Photochemical Processing of Electronic Materials

Photochemical Processing of Electronic Materials

Edited by
IAN W. BOYD and RICHARD B. JACKMAN

Department of Electronic and Electrical Engineering
University College London, UK

ACADEMIC PRESS
Harcourt Brace Jovanovich, Publishers

London San Diego New York
Boston Sydney Tokyo Toronto

This book is printed on acid-free paper

ACADEMIC PRESS LIMITED
24–28 Oval Road
LONDON NW1 7DX

United States Edition published by
ACADEMIC PRESS INC.
San Diego, CA 92101

A catalogue record for this book is available from the British Library
ISBN 0-12-121740-X

Typeset by Mathematical Composition Setters Ltd, Salisbury, Wiltshire, UK
Printed in Great Britain by Galliard (Printers) Ltd,
Great Yarmouth, Norfolk

Contents

7 Photo-Assisted II–VI Epitaxial Growth

S. J. C. IRVINE, H. HILL, J. E. HAILS,
G. W. BLACKMORE and A. D. PITT

8 IR and UV Laser-Induced Deposition of Hydrogenated Amorphous Silicon

PETER HESS

9 UV Lamp Deposition of a-Si : H and Related Compounds

W. I. MILNE and P. A. ROBERTSON

Contributors

T. S. Baller Philips Research Laboratories, P.O. Box 80 000, 5600 JA Eindhoven, The Netherlands.

Frank Beech Department of Electronic and Electrical Engineering, University College London, Torrington Place, London WC1E 7JE, UK.

G. W. Blackmore Royal Signals and Radar Establishment, St Andrew's Road, Malvern, Worcestershire WR14 3PS, UK.

J. E. Bourée CNRS—Laboratoire de Physique des Solides, 92195 Meudon, France.

Ian W. Boyd Department of Electronic and Electrical Engineering, University College London, Torrington Place, London WC1E 7JE, UK.

Gabriel M. Crean National Microelectronics Research Centre, Prospect Row, Cork, Ireland.

J. Dieleman Philips Research Laboratories, P.O. Box 80 000, 5600 JA Eindhoven, The Netherlands.

P. E. Dyer Department of Applied Physics, University of Hull, Hull HU6 7RX, UK.

J. Flicstein CNET—Laboratoire de Bagneux, 92220 Bagneux, France.

Eric Fogarassy Laboratoire PHASE, Centre de Recherches Nucléaires, 23 rue du Loess, 67037 Strasbourg, France.

J. S. Foord Physical Chemistry Laboratory, University of Oxford, South Parks Road, Oxford OX1 3QZ, UK.

François Foulon Department of Electrical Engineering, Imperial College of Science, Technology and Medicine, Exhibition Road, London SW7 2BT, UK.

F. N. Goodall Rutherford Appleton Laboratory, Chilton, Didcot, Oxfordshire OX11 0QX, UK.

J. M. Green Laser Technology Centre, AEA Technology, Culham Laboratory, Abingdon, Oxfordshire OX14 3DB, UK.

Mino Green Department of Electrical Engineering, Imperial College of Science, Technology and Medicine, Exhibition Road, London SW7 2BT, UK.

J. E. Hails Royal Signals and Radar Establishment, St Andrew's Road, Malvern, Worcestershire WR14 3PS, UK.

Peter Hess Institut für Physikalische Chemie, Universität Heidelberg, Im Neuenheimer Feld 253, D-6900 Heidelberg, Germany.

H. Hill Royal Signals and Radar Establishment, St Andrew's Road, Malvern, Worcestershire WR14 3PS, UK.

Patrik Hoffmann Laboratoire de Chimie Technique, École Polytechnique Fédérale de Lausanne, CH-1015 Lausanne, Switzerland.

S. J. C. Irvine Royal Signals and Radar Establishment, St Andrew's Road, Malvern, Worcestershire WR14 3PS, UK.

Richard B. Jackman Department of Electronic and Electrical Engineering, University of College London, Torrington Place, London WC1E 7JE, UK.

J. C. S. Kools Philips Research Laboratories, P.O. Box 80 000, 5600 JA Eindhoven, The Netherlands.

Ronald A. Lawes Rutherford Appleton Laboratory, Chilton, Didcot, Oxfordshire OX11 0QX, UK.

Baudouin Lecohier Laboratoire de Chimie Technique, École Polytechnique Fédérale de Lausanne, CH-1015 Lausanne, Switzerland.

W. I. Milne Cambridge University Engineering Department, Trumpington Street, Cambridge CB2 1PZ, UK.

M. R. Osborne Laser Technology Centre, AEA Technology, Culham Laboratory, Abingdon, Oxfordshire OX14 3PB, UK.

Jean-Michel Philippoz Laboratoire de Chimie Technique, École Polytechnique Fédérale de Lausanne, CH-1015 Lausanne, Switzerland.

A. D. Pitt Royal Signals and Radar Establishment, St Andrew's Road, Malvern, Worcestershire WR14 3PS, UK.

P. A. Robertson Cambridge University Engineering Department, Trumpington Street, Cambridge CB2 1PZ, UK.

Herbert Solka   Laboratoire de Chimie Technique, École Polytechnique Fédérale de Lausanne, CH-1015 Lausanne, Switzerland.

Hubert van den Bergh   Laboratoire de Chimie Technique, École Polytechnique Fédérale de Lausanne, CH-1015 Lausanne, Switzerland.

Marcel Widmer   Laboratoire de Chimie Technique, École Polytechnique Fédérale de Lausanne, CH-1015 Lausanne, Switzerland.

Preface

The rapid and efficient physical processes that lasers are capable of performing have already resulted in a steady increase since the early 1970s in applications on various production lines worldwide. These include machining, welding and alloying, surface hardening, insulation stripping, and impurity or defect gettering. It is the novel and unique processing characteristics presented by lasers that give rise to operating conditions never previously available, and thus the great variety of applications.

Lasers, and more generally lamps, can also give rise to a myriad of chemical reaction steps. The nature of light is such that the discrete quanta associated with particular emissions can selectively drive a key step in the reaction ladder leading to some surface modification or material removal, or to the formation of a thin film. The radiation can also be transformed into heat and drive a range of thermally controlled phenomena. These regimes of photochemical processing of electronic materials, have historically been largely unexplored, and only within the last decade have they begun to attract sizeable worldwide interest. The burgeoning of the field has been most evident at international level during events organized under the auspices of the Materials Research Society (MRS) and the European EMRS—societies whose remit is to foster and encourage interdisciplinary science.

It is intended that this new volume complement and update the existing work already published in this field. In this respect, the main concepts and basic principles are restricted to one introductory chapter. The following 17 chapters tackle specific and specialized areas of research that are currently being studied on a worldwide scale. This book contains reviews on UV lens design, laser lithography with respect to other lithographic techniques, photo-nucleation, excimer laser development, incoherent lamp applications, and *in situ* laser characterization; topics that have not been reviewed or assessed before in such detail. Together with articles on precursors for direct writing, laser ablation (of polymers and of superconductors), laser doping, etching, deposition and growth, it is hoped that this is a valuable and unique contribution to the field. It is aimed not only at both active and new researchers in the area, but also at those who feel they could be potential users of the technology.

This volume evolved from the 5th UK Photochemical Processing Workshop, a three-day meeting held at University College London (UCL) that was strongly supported by the EMRS/EEC Network on Laser Chemistry. The workshop was co-sponsored by several other learned societies, namely the British Association of Crystal Growth, the Institute of Physics, the Institution of Electrical Engineering, and the Royal Society of Chemistry. It additionally benefited from support from the EEC/Science programme on Photoprocessing of GaAs, the European Office of the United States Air Force, and several industries, whose inspiration and generosity is gratefully acknowledged.

The editors would like to take this opportunity to thank many people who have assisted, not only towards this volume, but also in the development of understanding in the field over the years. In particular, we are grateful to the authors of the chapters for delivering their erudite contributions (essentially) on time! Very special thanks go to our closest colleagues in the laboratory at UCL, F. Beech, B. Bradley, T. Kerr, P. Patel and G. Tyrrell, without whose help this project would have floundered at the first hurdle.

Finally, we should both like to take this opportunity to admit our indebtedness to our wives and families, who have so patiently tolerated the eccentricities of the British academic over the past years.

Ian W. Boyd
Richard B. Jackman

University College London
July 1991

1 Photochemical Processing: Fundamental Mechanisms and Operating Criteria

IAN W. BOYD

*Department of Electronic and Electrical Engineering,
University College London, UK*

1.1 INTRODUCTION

Photo-induced chemical reactions are not only central to everyday events such as photography, photocopying, polymer degradation and the bleaching of dyes, but they also play a vital role in the life support system of the planet. For example, virtually all forms of life are either directly or indirectly dependent upon the photosynthesis of organic compounds from CO_2 and H_2O by

$$CO_2 + 2H_2O \rightarrow [CH_2O] + H_2O + O_2 \tag{1.1}$$

as it is instrumental in initiating the growth of vegetation and producing the O_2 that makes up our atmosphere.

While such reactions must occur in the neighbourhood of light-absorbing pigments such as chlorophyll, many other photoreactions in nature do not require intermediaries to host the event. For example, O_2 within our atmosphere also plays a vital role in shielding us from harmful ultraviolet (UV) radiation from the sun through the following photo-induced reactions:

$$O + UV \rightarrow O^+ + e^- \tag{1.2}$$

$$O_2 + UV \rightarrow O_2^+ + e^- \tag{1.3}$$

$$O_3 + UV \rightarrow O_2 + O \tag{1.4}$$

Nitrogen ions, atoms and molecules similarly undergo a range of photon interactions, by themselves, and also in conjunction with various forms of oxygen.

In reality, a myriad of photo-induced chemical reactions can be initiated, either in the gas phase or within a host solid or liquid. Several of these have

Photochemical Processing of Electronic Materials
ISBN 0-12-121740-X

been studied for centuries, while most have only received attention since various intense artificial sources of light have become accessible. The availability of flash lamps not only enables new chemical reactions to be induced, but also allows reaction kinetics to be studied on timescales much less than a second.

The advent of the laser has stretched the list of photo-induced reactions considerably. The range of wavelengths and intensities now available have allowed access not only to new reactions but also to completely new non-linear reaction mechanisms as well as unexplored fields of study. Reaction events down to tens of femtoseconds (1 fs $= 10^{-15}$ s) can at present be chronicled using laser pulses.

While the vast majority of pioneering studies were performed within the gas phase, around the beginning of the 1980s the use of photons to stimulate reactions that formed thin film layers or delineated patterns on solids began to attract considerable attention [1–4]. Over the past decade, the field has burgeoned and currently sustains several international conferences and workshops each year. Interest in the subject can be found in a wide variety of disciplines, the basic sciences of chemistry and physics, the applied areas of materials and engineering, and the fundamental areas of surface science and quantum electronics. The potential applications are diverse, as the content of this volume testifies.

In this chapter, the reader will be introduced to the basic concepts behind photochemical materials processing. The photon sources used will be described, the underlying mechanisms reviewed, and the processing techniques discussed.

1.2 PHOTON SOURCES

1.2.1 Incoherent sources

For many photo-induced reactions, it is not necessary for the photons to be coherent, i.e. in phase with each other. Nor is it necessary for them to be monochromatic (of a single wavelength). In these instances, photons from a variety of lamps can be used. Such sources may be resistively heated wires (as in a light bulb) or gas discharges (as in a sodium vapour street light). By electrically exciting specific gases or vapours, a range of characteristic discharge spectra can be produced.

Where large continuous powers are desirable, such as for photothermal reactions, current-heated tungsten filaments are commonly used. Arranged in banks, these can produce a steady output at levels approaching 100 kW in a continuum stretching from the near-UV through the visible, peaking in the near-infrared (IR) and falling off into the IR. Indeed, the spectrum is close to that expected from a high-temperature black body. Unavoidable but gradual evaporation of tungsten atoms can be reduced by adding trace amounts of iodine, which also helps to improve the UV output.

Equally powerful light containing more UV can be acquired from arc lamps, where a gas, or mixture of gases, at medium pressures (450–1500 Torr) is subjected to a large electrical discharge from a capacitor bank. Often, the discharge has to be initiated by pre-ionizing the gas with a very high voltage "trigger". The light emitted in this case is more characteristic of energy levels available to the ions and atoms of the discharged species. Hg-based vapours are the most widely used, and the greatest emitted intensity usually appears at 365 nm (at efficiencies of up to 6%) along with subsidiary peaks around 300 and 313 nm. Possible lower wavelengths are masked because of self-absorption at these pressures. Such lamps are available commercially at ratings up to 60 kW over several metre column lengths. Higher photon fluxes can be achieved by high-pressure (up to 100 kTorr) short arc lamps. Several kilowatts can be obtained from arc lengths of only centimetres, although the spectrum at these pressures is characterized by only minor features superimposed upon a broad continuum. Hg/Xe mixtures give more spectral detail, especially in the UV; a typical spectrum is shown in Fig. 1.1. Efficiencies even of the strongest lines, however, are no more than 1–2% at best.

Low-pressure lamps (several torr and less) are among the most often used sources today for high-intensity UV light. At these pressures, under a high-voltage discharge, Hg can emit several distinct wavelengths previously self-absorbed. These characteristic lines arise from different atomic transitions:

$$^3P_1 \rightarrow {}^1S_0 + 253.4 \text{ nm,} \qquad (1.4)$$

$$^1P_1 \rightarrow {}^1S_0 + 184.9 \text{ nm} \qquad (1.5)$$

and appear at power ratios of about 7:1 for the longer to shorter wavelengths. Under optimum conditions, they can be up to 80% efficient in their conversion of electrical to optical energy. If one retains low-pressure operation, but increases the discharge current to more than 1 A cm^{-2}, the two characteristic lines saturate, and another line, previously weaker, at 194.2 nm becomes dominant and continues to increase up to current densities beyond 20 A cm^{-2}. A 1.4 kW Magnetron system can heat Ar atoms, which vaporize Hg to a pressure of 1–2 atm and produce, at efficiencies of greater than 9%, more than 100 W of UV light in each of the 200–260 nm and 260–400 nm regions, as well as 225 W between 400 and 700 nm [5].

Incoherent sources of high intensity deep UV light (<180 nm) are not currently readily available. Hydrogen, deuterium (D$_2$) and noble gas lamps emit in this region only at modest levels, where, at pressures of less than 2 Torr, several discrete spectral lines can be obtained preferentially over the

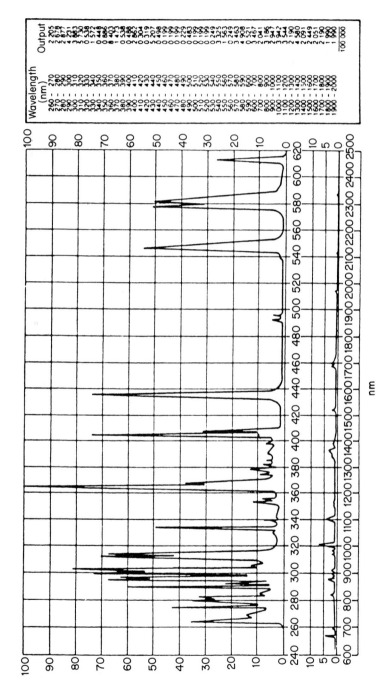

Fig. 1.1 Emission spectrum of an HG/Xe arc lamp.

low-level background continuum. D_2 lamps have been most widely used, emitting from atomic levels in the 120–200 nm region, while at higher wavelengths a high continuum develops from molecular electronic levels. The overall UV efficiency of these systems, however, is relatively low.

For arc lamps, then, high lamp output powers are typically associated with a wide continuum of wavelengths, while narrower linewidth outputs are unavoidably generated at much more modest power levels. A broadband lamp output can often be more efficiently absorbed in a photochemical reaction, however, than a single wavelength source in the same region. The incoherence and solid-angle emission of 4π steradians from most lamps ultimately means that, for large-area irradiation applications, lamp sources will also be extremely useful.

An alternative to the traditional arc lamp is the miniature dielectric discharge system, which has been shown to be useful for producing deep-UV light. Also known as the "silent" discharge, it operates at atmospheric pressures in a sandwich structure consisting of a dielectric barrier and electrodes, across which is applied a high-voltage alternating at 50–5 kHz [6]. Current flow is initiated through a large density of randomly distributed microdischarge filaments, each of submicrosecond duration. Various gases and gas mixtures can be used to obtain conversion efficiencies of typically 10%. Figure 1.2(a) shows one of many possible geometric configurations used to generate the radiation, while Fig. 1.2(b) demonstrates the spectral output obtained when different gases are present in the discharge. The flexibility of both the gas-fill (many inert and excimer-forming gas mixtures can be used) and the geometry controlling the emission mean that this type of lamp should find many applications in an advancing photochemical processing field.

In terms of capital cost, lamp sources are significantly cheaper to buy than most coherent laser sources, and are also much less expensive to operate. In situations where near-diffraction-limited photon definition, or extremely high intensities, are desirable, more coherent light, such as from lasers, will be required. The general properties of lasers are outlined below.

1.2.2 Coherent sources

The most striking property of the laser beam is its directionality and highly collimated nature. Unlike lamps, laser beams have a very low angle of divergence, typically 0.2–10 rad. This means they can be propagated very large distances efficiently without any energy being lost through divergence out of the path, and thus that large amounts of energy can be readily manipulated and directed into almost any location. By comparison, conventional sources of radiation such as the lamps mentioned above require

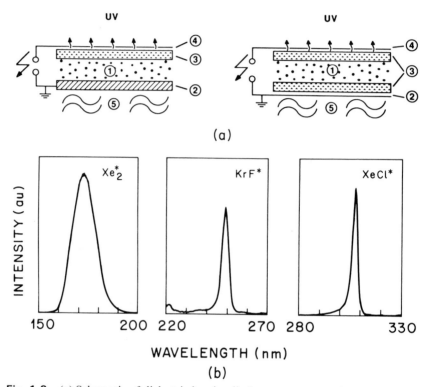

Fig. 1.2 (a) Schematic of dielectric barrier discharge systems used to generate UV radiation: (1) discharge gap; (2) ground electrode; (3) dielectric barrier; (4) UV-transparent high-voltage electrode; (5) coolant [6]. (b) Typical spectra obtained for Xe, KrF and XeCl [6].

very elaborate collection geometries to minimize energy losses, which nevertheless still tend to be quite significant.

Another property of laser radiation, governed by the fundamental laws of diffraction, is its ability to be focused to extremely small dimensions. A single-mode beam of laser light can, depending on the quality of the optics used, be focused to the dimensions of its own wavelength. This is a fundamental limit, which cannot be surpassed irrespective of the lens quality used. Traditional light sources cannot compete in this regime. It is possible with lasers, however, to define surface relief that is smaller than the diffraction limit of the beam. The chemical reaction induced by the photons need only be a strong function of the incident fluence or the induced temperature profile for this to be possible. The more strongly nonlinear this relationship is, the easier it will be to "defeat" the lower limits of resolution dictated by diffraction, as is discussed in Chapters 3 and 4 of this volume.

Table 1.1 Properties of some of the commercially available higher-power/energy lasers.

Laser type	Wavelength (μm)	Power (W)	Energy (mJ)	Pulsewidth (ns)	Beam size (mm)
CO_2	9–11	500			20
			5000	200	20
CO	5–7	100			10
			1000	10	
Nd:YAG	1.064	500			10
			1200	4–20	10
Nd:glass	1.054		5000	4–50	10
He:Ne	0.63	0.01			1.5
Nd:YAG	0.532		600	4–20	
Cu vapour	0.51–0.58		10	20	20
Ar ion	0.48–0.51	20			2
	0.35	5			1.5
Nd:YAG	0.355		300	4–20	
Kr ion	0.33–0.8	5			1
XeF	0.351		500	20	20 × 10
XeCl	0.308		1000	20	20 × 10
Nd:YAG	0.266		200	4–20	
KrF	0.248		1000	20	20 × 10
ArF	0.193		400	20	20 × 10

Advances in digital electronic control techniques, together with the ease of manipulation of laser beams, is conducive to novel direct-writing applications. In particular, by controlling the velocity and direction of the reaction-controlling laser beam, lines, curves or almost any pattern can be defined directly on a substrate, through localized deposition, growth, etching, ablation, or whatever method is desirable. This offers a single-step method for engraving micrographic structures as used in micromechanics or microelectronics manufacturing. Standard present-day techniques require several serial production steps to achieve essentially the same result. Details of this technique will be discussed in more detail by Van den Bergh in this book.

Although laser light is essentially monochromatic, this does not imply that each laser system can only emit a single frequency. Some lasers can be readily tuned over a considerable range of discrete wavelengths, they can be operated in a multiline mode, or they can be used to pump a range of commercially available dyes. Therefore the choice of colours available is much more than the number of different systems on the market, although it must be said that the spectral availability at certain power levels is not yet close to what would ideally be desired by spectroscopists. Nevertheless, tunability and monochromatic properties present a unique method of reaction control never encountered before the emergence of the laser. Table 1.1 shows a selection of the laser wavelengths commonly used for materials processing. With this spread of frequencies available, a variety of physical and chemical mechanisms are on hand to induce many kinds of reaction phenomena for thin-film processing.

A wide range of laser pulse durations are currently available for processing and reaction monitoring. Laser pulses as short as 8 fs can now be generated. Although such pulsewidths are not readily available for photochemical processing (and it is not obvious that they will ever be required for such applications), the generation and application of picosecond and nanosecond laser pulses allows novel and unique processing conditions to be accessed. We shall see later in this volume (see the sections on ablation in Chapters 14 and 15, for example) that nanosecond irradiation procedures are quite commonly used to induce thin-film growth and delineate microstructures.

1.3   PHOTO-EXCITATION OF MATTER

Radiation from the UV to the IR interacts primarily with the outer electrons of an atom, and sometimes, though less commonly, with a particular set of atomic vibrations, or phonons. The frequencies and energies associated with this band of electromagnetic radiation do not induce nuclear disturbances, nor do they affect the energy levels of the inner core electrons of an atom. The optical properties of any material will therefore be mainly affected by the nature of its outermost electron(s) and thus be predicted by knowing their electronic configurations. Moreover, the behaviour of the optically excited system after absorbing the photons will also depend on the atomic arrangement, as well as the nature of any other nearby species.

It is important at this stage to introduce the concepts of pyrolytic and photolytic reactions. The former are ideally controlled by *thermally stimulated* mechanisms, while the latter are initiated by direct *photonic stimulation* of a particular bond, after which a reaction will occur either as a result of the disruption of the bond, or simply as a consequence of the atom being in an excited state. In many of the photo-induced reactions recorded, both thermal and nonthermal processes are present. There are nevertheless sometimes quite obvious and discernible features that distinguish between the two mechanisms and lead to particular limitations and/or advantages of either process.

1.3.1   Excitation of the gas species

It is well known, that compared with solids, atomic and molecular species only exhibit a relatively small number of excited states accessible by optically induced transitions. The statistically allowed transitions between rotational, vibrational and electronic levels, within the selection rules of quantum mechanics, result in a series of discrete and sometimes extremely narrow lines in the absorption (and emission) spectra of most gases. These are often used as an identifiable fingerprint for the species.

The energy ratios associated with electronic, vibrational and rotational transitions are about $10^6 : 10^3 : 1$. Pure rotational spectra are exhibited only by molecules possessing a permanent dipole moment, and the vibrational spectra require a dipole change during the motion, while electronic spectra are exhibited by all molecules since changes in the electronic distribution in a molecule are always accompanied by simultaneous dipole changes.

1.3.1.1 Selective vibrational excitation

Atoms form molecules by arranging themselves to participate in some specific internal electronic timesharing. The nuclei of all the atoms attract electrons surrounding the complete molecule while simultaneously repelling each other. This results in a balanced equilibrium of interatomic spacings and charge distribution. External influences may disturb this equilibrium, resulting in bond compressions and extensions. Real molecules, however, do not follow the rules of simple harmonic motion exactly, and, when stretched, tend to dissociate eventually into their component atoms. When the Schrödinger equation is solved for a diatomic molecule, it gives rise to a set of allowed vibrational energy levels for this anharmonic oscillator as shown in Fig. 1.3.

Vibrational states of polyatomic molecules are quite different. A nonlinear molecule can have $3N - 6$ different internal vibrations (linear polyatomic species have $3N - 5$). For the diatomic molecule described above, there can be only one fundamental vibration, which is complicated by overtone vibrations governed by anharmonicity. Each internal vibration of a polyatomic molecule is also affected by these phenomena, and, furthermore, there are additional effects arising from combinations of vibrations as well as overtones, and the intensities of these can be considerably enhanced by resonance phenomena. An important consequence of this is that many discrete energy levels now exist at lower energies; these rapidly converge into a quasicontinuum at higher energies and eventually into a true continuum above the dissociation energy. Understandably, the complexity of the situation is proportional to the number of atoms in the molecule. Therefore, although only one or two modes of vibration actually absorb the radiation, the energy so absorbed can be coupled rapidly into the quasicontinuum, thereby heating up the molecule towards a state of dissociation via a strictly multiphoton process. There is evidence [7] that Si can be deposited in this manner, by the vibrational excitation of SiH_4 with the 10P20 line of a CO_2 laser, as is discussed by Hess in Chapter 8.

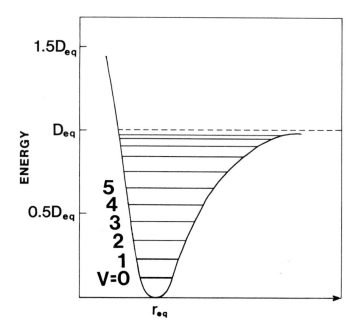

INTERNUCLEAR DISTANCE

Fig. 1.3 The allowed vibrational energy levels for a diatomic molecule undergoing anharmonic oscillations [3].

Another mode by which energy is absorbed into vibrational states of a polyatomic or diatomic molecule is collisional transfer. This process could clearly compete with that described immediately above, making it difficult to distinguish between the two, or indeed to purposely select the former mode of operation. However, collisional transfer can introduce interesting alternatives to direct photochemistry. It is possible to absorb energy into an intermediate species which then transfers it by collisional scattering to the desired molecule, which subsequently undergoes the desired chemical reaction. This technique, known as photosensitization, is quite common in the field of photochemistry. For example, Hg atoms have been used to assist in the photodeposition of Si and SiO_2 when using UV light (see Chapters 9 and 10). It is important, however, that these atoms do not contaminate the film that is actually being grown.

Vibrational excitation from one state to the next highest, or to another not too far removed from the ground state, can enhance the sticking coefficients of particular molecules on certain surfaces. N_2O molecules, excited from the (001) to the (100) vibrational level by photons at 10.8 μm, react with Cu (most likely forming N_2 and CuO) at a rate enhanced by a factor of 5000 [8]. For many chemical reactions, however, the activation energies are much higher than a single vibrational excitation. To initiate these, it is often necessary to employ multiphoton excitation, leading either to a highly excited reactive species, or to complete rupture of the bonding and dissociation of the molecule.

Polyatomic molecules in their ground electronic states can also be dissociated by multiple photon absorption. The decomposition of SF_6 with a high-power CO_2 laser has been studied [9], and it is found that the F atoms, formed by photodissociation of SF_6 into $SF_5 + F$ and the subsequent dissociation of the unstable SF_5 into $SF_4 + F$, can diffuse in their reactive state over appreciable distances to interact with a nearby Si surface. CF_3 radicals can similarly be generated from CF_3Br to react with SiO_2 [10]. This laser-induced process is expected to be more widely used in the future for etching studies, since many halogen-containing polyatomic molecules such as CF_3X, CF_2X_2, where $X = Cl$, Br, I, etc., can be readily dissociated to produce radicals [11].

1.3.1.2 Selective electronic excitation

The energy levels associated with electronic excitation are almost identical to the Morse potential shown in Fig. 1.3, but are displaced vertically in energy above the set of levels associated with the ground state. Quantum theory predicts that the most probable transition of single-photon electronic excitation from one set of levels to the other is where the internuclear distances of the upper and lower excited states are essentially equal. Transitions to neighbouring vibrational states will occur with a much reduced probability. Consider the case where the upper electronic state gives rise to a larger separation of the atoms. A vertical transition from the ground state will access one of the higher vibrational levels in the ladder of the upper electronic branch. If the upper-state separation is very large, excitation to a vibrational state far into the upper level could exceed the molecule's dissociation energy.

Another possibility for dissociation by electronic excitation occurs when the upper electronic state, unlike in the previous case, is unstable. This happens when there is no definable energy minimum in the upper electronic state. Here the whole spectrum for the system will exhibit a continuum, the lower limit of which will be the energy difference between the lowest vibrational level of the lower electronic state and that just needed to dissociate the molecule. Occasionally, the usual stable excited states can be intersected by another set of unstable continuous states, at some energy level well below the expected dissociation value. In this case, there is a finite probability of radiationless transfer between the curves, so that "predissociation" can occur.

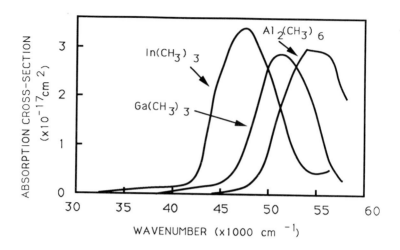

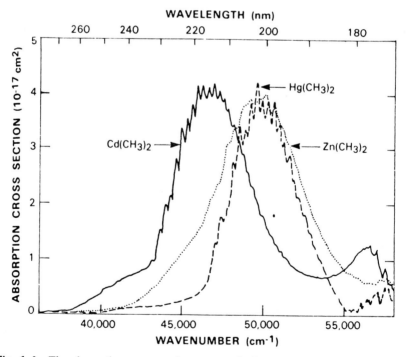

Fig. 1.4 The absorption spectra of some metal alkyls. (Adapted from [81], and courtesy of R. M. Osgood, Columbia University.)

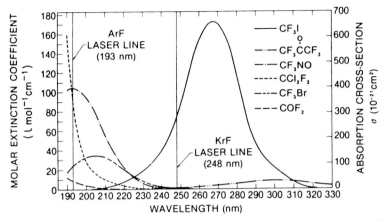

Fig. 1.5 The absorption spectra of various radical precursor compounds [70].

One of the most widely studied reactions using electronic excitation and dissociation of molecules to form etch-patterned microstructures involves Cl_2. In fact, single-photon photofragmentation is central to laser-assisted etching of many materials involving not only Cl_2 but also Br_2 and I_2. The Ar-laser radiation used in these applications can readily excite the Cl_2 molecules into their first dissociative state [12, 13]. Details of these reactions will be treated later in this volume in Chapters 12 and 13.

Photolysis of $Al_2(CH_3)_6$ and many other metallo-organic molecules by UV radiation is also possible through laser-induced dissociative electronic transitions. Figure 1.4 shows the absorption spectra of several such molecules, which indicate that they can be readily dissociated by the photons at 193 and 248 nm available from excimer lasers. Figure 1.5 shows the vibrational absorption spectra of a range of halogen-containing precursors for laser-induced etching [14]. The detailed spectroscopy and photochemistry of a wide range of molecules are excellently reviewed by Rothschild [1, 15].

1.3.2 Interaction of laser radiation with solids

1.3.2.1 Metals

Metals are characterized by their loosely bonded outermost electrons, which can travel freely from atom to atom. Their optical properties can be described within the framework of the highly successful "free electron gas" model. Calculations on the dielectric response of such an electron gas reveal a plasma frequency w_p at which an undamped plasma of electrons can resonantly oscillate. For low-frequency radiation, much lower than this plasma edge, absorption is dominated by the collective properties of the free carriers. Many metals therefore exhibit almost identical spectral behaviour in the IR, and the absorption coefficient is inversely proportional to the square root of the frequency of the radiation. In this regime, however, the light is strongly reflected. Thus, to make use of the efficient absorption processes, extremely large laser beam intensities must be used to compensate for the large reflectivity losses.

As the wavelength of the incident radiation decreases, both absorption and reflectivity also decrease. The precise locations of their minima in this region depend uniquely on the band structure of the material in addition to the plasma resonance mentioned above. At high frequencies, the free carrier contribution to the optical properties becomes extremely small, the metal exhibits characteristics related to the lattice properties, and eventually, for most metals, interband absorption will occur beyond the minimum. The characteristic colours of gold and copper are a consequence of an absorption minimum in the visible.

The energy decoupled from a laser beam into the electronic system of the metal is rapidly redistributed among the electrons via collisions, and eventually to the lattice by various scattering mechanisms. Typically, times of the order of picoseconds or less are required to convert the photonic energy into random atomic motions. Thus the metal can be rapidly heated up within the duration of a laser pulse. In this way, the rate of supply of energy (proportional to the beam flux and duration of the exposure) is balanced by heat losses due to thermal diffusion into the bulk, radiation losses from the surface, or convective losses. The actual temperature attained in this way can be determined using the usual heat diffusion equation.

1.3.2.2 Insulators and semiconductors

Unlike metals, insulators and semiconductors do not contain appreciable numbers of free carriers in their conduction bands. Each material, however, can be characterized by a parameter, known as the bandgap, that defines the amount of energy E_g required by an electron in the highest valence state to cross into the lowest level in the conduction band. For large E_g values of the order of 5–9 eV (i.e. into the UV), the material is a very strong insulator. If E_g is only in the region of a fraction to a few electronvolts then it is a semiconductor.

Above and below their bandgap energies, insulators and semiconductors exhibit similar optical properties. For radiation whose wavelength is shorter than E_g, i.e. $E_p = h\nu > E_g$, electronic transitions between the valence and conduction bands are induced. This occurs in the near IR for silicon (1.12 eV), germanium (0.67 eV) and gallium arsenide (1.42 eV), the

IR for indium antimonide (0.18 eV), the green (2.26 eV) for gallium phosphide, just into the UV (3.5 eV) for the chalcopyrite $CuAlS_2$, and the vacuum UV (8 eV) for silicon dioxide. Across the band edge, the absorption coefficient increases from typically less than 1 cm^{-1} to around 10^6 cm^{-1}, i.e. approaching that usually exhibited by metals in the IR. In the strongly absorbing region of the spectrum, these materials also exhibit their highest reflectivity.

At energies below the bandgap, optical properties are determined by much weaker intraband electronic transitions and by excitation of specific vibrational modes within the lattice itself. The frequency and precise location, as well as the overall strength of these modes, are strongly dependent upon the exact nature of the bonding arrangements and the type of atomic species involved. The spectral features in this region are often exceedingly complex. More details of the band structure and the optical properties of semiconductors can be found in [3, 16].

1.3.2.3 Nonlinear optical absorption

In most photochemical reactions, the optical absorption is linear, i.e. proportional to the number of absorbing centres within the path of the radiation and the scattering cross-section. However, this linear absorption regime, characterized by the well-known Beer Law

$$I/I_0 = \exp(-\alpha d), \tag{1.6}$$

can be disturbed in several ways. For example, if a new absorbing state is created once the absorption process has occurred then the absorption can behave nonlinearly. This will happen when free carriers are generated across the bandgap of a semiconductor or insulator that can subsequently contribute to enhanced absorption, i.e. free carrier absorption (FCA). Multiphoton absorption can also occur, particularly in the presence of an intense laser beam. The simultaneous absorption of two photons is not uncommon in semiconductors, and can also be found in gases. For excitation of valence electrons to the conduction band via two-photon absorption (TPA), energy conservation requires

$$E_g \leqslant 2h\nu. \tag{1.7}$$

The probability of the occurrence of such absorption is proportional to the square of the intensity of the laser beam. For pulses shorter than the carrier lifetime, TPA is always intensity-dependent, while FCA depends only upon the number of free carriers created, and is approximately fluence-dependent.

1.3.3   Plasma formation

Extremely high intensity and ultrashort duration laser pulses can generate
a dense plasma above the surface of many types of materials. Indeed,
plasma formation by laser-induced dielectric breakdown at a gas–solid
interface has attracted considerable attention over the years [17, 18].
This optically induced breakdown involves the radiative heating of free
electrons within the solid by the intense laser pulse. These electrons couple
very strongly to the incident radiation, and, by absorbing the energy,
accelerate and collide with ions and various gas phase atoms (inverse
bremsstrahlung). In this way, more atoms are ionized, and they in turn
provide more electrons, which further increase the rate of ionization. This
electron gas, rapidly formed above the surface by impact (sometimes
known as "cascade") ionization, absorbs the remainder of the laser energy,
thereby leaving the sample relatively cool and shielded. Therefore, in an
oxygen ambient medium, heterogeneous chemistry can be initiated by the
laser beam. This technique has been used to induce the growth of thin oxide
layers on the surface of niobium films [19], and has been called "laser
pulsed chemistry" (LPPC).

Plasma formation also occurs during laser ablation—a term applied to
the photonic removal of polymeric materials, such as photoresists, plastics
or biological specimens, and to the growth of multicomponent layers such
as superconducting, magnetic, ferroelectric or optical films. The strongly
absorbed radiation, which ideally must be supplied in short (<50 ns)
pulses, drives material from the surface at supersonic velocities. The ejec-
tant is known to consist of atoms, neutrals, ions clusters and even microscopic droplets, and these can react with the remnants of the radiation
pulse as well as any gaseous phase in the immediate environment. Under
optimum conditions, the particles removed can either by pumped away as
volatile by-products, or directed on to an appropriate substrate to engineer
thin-film growth. Both of these techniques are discussed in detail later in
this volume in Chapters 14 and 15.

1.3.4   Interactions with surfaces and adsorbates

The solid surface serves as the rendezvous for many gas phase species and
the host to many of the subsequent chemical reactions. Because of the trun-
cation of the atomic lattice, the bonding and electronic structure of surfaces is often exceedingly complex, and often it is the special properties of
the surface that induce or enhance a particular mechanism. The optical
properties of surfaces are thus not only inherently elaborate, but can be significantly modified by the presence of impurity species attached to them.

The effect of adsorbates and the phenomenon of desorption are described briefly below.

1.3.4.1 Adsorbates

When an atom or molecule interacts with a clean surface, it perturbs its already complex electronic structure. If this adsorbate remains relatively inert such that the attractive atomic interaction is very weak then the species is *physisorbed*. If the foreign material is reactive and a new chemical bond is formed then the species is *chemisorbed*. While both mechanisms effect a change in the electrostatic potential of the surface, only the chemisorption influences the valence levels of the substrate. In fact, it introduces a completely new set of modified electronic energy levels locally within the surface-adsorbent complex. Chemisorption is highly specific in character, and often (though by no means always) only single monolayers can be formed. However, physisorption involves van der Waals type forces, and, although these are significantly weaker, they can extend from one monolayer to another. It can therefore by anticipated that several (or many) monolayers can occasionally be physisorbed on to a surface (see Section 1.4).

Once adsorption has occurred, it presents an additional set of possibilities for laser-assisted surface interactions. Photochemical coupling of the radiation can now occur where before it was not possible because of the lack of appropriate absorption sites. Consequently, a new avenue is now available for deposition, etching and film growth. The interaction of laser radiation with even a lightly contaminated surface can also give rise to the complimentary process of desorption. Here the photons directly, or indirectly, encourage bond breaking between adsorbate and surface, with the impurity then leaving the surface.

For many molecules, consecutive excitations up the vibrational ladder cannot be practically achieved using lasers because of the usual high degree of anharmonicity (Section 1.3.1.1). For example, to reach the nth level of vibrational excitation, n different photons of specific frequency would be required. However, when these molecules are adsorbed on to a surface, the number of available excitational modes is increased significantly. In isolation, a diatomic molecule has only one mode of vibration, whereas upon adsorption it can have up to six modes, including hindered translational and rotational motions. As with the polyatomic molecules described earlier, this increase in the density of states brings the level of the quasi-continuum lower down into the Morse potential and therefore makes it more easily accessible to smaller numbers of multiple photon absorption.

The vibrational activation of adsorbate–surface complexes is damped rapidly by the surface itself. Nevertheless, it should be possible to excite

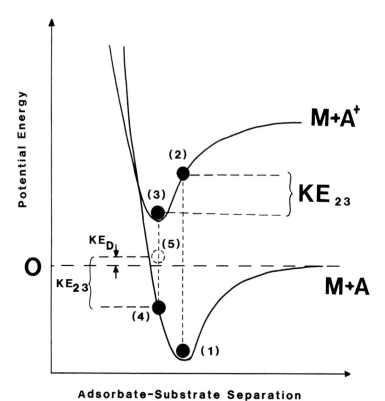

Fig. 1.6 A possible mechanism for desorption via intramolecular energy transfer from a photo-active mode to an adsorbate stretching mode [20].

other intramolecular vibrational modes during this relaxation, such that, before the energy is transferred to bulk phonons and manifested as heat, indirect excitation of a weaker bonding unit can occur. Clearly, the various timescales involved must be carefully summed up and compared with competing processes, and it is often quite difficult to identify the more complex polyatomic intermolecular energy transfer pathways. Figure 1.6 gives a good picture of an intramolecular energy transfer mechanism by which excitation of a mode not even directly connected with the adsorbate vibrational modes results in the eventual excitation of the modes after some period of time [20]. This can often lead to desorption of a species from the surface, as will be discussed in the following section.

1.3.4.2 Desorption

Desorption can be governed by the changes in the thermal properties of the material resulting from heating, or by any one of several photo-induced absorption mechanisms. In optically assisted thermal desorption, the foreign species is released from the surface when energy is transferred by the energetic phonons from the hot substrate into the surface–adsorbate bond. The molecular desorption occurs with a characteristic translational temperature of the desorbed species. In this case, the photons are absorbed by the substrate, and the modified electronic structure does not play a significant role in the process.

Photodesorption can occur in several ways, by radiation coupling into the vibrational modes of the system or into electronic transitions. When the energy is resonantly absorbed into the internal vibrations of the adsorbate–surface system, IR wavelengths are involved. Laser-assisted desorption of CH_3F from NaCl, recently investigated using a CO_2 laser, is a multiphoton excitation process [21]. Time-of-flight (TOF) diagnostics of CO_2 laser-induced desorption of CH_3F, CH_3COOH and CCl_4 on various substrates lead to the conclusion, from correlation of the heat of vaporization and the strength intermolecular interaction in the van der Waals films [22], that the observed wavelength dependences may be explained by resonant heating via selective excitation of the internal modes of the condensed molecules. UV photodesorption is initiated by promoting electronic excitations. However, the energy absorbed in this mode may not remain localized for long times within the excited adsorbate–substrate complex, and thermal redistribution may take place, thus making the dominant desorption process thermal [23]. In general, photodesorption dynamics are not well understood. Since the migration of reactants adsorbed on the sample can be a limiting step in the overall rate process of a surface chemical reaction, photostimulated surface motion of adsorbed atoms may be important. A theoretical approach has found that the photons could enhance diffusion-type behaviour under conditions of steady state excitation [24]. Laser stimulation of electron migration between surface groups has also been suggested [25] to explain the increased rate of decomposition of NH_2 groups on SiO_2 during CO_2 laser irradiation. Desorption pathways involving chemical reactions as essential components of a multistep process have also been proposed for desorption from oxide and sulphide surfaces exposed to UV–visible radiation [26].

1.4 ADSORPTION

Prior to any photo-induced film growth or etching, chemical species must interreact with the sample surface. In Section 1.3.4.1, photon interactions with physisorbed and chemisorbed species have already been addressed. Adsorption is most easily identified in porous substances that exhibit large surface areas, such as aluminium oxide and zinc oxide. Although a large surface area is important, many other properties of the adsorbate and the surface determine the extent of adsorption. A rise in pressure and a decrease in temperature usually increase adsorption, and adsorption isotherms show this relationship for different pressures at a specific temperature T. For a limited pressure range at constant T, the variation of adsorption with pressure P can usually be represented empirically by

$$a = mP^n, \tag{1.8}$$

where m and n are constants for a given surface–adsorbate combination at a particular T, and a is the amount of gas adsorbed per unit mass of adsorbent. A more accurate relationship, known as the Langmuir adsorption isotherm, relating a to P for a given temperature is given by

$$a = \frac{k_1 P}{k_2 P + 1}, \tag{1.9}$$

where k_1 and k_2 are constants. This formula is strictly only applicable to monolayer coverage, and numerous instances of this relationship have been reported over the years—understandably, primarily for examples of chemisorption.

At very low temperatures, the value of $k_2 P$ becomes negligible compared with unity, and adsorption is directly proportional to the pressure. At higher pressures, unity becomes insignificant compared with $k_2 P$, and the adsorption is determined by the ratio k_1/k_2. This limiting adsorption with increasing pressure is applicable at any temperature, and is a consequence of one-monolayer surface coverage. At intermediate pressures, expressions of the form (1.8) would become quite applicable with reasonable accuracy.

Although physical adsorption can also result in only single-layer coverage of the surface (Type I), it is usual to find a *multilayered* structure built up on the adsorbent. In fact, five types of physisorption isotherms can be identified. By applying the Langmuir approach to such multilayered physisorption, Brunauer, Emmett and Teller (BET) [27] derived the equation

$$\frac{v}{v_m} = \frac{PP_0}{P_0 - P} \frac{c}{P_0 + P(c - 1)}, \tag{1.10}$$

where P_0 is the pressure required to condense the gas at the operating temperature (i.e. the vapour pressure), v is the volume of gas adsorbed at pressure P, v_m is the volume adsorbed when the adsorbent is covered by a *single* layer or molecules, and c is a constant approximately equal to $\exp(\delta R/RT)$, where δE is the energy associated with the adsorption of the first monolayer. When $c > 1$ and $c < 1$, adsorption characterized as type II and type III respectively is achieved. Types IV and V occur as modifications of types II and III if gas condenses inside some of the small pores and capillaries of the surface structure at pressures below P_0. From a plot of $P/v(P_0 - P)$ against P/P_0, the slope of the straight line isotherms should give, according to the BET equation (1.10), the ratio $(c - 1)/v_m c$ and the intercept $1/v_m c$. From these, v_m can be calculated, and knowing the molecular diameters and assuming close packing, the surface area of the adsorbent surface can be estimated. The ratio v/v_m is sometimes defined as θ, the fractional surface coverage. Several aspects of physisorbed alkylmetal molecules, in relation to photodeposition of metals, are addressed in [28, 29].

The large number of the vibrational modes characterizing adsorbed species, together with the high molecular densities possible with adlayers, enable film nucleation to be precisely pinned by *direct* photolysis of the surface-adsorbed species. In this way, ultra-high resolution is possible, because gas phase diffusion is no longer important. Once a few nucleation sites have been established at a particular site, deposition could in principle be continued via normal gas phase pathways.

An example of a surface-catalysed process involves adsorbed layers of the organic molecule methyl methacrylate (MMA) being polymerized into films of poly-MMA, or PMMA, by absorption of UV laser radiation [30]. Molecules of MMA from the ambient medium form on the surfaces exposed to the vapour, and the free radical catalysed polymerization is initiated while rapid collisions of further vapour phase species continually replenish the adlayer. Polymer growth, initiated by a weak photodissociative absorption at 257.2 nm, can be increased substantially using a compound adsorbed-layer structure. This sensitizing layer is formed by exposing the original SiO_2 substrates to $Cd(CH_3)_2$ vapour, then evacuating the cell and introducing the MMA. Although the technique leaves a single chemisorbed monolayer under the PMMA, an enhanced preparation rate of nearly two orders of magnitude (up to $1\ \mu m\ s^{-1}$) can be achieved.

As the level of materials technology increases and surface structures become progressively smaller and more complex, the importance of surface structure and quality will become more evident. As we have seen in this section, the interaction of gaseous species and surfaces is an extremely complex situation. The presence of energetic photons, although sometimes providing an important clue towards the understanding of some reactions, can make it even more difficult to interpret the basic mechanisms involved. Although most photoprocessing studies are very much application-driven, it is important that there remains a healthy degree of fundamental investigation of the mechanisms involved.

1.5 LASER-INDUCED HEATING

1.5.1 Thermalization

Photonic absorption in solids is associated in most cases with an instantaneous increase in the energy of the carrier system and to a lesser extent
the lattice itself. Where free carriers have been photo-excited or photogenerated, they immediately assume the full energy of the photon and
increase their kinetic and potential energy levels. The energy is then rapidly
redistributed among the carriers through carrier–carrier interactions,
between carriers and the lattice through carrier–phonon scattering, and
also within the lattice through phonon–phonon interactions. Ultimately,
the energy will be degraded to atomic vibrations, and the sample will heat
up.

Within the electronic system, scattering processes involve carrier–carrier
and carrier–plasmon (oscillation of the carrier plasma) collisions. These

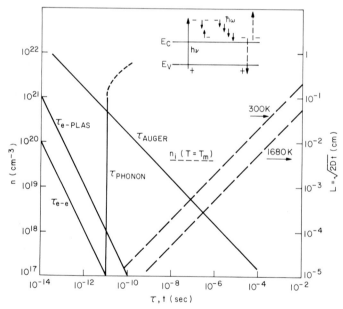

Fig. 1.7 A summary of the various electron scattering times in silicon, which also
applies in general to most other semiconductors, and in some instances to metals
and insulators. Also included are the characteristic heat diffusion lengths for silicon
at two temperatures in the solid phase. The insert is a schematic of some of the
important scattering processes. The vertical axes show the carrier density on the
right, and the thermal diffusion length on the left. (After Brown [31].)

depend upon the density of the carriers. As shown in Fig. 1.7, these processes are very rapid, occurring in 10 fs for densities around 10^{20} carriers
cm^{-3} [31]. Another process that assists in the redistribution of energy
within the carrier system is the Auger process whereby an electron and hole
recombine and transfer their recombination energy to a third carrier. In
this way, the density of the plasma can be significantly reduced, although
the total energy of the system will remain approximately constant. (The
opposite mechanism of impact ionization can also occur through the creation of an electron–hole pair by a sufficiently energetic third carrier.) This
thermalization process may take only a few picoseconds to complete.

Energy can escape from this quasithermalized system via carrier–phonon
scattering, and in this way the lattice will heat up according to the number
of lattice phonons created. Typically, about 20 phonons must be generated
to transfer 1 eV from the carriers to the lattice, and the time required for
this has been estimated to be less than 10 ps [32] for $n \approx 10^{21}$ cm^{-3}. Hence
it is appropriate to consider that for nanosecond and picosecond irradiation the lattice is heated instantaneously.

1.5.2 Heating models

A large number of models have successfully modelled the temperatures
produced by laser-induced heating of solids. Frequently described as
"simple" thermal models, these are often quite complicated, involving
nonlinear optical and thermodynamical processes such as absorption,
reflection, carrier and thermal diffusion, melting and evaporation, as well
as the energy of the reaction. Strictly, there does not exist a universal model
describing laser-induced heating, and many research groups develop
models specific to their individual needs. In many cases a few simplifying
assumptions, with little loss in accuracy, can provide analytical rather than
numerical solutions to each problem. Constants can be given an average
temperature-dependent value throughout the calculation if they do not vary
much during the heating cycle, and, in many CW modelling situations,
only the strongest nonlinearity (usually thermal conductivity) is given an
empirical temperature dependence. For detailed models of specific
irradiation conditions, the reader is referred to [1–6, 33–39].

If the energy from the laser beam induces a phase change and if the
deposition of energy continues after melting or evaporation has occurred,
the heating models must include the various latent heats of reaction, as well
as the altered optical and thermal properties of the system. There are a
large number of so-called "melting models" in the literature, only a fraction of which are referenced here [40–47]. When significant evaporation
occurs near the boiling temperature of the material, modelling becomes

appreciably more difficult. The energy that would otherwise be incident on the sample surface will now be scattered away from it by the evaporant rather than being coupled into it. Therefore the nature of the evaporant, as well as any interaction with the environment, must be included in any detailed model of the system.

1.6   NUCLEATION AND GROWTH

Three principal mechanisms of thin-film formation have been identified over the years. The first, the Van der Merwe mechanism, is characterized by a layer-by-layer growth, the second is a three-dimensional nucleation, growth and coalescence of islands (the Volmer–Weber mechanism), and the third, operating via the adsorption of a monolayer and subsequent nucleation on top of this layer, is known as the Stranski–Krastanov mechanism (see for example [48]). In the majority of cases, the second process occurs, and this is the one thought to be closest to that operable in many of the laser-induced deposition techniques.

The individual steps comprising the growth of a thin film from the gas phase include impingement and sticking of the components onto the surface, surface migration of the adsorbed species, nucleation, clustering of individual nuclei into interconnecting islands, and then filling in of the network spacings to form uniform layers. Some films may be tens to hundreds of nanometres thick before this final stage is considered complete. Nucleation can be influenced by surface structure, such as point defects or crystal dislocations; in fact, this property is often used to decorate such features on crystal surfaces in order to detect them. Clearly, the initial stages are crucial to the quality of the film structure, and a photonic presence during the process could be beneficial or disastrous.

Most theoretical treatments of nucleation [49, 50] are based on deterministic rate equations first derived by Zinsmeister [51], and yield a unique solution for each set of initial conditions. Particles in a vapour will impinge upon a surface in a random fashion, and, assuming a high sticking coefficient s, will adsorb. (In practice, $s \approx 1$ if the desorption energy of the particle on the substrate is less than 4% of its incident energy [48].) If a particular species is attracted by the fields at the surface, it may remain there for some time, able to move around until either losing excess energy to a defect, thereby becoming temporarily bound, or gaining sufficient energy to escape. The time spent on any weakly bonded sites, t_b, can be written as

$$t_b = t_0 \exp(\varphi_m/kT), \qquad (1.10)$$

where φ_m is the activation energy for migration and t_0 is the inverse surface vibrational frequency of the adatoms. The total time spent on the surface is determined by the desorption activation energy φ_d, and is of the form

$$t = t_0 \exp(\varphi_d/kT). \tag{1.11}$$

The value of t_b is largest at low sample temperatures, where the particles lose all their excess energy very quickly and become bonded. At increased temperatures, the species may occasionally move from site to site in a random fashion, until, at very high temperatures, t_b is so small that the adsorbed particles act as a two-dimensional gas on the substrate. Nucleation can only occur if sites exist where the particles can lose their excess energy. The precise nature of these preferred sites still remains a relatively unexplored regime of thin-film preparation, and even more so in photo-assisted growth (see Chapter 5).

The re-evaporation of desorbed species also affects the rate of nucleation. If the adsorption rate exceeds the desorption rate then the surface will become supersaturated and nucleation more likely. The controlling factor is the ratio φ_d/φ_s, where φ_s is the heat of sublimation of the condensed film. For small ratios, nucleation occurs without supersaturation, and the coverage can be quite high, whereas if $\varphi_d > \varphi_s$ then condensation will only occur if a high degree of supersaturation is achieved. At a value around 1, the classical theory of nucleation, based on thermodynamic concepts (capillary theory) can be readily applied [48].

Surface migration (or diffusion) can influence the nucleation rate. This accelerates cluster formation and subsequent film growth. Without migration, cluster formation will be slow, determined only by direct impingement of particles on the surface. Cluster density will also be strongly determined by the diffusion processes. After some density of clusters has been formed, either directly or by migration, the diffusion of species serves a dual purpose of not only finding new nucleation centres, but also supplementing the growth of existing clusters and assisting in the coalescing of neighbouring clusters.

Cluster growth relies upon the formation of bonds when impinging or migrating atoms encounter a cluster. Conversely, aggregate disintegration may occur instead. The cluster stability will depend upon many parameters such as surface tension forces at the substrate, the bonding relationship between aggregate atoms and substrate, and the energy difference between bonded and unbonded species, as well as ambient pressure and temperature. Under optimum conditions, clusters, acting as established seeding sites for incident particles, will continue to grow in three dimensions, eventually forming large "islands", which will coalesce when their boundaries overlap. In many cases, sufficient heat may be liberated upon coalescence to melt the islands (being microscopic clusters, their melting point could be significantly lower than that of the bulk film). The structural quality of the grown film will be dictated not only by such processes, but also by the relationship between the thermal properties of the film and substrate.

The overall nucleation rate J, determined by classical capillary theory, will depend upon N^*, the concentration of aggregates over a critical size, given by

$$N^* = n_0 \exp\left(-\delta G^*/kT\right), \tag{1.12}$$

where δG^* is the corresponding critical energy of nucleus formation and n_0 is the density of adsorption sites. It will also depend on Γ, the rate at which molecules join the critical nucleus by surface diffusion (those joining by direct impingement are negligible at this stage). Thus the nucleation rate is

$$J = 2\pi r^* Z \Gamma N^* \sin\theta, \tag{1.13}$$

where r^* is the radius of the critical nucleus, θ is the contact angle of the aggregate with the substrate and Z is Zeldovich's constant (≈ 0.01). While this theory provides a qualitative view of the main mechanisms of film formation, it uses thermodynamic concepts that more correctly apply to macroscopic systems. For submicron applications, more specialized alternative theories may be required (see Chapter 5).

1.7   CHEMICAL REACTIONS AND GROWTH RATES

The range of mechanisms by which photons can initiate chemical reactions has already been addressed in this chapter. However, the preparation of the substrate or the gas phase species for reaction is only one step in a chain of events that must be rigorously followed before the desired chemical interactions occur at all. The rate at which the photo-induced reaction can proceed can depend on

(a) the irradiation geometry;

(b) the properties of the laser radiation, such as wavelength, pulsewidth, intensity and spot size;

(c) the composition of the gaseous environment;

(d) the properties of the gaseous mixture, such as pressure, temperature and flow rate;

(e) transport of the species, which may be affected by diffusion, molecular flow, convection, etc.;

(f) the properties of the substrate, such as temperature, surface structure and preparation.

It would be a monumental task to calculate the reaction rate of photo-induced reactions on a purely theoretical basis. However, if one or two limiting steps could be identified, a good approximation for the expected rate might be possible. For low incident photon flux, Ehrlich and Tsao [52] have shown that at the centre of a laser beam of intensity I, on the sample surface, surrounded by a density N_p of parent molecules, the reaction rate due to gas phase processes, $j_{\text{gas phase}}$, can be written as

$$j_{\text{gas phase}} \approx \tfrac{1}{2} k_g I N_p (P) \sigma w_0 \beta, \tag{1.14}$$

where σ is the photodissociation cross-section. The reaction rate for gas phase molecules is proportional to w_0^{-1}, where w_0 is defined as the beam waist; β is the sticking probability and k_g is a geometry-related constant.

For processes in which the controlling photochemistry occurs in a surface-adsorbed layer, the reaction rate for the adsorbed phase, $j_{\text{ads phase}}$, at the beam centre is similarly shown to be

$$j_{\text{ads phase}} \approx I\sigma\Theta(P, T), \tag{1.15}$$

i.e. it is proportional to w_0^{-2}, assuming that the photochemical cross-sections per molecule are independent of surface coverage $\Theta(P, T)$. Although not always true, this is a useful simplifying assumption.

The gas phase reaction depends on the power/radius ratio since the re-actant species are created in three dimensions above the surface, rather than in the "two" dimensional monolayer. It has been shown that, when the incident beam is approximated by an infinitely long cylinder of radius w_0, the geometrical distribution of excited molecules arriving at the surface is [52]

$$k_g = \frac{\text{solid angle}}{4\pi} = \left(1 - \frac{2}{\pi}\right)\tan^{-1}\left(\frac{1}{8}\frac{r}{2w_0}\right). \tag{1.16}$$

At the beam centre ($r = 0$), this tends to $k_g = \tfrac{1}{8}$. For gas phase reactions, the rate depends linearly on the density of the parent molecules, which in turn is directly proportional to the pressure. Therefore a linear increase in pressure will result in a corresponding increase in reaction rate. While the reaction is controlled by this mechanism, and (1.16) remains valid, there should not therefore be any strong dependence on temperature.

By contrast, reactions dominated by adsorption mechanisms are quite strongly dependent on the degree of coverage of the adsorbent, and this in turn is strongly influenced by both temperature and pressure. As already pointed out in Section 1.3.4, except for very low pressures and monolayer adsorption or chemisorption, the coverage of molecules onto the surface is very often a nonlinear function of pressure. Adsorption is by all accounts an exothermic reaction. In contrast, desorption is endothermic, requiring energy to function. Thus a reduction in temperature will favour the forward reaction of adsorption.

Rapid reaction rates can be achieved using extremely high levels of irradiance from laser systems. In fact, local depletion of donor molecules supplying the reactive species can occur. In this regime, it is not the microchemical reaction rate that limits the process, but the availability of reactive species in the processing volume, i.e. mass transport into the reaction zone. In extreme cases where very efficient radiation coupling to the precursor prevents surface reactions, copious quantities of undesirable powdery deposits may be produced. The most usual way of avoiding this is to introduce an inert buffer gas such as N_2, He or Ar. This, however, can also indirectly reduce the deposition rate of the most desirable film structure.

If the laser-induced reaction is limited to a small area on the surface, the mass transport must be modelled by a full three-dimensional diffusion equation, since the reaction now enjoys a supply of reactant from a small zone extending radially outward from its centre [53]. Conventional large-area planar reactions that are diffusion-controlled by the reactant supply or release of products use the one-dimensional diffusion equation. Considering only concentration (and not temperature) gradients in the external medium, leading to a constant diffusivity D, and assuming the reaction rate to be proportional only to the surface concentration of reactants within a small hemispherical zone of radius w_0, the steady state surface reaction rate j_{surf} is given by [53]

$$j_{surf}(t \rightarrow \infty) = \frac{Dn_\infty}{r_0 + w_0}, \qquad (1.17)$$

where $r_0 = 2D/fv$ is related to the mean free path of molecules in the gas, v is the r.m.s. velocity of the gas molecules, f is the fraction of collisions of the surface that result in a reaction and n_∞ is the fixed molecular density away from the film. This equation is plotted in Fig. 1.8 as a function of pressure for various values of w_0. Under most conditions, the steady state conditions were reached within a millisecond for $w_0 = 100$ μm, within a microsecond for $w_0 = 10$ μm, and significantly more rapidly as beam dimensions approached 1 μm. In fact, the steady state time t_{ss} can be written as

$$t_{ss} \approx \frac{1}{D} \left(\frac{1}{r_0} + \frac{1}{w_0} \right)^{-2}. \qquad (1.18)$$

For small spots, $r_0 \gg w_0$, the reaction rate $j_{surf}(t \rightarrow \infty)$ is

$$j_{surf}(t \rightarrow \infty) \rightarrow \frac{Dn_\infty}{r_0} = \tfrac{1}{2} fvn_\infty, \qquad (1.19)$$

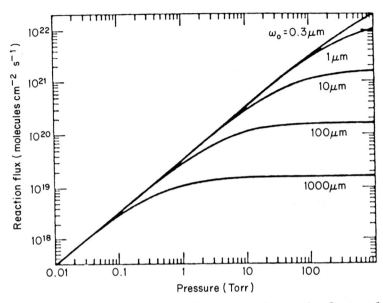

Fig. 1.8 Computed steady state values for the molecular reaction flux as a function of pressure given by (2.51) for various values of spot size. The curves are typical of thermally activated laser etching. (See the work of Ehrlich and Tsao [54] for more details.)

and is limited by interfacial kinetics, is almost linear with pressure and is independent of w_0. For the other extreme, where $r_0 \ll w_0$, the surface reaction rate can be written as

$$j_{\text{surf}}(t \to \infty) \to \frac{Dn_\infty}{w_0}. \qquad (1.20)$$

Here it is limited by diffusion kinetics, but independent of pressure since D is proportional to $1/P$ and inversely proportional to w_0. A comparison of planar reactions and localized laser-induced reactions in the high-pressure limit leads to a predicted enhancement of the latter over the former by factors approaching infinity! A more realistic comparison must recognize that most planar film growth occurs in flowing reactors, where diffusion of the necessary species only need occur across a thin boundary layer, which is typically 3 mm in an atmospheric reactor. Therefore, using $w_0 \approx 0.3$ μm, the expected geometric enhancement would be four orders of magnitude. Growth rates approaching 1 mm s^{-1} have been achievable for some thin films by taking advantage of this possibility in pyrolytic deposition.

Just as gas phase diffusion to the surface can be a mass transport limiting case, surface migration can similarly hold up the laser-induced reaction. At high laser intensities, the adsorbed molecules are rapidly transformed into their relevant products. Nucleation, clustering and film growth, however, depend upon diffusion inside the laser-influenced zone. At lower intensities, the reaction will simply be photon-limited [52].

Many thin-film reactions can be characterized by one of two types of growth process: by interface or bond-breaking kinetics, which tend to represent linear growth rates, or by a diffusion-controlled reaction, where the transport of reactant to the reaction site exhibits parabolic-type behaviour. Diffusion-limited reactions can be represented by

$$d^2(t) = \beta t \, \exp\left(-E_a/kT\right), \qquad (1.21)$$

while film growth controlled by a rate-limited process is

$$d(t) = Bt \, \exp\left(-E_a/kT\right), \qquad (1.22)$$

where $d(t)$ is the reacted thickness at time t, B and β are constants, E_a is the activation energy, and k and T have their usual meanings. For a stationary laser beam, when growth during the thermal equilibration time is negligible, the above equations can be readily applied. However, if a laser beam is scanned over a larger area such that any point on the workpiece is subjected to temperature fluctuations as the beam approaches and leaves the point then (1.22) is more accurately written as

$$d(t) = B \int_0^t \exp\left[-\frac{E_a}{kT(t')}\right] dt', \qquad (1.23)$$

where $T(t')$ is the time-varying temperature induced at the point of interest by the moving laser beam, which may be measured or calculated [54–56]. This relationship will also apply for (stationary) pulsed irradiation situations.

Where multiple overlapping scans are used, the reacted layer at a given time is a summation of all the scans, fully and partially, passing through each point. The average reacted thickness grown, $\langle d \rangle$, over an area A scanned by the beam in time t is given by

$$\langle d \rangle = \frac{1}{A} \int_0^\infty d(r) 2\pi r \, dr = \frac{2BT}{A} \int_0^\infty \exp\left[-\frac{E_a}{kT(r)}\right] dr \qquad (1.24)$$

for reaction-limited kinetics. For both (1.23) and (1.24), the temperature is the maximum induced temperature T for an "effective" time t_{eff} [55, 56]. Similar derivations can be employed for other types of reaction, governed by, for example, logarithmic and/or power laws as found in metal oxidation.

1.8 MODES OF PHOTOCHEMICAL PROCESSING

1.8.1 Geometrical configurations

Many geometric configurations for photochemical processing have been conceived. However, two distinct classes of photoprocessing can be defined, depending upon whether one requires discrete and localized reactions or large-area uniform coverage. Large-area processing is principally the domain of incoherent photon systems. Large banks of lamps, or single large-intensity lamps, in conjunction with specially designed housings, maximize the photon collection efficiency and optimize the directionality of the output. These may be external to the reaction chamber, or an integral part of it, depending upon the nature of the reaction and the wavelength of light needed. Figure 1.9 shows an example of a processing system comprising a large-area Hg lamp. This is designed to operate in conditions where the lamp itself is not subjected to any destructive processing gases or by-products (e.g. from photo-oxidation). Figure 1.10 is a schematic representation of another system using an internal lamp, where there is no window barrier between the photon source and the processing gases. Deep-UV photons are generated by the discharge and are directed into the reaction zone, which is differentially pumped to help prevent intermixing of the species. The temperature of the sample can in both cases be varied independently using a resistively heated stage. Milne and Robertson describe other lamp systems in detail in Chapter 9. The use of lasers in this regime is limited only by the photon flux levels available from present day systems (see Table 1.1) and the ability to produce geometrically uniform beams.

For pyrolytic reactions, the radiation must impinge upon the sample surface and be absorbed, as shown in Fig. 1.11, using a laser beam. For

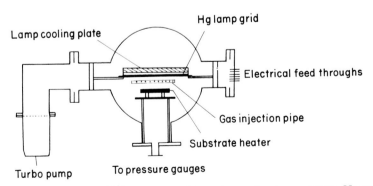

Fig. 1.9 A large-area UV lamp processing system using low-pressure Hg vapour capable of producing 185 nm light [85].

photolytic processing, the photons need not be incident upon the sample surface. The photo-excited species must only be generated close to it. The beam that controls the reaction travels some distance above and parallel to the substrate. Close proximity ensures that the photostimulated products condense and nucleate on the material in that neighbourhood and do not inefficiently coat other parts of the chamber, as is the case in conventional deposition. Even so, it is often necessary to continuously flow inert gas across the window of the chamber in order to eliminate unwanted deposition in that region. With this configuration, material can be deposited in a line along the path of the beam. Sometimes a second beam may be introduced into the chamber, incident directly onto the sample.

For microphotochemistry, maskless direct writing or pantography [57, 58] or pattern projection, which requires masks, can be performed. The concept of maskless patterning originated in the mid-1960s when a patent [59] described the use of a directed laser beam focused onto a sample to initiate certain pyrolytic chemical reactions in the presence of

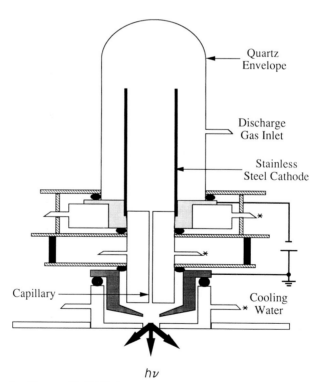

hν

Fig. 1.10 An "internal" VUV lamp system used to produce vacuum UV light applicable to direct photochemical reactions [83].

particular vapours and discussed possibilities for complex relief definition by means of movable stages. Since then, these ideas have been applied to many physical applications in laser cutting, welding, scribing and trimming of a wide variety of materials [60−63].

There are essentially two choices for direct writing. The sample can be moved by means of a programmable motorized stage, enabling positioning of the workpiece with respect to the laser beam in two, and sometimes three, dimensions. Practical limitations include speed of processing, accuracy of alignment for single layers and registration for multilayers, and resolution. There is presently much activity in the development of high-speed positioning equipment capable of 0.1−1 μm resolution. Another possibility for direct writing applications is to keep the sample stationary and to move the laser beam across the surface of the workpiece. Since lower loading of the motors is involved and angular movements can be employed, processing speeds can be considerably faster. In this mode, the sample can be held in position by a vacuum stage and conveniently pre-heated without disturbing the positioning capability of the system. The

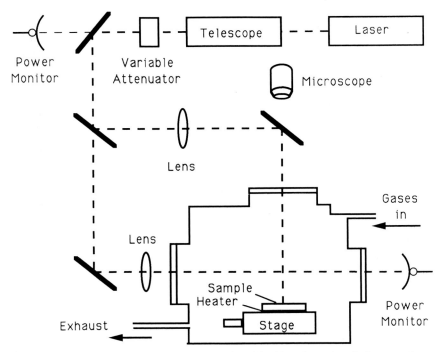

Fig. 1.11   A common experimental set-up for thin-film photochemical processing. Light can be incident parallel or perpendicular to the substrate, or both [3].

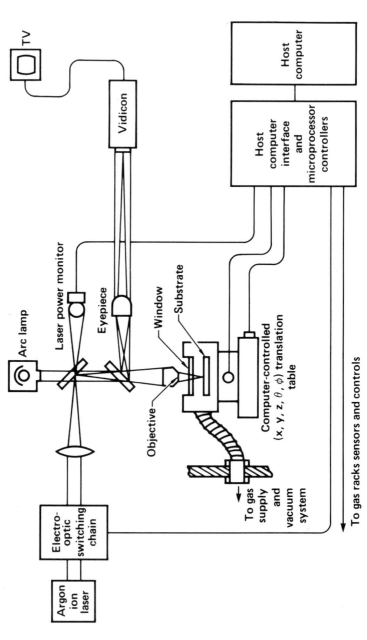

Fig. 1.12 Arrangement used for laser direct writing of microstructures [58]. (Courtesy of Lawrence Livermore Laboratory.)

optics required in this case inevitably need to be quite complicated, particularly in applications where spot-size stability is required. This problem becomes severe as one attempts to cover increasingly larger areas on the sample. Specialized optics are being developed to improve this limitation, but at present sample movement is preferable. One of the more comprehensive experimental set-ups used is shown in Fig. 1.12 [58]. The essential elements of this application include a laser with the capability of temporal modulation, an optical microscope to enable processing and simultaneous monitoring, real-time beam characterization apparatus, and a reaction chamber mounted on a computer-controlled (x, y, z, θ, ϕ) translation stage. The laser wavelength must be chosen to suit the requirements of the chemical reaction, and enable sufficiently lossless transmission of the beam through the system to the reaction zone. Different gas mixtures and optical wavelengths can readily be incorporated into the type of set-up.

The direct writing, or laser pantography, concept is an innovation in microfabrication technology, and can find applications where limited drawing or restructuring are required in low volumes (i.e. new designs, customized or non-standard devices, etc.). The technique has drawbacks in its processing time limitations [3]. It is therefore widely recognized that large-area pattern replication rather than line-by-line sequential definition is potentially more useful for extended-coverage pattern generation.

1.8.2 Pattern generation

Early work in this field used contact lithography apparatus to etch submicron patterns in various materials with feature sizes down to 0.3 μm [64, 65]. Stencil masks made by electron beam lithography in a similar contact mode, but in conjunction with an F_2 excimer laser (157 nm), have reproduced etched features as small as 200 nm in self-developing resist [66], while the same laser has reproduced mask patterns as narrow

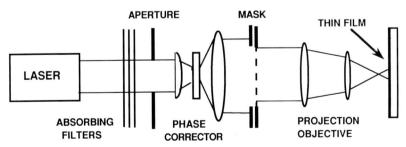

Fig. 1.13 One of the earliest reported laser projection systems for directly patterned film growth [70].

as 150 nm in conventional resists [67, 68]. These are among the smallest features yet delineated by optical lithography. Figure 1.13 shows an early Nd : glass laser projection system used by Metev *et al.* in 1980 [69] to selectively oxidize chromium, while an arrangement more akin to proximity printing has been used for etching by Loper and Tabat [70]. More advanced image projection methods have recently been applied to selective film growth. For more complete and detailed accounts of such laser lithography, the reader is referred to Chapters 3 and 4.

1.8.3 Process uniformity

Demand for process reproducibility and uniformity requires that the optical and thermal properties are uniform over the entire sample area and also from sample to sample. With respect to the radiation, it is also desirable to use a beam whose power level, profile and temporal duration are stable and repeatable. Presently, the shot-to-shot reproducibility of the best Q-switched lasers is better than 5% (except perhaps for the excimer—see Chapter 2) but closer to 20% for most high-power mode-locked lasers. Pulsed lamps exhibit much better stability. On the whole, the CW sources produce a very stable output—certainly to within better than 2%—especially when various forms of feedback mechanisms are applied.

The beam profile can often be a problem in laser systems, and, although it is always advisable to optimize this at source, it is usually necessary to spatially filter the high-frequency noise content. The clean TEM_{00} Gaussian beam retains its form after focusing, so, by Fourier transforming (focusing) a noisy beam, the higher frequencies will be displaced to points further out in the transform plane away from the focused Gaussian beam. By inserting an appropriately sized aperture at the focal point, the noise will be blocked and only the central Gaussian will pass down the optical axis. A second lens is often used to rescue the cleaned-up beam from diverging rapidly out of control.

As mentioned earlier, a uniform beam profile is often necessary for beam processing. Some laser systems are notorious for producing beams with very large intensity inhomogeneities across them (e.g. hot spots). Therefore a device that spatially homogenizes laser beams would be highly attractive. A quartz rod with a right-angle bend and a rough entrance surface can cause sufficient multiple internal reflections in a propagating beam to mix up its spatial and temporal intensity and phase, and present at the polished exit surface an essentially uniform (spatially and temporally) beam profile [71]. Modifications to this design have appeared over the years, and the better efforts give fluence uniformities within 5% over the output face of the quartz waveguide. A drawback of this method is the rapid divergence of the outcoming beam, which always demands that the sample be placed within a few millimetres of the output face of the homogenizer. Another device, based upon the principle of the kaleidoscope, has been shown to be capable of homogenizing the profile of a CO_2 laser beam to within 8%, with an overall energy loss of only 14% [72]. Lacombat et al. [73], and later Kawamura et al. [74], reported on the use of wedge prisms, while a patent by Brunsting [75] and an optical integrating device manufactured by SPAWR Optical Research [76] use segmented mirrors to achieve the necessary uniformity. Both the prism- and the mirror-based devices, which can comprise up to 25 segments [76], provide a uniform beam by optimized overlapping of the individual segments only at one single focal plane in the system. An improved version of the folding mirror system, by using a wedged tunnel configuration, produced a superior uniformity and a collimated beam [77]. With a two-stage device, an intensity uniformity of approximately ± 1.5% was achieved over an area containing around 75% of the beam energy. Both a patent [78], and a publication [79] describe the use of specially designed lenses to redistribute the peak central intensity of the laser beam around the profile edges. Here, unlike in the applications mentioned above, the phase of the wavefronts is not disturbed, and spatially coherent monomode beams may be homogenized.

1.9 SUMMARY

The aim of this chapter was to briefly set a foundation from which many of the subsequent chapters could be more readily digested. In it, the concept of photochemical processing, applied to thin films and surfaces, has been introduced. The current state-of-the-art development of the wide variety of photon sources available, including lasers and lamps, has been briefly reviewed. The fundamental mechanisms behind photon–atom interactions have been introduced, including absorption mechanisms in solids and in gases, energy redistribution and heating, the interaction of species with surfaces, and different modes of film growth and particle removal. Several important practical aspects of photoprocessing have been described, such as geometrical styles and methods for obtaining uniform beams and producing localized patterns on surfaces.

REFERENCES

[1] D. J. Ehrlich and J. Y. Tsao (eds), *Laser Microfabrication: Thin Film Processes and Lithography* (Academic Press, San Diego, 1989).

[2] K. G. Ibbs and R. M. Osgood (eds), *Laser Chemical Processing for Microelectronics* (Cambridge University Press, 1989).

[3] Ian W. Boyd, *Laser Processing of Thin Films and Microstructures* (Springer-Verlag, Berlin, 1987).

[4] D. Bäuerle, *Chemical Processing with Lasers* (Springer-Verlag, Berlin, 1986).

[5] J. C. Mathews, M. G. Ury, A. D. Birch and M. A. Lashman, *Proceedings of SPIE Conference on Microlithography, 1983.*

[6] B. Eliasson and U. Kogelschatz, *Appl. Phys.* **B46**, 299 (1988).

[7] M. Snells, E. Borsella, R. Fantoni and A. Giardini-Guidoni, in *Laser Processing and Diagnostics* (ed. D. Bäuerle), p. 210 (Springer-Verlag, Berlin, 1984), and references therein.

[8] M. Bass and J. R. Franchi, *J. Chem. Phys.* **64**, 4417 (1976).

[9] T. J. Chuang, *J. Vac. Sci. Technol.* **21**, 798 (1982).

[10] J. I. Steinfeld, T. G. Anderson, C. Reiser, D. R. Denison, L. D. Hartsough and J. R. Hollahan, *J. Electrochem. Soc.* **127**, 514 (1980).

[11] T. J. Chuang, *J. Chem. Phys.* **74**, 1453 (1981).

[12] F. A. Houle and T. F. Chuang, *J. Vac. Sci. Technol.* **20**, 790 (1982).

[13] T. J. Chuang, *IBM J. Res. Develop.* **26**, 145 (1982).

[14] G. E. Jellison and F. A. Modine, *Appl. Phys. Lett.* **41**, 180 (1982).

[15] M. Rothschild, in [1], p. 163.

[16] R. A. Smith, *Semiconductors* (Cambridge University Press, 1979).

[17] N. Bloembergen, *IEEE J. Quantum Electron.* **10**, 375 (1974).

[18] J. A. McKay and J. T. Schriemf, *IEEE J. Quantum Electron.* **10**, 2008 (1981).

[19] R. F. Marks, R. F. Pollak, Ph. Avouris, C. T. Lin and Y. J. Thefaine, *J. Chem. Phys.* **78**, 4270 (1983), and references therein.

[20] T. J. Chuang, in [1], p. 87.

[21] J. Heidberg, H. Stein and E. Riehl, *Phys. Rev. Lett.* **49**, 666 (1982).

[22] B. Schafer and P. Hess, *Appl. Phys.* **B37**, 197 (1985).

[23] R. S. Lichtman and D. Shapira, *CRC Crit. Rev. Solid State Mater. Sci.* **8**, 93 (1978).

[24] M. S. Slutsky and T. F. George, *J. Chem. Phys.* **70**, 1231 (1979).

[25] M. S. Dzhidzhoev, A. I. Osipor, V. Ya. Panchenko, V. T. Platonenko, R. V. Khokhlow and K. V. Shaitan, *Sov. Phys. JETP* **47**, 684 (1978).

[26] S. Baidyaroy, W. R. Bottoms and P. Mark, *Surf. Sci.* **28**, 517 (1971).

[27] S. H. Brunauer, P. H. Emmett and E. Teller, *J. Am. Chem. Soc.* **60**, 309 (1938).

[28] C. J. Chen and R. M. Osgood, *Appl. Phys.* **A31**, 171 (1983).

[29] D. J. Ehrlich and R. M. Osgood, *Chem. Phys. Lett.* **79**, 381 (1981).

[30] D. J. Ehrlich and J. Y. Tsao, *Appl. Phys. Lett.* **46**, 198 (1985).

[31] W. Brown, in *Laser and Electron Beam Solid Interactions and Material Processing* (ed. J. F. Gibbons, L. D. Hess and T. W. Sigmon), p. 20 (North-Holland, Amsterdam, 1981).

[32] E. J. Yoffa, *Phys. Rev.* **B26**, 2415 (1981).

[33] N. Bloembergen, in *Laser–Solid Interactions and Laser Processing* (ed. S. D. Ferris, H. J. Leamy and J. M. Poate), p. 1 (AIP, New York, 1979).

[34] H. S. Carslaw and J. C. Jaeger, *Conduction of Heat in Solids* (Clarendon Press, Oxford, 1959).

[35] M. Lax, *J. Appl. Phys.* **48**, 3919 (1977).

[36] Y. I. Nissim, A. Lietoila, R. B. Gold and J. F. Gibbons, *J. Appl. Phys.* **51**, 274 (1980).

[37] J. E. Moody and R. H. Hendel, *J. Appl. Phys.* **53**, 4364 (1982).

[38] F. Ferrieu and G. Auvert, *J. Appl. Phys.* **54**, 2646 (1983).
[39] P. Schvan and R. E. Thomas, *J. Appl. Phys.* **57**, 4378 (1985).
[40] A. E. Bell, *RCA Rev.* **40**, 295 (1979).
[41] P. Baeri, S. U. Campisano, G. Foti and E. Rimini, *J. Appl. Phys.* **50**, 788 (1979).
[42] J. R. Meyer, M. R. Kruer and F. J. Bartoli, *J. Appl. Phys.* **51**, 5513 (1980).
[43] M. Bertolotti and C. Sibilia, *IEEE J. Quantum Electron.* **17**, 1980 (1981).
[44] R. F. Wood, *Phys. Rev.* **B25**, 286 (1982), and references therein.
[45] D. L. Kwong and D. M. Kim, *J. Appl. Phys.* **54**, 366 (1983).
[46] D. Agassi, *J. Appl. Phys.* **55**, 4376 (1984).
[47] M. O. Thompson, G. J. Galvin, J. M. Mayer, P. S. Peercy, J. M. Poate, D. C. Jacobson, A. G. Cullis and N. G. Chew, *Phys. Rev. Lett.* **52**, 2360 (1984).
[48] L. Eckertova, *Physics of Thin Films* (Plenum, London, 1984).
[49] J. A. Venables, *Phil. Mag.* **27**, 697 (1973).
[50] K. Takeuchi and K. Kinosita, *Thin Solid Films* **90**, 31 (1982).
[51] G. Zinsmeister, *Vacuum* **16**, 529 (1966).
[52] D. J. Ehrlich and J. Y. Tsao, *Proc. SPIE* **459**, 2 (1984).
[53] D. J. Ehrlich and J. Y. Tsao, *J. Vac. Sci. Technol.* **B1**, 969 (1981).
[54] A. Lietoila, R. Gold and J. F. Gibbons, *Appl. Phys. Lett.* **39**, 810 (1981).
[55] R. B. Gold and J. F. Gibbons, *J. Appl. Phys.* **51**, 1256 (1980).
[56] Z. L. Liau, *Appl. Phys. Lett.* **34**, 221 (1979).
[57] D. J. Ehrlich, R. M. Osgood and T. F. Deutsch, *Appl. Phys. Lett.* **39**, 957 (1981).
[58] B. M. McWilliams, I. P. Herman, F. Mitlitsky, R. A. Hyde and L. L. Wood, *Appl. Phys. Lett.* **43**, 946 (1983).
[59] R. Solomon and L. F. Mueller, *US Patent* 3 364 087 (January 1968).
[60] S. S. Charschan, *Lasers in Industry* (Van Nostrand, New York, 1972).
[61] W. W. Duley, CO_2 *Lasers. Effects and Applications* (Academic Press, New York, 1976).
[62] J. F. Ready, *Industrial Applications of Lasers* (Academic Press, New York, 1978).
[63] H. Koebner, *Industrial Applications of Lasers* (Wiley, Chichester, 1984).
[64] K. Jain, C. G. Wilson and B. J. Lin, *IBM J. Res. Dev.* **26**, 151 (1982).
[65] K. Jain, *Lasers & Applics* **2**(9), 49 (1983).
[66] D. Henderson, J. C. White, H. G. Craighead and I. Adesida, *Appl. Phys. Lett.* **46**, 900 (1985).
[67] H. G. Craighead, J. C. White, R. E. Howard, L. D. Jackel, R. E. Behringer, J. E. Sweeney and R. W. Epworth, *J. Vac. Sci. Technol.* **B1**, 1186 (1983).
[68] J. C. White, H. G. Craighead, R. E. Howard, L. D. Jackel, R. E. Behringer, R. W. Hepworth, D. Henderson and J. E. Sweeney, *Appl. Phys. Lett.* **44**, 22 (1984).
[69] S. M. Metev, S. K. Savtchenko and K. Stamenov, *J. Phys.* **D13**, L75 (1980).
[70] G. L. Loper and M. Tabat, *Proc. SPIE* **459**, 121 (1984).
[71] A. G. Cullis, H. C. Weber and N. G. Chew, *J. Phys.* **E12**, 688 (1979).
[72] R. E. Grojean, D. Feldman and J. F. Roach, *Rev. Sci. Instrum.* **51**, 375 (1980).
[73] M. Lacombat, G. M. Dubroeucq, J. Massin and M. Brevignon, *Solid State Technol.* August, 115 (1980).
[74] Y. Kawamura, Y. Itagaki, K. Toyoda and S. Namba, *Optics Commun.* **48**, 44 (1983).

[75] A. Brunsting, *US Patent* 4 327 972 (May 1982).
[76] SPAWR Optical Research, 1527 Pomona Road, Corona, CA 91720, USA.
[77] M. R. Latta and K. Jain, *Optics Commun.* **49**, 435 (1984).
[78] H. J. Kreuzer, *US Patent* 3 476 463 (1969).
[79] D. Shafer, *Optics & Laser Technol.* **14**, 159 (1984).
[80] H. H. Gilgen, C. J. Chen, R. Krchnavek and R. M. Osgood, in *Laser Processing and Diagnostics* (ed. D. Bäuerle), p. 225 (Springer-Verlag, Berlin, 1984).
[81] C. J. Chen and R. M. Osgood, *J. Chem. Phys.* **81**, 327 (1984).
[82] V. Nayar, P. Patel and I. W. Boyd, *Electron. Lett.* **26**, 205 (1990).
[83] P. Patel and I. W. Boyd, *Appl. Surf. Sci.* **46**, 352 (1990).

2 Developments in Excimer Lasers for Photochemical Processing

J. M. GREEN and M. R. OSBORNE

Laser Technology Centre, AEA Technology, Culham Laboratory, Abingdon, Oxfordshire, UK

2.1 INTRODUCTION

Lasers are an enabling technology—and nowhere more so than in the application of excimer lasers to the field of photochemical processing, where the availability of powerful ultraviolet (UV) laser radiation is inspiring a technological revolution. After over a decade of intense development, one class of excimer laser, the discharge pumped rare gas halide (RGH) laser, has achieved an industrial level of reliability and ease of operation, and is making rapid progress towards an output of over one kilowatt power. With carbon dioxide and neodymium–YAG, it will complete a suite of lasers offering high powers from the ultraviolet through to the far-infrared, for exploitation in a wide range of materials processing applications.

The aim of this chapter is to review the present state of the art of commercial excimer lasers (Section 2.3) and to indicate the improvements that the current worldwide programme of laser development is likely to effect (Section 2.4). To put these sections into context, and to understand better the inherent strengths and weaknesses of excimer lasers, Section 2.2 gives an outline of the basic physics of excimer lasers, and a brief account of their historical development. The complex relationship between laser parameters and photochemical processing is also described in this section.

2.2 BACKGROUND

2.2.1 Excimer species

Excimer molecules have dissociative ground states, which make them ideal lower laser levels. Many species have been examined as potential high-

Photochemical Processing of Electronic Materials
ISBN 0-12-121740-X

Table 2.1 Partial listing of lasing excimer species. Those in the top section can all be efficiently excited in a common discharge laser system, and are listed in order of performance. The excimers in the middle section can also be made to lase in discharge systems, whereas those in the bottom section require intense e-beam excitation.

Lasing molecule	Wavelength (nm)	Photon energy (eV)
KrF^*	248	5.0
$XeCl^*$	308	4.0
ArF^*	193	6.4
XeF^*	351	3.5
F_2^*	157	7.9
$KrCl^*$	222	5.6
Xe_2F^*	610 ± 65	1.8–2.3
$HgBr^*$	about 500	2.5
$HgCl^*$	558	2.2
HgI^*	444	2.8
Hg_2^*	335	3.7
Xe_2Cl^*	490 ± 40	2.3–2.8
XeO^*, KrO^*	558	2.2
Ar_2^*	126	9.9
Kr_2^*	147	8.5
Xe_2^*	172	7.2

power laser systems, and some of them are listed in Table 2.1, together with their lasing wavelength. (Most of these species are strictly exiplexes.) The reader is referred to any of several reviews [1–4] for a more comprehensive listing. The laser physics of these species dictates that very high instantaneous excitation intensities of a multiatmospheric gas mixture are required to reach lasing threshold: typically of order $100 \, kW \, cm^{-3}$ for the longer wavelengths (e.g. XeF^*), and up to of order $10 \, MW \, cm^{-3}$ at short wavelengths (e.g. Ar_2^*). This requirement is at the heart of many of the technological challenges of excimer lasers. The first technique able to provide such intensities was the electron beam. Injecting a high-energy electron beam (e-beam) through a foil window into the laser gas mixture, all of the species in Table 2.1 can be made to lase, and this remains the method of choice for the largest laser pulse energy systems. However, heating of the foil window, which must bear the pressure differential of several atmospheres between the laser gas and the vacuum environment of the e-beam source, severely limits the pulse repetition frequency (typically $\leqslant 1 \, Hz$) and results in a comparatively short lifetime (10^2–10^4 shots), making these systems unsuitable for industrial applications.

The major breakthrough in producing industrially acceptable excimer lasers was the development of avalanche discharge excited systems in 1976. The technology of such systems is much more rugged than for e-beams, but imposes several restrictions on the composition and pressure of the gas mixture. Only those species that lie in the upper two sections of Table 2.1 have been made to lase reliably under discharge excitation. Of this subset, only those in the upper section lase readily in a common laser system, and hence these are the most frequently exploited lasing species—the so called rare gas halide (RGH) group, with the exception of F_2, which is clearly not a rare gas halide, nor even an excimer in the strict sense. In general, KrF^* and $XeCl^*$ give the best laser performance, with KrF^* producing the highest powers (usually), and $XeCl^*$ always giving the longest laser gas lifetime. ArF^* and XeF^* are close behind KrF^* in terms of power, but further away from $XeCl^*$ in gas lifetime. $KrCl^*$ and F_2^* generally produce significantly lower powers, although the gas lifetime of the former is quite good, and the short wavelength of the latter attractive for some applications.

2.2.2 Laser design

The very high power deposition required in the laser discharge (typically of order 100 MW) has several important consequences for the laser design. The first is that the laser is operated in pulsed mode only, and relatively short pulses at that. Laser output pulse durations of 10–20 ns have been the rule until recently. The second is a requirement for very low inductance discharge circuitry and switching. This dictates the TEA geometry familiar from CO_2 lasers (Fig. 2.1), in which two extended electrodes are placed parallel to the optical axis. The third requirement is that the gas heated by the discharge (the so-called "active volume") be removed from the electrode region between laser pulses, to prevent thermal gradients causing discharge instability. A further requirement, brought about by the need to employ high (multi-atmospheric) gas pressures, is a need to preionize the active volume before voltage is applied to the electrodes. Without this, an arc rather than a glow discharge is produced, resulting in no laser output. This requirement is particularly severe in excimer lasers because the gas mixture includes a strongly electronegative halogen (e.g. F_2 or HCl), which reduces the lifetime of any free electrons to only of order 1 ns.

The basic scaling of excimer lasers is illustrated in Fig. 2.1. Increasing the discharge volume increases the pulse energy, while an increase in gas flow is necessary to increase the pulse repetition frequency (PRF).

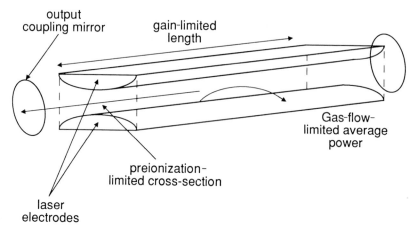

Fig. 2.1   Schematic of TEA (transverse electric atmospheric) RGH laser.

2.2.3   Photochemical application considerations

Pulse energy and repetition frequency are only two of many parameters that influence the commercial and scientific viability of a particular photo-chemical application. The considerations for a general photochemical process can be conveniently divided into three areas: first, the nature of the photochemical interaction, which is influenced primarily by the laser wavelength, pulse duration and pulse energy/peak power; secondly, the propagation, imaging and focusing properties of the laser beam, its size, uniformity, divergence and spectral bandwidth, which determine what may be termed the quality of the process; thirdly, considerations of commercial viability, which are influenced by such parameters as the capital and running cost of the laser and its reliability, and others related to the throughput of the process, principally average power and PRF. Clearly, particular laser parameters can influence more than one area of consideration. For example, while laser bandwidth is important in the context of chromatic aberrations, it can also influence the nature of the photochemical process: average power and PRF can influence the photochemical process through heating effects, as well as having an impact on the overall throughput of the process.

The technology of the RGH discharge laser can be similarly divided into three key technology areas, but, unlike the three areas considered above, the parameters of the lasers are almost invariably influenced by more than one key technology. This results in a rather complex relationship between laser application and laser system technology as summarized in Table 2.2. In describing the three key technologies, "discharge" refers to the tech-

Table 2.2 Applications considerations for excimer laser technological developments.

Laser beam utilization		Laser system technology		
Aspect	Considerations	Discharge	Gas	Optics
Beam interaction parameters				
Wavelength	Bond breaking		•	○
Pulse duration/peak power	X-ray generation, nonlinear effects	•		○
Pulse energy	Thermal effects	•		○
Propagation, imaging and focusing				
Beam size and uniformity	High uniformity for photolithography	•		○
Beam divergence	Time-varying			•
Spectral bandwidth	Chromatic aberration			•
Additional process requirements				
Capital cost	Depreciation and interest	•	○	○
Running cost	Switches, gas consumption	•	•	○
Reliability	Component lifetime, control system	•	•	•
Reproducibility	3% s-s jitter at present	•		
Average power	Parallel processing gives higher throughput	•	•	○
Pulse repetition frequency	Absorption depth limits material removed per pulse	•	•	○

•, strong dependence; ○, weak dependence.

nology of discharge excitation: preionization, high-voltage switching and electrode profiling. "Gas" refers to the composition and condition of the laser gas mixture, the control of impurities (which have a significant effect on laser performance), gas circulation and shock wave management. Finally, "optics" refers to the design of the laser resonator, including intracavity components, as well as the coupling of more than one laser in an oscillator–amplifier configuration, and wavelength shifting techniques, beam combining and pulse stretching/compression techniques.

2.2.4 Early excimer laser development

The early commercial excimer lasers, the first of which appeared in 1976, drew heavily on existing TEA CO_2 laser technology. However, the greater demands that the excimer lasers place on the preionization and the pulsed power and switching lead to less than optimum performance and unacceptably reduced component lifetimes. Incompatibility between the highly corrosive excimer laser gas mixture and some of the materials used in the early lasers also resulted in degradation of the laser components and, more immediately, poisoning of the expensive laser gas mixture. The study of the halogen compatibility of materials under the conditions existing inside the laser (high pressure, shock waves, heat, intense UV irradiation and electrical stress) has yielded many improvements in this area over the years, but remains an ongoing science.

From a very early stage, excimer manufacturers throughout the world adopted the so-called automatic preionization circuit developed in the UK [5]. This remained in almost universal use until the late 1980s. Over this period, although the basic laser design did not change, many refinements were incorporated.

Possibly greater emphasis has been placed on improving reliability and user friendliness than on outright laser performance. For example, in addition to the improved gas lifetime alluded to above, the high-voltage switching stage has seen constant development, from spark gaps, to thyratrons, and then with varying degrees of magnetic assistance and compression. Perhaps more than any other development, however, the incorporation of micro-processor control has revolutionized the operation of excimer lasers.

2.3 MODERN COMMERCIAL EXCIMER LASERS

2.3.1 Performance

Most manufacturers now offer lasers with some form of microprocessor control. This has finally rid excimer lasers of the reputation of "needing a PhD student to operate them". The microprocessor can take care of routine tasks (such as evacuating the laser and refilling with some preset gas mixture), and can also effectively enhance laser performance (such as by keeping the pulse energy/average power constant by a combination of varying the applied voltage and adjusting the gas mixture). With the addition of the microprocessor, the excimer laser now closely approximates a reliable "black box" source of UV photons. The performance available from commercial excimer lasers is improving constantly, and any figures given here will date rapidly. However, to give some idea of orders of magnitude, Table 2.3 presents, in the first column, the specification of a typical (hypothetical) 1990 commercial system. These specifications, which are all attainable simultaneously, are typical of the performance of a "middle of the range" laser operating on KrF^* from any of the major manufacturers.

Table 2.3 Performance summary of discharge excited rare gas halide lasers.

Discharge pumped rare gas halide laser performance	Typical commercial	Extreme commercial	Extreme research
Applicability of specifications	Simultaneous	Not simultaneous	Not simultaneous
Average power (W)	80	160	750
Pulse energy (J)	0.4	2.0	80
Pulse repetition frequency (Hz)	200	1000 (2000)[a]	4000, 3500 (12 000)[a]
Pulse duration (ns)	15	250 (0.03)[b]	1500 (0.0002)[b]
Peak power (W)	5×10^7	10^8	10^9 (10^{15})[b]
Spectral bandwidth (GHz)	900	5	0.15
Beam divergence (mrad)	3	0.2	0.012

[a] excited by microwave discharge.
[b] system employing more than one laser.

Lasers of this specification have existed for a number of years now, and are likely to continue to be sold for many years to come.

Each of the specifications in the first column can be bettered by other commercial systems, although there is often a trade off (for example, higher PRF tends to imply lower pulse energy). The second column indicates the extreme values of each parameter currently available in a commercial system. It must be emphasized that these specifications are *not simultaneous*: no laser exists that can meet all of these parameters at the same time. Furthermore, it must be remembered that these figures apply to avalanche and microwave discharge excited lasers only. Electron-beam excited lasers are commercially available that can exceed some of these figures, pulse energy in particular.

The final column in Table 2.3 represents the extreme parameters that have been obtained in research laboratories (as of July 1990), and will give some indication of what may become commercially available in the future.

The parameters listed in Table 2.3 are those that have the most obvious influence on any photochemical application of excimer lasers. However, there are several other parameters that should not be ignored. Perhaps the most important of these is stability. Shot-to-shot energy fluctuations are generally guaranteed to about ±3%, but many manufacturers add the caveat that *all* of their individual pulse specifications are based on 90% of pulses! The remaining 10% are presumably totally unspecified. This unsatisfactory situation is a throwback to the days when excimer lasers were far less well developed than they are today, and does a severe disservice to the credibility of the modern laser. Experience shows that pulses lying out of the specified energy band are very rare indeed on the current generation of lasers, and one can only hope that the specifications are changed to acknowledge this fact in the near future.

Another parameter of importance in some applications is timing jitter with respect to an external event. This may be as good as about ± 2 ns on a new laser, but will degrade with age. Manufacturers may not specify jitter unless asked.

2.3.2 Reliability and running costs

All of the elements in the laser pulsed power circuit are subject to considerable electrical stress, and have finite lifetimes. In some early excimer circuits, the thyratron was clearly the weakest link, and one of the most expensive to replace. However, with circuits designed to stress this component less, its lifetime has been extended to meet that of other components. Several manufacturers now recommend a "scheduled overhaul" after something like 10^9 shots, at which time the laser electrodes, preionizer pins and thyratron will be refurbished or replaced on exchange. The exact interval and the work carried out varies between manufacturers, and is under constant review, but it is necessary to budget some fraction of the capital cost of the laser (maybe 20–40%) for this service. Failure of all other components is unlikely, although in such a complex system unscheduled down time cannot be ruled out. Some manufacturers may provide MTBF figures, or may guarantee laser availability.

Routine maintenance is limited to cleaning of the laser optics. This can mean total loss of the laser gas mixture unless gate valves are fitted. Cleaning intervals vary according to use (and are of order 10^6–10^8 shots), and in practice it is difficult to thoroughly clean the optic (inside and out) without causing some damage. Somewhat degraded laser performance may be expected after, say, ten cleaning operations.

The economics of running an excimer laser depend on the way in which it is used. The reader is referred to the review by Klauminzer [6] for one summary. In almost all cases, capital cost/depreciation and interest will be significant. If the laser is run on a production line basis (i.e. two-shifts operation at high repetition rate, constant operation with the same excimer species) then the maintenance, gas consumption and electrical consumption costs may be comparable to the depreciation and interest. If used in a scientific application (intermittent operation at lower repetition rate, and changing excimers) the gas costs will dominate.

The lifetime of the excimer laser gas mixture has been the subject of a great deal of confusing information in the past. The gas mixture is degraded either by loss of one (or more) of the gaseous components (usually the halogen) by reaction with some material inside the laser, or by the production of species that "poison" the gas and spoil the laser kinetics. Both of these processes can occur when the laser is running, and also when it is turned off. Hence there are both static and dynamic gas lifetimes. Lasers that are good in the latter respect (for example, with efficient discharge leading to little erosion of the electrodes and preionizer pins) need not be good in the former (with regard, for example, to choice of materials and size of reservoir) and vice versa. Manufacturers can therefore choose to represent their laser in the most attractive way. The introduction of on-line gas processors to improve gas lifetime by removing impurities has allowed further scope for confusion, and finally the advent of microprocessor controlled gas renewal and replenishment systems has made a simple gas lifetime impossible to define. However, although "gas lifetime" is no longer meaningful, "gas consumption" (defined as the gas consumed, under the control of the microprocessor, to keep the laser output at a specified level) certainly is. As the trend towards quoting gas consumption (and static lifetime) continues, the potential user will, for the first time, be able to make reliable gas cost comparisons.

2.3.3 Future trends

Since the mid-1980s, the major manufacturers have sought to develop lasers aimed specifically at the industrial market place, and sell these in parallel with their existing "scientific" models. The industrial lasers typically employ heavily overrated components, and may be optimized for a single gas mixture. The differing needs of the shop floor and the laboratory will lead to further diversification of the two strands, although some of the convenience features of the industrial models are optional on scientific lasers.

As noted earlier, the basic design layout of most of today's lasers was settled in the mid-1970s. However, within the past few years, several new approaches have appeared. In addition to the commonly used spark preionization, lasers using corona (for longer life and lower gas contamination) and X-ray (for larger volumes and pulse energies) preionization have reached the market place. Also, a "photoswitched" laser, eliminating the expensive conventional switch in the main discharge circuit, and lasers excited by microwaves offering very high pulse repetition rates are targetting lower average power applications.

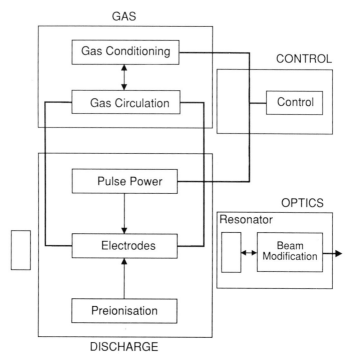

Fig. 2.2 Interaction of the various laser components, divided into "discharge", "gas", "optics" and "control". The modest control shown is typical of that provided automatically in commercial systems.

2.4 EXCIMER LASER DEVELOPMENT

There is currently intense interest in Japan, the USA, the USSR and Europe in developing high-average-power excimer lasers. The efforts may conveniently be divided into five areas—sensing and control, preionization, pulse power, gas flow and optics—as shown in Fig. 2.2. These will be discussed in turn.

2.4.1 Sensing and control

Although laser manufacturers have greatly simplified the operation of their lasers using microprocessor based control, there is at present only the minimum of monitoring of the actual laser performance. Only the laser pulse energy is directly measured, and any decrease initiates replenishment or partial exchange of the laser gas, degradation of which is assumed to be the cause of the energy decrease. This ignores all of the other possible causes of decline (e.g. dirty or misaligned optics, failing capacitor or thyratron, and worn electrodes or preionization pins). Systems to constantly monitor these features (and automatically correct where possible) are currently under development, the goals being improved system stability, and prior warning of imminent component failure. At present the only on-line guide offered on commercial systems is the total shot count of the laser.

2.4.2 Preionization

It has long been realized that a uniform distribution of free electrons (of order 10^8 cm^{-3}) within the discharge region prior to the application of the voltage is essential to the formation of a uniform and stable laser discharge in an excimer laser gas mixture. Taylor [7] provides a good introduction to this complicated subject.

A very specific problem posed by RGH lasers is the rapid dissociative attachment of electrons to the halogen donor molecule, creating negative ions. The accumulation of low-ionization-potential negative ions may in principle be exploited as a ready source of electrons (see below). In practice, however, a "flash" preionizer creates electrons at a sufficient rate to compete directly with dissociative attachment, to create an instantaneous free electron density of 10^6–10^8 cm^{-3}.

"Flash" UV sources include corona types and the commonly used sparks. Sparks are undoubtedly the more powerful ionization source, but are localized, and result in erosion of the pin electrodes (low lifetime). Furthermore, the sputtered materials contaminate the gas mixture. Corona discharge over the surface of a dielectric produces much less contamination. The recent results of a record pulse duration (> 1 µs) in XeCl at good efficiency with simple corona preionization [8] give hope that this rugged and simple technique can be exploited further. Plots of laser output pulse energy versus preionization energy for different types and arrangements of preionizers but the same electrode configuration show a saturation of laser output power at high preionization, but at different values for each [9]. This implies that preionization uniformity rather than density is the ultimate issue.

Extended, robust UV preionization is provided by the plasma electrode, in which a quasi-CW surface (sliding) discharge across a dielectric provides UV preionization as well as acting as an electrode. Microsecond XeCl pulses have been achieved in this way, but the lifetime of the dielectric has so far been found to be severely limited by thermal loading [10].

The most significant development in excimer laser research in the last decade has been the development of the X-ray preionizer. The major attraction of X-ray preionization is its greater range. UV radiation has a useful preionization range of a few centimetres only (which limits the performance of commercial devices), whereas X rays have useful ranges of metres, and can be located sufficiently far from the discharge region to provide highly uniform and well defined preionization. Other benefits of X-ray preionization are that the preionizer can be located behind a solid electrode (as opposed to a mesh for some UV schemes) and that it gives rise to no contamination of the laser gases.

A trend towards improved energy extraction and pulse duration has been demonstrated for X-ray preionization over UV alternatives [11]. However, from a practical point of view, the complex and inherently inefficient preionizer with a relatively large and separately pumped vacuum chamber needing high voltage (100–150 kV) and X-ray screening is a severe disadvantage. For large-cross-section discharges there is no practical alternative to X rays (apart from an auxiliary UV laser [7]), but the reliability of X-ray sources and their performance at high PRF remains to be demonstrated experimentally, and for small-cross section discharges the corona UV source holds particular attractions, especially for high (>1000 Hz) PRF operation.

The problem of X-ray preionizer operation at PRFs in excess of 1 kHz rests in the development of a reliable electron beam cathode with long lifetime. Field emission cathodes (such as carbon whiskers) traditionally used in large electron beam sources have a very limited life. The corona plasma source [12] gives a plentiful (space-charge-limited) supply of electrons. However, because it operates in a vacuum, the corona discharge necessarily requires the vaporization of some dielectric material, the bulk of which is not easily cooled (these drawbacks do not apply to the situation of the corona preionizer, where the UV-emitting discharge takes place in a surrounding atmosphere of gas).

Thermionic cathodes based on low-work-function coatings on tungsten filaments can in principle supply a sufficient electron density, and at high PRFs, but suffer from the major operating difficulty that high-vacuum conditions must be maintained to prevent contamination of the cathode surface.

It is arguable from the above that a technology breakpoint may exist for preionizers at a PRF of about 1 kHz for a laser average power of about 1 kW, since X-ray preionizers favour both higher laser pulse energies (i.e. larger discharge cross-sections) and lower PRFs, while UV sources are limited to lower laser pulse energies but can operate at substantially higher PRFs.

The major thrust in preionizer development remains in the field of flash sources. However, it has been demonstrated that the electrons lost by dissociative attachment can be liberated from the accumulating negative ions some time later, and these ions could be produced by a relatively long preionization pulse. Careful analysis of the mechanism for electron release [13] favours UV over collisional detachment to explain the observation that uniform discharges can be produced many attachment times after termination of the preionization pulse. The technique of negative-ion assisted preionization (NIAP) may have particular relevance to high-PRF laser operation, since in principle it reduces the peak power loading on the preionizer [14].

2.4.3 Pulsed power

Three areas of development merit special attention: magnetic techniques to reduce the requirements on the primary switch, the use of pulse forming lines or networks (PFLs or PFNs) to improve energy transfer and laser pulse duration, and preionization switching as a means of further simplifying (and reducing) switching requirements.

The saturable inductor magnetic switch is passive (the magnetic material carries only a fraction of the current it switches) and has an almost infinite lifetime. However, it is not a true on/off switch, and requires a small magnetizing current to achieve its switching action. "Magnetic assist" (to reduce the current rise time through the thryatron switch) and "magnetic switch control" (to prevent current reversal and provide some pulse sharpening) are presently employed in commercial excimer lasers. However, more ambitious magnetic pulse compression using a chain of coupled capacitor/saturable inductor resonant circuits (a Melville line) can give orders-of-magnitude enhancement in current rise times as well as optional voltage multiplication. Of particular importance in this respect is the possibility of using all solid state (thyristor) switching of excimer lasers. This would represent a major technological breakthrough for high-reliability high-PRF performance. Its feasibility for high-power excimer lasers has recently been demonstrated [15a, b].

Commercial excimer lasers largely favour simple capacitor discharge circuits [16], primarily because these circuits are most tolerant to differences in gas composition. Pulse forming lines can be more efficient, and can transfer 100% of their stored energy to the laser gas at constant voltage (i.e. with no ringing), but only provided that the PFL and discharge have matched impedances. These requirements define specific operating conditions. The initial voltage on the PFL under these conditions is twice the self-sustaining voltage of the discharge [17–19] and is generally several times too low to achieve the necessary fast gas breakdown, so a separate higher-voltage ("spiker" or prepulse) circuit is needed to initiate the discharge. This voltage may be applied to one of the electrodes, in which case a saturable inductor [18, 20] provides the necessary isolation between it and the PFL. A recently reported alternative uses a midplane grid to initiate the discharge [21].

The spiker–sustainer (PFL) circuit described above has produced the highest efficiencies and the longest pulses from discharge-excited lasers [8, 22]. Its use also greatly relieves the switching requirements, since in principle only the spiker need be switched—and this carries only a small fraction of the total deposited energy. The PFL is connected directly to the electrodes, so that during its charging (before the spiker pulse is initiated) a (low) voltage is also present across the discharge electrodes. To reduce the probability of premature breakdown of the discharge, the charging circuits for the PFL may be switched, but the requirements on a second thyratron would be quite modest. As an alternative to the spiker circuit, UV (laser, spark and corona) and X-ray triggering of the discharge have been used successfully [23, 24]. In this way, the switching requirements are still further reduced. The drawback of the technique is that the PFL must be charged to a substantially higher voltage than required for spiker–sustainer operation in order to achieve sufficiently rapid gas breakdown, which means that the most efficient energy transfer conditions cannot be satisfied. Use of this technique also greatly increases the probability of premature breakdown, partly in high-PRF operation.

Radiofrequency excitation of gas lasers has recently made a comeback in the infrared, but the potential of this technique for UV lasers remains to be fully demonstrated. Less contamination from electrodes in contact with the gas is obviously of particular relevance in the context of RGH lasers. In addition, results to date show that the stability of the RF excited discharge is considerably better than that of its DC counterpart: laser pulse durations compare well with the longest that have been achieved by other discharge techniques (about 500 ns, and limited by halogen donor depletion) [25], and high-PRF laser operation (up to 8 kHz) has been achieved without gas flow [26]. Yet, although volume efficiencies of RF-excited XeCl and XeF lasers compare favourably with their DC counterparts, the pulse energy obtained from these devices is disappointingly small. This is because of the rapid collapse of the discharge (within 200 ns) into a submillimetre-thick region around the walls of the containing waveguide structure. This phenomena has been attributed to the generation of surface waves [27]. An explanation of why this occurs has been given in terms of an increasing plasma absorption of microwaves with increasing plasma density [28].

To minimize the effect of discharge collapse onto the wall of the ceramic tube, a small-diameter (about 1 mm) tube is typically used, but the useful straight length of such a tube is limited to 30 cm in practice. The small active volume and high intracavity losses ensure a low laser pulse energy, but it is obviously possible to improve the situation by increasing the surface area of the ceramic container (for example, by simultaneously exciting gas in a bundle of ceramic tubes). It remains to be seen whether by such techniques as externally applying a magnetic field, profiling the microwave pulse shape, or working in an entirely different regime of laser gas pressure and composition, the laser performance can be significantly improved.

2.4.4 Gas flow and shock waves

The greatest thrust in excimer laser development at the present time is towards higher average powers, and the overriding considerations relate to the efficient generation of high volume flow rates, the aerodynamics of the discharge region and the management of shock waves.

At a typical specific energy extraction of 2 J l^{-1} and a clearing ratio (defined as the volume of gas passing through the discharge region between successive discharges, divided by the volume of the discharge) of 2, a 1 kW excimer laser would require the high gas volume flow rate of $1 \text{ m}^3 \text{ s}^{-1}$. For a 1 kW, 1 kHz laser, with a neon buffer gas, this flow rate corresponds to a kinetic energy per unit time through the discharge region of the order of 10 kW, and, depending on the losses in the gas flow loop (due particularly to the creation of eddy currents in detached boundary layers), a fraction of this power has to be provided by the gas circulator. This circulating gas power scales as the cube of both laser PRF and clearing ratio. Minimizing the latter involves achieving a uniform gas flow throughout the discharge region, but the boundary layers at the (curved) electrode surfaces can make this difficult. Furthermore, the preionized region may extend well beyond the boundaries of the discharge region, and this too can adversely affect the minimum achievable value of the clearing ratio. Both factors have significant effects on the geometry of the discharge region, especially around the edges of the electrodes.

Gas circulator power loss considerations also highlight a difficulty of high-power laser operation at high PRFs. Interestingly, the 1 kW, 1 kHz preionizer technology breakpoint noted in Section 2.4.2 may also exist in relation to gas technology, since at 1 kW average power an increase in PRF above 1 kHz will rapidly raise the circulating gas kinetic energy per unit time to the same magnitude as the electrical power fed into the laser discharge (50 kW at 2% electrical efficiency). Unless this kinetic energy can be largely preserved as the gas flows around the flow loop, the gas circulator and the laser high-voltage power supply will consume comparable amounts of power. But it is not only the preionization and circulator technology that may set a practical limit to the laser PRF. Shock waves caused by the rapid heating of the gas in the discharge region [29] take away about half the energy deposited by the discharge, and such waves must be effectively dissipated between discharge pulses if discharge uniformity is to be preserved [30, 31].

Since, by definition, shock waves travel at speeds above Mach 1, only waves with small velocity components in the direction of gas flow (e.g. waves bouncing between the electrodes and laser windows/mirrors) can be "blown out" of the discharge region. Other shock waves arrive back in the discharge region by reflection off the contact surface of the expanding cloud of discharge-heated gas, off obstacles in the gas stream close to the electrodes such as current returns or a UV preionizer structure, and perhaps even from walls of the flow loop further away from the discharge region.

The detrimental effect of increasing PRF on pulse energy has been investigated. In general, serious effects are observed at PRFs exceeding 1 kHz. Careful investigation of shock wave effects in the LUX laser [31] have revealed the persistence of transverse shocks bouncing between the solid electrodes as the dominant PRF-limiting effect (despite a clearing ratio exceeding unity)—an effect that has been reduced but not eradicated by incorporating nickel sponge at the edges of the discharge region. In shock wave studies on Culham's CHIRP laser [29], these transverse shocks were effectively damped by a mesh electrode without the need for further shock damping material. In all the above high-power lasers, clearing ratios of between one and two were found adequate.

Downstream arcing at multikilohertz PRF has also been reported in UV-preionized devices, despite a high clearing ratio [32]. The persistence of discharge products in slow-moving boundary layers adjacent to the electrodes may be responsible for this phenomenon.

The level of density perturbation that significantly influences discharge uniformity is thought to be an order of magnitude or more greater than that which seriously affects the phase front of the propagating laser beam. Techniques to improve discharge stability and laser beam quality may be therefore necessary in future devices.

2.4.5 Optics

The output from the laser is determined primarily by the choice of gas and the dimensions of the discharge, typically 2 cm × 1 cm × 50 cm. The high gain of the short-pulse laser (5–15% cm^{-1}) means that little feedback is required, and a standard resonator consists of an aluminized or dielectric-coated magnesium flouride total reflector and a calcium or magnesium flouride uncoated (8%) reflector. The combination of high-gain, low-Q cavity and low number of round trips means that the mode competition is low, and the laser output consists of a large number (of order 10^7) of spatial modes with a fairly high divergence of 2–4 mrad, different in the orthogonal axes of the approximately $2 \text{ cm} \times 1 \text{ cm}$ rectangular output beam.

Improvements in beam divergence can be achieved in longer-pulse excimer lasers [33] and those employing a resonator with unstable geometry. In the latter case, the beam divergence can be reduced by an order of magnitude with little (about 25%) loss of pulse energy. Where for particular applications (e.g. ophthalmology) a more uniform spatial intensity distribution is required, this can be achieved by increasing the dimensions of the discharge. Most manufacturers also offer oscillator-injection locked amplifier systems, which can be used, for example, to provide beams at greatly reduced bandwidth and divergence.

As an alternative to looking to resonator and discharge design to improve laser beam parameters, passive external components such as beam homogenizers, pulse stretchers and chromatic-aberration-free lenses are available. One attraction to this alternative is that it preserves the poor spatial coherence of the laser beam and so minimizes fringing effects resulting from optical inference.

Laser cavities employing variable-reflectivity mirrors (VRM) and novel unstable resonators [33], such as the self-filtering unstable resonator [34] are designed to optimize the (low) energy of the diffraction-limited beam that can be extracted from an excimer laser. A flowing-liquid stimulated Brillouin scattering (SBS) phase conjugate mirror (PCM) resonator has also been developed for high beam quality at high average power, but the greatest use of phase conjugation, which like other nonlinear processes is more easily realized at the shorter wavelengths of excimer lasers, is in the double-pass amplifier scheme, where the PCM corrects for the optical inhomogeneities of the amplifier [35, 36].

Another nonlinear process usefully exploited by excimer lasers is stimulated Raman scattering (SRS). A good spatially and temporally coherent beam at a Raman-shifted wavelength can efficiently extract power from a relatively incoherent excimer laser beam when the two are present simultaneously in the SRS medium. This phenomenon can be used for beam clean-up, pulse shortening and lengthening, and even for combining the output of several excimer lasers to achieve higher powers or higher PRFs in a modular fashion [37].

Intracavity line-narrowing components have been used in commercial photolithographic excimer laser systems. Frequency locking to an absorption feature for high wavelength stability is possible.

The KrF and ArF lasers have particularly broad gain bandwidths

(50 cm^{-1} and 200 cm^{-1} respectively), which are ideal for amplifying pico-second pulses. Commercial excimer oscillator–amplifier systems can be utilized to provide tens of millijoules in subpicosecond pulses, with the excimer oscillator pumping a dye chain that generates the short "seed" pulse, which is then frequency-doubled and double-passed through the excimer amplifier. Such pulses are very suitable for X-ray generation, making compact sources for higher-resolution photolithography.

2.4.6   High-average-power systems

In designing high-average-power systems, advances have to be made in each of the five areas described above. The differing pulse energies and repetition rates under investigation lead to a wide variety of high-power designs.

The performance of a number of high-average-power lasers are plotted in Fig. 2.3. Excluded from this figure are a number of e-beam pumped lasers that can provide pulse energies at 10^2–10^4 J, and are, in principle, capable of repetition rates of the order of 0.1 Hz [38]. With discharge exci-

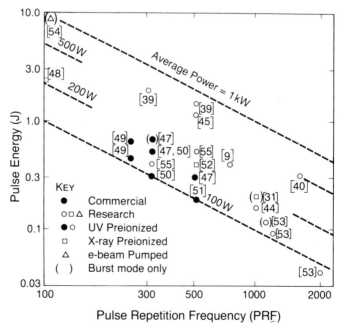

Fig. 2.3 Performance summary of several high-average-power excimer lasers (single modules). Square-bracketed numbers adjacent to points refer to references.

tation, average powers of 500 W have been achieved at PRFs between 300 and 1600 Hz [39, 40], with a maximum of 800 W being reported at 500 Hz [39]. There is currently intense activity on four continents to increase the average power available, particularly at higher PRFs. Powers of the order of 1 kW are expected to be reported by several groups by 1992.

Figures 2.4 and 2.5 show cross-sections of several high-power lasers. From the wide range of devices produced, some general trends may be tentatively identified. It is certainly the case that earlier work (Fig. 2.4) tended

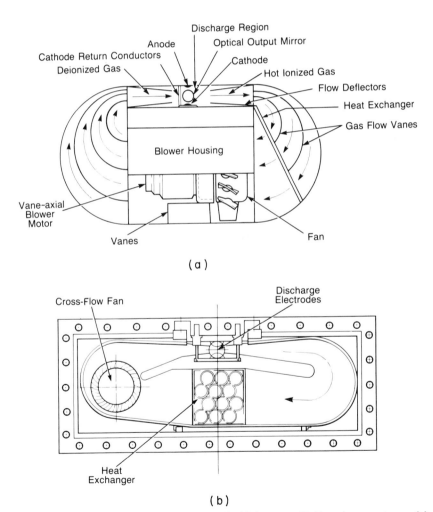

Fig. 2.4 Cross-sectional view of two early high-power KrF excimer systems: (a) 200 W, 1000 Hz (1980) [44]; (b) 300 W, 800 Hz (1988) [9].

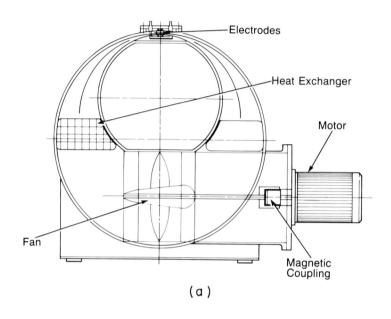

(a)

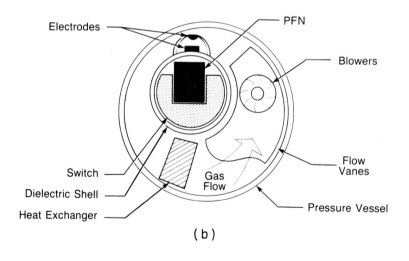

(b)

to concentrate on the KrF excimer, whereas more recently attention has turned to XeCl (Fig. 2.5). This is due to a combination of factors, including the advances in efficiency [18, 22] and pulse duration [8, 10] achieved with XeCl, the reduced materials compatibility problems with the chloride gas mixture [41] and the greater resilience of the XeCl discharge to perturbations in gas density [42, 43] that occur at high PRFs [29, 31]. A second trend observable between Figs 2.4 and 2.5 is a move away from a fabricated gas flow loop towards a system formed largely by the cylindrical shape of the main pressure vessel. This may be attributed to the requirements of structural integrity as the systems become larger [40]. However, for several of the largest lasers now under construction, the power requirements for the gas flow are so high that a more radical, aerodynamically profiled "wind tunnel" type flow loop, such as that shown schematically in Fig. 2.6, begins to look attractive. Such flow loops are ideal for laser development work in so far as each component is readily accessible for modification, but it remains to be seen which flow loops are adopted for the next generation of commercial high-power systems.

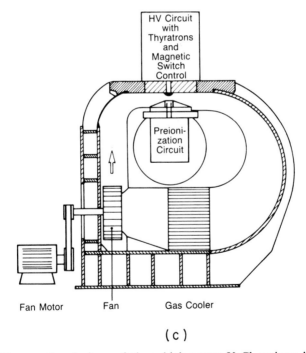

(c)

Fig. 2.5 Cross-sectional views of three high-power XeCl excimer lasers, illustrating a near cylindrical closed flow loop (i.e. contained within a single pressure vessel) design: (a) 500 W, 1600 Hz (1990) [40]; (b) 520 W, 500 Hz (1990) [45]; (c) 800 W, 500 Hz (1990) [39].

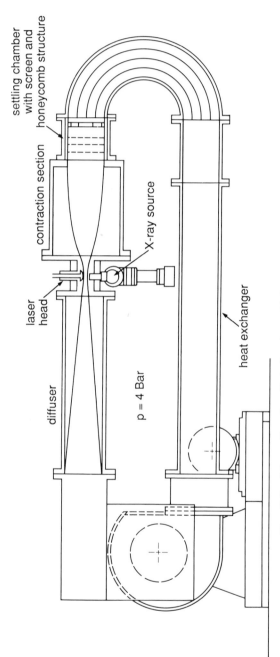

Fig. 2.6 "Open flow loop" design for a high-power excimer laser [46].

2.5 CONCLUSIONS

The story of excimer laser development has been one of meticulous attention to detail. At various times, each aspect of the complex laser system has come under the closest scrutiny, and has been optimized for performance, lifetime and reliability. This is a continuing process, the weaker links in the chain are being strengthened, and the results are impressive. Within the past 15 years, excimer lasers have been developed from a scientific curiosity to a laboratory tool, and are now approaching industrial maturity. With the immense applications potential now driving worldwide development, it will not be long until industrially rated kilowatt-level lasers are available.

REFERENCES

[1] C. K. Rhodes (ed.), *Excimer Lasers*, 2nd edn (Springer-Verlag, Berlin, 1984).
[2] J. M. Green, *Optics and Laser Technol.*, 289 (December 1978).
[3] M. J. Shaw, *Prog. Quantum Electron.* **6**, 3 (1979).
[4] M. H. R. Hutchinson, *Appl. Phys.* **21**, 95 (1980).
[5] A. J. Kearsley, A. J. Andrews and C. E. Webb, *Optics Commun.* **31**, 181 (1979).
[6] G. Klauminzer, *Laser Focus* **21** (12), 108 (1985).
[7] R. S. Taylor, *Appl. Phys.* **B41**, 1 (1986).
[8] R. S. Taylor and K. E. Leopold, *J. Appl. Phys.* **65**, 22 (1989).
[9] H. M. von Bergmann, G. L. Bredenkamp and P. H. Swart, *Proc. SPIE* **1023**, 20 (1988).
[10] S. V. Mel'chenka, A. N. Panchenko and V. F. Tarasenko, *Sov. J. Quantum Electron.* **14**, 1009 (1984).
[11] M. Steyer and H. Voges, *Appl. Phys.* **B42**, 155 (1987).
[12] T. Letardi, P. DiLazzaro, G. Giordano and C. E. Zheng, *Appl. Phys.* **B48**, 55 (1989).
[13] M. R. Osborne, R. J. Winfield and J. M. Green, *J. Appl. Phys.* **65**, 5242 (1989).
[14] M. R. Osborne, R. J. Winfield and J. M. Green, in *Proceedings of 9th National Quantum Electronics Conference, Oxford*, Paper 8. Unpublished.
[15a] T. Shimada, M. Obara and A. Noguch, *Rev. Sci. Instrum.* **56**, 2018 (1985).
[15b] H. M. von Bergmann and P. H. Swart, in *Proceedings of Conference on Lasers and Electro-optics, Baltimore, 1991*, Paper WF3, p. 244 (Optical Society of America, Washington DC, 1991).
[16] T. J. McKee, G. Boyd and T. A. Znotins, *IEEE Photon Technol. Lett.* **1**, 59 (1989).
[17] R. S. Taylor, P. B. Corkum, S. Watanabe and A. J. Alcock, *IEEE J. Quantum Electron.* **19**, 416 (1983).
[18] W. H. Long, M. J. Plummer and E. A. Stappaerts, *Appl. Phys. Lett.* **43**, 735 (1983).
[19] M. R. Osborne, P. W. Smith and M. H. R. Hutchinson, *Optics Commun.* **52**, 415 (1985).

[20] C. H. Fisher, M. J. Kushner, T. E. DeHart, R. A. Petr, J. P. McDaniel and J. J. Ewing, *Appl. Phys. Lett.* **48**, 1574 (1986).

[21] H. Reiger and M. Cornell, in *Proceedings of Conference on Lasers and Electro-Optics, Baltimore, 1989*, Paper WF26, p. 190 (Optical Society of America, Washington DC, 1989).

[22] J. W. Gerritsen, A. L. Keet, G. J. Ernst and W. J. Witteman, *J. Appl. Phys.* **67**, 3517 (1990).

[23] B. Lacour and C. Vannier, *J. Appl. Phys.* **62**, 754 (1987).

[24] S. Bollanti, P. Di Lazzaro, F. Flora, G. Giordano, T. Hermsen, T. Letardi and C. E. Zheng, *Appl. Phys.* **B50**, 415 (1990).

[25] C. P. Christensen, R. W. Waynant and B. J. Feldman, *Appl. Phys. Lett.* **46**, 321 (1985).

[26] C. P. Christensen, C. Gordon, C. Moutoulas and B. J. Feldman, *Optics Lett.* **12**, 169 (1987).

[27] R. Waynant, W. M. Bollen and C. P. Christensen, in *Proceedings of Conference on Surface Waves in Plasmas and Solids, Ohrid, Yugoslavia, 5–7 September 1985*, p. 674 (World Scientific, Singapore, 1986).

[28] J. M. Green, Microwave requirements for a rare gas halide laser (personal communication).

[29] J. Fieret, in *Proceedings of OE Lase '90, Technical Conference on High Power Lasers, Los Angles, 1990*.

[30] V. Yu Baranov, V. M. Borisov, A. Yu. Vinokhodov, F. I. Vysikailo and Yu. B. Kiryukhin, *Sov. J. Quantum Electron.* **14**, 558 (1984).

[31] M. L. Sentis, P. Delaporte, B. M. Forestier and B. L. Fontaine, *J. Appl. Phys.* **66**, 1925 (1989).

[32] K. Hotta, S. Ito and M. Arai, in *Proceedings of Conference on Lasers and Electro-Optics, Anaheim, California, 1990*, Paper CTHB1, p. 346 (Optical Society of America, Washington DC, 1990).

[33] T. J. McKee, *Optics Lett.* **15**, 795 (1990).

[34] V. Boffa, P. DiLazzaro, G. P. Gallerano, G. Giordano, T. Hermsen, T. Letardi and C. E. Zheng, *IEEE J. Quantum Electron.* **23**, 1241 (1987).

[35] M. Sugii, O. Sugihara, M. Ando and K. Sasaki, *J. Appl. Phys.* **62**, 3480 (1987).

[36] M. R. Osborne, W. A. Schroeder, M. J. Damzen and M. H. R. Hutchinson, *Appl. Phys.* **B48**, 351 (1989).

[37] M. J. Shaw, J. P. Partanen, Y. Owadano, I. N. Ross, E. Hodgson, C. B. Edwards and F. O'Neill, *J. Opt. Soc. Am.* **B3**, 1374 and 1466 (1986).

[38] M. Obara, in *Proceedings of European Conference on Laser Treatment of Materials (ECLAT '90), Erlangen, Federal Republic of Germany, 1990*, p. 799 (Sprechsaal Publishing, Coburg, 1990).

[39] P. Oesterlin and D. Basting, *Physics World*, **3** (7), 43 (July 1990).

[40] H. M. von Bergmann and P. H. Swart, in *Proceedings of International Conference on Lasers '89*, p. 80 (STS Press, McLean, 1990).

[41] G. M. Jursich, W. A. von Drasek, R. Brimacombe and J. Reid, in *Proceedings of Conference on Lasers and Electro-Optics, Anaheim, 1990*, Paper CTUI2, p. 180 (Optical Society of America, Washington DC, 1990).

[42] M. R. Osborne, *Appl. Phys.* **B45**, 285 (1988).

[43] M. M. Turner, Modelling of the self sustained, transverse discharge excited xenon chloride laser. PhD Thesis, University of St Andrews (1990).

[44] T. S. Fahlen, *IEEE J. Quantum Electron.* **16**, 1260 (1980).

[45] T. Fahlen, XMR Inc., Personal communication (1990).

[46] W. H. Witteman, F. A. van Goor, G. B. Ekelmans, M. Trentelman and
     G. J. Ernst, in *Proceedings of 1st International Congress on Optical Science
     and Engineering, Hamburg, 1988* (SPIE, Bellingham, WA, Vol. 1023, 1988).
[47] XMR Inc, Santa Clara, California, Commercial device.
[48] U. Schmidt and W. Rath, in *Proceedings of Conference on Lasers and
     Electro-Optics, Anaheim, 1990*, Paper CTHB4. p. 348 (Optical Society of
     America, Washington DC, 1990).
[49] Lambda Physik, Göttingen, Germany, Commercial device.
[50] Lumonics Inc., Kanata, Canada, Commercial device.
[51] Questek Inc., Billerica, Massachusetts, Commercial device.
[52] R. R. Butcher, S. Swisher, G. F. Erickson, R. Tennant and W. Willis, in
     *Proceedings of Topical Meeting on Excimer Lasers, Incline Village, Nevada,
     1983*, Paper MB5, p. 5 (Optical Society of America, Washington DC, 1983).
[53] S. Ito, M. Arai and K. Hotta, in *Proceedings of Conference on Lasers and
     Electro-Optics, Baltimore, 1989*, Paper THF2, p. 268 (Optical Society of
     America, Washington DC, 1989).
     S. Ito, M. Arai and K. Hotta, in *Proceedings of Conference on Lasers and
     Electro-Optics, Anaheim, 1990*, Paper CTHB1, p. 346 (Optical Society of
     America, Washington DC, 1990).
[54] Maxwell Inc., San Diego, California.
[55] V. M. Borisov, A.-Yu. Vinokhodov and Yu. B. Kiryukhin, *Sov. J. Quantum
     Electron.* **17**, 595 (1987).

3 Deep-UV Optics for Excimer Lasers

F. N. GOODALL

Rutherford Appleton Laboratory, Chilton, Oxfordshire, UK

3.1 INTRODUCTION

The drive for increasing device performance in terms of speed and power consumption in semiconductor devices demands ever-decreasing linewidths. To obtain large fields with enhanced resolution requires either higher numerical apertures (NAs) or alternatively shorter wavelengths. The lens design for such field sizes ($>1 \text{ cm}^2$) with NAs greater than 0.5 is extremely difficult and furthermore results in reduced depth of focus. This makes the shorter-wavelength option more attractive, and the increasing reliability of excimer lasers capable of high fluence at 248 and 193 nm has given rise to a great deal of research effort to develop suitable optical systems at these wavelengths, principally at Zeiss, Tropel, Ultratech and RAL.

Continuing improvement in excimer laser reliability has enabled the stage to be reached where they may be realistically used in industrial processes. At the present time there are two principle uses in industry. The first is for conventional lithography, and complete excimer wafer steppers are already in use at 248 nm [1]. The second is the micropatterning of surfaces, where a projected image causes material to be selectively removed by direct ablation. This technique can be used to create submicron patterns in a variety of materials such as ceramics, silicon and polymers [2].

Recently, a great deal of effort has been expended on studies of laser chemical processing [3]. The principle here relies upon the high photon energy available at excimer wavelengths, causing localized chemical reactions that can be used to deposit, dope and etch submicron linewidths. If suitable imaging optics can be developed, this could eventually lead to the replacement of conventional semiconductor processing, either in whole or in part.

Two principal optical techniques can be used. One is to focus the laser

Photochemical Processing of Electronic Materials
ISBN 0-12-121740-X

energy into a spot that can be scanned to write the required pattern (laser direct writing or laser pantography). This is particularly useful at the experimental stage, where the fundamental physics can be established using relatively simple optics. However, such an approach would prove cumbersome during real device fabrication, and if the full potential of this technology is to be realized then full-field high-NA lenses must be developed. These will then allow direct imaging of the desired circuit layout to be achieved in a single step. This of course assumes that the desired reaction only occurs where the light strikes the semiconductor surface.

3.2   THE OPTICAL DESIGN PROBLEM

The excimer laser wavelengths at 248 and 193 nm have relatively poor transmission in most materials. Further, the bandwidths at each of these wavelengths are sufficiently large to cause chromatic aberration problems. Unlike the longer UV wavelengths (>360 nm) presently used for lithography, where a large selection of glasses with different refractive indices and dispersions are available, the situation in the deep UV is much more difficult. Only fused silica, calcium fluoride and lithium fluoride have good transmission properties, and even these can suffer from the formation of colour centres, particularly at 193 nm [4]. Consequently, the lens designer is forced to consider novel approaches to the problem, using for example catadioptric systems, which consist of a mixture of refracting and reflecting elements, and possibly all-reflecting systems, to utilize the full laser bandwidth. Alternatively, the frequency of the laser emission can be narrowed sufficiently so that a more conventional all-refracting system can be used. There are advantages and disadvantages in each approach, and these are considered in detail below. Systems that have been designed, fabricated and tested are discussed, and some results on conventional lithography and micromachining of polymers and ceramics are presented.

The requirements of complete optical systems for laser chemical processing are also considered here. The development of such systems capable of semiconductor device fabrication presents a challenge to the optical designer that is different to both conventional lithography and micromachining. In this case the imaging has to take place in a gas cell through a window onto the underlying substrate. This poses unique problems where the traditional application of conventional lithography will not be sufficient. For example, the final lens-to-substrate distance must be much greater than is normally the case, being typically 2 cm, which places constraints on an already difficult lens design. The wafer metrology, alignment of the multilayer patterns, maintenance of focus and total system design are formidable problems, which must be solved if this technology is to have a significant chance of replacing conventional resist techniques.

Finally, although an increasing amount of information is available on the basic physics and chemistry of the processes likely to be involved, the use of full-field imaging may give rise to unexpected local interactions, similar in nature to the proximity effect in e-beam lithography, particularly when submicron dimensions are involved.

3.3   GENERAL OPTICAL DESIGN CONSIDERATIONS

3.3.1   Resolution and depth of focus

Initial parameters needed to define the optical system are resolution R and depth of focus D. For a diffraction-limited system these are given by

$$R = 0.61\lambda/\text{NA}, \tag{3.1}$$

$$D = 0.61\lambda/\text{NA}^2, \tag{3.2}$$

where λ is the illumination wavelength and NA is the numerical aperture. By plotting λ as a function of NA, the available range of NA values is shown for the particular value of operating wavelength chosen (see Chapter 4, Fig. 4.1). Initial work at RAL was based on the assumption that a depth of focus of about 1.5 µm was a practical value on a production machine at 249 nm, giving an NA of about 0.3. However, this is somewhat conservative, and it is possible to work with depth-of-focus values somewhat smaller than this, giving NAs of 0.4–0.45. Hence the window shown in Chapter 4, Fig. 4.1 can be enlarged, as can be determined by plotting (3.1) and (3.2) again for the larger NA.

Since no lens design is perfect, it is important to indicate the total performance resulting from aberrations and diffraction. This is demonstrated using the diffraction modulation transfer function, where the amplitude ratio of light to dark for a grating structure imaged through the lens is shown as a function of spatial frequency. For photoresist materials used in conventional lithography (novolacs) it is normal to work with a modulation ratio of approximately 60%.

3.3.2   Aberration correction

To produce aberration-free lenses that are diffraction-limited at excimer wavelengths is a formidable task, particularly with the limited number of materials available. The problem may be somewhat reduced if distortion in the image can be tolerated. This particular aberration does not affect resolution, and for experimental purposes is unlikely to matter. It is also possible that for the cases where it does matter it can be corrected for in the data by predistorting in the opposite sense. The essential requirement when lenses are fabricated for production is that the distortion in each is identical. It need not necessarily be zero, and this somewhat eases the lens design problem. Other types of aberration such as spherical, coma, astigmatism and chromatic affect resolution, and must be minimized by careful choice of initial lens type and subsequent design work.

3.3.3 Telecentricity

One other lens parameter that is important is the demagnification factor. It is essential to hold this constant at the correct value, especially when stepping and repeating an image to create a single large area. In this way small variations in the image distance can be tolerated—as long as they remain within the depth of focus. If the system is not telecentric, small variations in the magnification occur through the focal range. Consequently, if a large final image is to be achieved by stepping and repeating a smaller subfield then the boundaries of each subfield will not necessarily be coincident, giving rise to either a gap or an overlap (i.e. stitching errors).

3.3.4 Illumination optics

In considering the excimer laser as a light source, the intrinsic coherence of the radiation must be taken into account. The natural linewidth is typically about 0.3 nm at 248 nm, but is decreased by a factor of $\frac{1}{100}$ in a typical line-narrowed laser used for lithography. Normally the high NA of the imaging lens and the magnification of typically $5\times$ would lead to a value of partial coherence of the system of about 0.2, and for line-narrowed systems the situation is much worse, whereas it is normal in lithography to work with a value of about 0.6. The procedure adopted at RAL to achieve this is to use a fly's eye imaged into the entrance pupil of the lens, which also achieves beam homogenization. This technique has also been used by Horiike and co-workers [5], while other methods have also been successfully applied [1, 6]. This reduces the problem of speckle, which becomes severe in line-narrowed systems. For a system consisting of a small primary mirror and a large secondary mirror such as that shown in Fig. 3.1 (called a Cassegrain system) a ring field was used to give greater efficiency.

3.4 LENSES FOR LASER CHEMICAL PROCESSING

When considering wavelength selection for laser chemical processing (LCP) 193 nm is of much greater interest than 248 nm. Chromatic correction for the full laser bandwidth makes the lens design for 193 nm much more difficult, since the dispersion at this wavelength is much greater than at 248 nm. Damage, manifested through the formation of colour centres, is known to develop in both materials, particularly at 193 nm [7]. However, calcium fluoride which is free from this disadvantage, is available, while progress is being made by Hereus in Germany and Shinetsu in Japan in the case of fused silica, and it seems likely that this problem can be solved.

One approach is to consider catadioptric systems, which consist of refracting and reflecting elements combined. These can be designed with most of the power in the reflecting surface, and hence have relatively broad bandwidths. Such a system was designed and built for MIT Lincoln Laboratories. It has a numerical aperture of 0.35 and a field size of 4 mm diameter. The final lens-surface-to-substrate distance was 1 cm. Subsequent design work has shown that this can be increased to 2 cm without much difficulty. The construction is of the Cassegrain type; the layout is shown in Fig. 3.1 with a photograph of the actual lens in Fig. 3.2.

Such systems normally use mechanical support for the primary mirror. This, however, increases the diffraction around the obstruction and leads to poorer performance. The RAL design gets round this problem by depositing the primary mirror onto the centre of a lens element of full aperture. The final diffraction performance of the system can be evaluated by inspection of the MTF curves in Fig. 3.3. As can be seen, the modulation is reduced compared with an unobstructed system, and this effect is unavoidable in such Cassegrain systems. Subsequent work on this type of

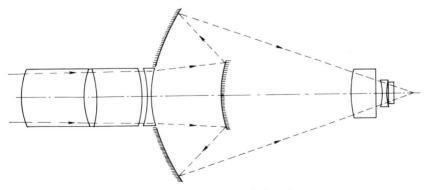

Fig. 3.1 193 nm catadioptric lens layout.

Fig. 3.2   193 nm catadioptric lens.

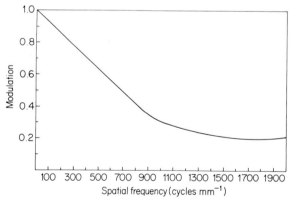

Fig. 3.3   MTF for 193 nm catadioptric lens.

design has shown that even higher NAs up to 0.4 and field sizes up to 6 mm diameter can be achieved.

Lens coatings are an extremely important consideration both in the case of the mirror and for the refractive elements. The mirror coatings are multilayer in nature and give high reflection coefficients (about 98%) at both 193 and 248 nm for the full range of ray angles incident on both the primary and secondary mirrors. The coatings for the refractive elements are colloidal silica and form a graded index match with efficiencies approaching 100%. In both the case of the reflection coatings and lens coatings there is an additional need to withstand the high energy densities of up to 1 J cm^{-2} and high peak pulse powers. Experience with the two types of coating has proved to be very satisfactory in this respect with up to 2 J cm^{-2} and lifetimes as long as the lens material itself.

In order to construct a complete system, many other factors must be taken into account. The wafer in the gas cell must be maintained orthogonal to the lens axis. The cell window is part of the lens system and must be included at the design stage. It does, however, have the advantage that it can be moved along the axis without changing the lens performance. However, any tilt will produce nonsymmetrical aberrations and must be avoided.

It is not certain at this stage whether die-by-die alignment is necessary. One major advantage of laser chemical processing is that it is a moderately cool process, and the relative shift that takes place between die due to the high-temperature processing when conventional methods are used will be absent. Thus it is possible that only global alignment will be necessary. This would be a major bonus, and such a system would then have a decisive advantage over wafer stepper technology. However, other problems will need to be tackled. The maintenance of focus will be a problem. If the wafer remains in the cell for several processes that use different gases then the refractive index will be changing in the final image space, thus causing the lens to go out of focus, especially for high numerical apertures. Temperature effects will also be present, especially for parts of the process that need heated substrates. If we assume that the above problems can be solved, it will then remain to be seen whether there is a serious proximity effect when full-field imaging is attempted.

Some work has been done using the all-refractive lens systems for laser oxidation of silicon using 248 and 193 nm full-field imaging, resulting in patterns over a relatively large area with linewidths down to around 1 μm [8].

Using a focused excimer laser beam and irradiation of Ti films on LiNbO$_3$ substrates, the diffusion of Ti into the substrate has also been studied. The observations suggest that lateral diffusion of Ti can be reduced, giving more-rectangular doping profiles. This could in turn lead to a process whereby complete waveguide structures could be imaged and implanted into the substrate, thus leading to a considerable simplification in manufacture [9].

3.5 MICROMACHINING

Excimer lasers are capable of delivering high fluences, which can cause direct photoablation of a wide variety of materials. Suitable lenses can be designed to project patterns directly, thus avoiding the use of photoresist.

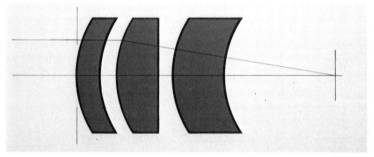

Fig. 3.4   193 nm triplet lens layout (NA = 0.17, magnification = 0.06).

Fig. 3.5   Ablation results in PM20.

By suitable choice of energy and using diffraction-limited lens designs, micron and submicron features can be formed with vertical walls.

The first work done used a triplet lens design with NA of 0.17 at 193 nm wavelength [10]. The aberration correction was for spherical aberration and coma, similar to any ordinary non-flat-field microscope objective. This gave only a 500 μm field size, but was adequate for initial tests in image projection. These initial experiments were in cross-linked PMMA type PM20, and gave line pattern images about 0.5 μm wide with 0.5 μm spaces. This resolution is better than expected from the Rayleigh limit, but is accounted for by the extreme nonlinearity of the process. The lens layout is shown in Fig. 3.4, while the etching results are shown in Fig. 3.5.

Following on from these experiments, a great deal of theoretical design

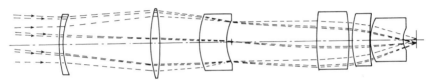

Fig. 3.6   248 nm all-refracting lens layout.

Fig. 3.7   248 nm all-refracting lens.

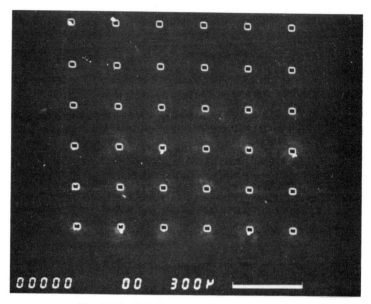

Fig. 3.8 Ablation results in polymer film.

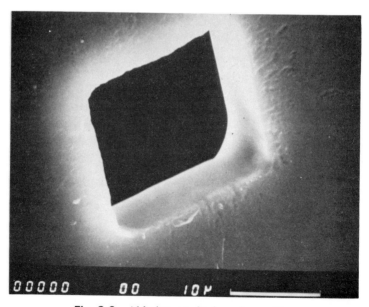

Fig. 3.9 Ablation results in polymer film.

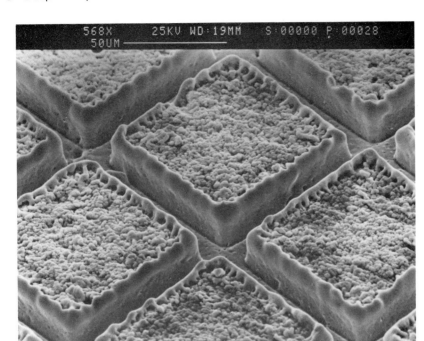

Fig. 3.10 Ablation results in ceramic.

work was done to explore the potential for this type of lens system. The
results at 248 nm were particularly encouraging [11].

Most work has been at 248 nm, using all-refractive lens systems with
typical NAs up to 0.3 and 1 cm diameter fields for high resolution and NAs
of 0.1 and field sizes up to 4 cm diameter for moderate resolution where
large-area coverage was the prime consideration. The layout of a typical
lens design of the type used is shown in Fig. 3.6, with the lens itself shown
in Fig. 3.7. Figure 3.6 shows a typical lens layout with six quartz elements,
an NA of 0.3, image size 1 cm and final lens-to-substrate distance of
1.4 cm. Figure 3.7 is for a similar all-quartz lens, but with an NA of 0.1
and a field size of 4 cm. The aberrations are reduced to below the diffraction limit, and field curvature is kept below the depth of focus. Results
using this later design of lens were obtained in polymers and ceramics
(Figs 3.8–3.11). It can be seen that vertical walls can be obtained for both
the ceramics and polymers by suitable choice of fluence.

Some work has been done at 193 nm, where 1–2 μm features have been
produced in PMMA using such lenses, but the chromatic aberrations preclude resolution in the submicron range with large fields >1 cm. Recent

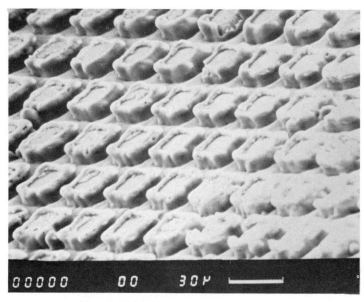

Fig. 3.11 Ablation results in sapphire.

design work has shown, however, that NAs up to 0.2 and 4 mm diameter field-size lens systems of the all-refracting type using fused silica and calcium fluoride can be realized, and hence submicron medium-sized fields can be achieved, which should prove useful for experimental work.

One of the most important experimental results that has been consistently observed is that, even for lenses of high numerical aperture and hence small depth of focus, the lens does not have to be refocused when ablating structures even up to 50 μm deep. This considerably simplifies implementing laser micromachining systems in an industrial process. So far, this has been found to be independent of the material used.

3.6 CONVENTIONAL LITHOGRAPHY

For conventional lithography the need is for high numerical apertures and large field sizes. The additional complexities of gas cell design, high power and large image distances are of course absent. Two main approaches have been adopted at 248 nm. One is to use a catadoptric Wynne Dyson lens, which gives excellent performance over a wide field [12]. The choice of unit magnification and concentric elements gives good inherent correction of the

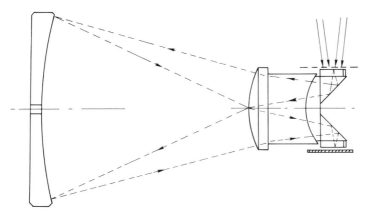

Fig. 3.12 1 : 1 Wynne–Dyson lens layout.

monochromatic aberrations. It is free of distortion, and doubly telecentric. Furthermore, it is chromatic not just over the laser bandwidth but even over the 254 nm line from a mercury arc lamp; it is described in Chapter 4. A similar system is currently used by Ultratech Stepper. The lens layout and a photograph are shown in Fig. 3.12 and Fig. 4.2 in Chapter 4.

Recent work has indicated that the refracting components, which were originally lithium fluoride and fused silica, can be replaced with all fused silica elements that are air-separated, thus obviating the need for the oil couplant used in the original design. The final image distance is small, usually of order 1 mm. This can be increased at the expense of field size, thus allowing the possible use of such a system for large-area chemical etching and laser chemical processing. In every respect this is an excellent system, because of its inherent high resolution, large field and relative ease of production. It suffers, however, from the one great disadvantage that final-size reticles are required. Although this has to be compared with the even greater problem of producing X1 masks for X-ray lithography, which requires a totally different mask technology, it is pertinent to consider the possibility of reduction systems.

As has previously been stated, there is great difficulty in producing all-refracting lens systems with large fields and high NAs at 248 nm, and the problem is even worse at 193 nm. Although the Cassegrain system goes some way towards reducing these difficulties, its field size is limited to approximately 6 mm unless aspherics are considered, and the central obstruction causes a deterioration in the MTF. A different approach has been considered at RAL; it gives a vastly improved performance, which is shown in Fig. 3.13. The results at 248 nm have NAs up to 0.5, with 3 cm total field diameter, and at 193 nm recent work has given 3 cm fields and

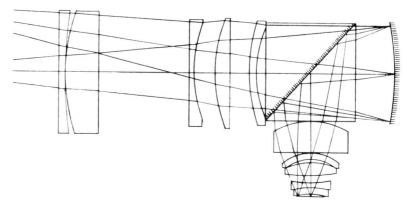

Fig. 3.13   Catadioptric lens using beamsplitter.

NAs of 0.3, with no line narrowing necessary at either wavelength. Such a design depends crucially on whether the beam splitter can be successfully fabricated. The coefficient of reflection of all rays incident on the 45° interface must be constant. This in turn requires multicoating techniques of high accuracy, which are relatively broadband, and this may prove very difficult to achieve in practice. However, much useful experimental work should be possible with such a system, even if it does not quite meet the more demanding specification needed for a commercial wafer stepper system.

Finally, the use of all-reflecting systems can be considered. Useful performance can be obtained, however, only with the use of at least four mirrors and aspheric profiles. The difficulty of fabricating these systems, together with the obstructions that cannot be avoided in implementing such a design, mean that this is not a practical approach for deep-UV lithography. Nevertheless, all reflecting systems may give good performance in the X-ray region, where high resolution can be achieved with the use of low numerical apertures and simple two- or three-mirror systems without the need for complex surface profiles such as paraboloid or even generalized polynomial shapes (aspherics).

One solution to the problem of all-reflecting systems that avoids the above difficulties is to consider the Offner system [13]. This is a unit magnification system that has good aberration correction due to symmetry. The scanning takes place over wafer and mask simultaneously, using a small annular portion of the field. In this way diffraction-limited performance may be obtained. Recently, a reduction system based on this approach has been made by Perkin Elmer. This system, while basically the same, is more difficult to engineer, since the scanning of the mask and wafer has to take place at different rates, differing in the reduction ratio required. Such a system could be used for most of the applications considered in this chapter, although its primary use has been for conventional lithography.

Finally, the use of line-narrowed systems may be considered. This solves the chromatic correction problem, but leads to a more complex laser system with a phase-locked oscillator. Furthermore, the increased degree of coherence that this implies can give rise to speckle problems, and care must be taken with the illumination optics to reduce this. However, most of these problems appear to have been solved, and such a system is now used successfully on the GCA Excimer Stepper [1].

Photoresist at 248 nm remains a problem, although much work is being done to rectify this, and already a negative resist is available. However, the most probable way forward lies in surface-modified resists, which work by having a silylation layer at the surface of a polymer. This layer is selectively removed by the imaging process to create the pattern, and subsequent reactive ion etching gives pattern transfer to the substrate, the silylation layer protecting the resist in the areas where it remains. This approach has two advantages: it acts as a bilevel resist with an extremely thin imaging layer, and it can be used at both 248 and 193 nm. For the high numerical apertures that will be used to take full advantage of excimer laser based lithography, it appears to be the most promising way forward.

3.7   CONCLUSIONS

There is a great range of activities requiring the use of excimer lasers for micromachining, photochemistry and conventional lithography, both for experimental purposes and in industry. This in turn has led to totally different approaches being taken, depending on wavelength (248 or 193 nm), working distance, resolution and field size. For each of these separate activities lens designs have been considered that avoid the use of line-narrowed lasers with their additional expense and complexity. However, for some purposes this may not be possible, and for systems using all fused silica optics strictly monochromatic designs already exist. There is therefore already a possible solution to the optics problems for use with deep-UV excimer lasers, and continuing progress should lead to a variety of total systems for micromachining, laser chemical processing and conventional lithography.

Although deep-UV optical lithography is expected to be the choice for replication down to 0.35 μm, below this it may be necessary to resort to different techniques entirely, and a comparison of the various options is given in Chapter 4. It is, however, dangerous to put a final figure on the lower resolution limit for optics, since a combination of resist chemistry, phase reversal masks and lens systems with very high numerical apertures may drive the figure lower than can be envisaged at the present time—it may yet prove equal to the challenge of X-ray lithography in resolution, and will be particularly competitive with synchrotrons in terms of cost.

REFERENCES

[1] V. Pol, J. H. Bennewitz, G. C. Escher, M. Feldman, V. A. Firtion, T. E. Jewell, B. E. Wilcomb and J. T. Clemens, *SPIE Opt. Microlithogr.* **633**, 6 (1986).
[2] F. N. Goodall, R. A. Lawes and G. Arthur, *Microelectron. Engng* **2**, 187 (1990).
[3] I. W. Boyd, *Laser Processing of Thin Films and Microstructures* (Springer-Verlag, Berlin, 1989).
[4] M. Rothschild, D. J. Ehrlich and D. C. Shaver, *Microcircuit Engng*, 167 (1989).
[5] Y. Horiike, in *Proceedings of Fall Meeting of the Materials Research Society, Boston, 1–6 December 1986*, Paper B&C1.2.
[6] V. Pol, J. H. Bennewitz, T. E. Jewell and D. W. Petero, *Opt. Engng* **26**, 311 (1987).
[7] D. J. Ehrlich and M. Rothschild, *Microcircuit Engng* **9**, 27 (1988).
[8] V. Nayar, I. W. Boyd, G. Arthur and F. N. Goodall, *Appl. Surf. Sci.* **36**, 134 (1989).
[9] S. A. M. Al Chalabi and F. N. Goodall, *Appl. Surf. Sci.* **36**, 408 (1989).
[10] F. N. Goodall, R. A. Moody and W. T. Welford, *Optics Commun.* **57**, 227 (1986).
[11] F. N. Goodall, R. A. Lawes and G. Arthur, *Microcircuit Engng* 187 (1989).
[12] F. N. Goodall and R. A. Lawes, *Microcircuit Engng* **6**, 1 (1987).
[13] Kingslake, R., *Lens Design Fundamentals* (Academic Press, New York, 1978).

4 Submicron Lithography for Semiconductor Device Fabrication

RONALD A. LAWES

Rutherford Appleton Laboratory, Chilton, Oxfordshire, UK

4.1 INTRODUCTION

For over two decades, the key lithographic technique of the semiconductor industry has been UV optical lithography. Progress in optical lithography, mainly through reductions in exposure wavelength, has enabled linewidths down to 1 μm to be defined routinely in production. Deep-UV optical systems using excimer lasers and associated processing technology are now being developed in research laboratories throughout the world to enable dimensions to be reduced to 0.3–0.5 μm. The design of such lenses is discussed in more detail in Chapter 3. Eventually, the feature size required by semiconductor circuits will be smaller than can be achieved using optical lithography due to limitations to resolution imposed by diffraction. Alternative lithographic techniques will then be required, and several possibilities are discussed in this chapter.

The use of electron beams to overcome diffraction effects in optical lithography was suggested more than 20 years ago, and scanning electron beam lithography has been routinely available for industrial mask making for the last decade. The major difficulty has been the relatively slow exposure speed and high system cost, and this has limited the use of electron beam lithography to mask making and to direct write on wafer for a few specialized applications, namely maskless lithography. Difficulties in achieving submicron resolution have been experienced due to the interaction of backscattered electrons from neighbouring exposed areas.

For some years, the search for an alternative technique for producing high-throughput low-cost submicron lithography for production manufacturing has centred on X-ray lithography, and major R&D programmes have been launched in the USA, Japan and Germany to find a suitable

Photochemical Processing of Electronic Materials
ISBN 0-12-121740-X

technology. X-ray lithography offers sub-0.5 µm resolution mainly because of the significantly shorter wavelengths compared with light optics. Unfortunately, with X-ray lithography there are major problems with mask fabrication and the provision of sufficiently bright X-ray sources.

The use of ion beams as an alternative to either electron beam lithography or X-ray lithography is also being pursued at R&D laboratories throughout the world. Scanning ion beam and projection ion beam lithography are currently at the research stage and have yet to reach the point where the techniques are useful for general semiconductor device fabrication. However, scanning ion beam machining and deposition is currently being developed for both optical and X-ray mask repair.

The review presented in this chapter will of necessity be selective and will concentrate on those applications of optical, electron beam and X-ray techniques currently in use in advanced industrial processes and those ion beam techniques that may be used within the next few years.

4.2 OPTICAL LITHOGRAPHY

Currently, high-resolution optical lithography is based upon the optical stepper, which offers submicron resolution, typically over a 20 mm × 20 mm field. In order to cover a semiconductor substrate of 100–200 mm diameter, the image in the optical field is repeatedly stepped in both X and Y directions. The step and repeat operation is performed by laser-interferometer controlled stages to a positioning accuracy of <0.1 µm, thereby allowing the registration of successive layers of a semiconductor device to a similar precision.

The resolution R of a projection optical system is given by

$$R = K_1\lambda/\text{NA}, \tag{4.1}$$

where λ is the wavelength of the radiation used and NA is the numerical aperture of the optical system. The depth of focus DF is given by

$$\text{DF} = K_2\lambda/\text{NA}^2. \tag{4.2}$$

The constants K_1 and K_2 are derived from experimental results and lie in the range 0.5–0.8. For systems with good aberration correction and high partial coherence, the normal Fraunhofer diffraction theory can be assumed. Hence

$$R = 0.6\,\frac{\lambda}{\text{NA}}, \tag{4.3}$$

$$\text{DF} = 0.6\,\frac{\lambda}{\text{NA}^2}. \tag{4.4}$$

It can be seen that high resolution can only be achieved at the expense of depth of focus, and therefore the design of optical systems will involve a trade-off between short wavelengths and large numerical apertures [1]. This is illustrated in Fig. 4.1 for $R < 0.5$ μm and $D > 1$ μm, where it can be seen that there is a limited envelope of λ versus NA where both conditions can be met simultaneously (namely the unhatched area).

Extensive experimental work is now in progress in many laboratories to develop suitable optical sources, lenses and resist technology applicable at deep-UV wavelengths. One major thrust involves the use of excimer lasers

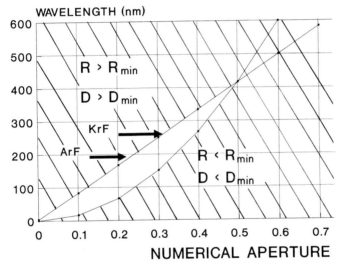

Fig. 4.1 Wavelength λ versus numerical aperture NA for an optical lens, illustrating the limitations imposed by a minimum resolution of 0.5 μm and a minimum depth of focus of 1 μm.

Table 4.1 Excimer lasers suitable for deep-UV optical lithography, together with the wavelength emitted.

Excimer laser		Wavelength (nm)
Fluorine	F_2	157
Argon fluoride	ArF	193
Krypton fluoride	KrF	249
Xenon chloride	XeCl	308
Xenon fluoride	XeF	351

as photon sources to provide appropriate wavelengths at suitable power levels. A key problem now being successfully overcome is the limited lifetime and reliability of the excimer laser itself. Table 4.1 shows the various excimer laser wavelengths that have been used to demonstrate lithography systems in the last few years.

The design of deep-UV lenses for optical lithography is well advanced, with two significantly different approaches being developed (see Chapter 3). One technique uses a lens made entirely from fused silica and requires a line-narrowed (e.g. 0.001%) excimer laser as a source [2, 3]. KrF lasers with specially designed cavities at $\lambda = 249$ nm have been used for this purpose, but the design, manufacture and maintenance of high-reliability lasers, suitable for operation in the production environment, remains a problem. This technique has the advantage that the lens can be designed with a demagnification ratio greater than unity (typically $5 \times$), and the required chrome-on-quartz masks can be obtained from existing mask-making equipment and processes. A second technique is to use a lens with chromatic correction and a much broader-bandwidth laser (e.g. 0.1%). Such lenses have been successfully constructed [4], using fused silica and

Fig. 4.2 A 1:1 magnification catadioptric lens designed for operation at KrF ($\lambda = 249$ nm) wavelengths and to resolve 0.5 μm features. The numerical aperture is 0.3 and the field coverage is a 60 mm diameter semicircle.

lithium fluoride, and employed in conjunction with a KrF ($\lambda = 249$ nm) or ArF ($\lambda = 193$ nm) laser source. The details of lenses published to date are mainly for 1:1 magnification catadioptric systems, which require the provision of high-quality masks.

This is a serious mask-making problem. For example, 0.5 μm lithography would require masks with features as small as 0.5 μm (cf. 2.5 μm for a 5× reduction lens) at a tolerance of 50 nm, which is beyond current technology. A 1:1 catadioptric lens, suitable for operation at KrF wavelengths, is illustrated in Fig. 4.2 (see also Chapter 3, Fig. 3.12 for more details). Typical submicron resist images obtained with the system are shown in Fig. 4.3.

The use of excimer lasers and the development of suitable lenses at deep-UV wavelengths offers the possibility of direct processing of substrates via photon-induced processes. In principle, the technology of laser chemical processing is capable of performing the essential device fabrication processes, such as etching, deposition of metals and semiconductor doping.

Fig. 4.3 Submicron resist images (approx. 0.8 μm) produced by the 1:1 catadioptric lens shown in Fig. 4.2. AZ2415 Novalac resist was used at a thickness of 0.6 μm. The resist is untreated after development in order to show the standing waves.

Fused silica lenses, such as that illustrated in Fig. 4.4, are capable of projecting complex submicron images, through a suitable gas cell and inducing various chemical reactions at the substrate surface. The energy of the laser (e.g. 1.0 J) is deposited onto the substrate in such a short time (10 ns) that very high irradiance levels are achieved (10^7 W cm^{-2}), and, at an appropriate wavelength, direct photoablation of material can take place. Figure 4.5 illustrates the direct photoablation of polyimide at the KrF wavelength $\lambda = 248$ nm. Experience with optical projection systems, under development, suggests that they can withstand high levels of irradiance

Fig. 4.4 A 10:1 fused silica reduction lens designed to operate at KrF wavelengths ($\lambda = 249$ nm). The NA of the lens is 0.3 and the field coverage is approximately 10 mm × 10 mm.

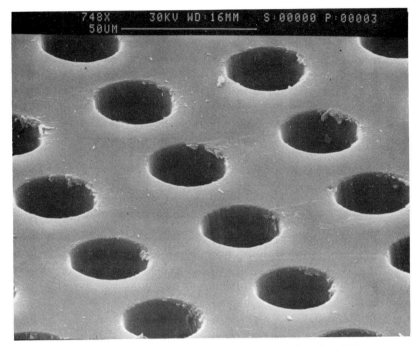

Fig. 4.5   Direct photoablation of polyimide film 50 μ thick.

without damage, provided that the materials used for the construction of the individual optical elements are defect-free.

As with many new techniques, laser chemical processing not only has to match the current and future resolution and process compatibility of existing methods of lithography, etching and deposition, but also has to offer novel and desirable capabilities.

4.3 X-RAY LITHOGRAPHY

X-ray lithography is a technique that "shadow-prints" a mask image onto a substrate, utilizing the very much shorter wavelengths of X rays in order to reduce diffraction effects to well below the deep-UV limits. X-ray lithography offers many advantages over optical lithography, notably high resolution, the ability to print very high-aspect-ratio patterns and low defect levels due to the relative insensitivity of X rays to any organic particulate contamination. In principle, sub-0.1 μm lines can be printed, but various effects conspire to degrade performance in a practical system [5, 6], as outlined below.

A simplified schematic of an X-ray lithography system is shown in Fig. 4.6, illustrating the relationship between the source, mask and substrate (wafer). It should be noted that, in any practical system, there must be a finite mask-to-wafer gap of several tens of microns in order to preserve the integrity of both the mask and the substrate during use.

The fundamental limits to resolution are set by Fresnel diffraction due

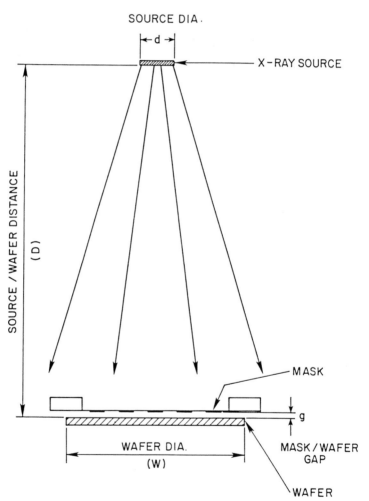

Fig. 4.6 Basic schematic of an X-ray lithography system. Geometrical distortion of the image is determined by the source diameter, mask-to-wafer separation and mask-to-source distance.

to the mask-to-wafer gap g and by the photoelectrons generated by the passage of the X rays through the resist. Some of these photoelectrons are released in a direction transverse to the direction of the X rays, and degrade resolution. The combined effect of diffraction and these spurious photo-electrons is shown in Fig. 4.7. Assuming that the effects of diffraction and photoelectrons can be added in quadrature, Fig. 4.7 shows that an X-ray system with $g = 20$ μm, requires an optimum wavelength $\lambda_{opt} = 1$ nm and should resolve 0.25 μm lines. In a practical system care must be taken to minimize the effect of penumbral blurring δ caused by the finite width of the X-ray source. The penumbral blurring may be estimated from Fig. 4.6 as $\delta = dg/D$ where d is the source diameter, D is the source-to-substrate distance and g is the mask-to-wafer gap as before. The X-ray source must be sufficiently bright to keep exposure times to a minimum and must be well collimated in order to reduce penumbral blurring. In practice this means a small source diameter (e.g. a few millimetres) and a large source-to-mask distance (e.g. 25–50 μm).

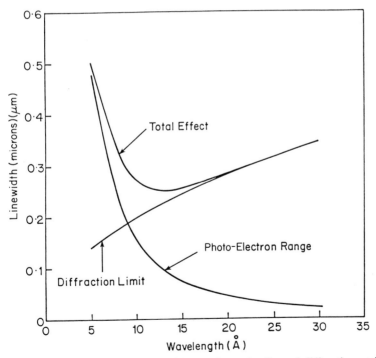

Fig. 4.7 Resolution limits for X-ray lithography set by Fresnel diffraction and the random direction of photo-electrons released by the resist and mask structure when absorbing the incoming X-ray photons. (The mask-transfer gap is 20 μm.)

Conventional e-beam generated X-ray sources have source sizes of a few millimetres, with output power levels of a few watts. Unfortunately, the large distance required for adequate collimation (e.g. several metres) results in prohibitively long exposure times (e.g. hours). Plasma X-ray sources, particularly those based upon lasers, have received much attention [7], but the only source so far able to meet all the requirements of X-ray lithography is the electron storage ring or synchrotron radiation source [8]. In the electron storage ring, electrons are constrained by a magnetic field to follow a circular orbit, and their associated radial acceleration causes electromagnetic radiation to be emitted in the forward direction. The result is a relatively wide-band source, with a spectral distribution similar to a black-body radiator and a critical wavelength λ_c below which 50% of the total energy is contained. The optimum energy for an electron storage ring for X-ray lithography is about 700 MeV. The radiation is strongly collimated in the forward direction and can be assumed to be parallel for lithographic applications. A very promising solution to developing an appropriate machine is the superconducting synchrotron storage ring, of which two are under construction in Europe and several in Japan and the USA. Figure 4.8 shows one of these machines, the "Helios" ring being constructed by Oxford Instruments for installation at IBM, East Fishkill. The

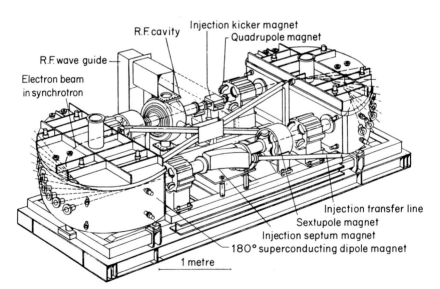

Fig. 4.8 The "Helios" superconducting electron storage ring. The electron energy is 700 MeV and the total emitted X-ray power 8 kW. The bending radius of the dipole magnets is 0.5 m and the magnetic field 4.5 T. (Courtesy of Oxford Instruments Ltd.)

electron energy is 700 MeV, the bending magnetic field is 4.5 T, the bending radius is 0.5 m, the total radiated power is 8 kW and the source size is 0.7 mm × 0.7 mm. While often referred to as a "compact" or "table-top" synchrotron, it nevertheless requires a small building to house not only itself but also ancillary equipment, such as the power supplies and other services.

Mask construction for X-ray lithography is fundamentally different and far more sophisticated than for conventional optical lithography, particularly in the provision of a supporting substrate for the absorber pattern. For conventional optical lithography, the supporting substrate is a relatively thick, near-optically flat piece of glass or quartz, which is highly transparent to optical wavelengths and provides a highly stable (2 parts in 10^6) basis for the thin (0.1 μm) chrome absorber pattern. In contrast, the X-ray mask consists of a very thin membrane (2–4 μm) of low X-ray absorbing material carrying a high X-ray absorbing pattern, typically made from a 0.5 μm thick layer of gold. The membrane must be transparent to the X rays and to light optics for alignment purposes and yet be stable to a few parts in 10^6. For example, the mechanical stress in the absorber pattern can cause in-plane distortion of the supporting thin membrane. This membrane can also be distorted due to environmental factors such as the X-ray radiation itself or humidity. A wide variety of single and multi-layer membranes have been examined, including those made from silicon, silicon nitride, boron nitride and silicon carbide. Each material has advantages and disadvantages, and further development work is necessary to determine which mask technology can be used for industrial wafer production.

4.4 SCANNING ELECTRON BEAM LITHOGRAPHY

Scanning electron beam lithography (SEBL) is based upon SEM technology, which uses the low equivalent wavelength of high-voltage electrons to eliminate diffraction effects for most practical applications.

In a practical system, resolution is limited by a number of aberrations inherent in any electron-optical system. An SEBL column exhibits both spherical and chromatic aberration and a number of deflection-induced aberrations. Deflection of the beam over a scan field of a few millimetres is possible, with the size limit being set by the magnitude of the deflection aberration coefficients. An electron beam machine will provide typically a 20 keV, 1–10 nA electron beam with a Gaussian-shaped beam of diameter of 0.05–0.1 μm, which can be deflected at speeds up to 10 MHz. In order to cover larger substrate areas, a laser interferometer controlled stage with a digitizing accuracy of typically 8 nm is provided and linked to the scanned e-beam field to a commensurate accuracy. A typical machine is shown in Fig. 4.9. Such a machine is very flexible and capable of undertaking applications ranging from 5 × reticle manufacture (e.g. 10^8 features with a minimum dimension of 1 μm) to nanonstructures with sub-0.1 μm features.

A typical submicron resist image produced by SEBL is shown in Fig. 4.10. The slight undercutting of the resist profile is caused by electrons backscattered from the resist–substrate interface. Backscattering of electrons from the solid substrate over distances of several microns can cause a significant problem for SEBL and give rise to "proximity" effects. When several shapes to be exposed lie within the range of the backscattered electrons, the energy absorbed at any point in the resist will be dependent on the proximity of adjacent structures. This is illustrated in Fig. 4.11, where several effects should be noted. The exposure of adjacent patterns that are close together (e.g. within 1–2 μm), results in a distortion of the shapes

Fig. 4.9 The EBMF-10.5 scanning electron beam lithography (SEBL) machine capable of supplying a 20–40 keV, 1–100 nA electron beam with a Gaussian profile and a spot size of diameter 0.025–0.1 μm. The machine is manufactured commercially by Cambridge Instruments.

Fig. 4.10 0.25 μm lines in a PMMA resist on a GaAs substrate. The beam current was 4 nA at an energy of 20 keV. The SEBL machine used for exposure was an EBMF-10.5. (Courtesy of Cambridge Instruments Ltd.)

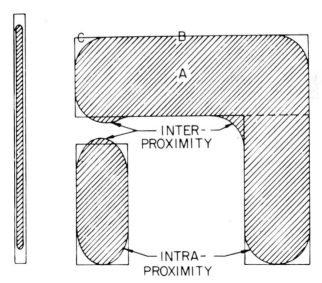

Fig. 4.11 Proximity effects with SEBL. Both interproximity and intraproximity effects can occur, as explained in the text. (After J. S. Greeneich [10].)

towards one another. This is termed the "interproximity" effect. At the corners and edges of isolated shapes, less dose is received than at the centre, and a variation in dimension occurs. This is known as the "intraproximity" effect [9].

There are three approaches to the elimination or reduction of proximity effects, particularly the interproximity effect. The first approach varies the electron dose using a software analysis of the global pattern and a known model for forward and backscattering [10]. One solution [11] attempts to provide constant average dose in each shape. This approach has received less attention in recent years due to the large computing resource needed to solve the mathematical formulae in a reasonable time (e.g. 1–2 h).

A second approach increases the voltage of the electron beam, thereby reducing the intensity of the backscattered electrons compared with the incident electrons. The proximity effect is dramatically reduced, as shown in Fig. 4.12, where sub-100 nm lines in resist have been defined. Using high voltages (e.g. 50–100 kV), in conjunction with a lift-off metallization technique [12], gold–palladium lines as small as 20–30 nm have been fabricated. The use of high voltages has contributed to the growth of interest in nanolithography, i.e. the range of applications where dimensions or toler-

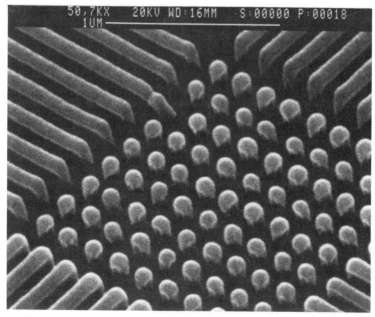

Fig. 4.12 Fabrication of nanostructures using high-keV-electron beams. The lines are approximately 70 nm wide in a PMMA resist and were exposed using a 40 keV electron beam. (Courtesy of Cambridge Instruments.)

ances are at 100 nm or below. However, not all substrates can withstand the bombardment of the energetic electrons that will normally pass straight through the resist with little energy loss, and so some applications cannot be undertaken with high-voltage electron beams. One such application appears to be silicon CMOS wafers.

Thirdly, a multilayer resist technique can be used to reduce the proximity effects by undertaking fine pattern definition on a very thin resist (e.g. 0.3 μm), which is separated from the device substrate by a thick (1–2 μm) intermediate layer. It is then possible to meet the normally conflicting requirements of high resolution, good linewidth control on nonplanar surfaces and resist compatibility with dry etching processes. Figure 4.13 shows a trilevel resist process consisting of a thin electron-sensitive resist on an aluminium pattern transfer layer on top of a thick polyimide planarizing layer. Submicron features were exposed in the top e-beam resist, using an SEBL machine, and the resultant developed image was etched into the aluminium using reactive ion etching (RIE). Using the patterned aluminium layer as an intermediate mask, the thick polyimide was then anisotropically etched using an oxygen-based plasma [13].

The major disadvantage of SEBL is the low exposure rate coupled with

Fig. 4.13 0.375 μm lines and spaces using a trilevel resist process. A 0.3 μm thick layer of Toya Soda CMS-EX(R) resist was spun on to a 0.1 μm thick layer of aluminium deposited on a 1.5 μm thick polyimide planarizing layer. (Courtesy of RSRE, Malvern.)

high system costs. At present, SEBL exposure is probably an order of magnitude more expensive than optical lithography for comparable tasks. The highest performance SEBL machines are currently those developed at IBM [14]. These novel machines use a variable shaped beam to provide beam shapes and sizes to approximate to the features to be printed. In principle, such machines offer submicron resolution at speeds 10–100 times faster than a Gaussian beam machine. However, interelectron effects in the high beam currents in the electron optical column limit the throughput of such machines, and it is unlikely that the promise of shaped SEBL technology will be fully realized for general semiconductor device fabrication.

4.5 FOCUSED ION BEAMS

Focused ion beam (FIB) equipment is similar in concept to that developed for SEBL, in that a computer-controlled scanning ion beam is produced with comparable beam dimensions and current levels, for example 50 nm beam diameter and $3\,\mathrm{A\,cm^{-2}}$ current density at beam energies of 50–150 keV. Most practical FIB systems for single-species ion beams use two lenses in a condenser–objective combination. A key component is the liquid metal ion source [15], which can be manufactured to produce ion beams from a number of different elements (e.g. gallium, boron and silicon).

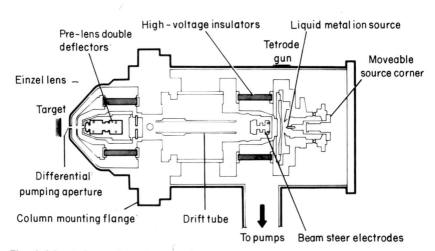

Fig. 4.14 A focused ion beam column capable of producing a 50 keV beam of ions with a beam diameter of 50 nm or less. The column forms part of the Oxford Instruments "MicroTrim" repair system for optical and X-ray masks. (Courtesy of Oxford Instruments Ltd.)

Figure 4.14 shows a typical ion column, designed for 50 keV operation, with the liquid metal ion source (typically gallium) and two accelerating gaps forming a tetrode gun/condenser and a final probe forming an einzel lens objective [16, 17]. An electrostatic pre-lens double deflector is positioned above the final lens, enabling field sizes of typically 0.5 mm to be readily achieved. FIB systems can be more complex than that shown in Fig. 4.14, depending upon the application. For example, a column for a mass-selected beam would use an alloy liquid metal ion source with either a magnetic [18] or an $E \times B$ mass-to-charge ion filter [19]. In general, a higher voltage would be used (namely 100–150 keV) and a third focusing lens employed.

FIB systems can be used for many of the processing steps common in semiconductor device fabrication, including lithography, ion implantation, etching and deposition. At one stage in the development of the technology, it was felt that scanning ion beam lithography (SIBL) might complement or even replace SEBL. For example, typical e-beam resists are up to 100 times more sensitive to ions than to electrons, and ion beams are less subject to the proximity effect than electrons. Unfortunately, the range of ions in resist is much shorter than for electrons, and hence much thinner resist layers (e.g. 0.1 μm) have to be used to ensure feature definition throughout the resist thickness (although doubly charged ions can be produced to increase the range). Thinner resist layers generally lead to more pinholes in the resist and more defects in the manufacturing process. In addition, the lower ion dose used to expose resists leads to random statistical noise, which has a deleterious affect on edge acuity and becomes critical as feature dimensions become submicron.

Focused ion beam implantation (FIBI) offers some novel fabrication technology when compared with the established flood ion implantation process. For example, the lateral grading of dopant profiles is relatively easy to achieve with FIBI but difficult to conceive with a flood system. FIBI necessitates the development of a reliable alloy source, and will require a mass-filtered column as discussed previously. For example, an Au–Si–Be alloy source would permit both p-type (Be) and n-type (Si) dopants for III–V compound semiconductor technology to be produced from a single source. However, it is unlikely that FIBI will be used to mass-produce semiconductor devices while the resolution of conventional flood implantation is adequate, since the scanning FIBI technique is many orders of magnitude slower in throughput.

The general use of scanning FIB in etching and deposition techniques for mass-produced semiconductor devices appears unlikely, given the apparent development potential of plasma-based systems to meet feature definition of 0.25 μm and below in silicon. The most successful application of FIB technology is currently the repair of very high-quality chrome-on-glass masks for optical wafer steppers. Furthermore, FIB repair techniques appear to be vital to the successful manufacture of defect-free X-ray masks [20].

Repair of optical masks is currently done using laser ablation to remove chrome defects and photo-induced deposition of carbon from a carbon bearing gas to repair pinholes. Both techniques suffer from the relatively poor resolution of the optics, which could be improved by using excimer lasers and deep-UV optics and with the after-effects of repair. It is the latter problem that is the most serious, and it has been shown [21] that there is considerable damage to the glass substrate, that the positional accuracy of the repair is limited to > 0.5 μm at best and that molten chromium debris are scattered and redeposited in the region of the repair.

In contrast, FIB milling can remove material virtually atom-by-atom, by sputtering with the scanning ion beam repeatedly and accurately over the chrome defect. The technique suffers from two problems. First, as the chrome is removed, ions from the primary beam (e.g. gallium) are progressively implanted into the glass substrate and cause a partially opaque stain. One technique used to overcome the problem is to bleed a gas (from a nozzle close to the ion beam) that recombines with the gallium and

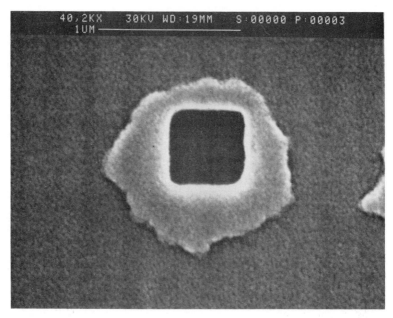

Fig. 4.15 A nominally square hole (0.4 μm × 0.4 μm) milled by FIB into a 1 μm isolated feature on a chrome-on-glass optical mask. The chrome thickness is approximately 0.1 μm.

substantially reduces its opacity [22]. Secondly, when the chrome defect under repair is isolated from ground, it will be charged positively by the incoming ion beam. The charge can build up sufficiently to make it impossible to image the feature using a secondary electron detector and to deflect the ion beam by several microns during the actual repair process. The effect can be neutralized by flooding the isolated feature with a beam of electrons from an electron gun, suitably placed near the final lens of the column. An excellent example of the "repair" of a submicron isolated chrome feature is shown in Fig. 4.15, where for demonstration purposes a 0.4 µm × 0.4 µm hole has been milled in an isolated 1 µm chrome feature.

Clear defects, such as pinholes, can be repaired by FIB-induced deposition from a hydrocarbon-bearing gas delivered from a nozzle close to the incoming ion beam. The ion beam decomposes the hydrocarbon molecules at the surface of the mask target and produces a tough highly adherent film of carbon with a similar resolution and accuracy to the milling operation. A 2 µm × 2 µm × 1 µm thick carbon patch deposited on to a chrome-on-glass optical mask is shown in Fig. 4.16 [23].

FIB mask repair is expected to become widespread during the next few years, even for optical 5 × reticle repair, as minimum features on wafer

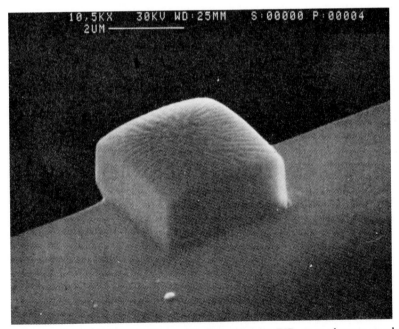

Fig. 4.16 A 2 µm × 2 µm carbon patch deposited by FIB on a chrome-on-glass optical mask to a thickness of approximately 1 µm.

Fig. 4.17 The FIB "MicroTrim" mask repair system, manufactured by Oxford Instruments. The FIB column shown schematically in Fig. 4.14 is fixed vertically above a vacuum chamber containing the stage mechanics and electron detectors. Operator interaction is via the keyboard, mouse and VDU.

approach 0.5 μm and below. Figure 4.17 shows an FIB machine, suitable for optical mask repair, which is commercially available and now undergoing trials at a UK mask facility.

FIB mask repair is seen as a key technology for X-ray lithography, although e-beam induced deposition [24] may offer a viable alternative because of the small features on the X-ray masks (i.e. 0.1–0.3 μm). The current capability of FIB systems will have to be improved to offer 20–30 nm beam diameters, 5–10 nm address increments, overall positional accuracy of a repair to better than 50 nm and higher currents. A stable FIB-induced metal deposition technique will be essential, along with a milling technique that avoids redeposition in what is essentially a three-dimensional structure, i.e. the X-ray absorber is typically 0.5 μm thick compared with an optical mask where the absorber is only 0.1 μm thick.

4.6 SUMMARY

Over the next few years, the use of excimer lasers will enable deep-UV optical lithography to reproduce features down to approximately 0.3 μm. The resultant small depth of focus will be overcome by various resist techniques, possibly those based upon the use of multilayer resists, to provide flat working surfaces. Laser chemical processes including ablation will find increasing usage, especially where large area, direct etching or material deposition of submicron features is required. It is likely that there will be a considerable number of applications in the mechanical engineering field (e.g. sensors) in addition to those in microcircuit fabrication technology.

X-ray lithography will not displace deep-UV optical lithography and become an accepted industrial technique until there is a need for sub-0.3 μm lithography, several major technical problems with the source and with mask making are overcome and the equipment costs are reduced. The major applications may well be for 0.25 μm lithography and for volume production in the late 1990s.

Scanning electron beam lithography will remain the accepted technique for high-resolution, high-complexity and high-quality mask making. For such applications, the problems with proximity effects and relatively low throughput have little relevance. For example, an acceptable e-beam exposure time for manufacturing a mask is 30–60 min. However, for direct-write applications, both problems must be solved before SEBL can displace optical lithography and/or meet the challenge of X-ray lithography for submicron applications. It is unlikely that this can be done in the near future, and hence direct-write scanning electron beam lithography will remain a technology for niche markets such as the quick-turnaround manufacture of application specific circuits.

Focused ion beam technology will play an important role in submicron lithography—but only in specific areas. Advanced optical masks will be repaired using FIB milling and FIB-induced deposition of carbon, although laser-based techniques may still be possible for less demanding applications. The superior resolution and placement accuracy of FIB milling and FIB-induced deposition of tungsten or gold will be essential if X-ray mask making is to succeed.

REFERENCES

[1] F. N. Goodall, R. A. Lawes and P. H. Sharp, *Microelectron. Engng* **5**, 445 (1986).
[2] V. Pol, J. H. Bennewitz, G. C. Escher, M. Feldman, V. A. Firtion, T. E. Jewell, B. E. Wilcomb and J. T. Clemens, *Proc. SPIE* **633**, 6 (1986).
[3] M. Sasago *et al.*, in *Technical Digest of International Electron Devices Meeting*, p. 316 (IEEE, New York, 1986).
[4] F. N. Goodall and R. A. Lawes, *Proc. SPIE* **922** (1988).
[5] A. Heuberger, *Microelectron. Engng* **5**, 3 (1986).
[6] A. Heuberger, *Microelectron Engng* **3**, 535 (1985).

[7] E. Turcu, G. Davis, M. Gower, F. O'Neill and M. Lawless, *Microelectron. Engng* **6**, 287 (1987).

[8] M. N. Wilson, in *Advanced Lithography* (ed. R. A. Lawes), p. 13 (IOPP, Bristol, 1988).

[9] J. S. Greeneich, in *Electron Beam Technology in Microelectronic Fabrication* (ed. G. R. Brewer) (Academic Press, New York, 1980).

[10] T. H. P. Chang, *J. Vacuum Sci. Technol.* **12**, 1271 (1975).

[11] M. Parikh, *J. Vacuum Sci. Technol.* **15**, 931 (1978).

[12] M. Hatzakis, *J. Electrochem. Soc.* **116**, 1033 (1969).

[13] A. Brown, S. H. Mortimer, S. J. Till and V. G. Deshmukh, *Microelectron. Engng* **3**, 443 (1985).

[14] H. C. Pfeiffer, in *Proceedings of 14th Symposium on Electron, Ion and Photon Beam Technology: J. Vac. Sci Technol.* **15**, 887 (1978).

[15] R. Clampitt and D. K. Jefferies, *Nucl. Instrum. Meth.* **149**, 739 (1978).

[16] P. D. Prewett, in *Advanced Lithography* (ed. R. A. Lawes), p. 61 (IOPP, Bristol, 1988).

[17] MicroTrim Mask Repair System, Oxford Instruments Ltd, Eynsham, UK.

[18] J. R. A. Cleaver, P. J. Heard and H. Ahmed, *Microcircuit Engng* **83**, 135 (1983).

[19] V. Wang, J. W. Ward and R. L. Seliger, *J. Vac. Sci. Technol.* **19**, 1158 (1981).

[20] U. Weigmann, *Microelectron. Engng* **6**, 617 (1987).

[21] P. J. Heard, P. D. Prewett and R. A. Lawes, *Microelectron. Engng* **6**, 597 (1987).

[22] P. J. Heard and P. D. Prewett, *J. Phys.* **D20**, 1207 (1987).

[23] P. D. Prewett, *Colloq. Phys.* **11**, 179 (1989).

[24] W. H. Brugner, *Microelectron. Engng* **9**, 171 (1989).

5 Promoting Photonucleation on Semiconductor Substrates for Metallization

J. FLICSTEIN

CNET — Laboratoire de Bagneux, France

J. E. BOURÉE

CNRS — Laboratoire de Physique des Solides, Meudon, France

5.1 INTRODUCTION

The potential of photo-induced metallization [1–5], for the processing of semiconductor materials for microelectronic devices [6], has been increasingly recognized over recent years. This interest has been reinforced by the availability of new commercial equipment dedicated to laser [7] or lamp [8] processing. A particular promising process is patterned laser metallization on various semiconductor surfaces from gas phase organometallic precursors. The advantages that laser processing offers are

(1) the possibility of high spatial resolution of the nucleation step;

(2) high selectivity to the wavelength of the formation and evolution of nuclei;

(3) high deposition rate,

(4) low-temperature alternatives to thermal processing.

Certain organometallic precursors for metallization (alkyls, carbonyls, halides and hydrides) have received attention because of their attractive intrinsic properties for processing [1]. The metal photonucleation stage is the creation of stable metal nuclei, induced by photons of different wavelengths, prior to metal deposition.

Why is understanding photonucleation necessary? Photonucleation and

Photochemical Processing of Electronic Materials
ISBN 0-12-121740-X

the initial stages of metal deposition are important because

(a) they control the main physical properties of the system: roughness and the adhesion at the interface;

(b) they can be the rate limiting step in deposition;

(c) they control the structural and electronic properties of the ultimate deposit.

Currently, several fundamental mechanistic aspects need to be clarified:

(i) the exact role of photon impingement on surface–adsorbate interactions in the metal photonucleation process;

(ii) detailed knowledge of fragments that may be produced as photo by-products, their nucleation on the substrate and their incorporation in the deposit.

Existing theoretical work shows [9, 10] two possible basic types of metal nucleation: (a) energetically more difficult, gas phase homogeneous nucleation, and (b) heterogeneous nucleation of isolated atoms and atomic clusters that is easily activated by substrate surfaces [11]. However, the objective here is limited to heterogeneous photonucleation from a gas phase on a semiconductor substrate. The recent availability of modular spectroscopic equipment [12], together with scanning tunnelling microscopes [13] (STM) based on ultrahigh vacuum (UHV), has made possible correlation between atomic-scale observations of surfaces, and photon-induced atomic interactions with surfaces, under *in situ* reproducible conditions. New ways to study the early stages in photonucleation process at nanometer level have therefore become possible.

This chapter is an overview of the advantages of photonucleation of metals on semiconductor substrates for metallization processing. The aim is to describe the state of our present understanding of nucleation phenomena assisted or induced by laser or lamp sources from gas-phase and absorbed precursor molecules.

5.2   BASICS: PHENOMENA AND MECHANISMS

On irradiating a silicon substrate with a low-fluence (5 mW cm^{-2}) UV mercury lamp emitting at 254 nm in a flow of trimethylaluminium (a precursor gas containing the metallic element to be deposited) some Al nuclei of different shapes are observed when the duration of exposure exceeds 30 min [5]. Thus, under some circumstances, it may be stated that metal photonucleation may be considered as a bottleneck between surface photodissociation and the beginning of growth. Metal is produced from the precursor by removing the ligands through photodissociation, which leads to a metal cluster. Therefore the concept of metal photonucleation must be considered as a sequence of steps performed under selective irradiation, leading from the precursor gas molecules to stable nuclei of metal on the surface concerned.

Thus the overall mechanism is normally complex and must be approached through the individual study of the interaction of light with matter (organometallic gas molecules, photoproducts and semiconductor surface) and through the study of the interaction of atoms, molecules and photodissociated fragments with the semiconductor surface. Then the elementary steps must be integrated in a general form. The interaction of light with gases and surfaces [14–19] has already been introduced in Chapter 1.

5.2.1 Overview of laser chemical vapour deposition

In a cold-wall CVD reactor, the sample susceptor is inductively heated while a precursor gas and an inert carrier gas flow through the reactor over the semiconductor substrate. Precursor molecules dissociate, either homogeneously or heterogeneously, and form the deposit material on the large-area substrate. At atmospheric total pressure, diffusion of precursor molecules to the surface and homogeneous gas phase decomposition dominate the kinetics of deposition. Therefore modelling of the flow for a reactor configuration is important [20, 21]. In low-pressure CVD (LPCVD), the total gas pressure is reduced below 1 Torr, there is no use of carrier gas, and the rate of the precursor dissociation, as controlled by the surface, becomes the limiting step.

In laser-assisted CVD, which means CVD enhanced by the use of a laser, photons of selected wavelength are absorbed by the semiconductor. Thus a localized hot reaction zone, determined by the photon energy distribution, is created (see Fig. 5.1). For typical Gaussian laser beam dimensions of 1 μm diameter, the hot reaction zone is 1 μm wide [22]. This is an estimate for the hot zone induced by a static beam at normal incidence. The value corresponds to a silicon substrate. At a low gas pressure (< 10 Torr), the mean free path of unreacted precursor in the photoreaction volume is equal to or larger than the zone lateral extension. When an initial metallic layer is formed on the semiconductor surface, this layer allows subsequent metallization to occur more freely, since there is an "autocatalytic" effect that enhances the ability of the precursor gas to be decomposed at these metallic surface sites [23]. Thus, this first layer is often referred to as a "prenucleation" layer [24] (see Fig. 5.1). Under these conditions, dramatic

LASER-ASSISTED METAL NUCLEATION

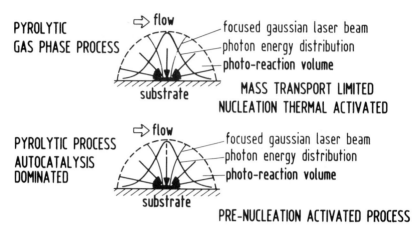

Fig. 5.1 Schematic of the principle of two processes of laser-assisted metal nucleation under static beam impact. The focused beam is impinging at normal incidence onto the substrate. The precursor flow is directed parallel onto the substrate. (a) Representation of the pyrolytic gas phase processes. The nucleation process is thermally activated. The nucleation contribution arrives mainly from the gas phase. The photoreaction rate, in the photoreaction volume, is possibly mass-transport limited for high power density. (b) Representation of the pyrolytic heterogeneous phase process. The laser heating is at normal incidence onto prenucleated substrate. The heterogeneous nucleation takes place onto an exclusively prenucleated site. A metal autocatalytic effect is possible [23a,b].

vertical deposition rates up to 10^8 Å s^{-1} [25] can be achieved; this protects the newly created metal–semiconductor interface since shorter-duration processing can be employed. Thus higher surface temperatures than in large-area CVD are allowed. Properties of thermally activated metal–semiconductor interfaces are known to exhibit certain differences, depending on the thermal regime.

An ideal metal should meet several criteria in a laser-assisted CVD process [22]:

(a) it should not interact chemically with the substrate surface;

(b) it should not permit the out-diffusion of the substrate constituent atoms;

(c) metal atoms should not (deeply) diffuse into the substrate;

(d) the stress at the created interface should be small, and the created interface should be smooth;

(e) the metal deposit should exhibit fast onset of nucleation;

(f) the metal should be structurally dense and compositionally pure;

(g) the metal deposit should exhibit mechanical resistance to cracking and peeling from the surface.

The extent to which these criteria can be achieved is limited by the threshold of the contribution to the deposit from homogeneous nucleation. Also, unwanted incorporation in the growing film of by-products from the thermal decomposition of the precursor can result from such rapid pyrolysis, as has been reported in [5, 26].

Several basic features [26b] differentiate laser or lamp-induced photolysis in CVD, from the previously presented laser-assisted pyrolysis processes, which can be summarized step by step as follows (see Fig. 5.2).

(i) When using UV photons (photon energy higher than 5 eV), the gaseous molecules or the adsorbed molecules are dissociated by one

LASER-INDUCED METAL NUCLEATION

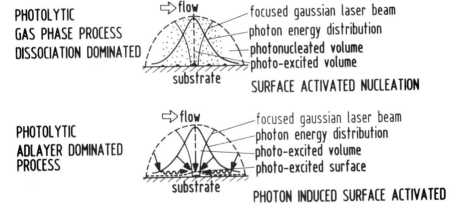

Fig. 5.2 Schematic of the principle of two processes of laser-induced metal nucleation under static beam impact. The focused beam is impinging at normal incidence onto the surface. The nucleation process is athermal. The precursor flow is directed parallel onto the substrate. (a) Representation of the photolytic gas phase process. The nucleation contribution arrives mainly from the gas phase, namely the photo-excited volume. The earliest clusters may assist subsequent nucleation onto the substrate, resulting in low deposit adhesion. The process is driven by the photo-dissociation rate. (b) Representation of the photolytic adlayer process (PAP). The nucleation is photon-induced and surface-activated onto the substrate adlayer. Nucleation takes place preferentially on a site induced by the photon beam. The continuity has to be ensured by the adlayer photodissociation and replenishment.

photon electronic excitation. This photolytic process is athermal and independent of the substrate nature, but depends selectively on the wavelength [27].

(ii) The lateral localization, in a microfocused beam-induced photolysis, is confined to a micron-dimensioned photo-excited volume where energetic photofragments and metal atoms are generated. This interaction limits the lateral spread of the photo-excited surface.

(iii) Decomposition by photolysis of the precursor molecule can now take place heterogeneously on the irradiated surface or homogeneously near the surface to yield adsorbed metal atoms. Contributions originating in the gas phase and in the adsorbed layer must be separated. It is experimentally found for metal alkyls that UV absorption curves relative to the adsorbed phase are broadened and shifted compared with the absorption curves relative to the gas phase [28]. Thus the rate of creation of photofragments and metal atoms is much greater in the adlayer than those associated with the photo-excited volume. Therefore this defines a photolytic adlayer-dominated process (PAP).

(iv) Heterogeneous nucleation on preferential sites of metal adatoms, created on the substrate through irradiation, may lead to the formation of stable nuclei and hence improve the rate at which subsequent thermal or pyrolytic film growth can occur [23]. This is a photon-induced surface-activated process.

(v)   There is desorption of photolysis secondary products (photo by-products) away from the reaction volume.

5.2.2 Preferential site of photonucleation

We shall treat heterogeneous photonucleation as two sequential processes:

(a) formation of an adsorbed layer on the semiconductor substrate;

(b) attainment of a critical nucleus size (from the supersaturated metallic atoms).

Additional photon effects (e.g. on surface mobility) can be subsequently included in this model. We are not considering here the sensitivity of heterogeneous photonucleation to structural defects.

(a) Chemisorption and physisorption [29–35]

Photon-activated adsorption on semiconductor surfaces has been briefly introduced in Chapter 1 and is rigorously discussed in [36]. One mechanism involves photo-induced adsorption or desorption applied to a semiconductor type of surface. Trapping of the precursor molecule in a chemisorption or a physisorption state (which serves as a precursor state) occurs as a result of the loss of the incident kinetic energy to the surface. The molecule preserves this state until enough energy is supplied, from the substrate or the impinging photon, to desorb without dissociating. Alternatively, (a) enough energy is supplied from the substrate or the impinging photon to reorient it into a more favourable configuration for dissociation, or (b) the molecule diffuses onto an energetically favourable site for further dissociation. Which mechanism is responsible for the dissociation of the molecule is dictated by (a) the photon energy and (b) the potential energy hypersurface of the interaction (which leads to the plot of energy changes against the bond distance during a successful dissociation).

Two mechanisms are used to describe the dissociative chemisorption of a precursor molecule on a surface [31]:

(1) direct dissociative chemisorption at the surface;

(2) precursor state dissociative chemisorption near the surface.

In the first case, the molecule dissociates into adsorbed fragments immediately upon its collision. In the second case, the molecule is adsorbed intact before dissociating. The following examples may be sufficient to demonstrate the extensive application of chemisorption to photonucleation.

Oxygen chemisorption on semiconductors is an example of the first case of dissociative chemisorption [37]. Here the temperature of the surface has no effect on the probability of dissociative chemisorption. In the primary contact of nascent metal atoms with this oxygen, oxidation phases are formed on the semiconductor. Therefore this may suggest an explanation of the successful use of oxygen in photoprocesses of metal oxidation.

It has been hypothesized that strong adsorption of methyl radicals on metals is responsible for the incorporation of carbonaceous photo by-products [38]. The activated dissociative chemisorption of CH_4 on metals is the most studied of all the systems using molecular beam techniques [31]. The effect of translational and vibrational energy on the dissociative chemisorption of CH_4 is discussed in [30b]. By analogy, for metal alkyl precursors, the picture of the effect of translational and vibrational energy might be significant for photonucleation of carbonaceous photo by-products in understanding the more general problem of carbon contamination and the pressure gap effect [39–41].

The pressure gap effect is a catalytic activity that is observed under high-pressure ($>10^{-4}$ Torr) conditions, common in heterogeneous processes, but is often absent in low-pressure ($<10^{-9}$ Torr) surface science experiments; for example, methane does not adsorb dissociatively on Ni as a room temperature gas at a pressure below 10^{-4} Torr. The reason for this lack of dissociative adsorption is the very low dissociative probability (of order 10^{-9}) due to the presence of an energy barrier along the dissociative reaction coordinate. However, an increase in pressure can increase the number of molecules per unit time able to overcome the barrier, and therefore leads to dissociation.

(b) Metal photonucleation on a semiconductor substrate

When two-dimensional nucleation occurs on a flat single-crystal substrate, the formation of the nucleus is the rate-determining step in the metal deposition process [10]. In order to promote a well compacted continuous deposition, the incident atoms must be able to rapidly diffuse until the preferential site for metal nucleation has been found. Direct condensation on the preferential site is possible if the supersaturation corresponds to $\Delta\mu > 0$, where μ is the chemical potential and $\Delta\mu$ is the thermodynamic driving force of the process (i.e. the supersaturation).

The probability of metal atoms sticking on a semiconductor flat surface is of particular importance in the early stage of metallization when the atom to be incorporated in the cluster is different from those belonging to the substrate lattice. In the case of a photon at normal incidence on the substrate, localized heterogeneous photonucleation can take place at the beam impact site if clusters form on the substrate within the matrix of an adsorbed gas precursor. At this step of photonucleation, there is no direct experimental evidence as to whether these clusters of the condensed phase are liquid-like or solid-like.

If the gas precursor has a greater electron affinity than the work function of the semiconductor, the precursor is an acceptor that may capture an electron from the surface state and induce band bending upwards. Therefore the adsorption process may be specifically induced by a photon, which makes more electrons available to an electronegative-type precursor, even in traces, like oxygen. We may consider the same mechanism for charging a semiconductor surface. In this case, the adsorption of stable donor or acceptor precursors onto the semiconductor surface leads to the formation of donor−acceptor (DA) complexes, which are neutral, if the process is carried out in darkness.

Under UV light, DA complexes may trap free carriers, which result from exciton dissociation [42]. Therefore local charges are produced on the surface. For simplicity, we can exclude the activity of local charges in the gas phase, contributing either to an increase in the surface density of nuclei or to their further development.

The energy of formation of a nucleus on a charged centre is lower than on a neutral one [43]. Hence, in the present case, this is the preferential site for photonucleation. In order to make a connection between charged centre density and density of nuclei, let us introduce a simple model. We suppose the nuclei to be generated only at an active charged centre (ACC). Classical nucleation theory [10, 44a] predicts that the onset of nucleation begins when the adatom surface density attains a critical value [44a]. Each nucleus builds up a circular diffusion zone of radius h. All other ACCs within a constant radius will be excluded from further activity. The maximum density of nuclei on the surface n_s is proportional to the nucleation rate and inversely proportional to the rate at which the diffusion zones spread. When saturation is reached in the unit area containing z ACCs, n_s nuclei will have formed. We should expect an increased density of nuclei when deposition takes place upon ACCs, relative to that for deposition upon a neutral surface, since there is a gain in the energy of formation of a nucleus on ACC. We can estimate the total area covered by diffusion zones $n\pi h^2$. An instantaneous increase of the density of ACCs by dz corresponds to an area increase of $n\pi h^2\, dz$, leading to formation of dn additional nuclei:

$$dn = (1 - n\pi h^2)\, dz. \tag{5.1}$$

Solving this for n, we obtain

$$n = \frac{1}{\pi h^2}\, (1 - e^{-z\pi h^2}). \tag{5.2}$$

This equation gives the connection between the density of ACCs and the density of nuclei. In the case of very dense active centres, the density of nuclei is determined by the zone radius only. Consequently, most of the ACCs are excluded from the photonucleation. This means, applied to the formation of nuclei, that correlation between ACCs and nuclei can be expected only in the case of small ACC density. However, z as well as h determines the nuclei density. Now we shall estimate the ratio of the density of nuclei for a charged surface n_s^* to that for nucleation on a neutral surface n_s:

$$\frac{n_s^*}{n_s} = \exp\left(\frac{\Delta E_q}{kT}\right),$$

where ΔE_q is the gain in energy of formation of a nucleus on a charged centre. Therefore, in the case of an acceptor-type precursor, with respect to the semiconductor substrate, UV photons lead to the formation of local charges on its surface. Subsequently, an increase in the rate of nucleation is manifested as compared with that registered in the absence of UV light.

5.2.3 Basic modes for photodeposition of initial layers

Experimental observations [45, 46] have led to the formation of three classical growth mode models of initial layers, illustrated in Fig. 5.3.

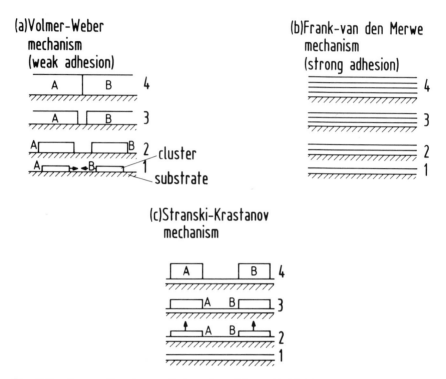

Fig. 5.3 Consecutive stages of photodeposition of initial layers: basic modes for cluster coalescence and aggregation [45]. A and B are clusters. The priority direction for rate of development is marked (see arrows). (a) Typical of weak adhesion of nucleus onto the substrate: island deposition mode (Volmer–Weber type). (b) The layer is more strongly bound to the substrate than to the next similar layer: layer deposition mode (Frank–van der Merwe type). (c) The first layer deposition is followed by island deposition: mixed deposition mode (Stranski–Krastanov type).

(a) The island deposition (or Volmer–Weber) mode (Fig. 5.3a)

This mechanism is typical of weak adhesion of a nucleus on the substrate, when the deposit is more strongly bound to itself than to the latter. The energy required to form a nucleus on the substrate is larger than in all the other cases. Accordingly, the role of the preferential sites of nucleation on the substrate is greatest. Weak adhesion also leads to a smaller interaction between the crystallographic orientations of the deposit–substrate interface. Quantitatively, the mutual nucleus–substrate orientation depends primarily on their interface energy. This interrelation appears in microscopic form (cf. the atomistic theory of nucleation [47, 48]) as the binding energy between nearest neighbours in a lattice, which corresponds to the adsorption energy of intrinsic or impurity entities at the nucleus–substrate interface Additional features come into play if the adsorbed layer consists of several components [49]. If two different entities A and B are present, their interaction energies ε_{AA}, ε_{BB} and ε_{AB} have to be taken into account. The sign and magnitude of

$$\Delta\varepsilon = \varepsilon_{AA} + \varepsilon_{BB} - 2\varepsilon_{AB}$$

determine the temperature versus coverage phase diagram. The following limiting cases may be distinguished:

(i) cooperative adsorption with the formation of a regular mixed phase, $\Delta\varepsilon \gg kT$;

(ii) competitive adsorption when both species will not be miscible but rather will coexist on the substrate surfaces in separate domains, $\Delta\varepsilon \ll kT$.

Obviously, the creation of an induced heterogeneity, by the interaction between adsorbed species, precursor intermediates or products, has consequences for the thermodynamic properties of precursor adsorbed layers on single-crystal substrate surfaces: the differential adsorption energy becomes dependent on coverage, and adsorption isotherms do not follow the simple Langmuir equation.

(b) The layer deposition (or Frank–van der Merwe) mode (Fig. 5.3b)

This mechanism will occur as an extreme case [50] when the deposit is more strongly bound to the substrates than to itself: $\varepsilon_s > \varepsilon_1$. This is a case of complete wetting, $\Delta\sigma < 0$, and experimental results [51] have shown that the formation of one or more adsorbate overlayers is energetically favourable even if the arrival rate is quite low ($\Delta\mu < 0$, and the gas is undersaturated). This appears explicitly as a macroscopic parameter $\Delta\sigma$, representing a measure of how much more difficult it is to split the deposit than to separate it from the substrate, or, more precisely, when

$$-\frac{\Omega\,\Delta\sigma}{a} \approx \varepsilon_s < \Delta\mu < 0, \tag{5.3}$$

where σ is the specific nucleus–substrate interfacial free energy, Ω is the specific volume per considered entity in the medium, a is the critical nucleus height, $\Delta\mu$ is the supersaturation value of the same considered entity, and ε_s is the adsorption energy of the same entity at a crystal interface, i.e. this is the "trivial case" when the deposited metal atoms fit into the substrate lattice, forming a "pseudomorphous overlayer".

(c) The intermediate deposition (or Stranski–Krastanov) mode
(Fig. 5.3c)

This is an intermediate situation between the cases already presented, in that layer deposition is followed by island deposition. In fact, the condition for good adhesion holds only at the beginning $2\sigma < \sigma_s$, while the adhesion of the bulk is weaker and subject to the opposite condition $2\sigma > \sigma_s$ (case of mixed adhesion; σ_s is the free energy of adhesion).

The fact that these models are oversimplifications is elegantly demonstrated by considering the photo-induced deposition of Al from trimethylaluminium (TMA) on a Si(100) surface. UV light has been found to decompose TMA to yield pure Al, but at only very slow rates. Visible light

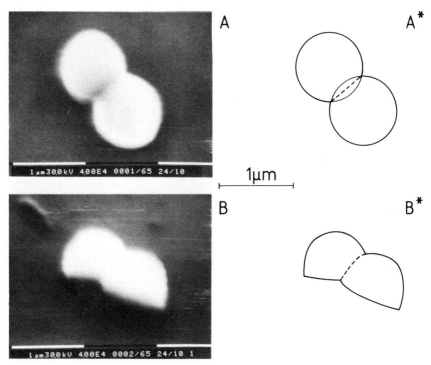

Fig. 5.4 Scanning electron micrograph presenting the coalescence of two solid Al islands to form a single solid island on silicon (100) [52]. (A) Planar view with corresponding scheme (A*) showing the spherical shape. Experimental conditions: TMA precursor gas irradiated during 30 min with a UV (254 nm, 5 mW cm^{-2}) lamp. (B) Same nuclei as above. Planar substrate was tilted by 15°, and revealed (see B*) a spherical cap characterizing a heterogeneous-type nucleation. The deposition mode suggested is the Stranski–Krastanov type.

(514 nm) is believed to promote a pyrolytic deposition step, leading to a rapid deposit of Al, which is highly contaminated [23a]. The combination of these two approaches can give clean deposits with reasonable deposition rates. We can begin to understand this by considering the type of photonucleation that each wavelength promotes. Figure 5.4 shows a scanning electron micrograph as an example of this case. Al nucleation and coalescence features included on Si(100) after 30 min irradiation of TMA gas with a UV lamp (5 mW cm^{-2}, 254 nm) at room temperature are clearly visible. The planar view provides evidence for the smooth spherical shape of the solid deposited nucleus. Tilting the substrate by 15° further reveals a spherical-cap-shaped nucleus. This geometry is most likely to arise from a heterogeneous nucleation process [44]. The adjacent illustration is helpful for the direct measurement of the contact angle, yielding a value of about 60°. Following conventional treatments, we can now evaluate the surface energy density and the absorption energy; the value of the latter indicates the tendency for development of the intermediate mode of deposition. The aggregation of solid Al nuclei is shown in a separate micrograph (Fig. 5.5). A round-shaped aggregate (800 nm × 900 nm) is displayed in (a). Then the same aggregate is presented (b) after geometrical decomposition into six spherical elementary nuclei, ranging from 100 to 400 nm each. Three additional nuclei (50 nm) are ready for incorporation in the main aggregate.

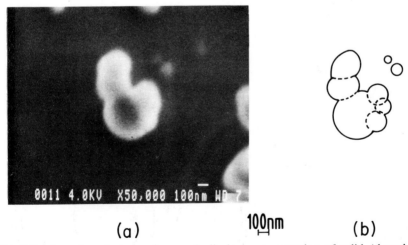

0011 4.0KV X50,000 100nm WD 7

(a)   100nm   (b)

Fig. 5.5 Scanning electron micrograph displaying aggregation of solid Al nuclei on silicon (100). After [52]. (a) A round-shaped aggregate (800 nm × 600 nm). Three stand-by nuclei are ready for incorporation into the main aggregate. (b) Decomposition of the aggregate into six spherical elementary nuclei, ranging from 100 to 400 nm. Experimental conditions. TMA precursor gas irradiated during 30 min with a UV (254 nm, 5 mW cm^{-2}) lamp.

Thus UV irradiation leads to the onset of surface nuclei, and can be thought of as a prenucleation stage. If we observe Al deposition assisted by visible (514 nm) laser light, subsequent to UV (254 nm) lamp-induced prenucleation, we see a process typified by the scanning electron micrograph reproduced in Fig. 5.6. This displays the annular zones obtained near the beam impact, on a (100)Si substrate. Starting from the outside zone, and going towards the beam impact, we distinguish gradually the following types of nucleation and deposition morphology (Fig. 5.7):

(a) spherical-cap-type nuclei;

(b) hemiellipsoidal-type nuclei;

(c) hexagonal-like cells; and finally

(d) layer-by-layer type deposition.

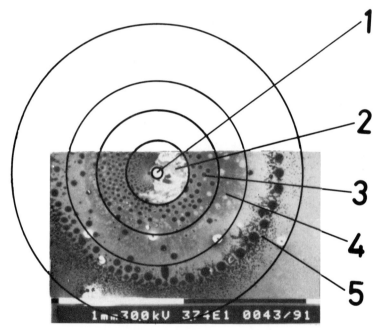

Fig. 5.6 Scanning electron micrograph showing the features of laser-assisted Al nucleation subsequent to UV (254 nm) prenucleation. The scheme emphasizes beam central impact, with the additional four regions radially disposed: (1) visible light (514 nm) laser beam impact accompanied by a characteristic crater; (2) layer deposition of Al; (3) mixed deposit of Al of layer and island type; (4) a few smooth and thin discs of C-containing species; (5) dense deposition of smooth and thin discs of C-containing species.

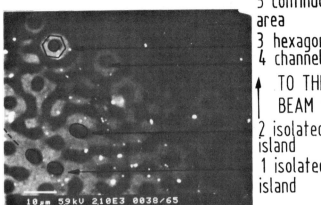

5 continuous deposition area
3 hexagonal-like cell
4 channel filling

↑ TO THE LASER
| BEAM CENTRE

2 isolated ellipsoidal island
1 isolated spherical island

Fig. 5.7 Scanning electron micrograph, zooming in to the near-by impact regions: (1) isolated spherical island; (2) isolated ellipsoidal island; (3) hexagonal-like cell; (4) channels of hexagonal-cell filling-up; (5) continuous deposition area. The deposition mode suggested is the Stranski–Krastanov type.

The spherical-cap type of nucleation on Si(100) can be easily understood at low UV fluence. For isotropic surface energies and a flat substrate, in classical nucleation theory, the equilibrium cluster shape is a spherical cap [44]. This results from the onset of heterogeneous nucleation during the lamp prenucleation step, and shows the threshold for laser-assisted nucleation (514 nm) [23]. It is believed that the hemiellipsoidal-type nuclei are a growth phenomena and result from surface plasmon interactions at 514 nm with the previous spherical-cap nuclei of Al [53, 54]. Then the continuous development of nuclei yields cells, which subsequently fill up to show the Stranski–Krastanov deposition mode.

This clearly demonstrates how the classification for basic modes of initial monolayer deposition is a helpful guide—but is an oversimplification of what actually happens in the photometallization process.

5.3 PHOTONUCLEATION FROM GAS PHASE AND SURFACE-ADSORBED PRECURSORS

The major processes responsible for photonucleation are reviewed in [11, 55, 56]. However, the behaviour of metallic adatoms under photon beams, directly related to the quantum yield of photodissociation, has been investigated only for isolated cases [57]. Many other factors contribute to photo-induced deposition, through poisoning of the photonucleation step and the subsequent photodeposition process, and by the formation of photo by-products, which have to be included in this picture. Moreover, to determine the chemical nature and concentration of very different photo by-products is difficult. The concentration of admolecules on the surface, their surface lifetime, desorption rate and surface diffusion rate are fundamental parameters under photon impingement.

5.3.1 Pioneering work in laser-assisted and laser-induced nucleation

The importance of photonucleation phenomena in photolytic deposition processes has been recognized ever since the first studies in this area. Ehrlich *et al.* [24] emphasized the surface activation contribution to metal deposition, with a surface-diffusion-dominated case [58], which was thoroughly analysed in [59, 60]. This concept has been extended by taking into account the gas phase contribution [61], which destabilizes the surface morphology with respect to deviations from uniformity. Concomitantly, the more conventional activated pyrolysis nucleation has been applied by several research groups [62–64].

Possible surface enhancement of photolytic processes has been a subject of much interest for the following reasons. This is the first step that one must study in order to control the phenomena of photonucleation on a rough surface. Although it is now clear that the surface plays a role in enhancing Raman scattering via its plasmon field, this is not specifically the case for the photodissociation of a gaseous precursor near or physisorbed onto a rough metallic surface [65, 66]. The reason for this is that, while the enhanced local field stimulates the absorption process, due to transitions to continuum states, the line-broadening effects play a competing role to suppress the dissociation. Ehrlich and Osgood [67] found that at 257 nm an important amount of dissociation occurs in the gas phase far above the substrate for $Cd(CH_3)_2$ but not for $Al_2(CH_3)_6$. Using a simple model [68], this is explained for TMA, where the dissociation occurs inside the chemisorbed layers, whereas plasmon resonance conditions are not fulfilled under the experimental conditions. From a practical viewpoint, this leads to the prenucleation concept [24], in activated pyrolysis [69] successfully applied to Al deposition, which was a breakthrough in metal deposition based on a two-step photonucleation process [5, 23].

The roles of the physisorbed layer and the gas phase contribution in a photodeposition process can be distinguished in a more global manner [70] using the dimensionless quantity G_1, defined by

$$G_1 = \frac{170\eta_p E_p}{\omega_0(\mu m) p_0(\text{Torr})},$$ (5.4)

where ω_0 is the laser beam radius, $E_p = A_{phy}p_0/A_{gas}p$ is the enhancement factor, η is the quantum efficiency or quantum yield, p_0 is the precursor vapour pressure at temperature T, p is the precursor partial pressure (Torr), A_{phy} is the optical absorption (physisorbed film) and A_{gas} is the optical absorption (precursor gas phase). A gas phase dominant contribution to the reaction leads to a value of $G_1 = 1$. This fast engineering approach can also enable us to gain initial insight into new photoprocesses that are to be considered.

5.3.2 The physical barrier to nucleation

The classical theory of nucleation considers the phase transition at constant composition [44a]. The transformation refers to a given set of thermodynamic conditions that lead to the birth of nuclei. If the number of atoms in the cluster is not very small then macroscopic concepts of surface energy and chemical potential can describe the system. The change in the thermodynamic potential of a system containing an aggregate has a maximum corresponding to the critical size of the nucleus. More details are outlined in Chapter 1. Since the new phase has a different specific volume and structure, the creation of a stable nucleus in the initial matrix causes interfacial stresses. The specific interfacial free energy contribution to the free energy of the system would play an ever increasing role with a spherical-type nucleus. Thus a physical barrier to nucleation exists when a heterogeneous nucleation process has a large maximum free energy ΔG^*. The rate of nucleation depends on the rates of surface diffusion and addition of new clusters to the nucleus. In such a case, this type of metal photonucleation is the rate-controlling process for metal deposition:

$$\Delta G^* \propto \Delta\sigma^3/\Delta G^2,$$

where ΔG is the free energy change of the metal and σ is the specific substrate–metal interfacial free energy. Thus at the nucleation temperature T established at the laser beam impact site, far enough from the equilibrium temperature T_e, we have

$$\Delta G \propto (T - T_e)^2. \tag{5.5}$$

The above relationship corresponds to what happens on a flat substrate in a pure pyrolysis case. As a whole, it is believed that the observed "volcano effect" during deposition [22] is due to the superheating $(T - T_e)$ required by the physical barrier of nucleation [71]. Furthermore, if we assume the expression for the rate of nucleation J for the spherical cap model given in [44] to be correct, we obtain

$$J \propto \exp(-1/T^3). \tag{5.6}$$

Fig. 5.8 Scanning electron micrograph displaying the "volcano effect" produced by a visible light (514 nm, 4 W) laser under static beam impact (diameter 50 μm). The focused beam is impinging at normal incidence onto the substrate. Well separated spherulitic particles are observed, which are ranged by size, in a decreasing symmetric manner around the "volcano".

Hence the nonlinearity of the temperature dependence of the nucleation rate may explain such an effect. A scanning electron micrograph displays this phenomenon (Fig. 5.8). It will be noted that this is a similar "volcano effect" to that observed (see Section 5.3.3) when the barrier role is played by a chemical energy barrier. The effect is reproduced under irradiation by a visible light laser beam at normal incidence. Well separated particles are displayed ranging over a narrow distribution, and seem to have been ejected from the "volcano". Thus the identifying feature for such a case is that the spherulitic particles next to the "volcano" are the heavier ones [72].

5.3.3 Photonucleation from a surface-reacted precursor

In real situations, the early stages of metal deposition may be far more complex than the simple planar processes previously developed by considering the basic modes for photodeposition of initial layers. In many cases, the photo by-products react in the adsorbed phase to form a species that then diffuses towards the preferential site for photonucleation and binds strongly to the substrate. One possibility is that the adsorbed photo by-product might only react with certain kinds of sites present on the surface. In this case, we may have an irreversible first-order process in order to bind to the preferential site for photonucleation. Another possibility, in the same case, is that the absorbed photo by-product reacts with a second species on the surface. The addition of new particles to the nucleus may require photochemical bond breaking, overcoming the chemical free energy barrier. This kind of photonucleation has been very poorly studied.

5.3.3.1 Low free energy change

The photonucleation of oxide(s) can be a rate-controlling factor [56] if the driving force ΔG is small enough [73]. How small are the values of ΔG of formation that we may obtain for metal oxide nucleation? The presence of adsorbed oxygen or oxygen-containing species is very common in photo-metallization experiments [74]. The various elementary mechanisms involved in the photoreactions of metals with traces of oxygen are transport in the gas, oxygen decomposition at the solid–gas interface, adsorption (physisorption and chemisorption) and desorption, diffusion in the metal oxide and at grain boundaries, chemical reaction, and nucleation of the oxide(s) [74]. Therefore IR laser or lamp assisted oxidation of metals [37, 38] can occur as a by-product during metallization. Significant photo-induced oxidation was observed for UV-irradiated samples of nickel [74b]. The mechanism involved is believed to be the electronic photo-excitation of adsorbed molecular oxygen. Thus it is possible that the oxygen dissociation is increased, and the atomic oxygen production leads to enhanced Ni oxidation.

In the case of Al deposition, prior to the formation of the Al_2O_3 cluster, oxygen will have been adsorbed at the surface of the "nascent" Al, so that the actual process will be more a transformation of the superficial Al with its adsorbed oxygen into Al_2O_3 [38]. The ΔG values should be decreased by the heat of adsorption of O_2 on Al; concomitantly, the adsorption of gases will lower the initial surface energy σ, which will increase $\Delta \sigma$(metal oxidation). Moreover, oxygen tends to dissolve into the Al, thus decreasing the free energy level of the metal and correspondingly decreasing the overall driving force for nucleation $\Delta G^* \propto \Delta \sigma^3 / \Delta G^2$.

5.3.3.2 High free energy change

In a different way, within the previously described system, we may have a high free energy change. That is possible if oxygen was drawn out of the silicon native oxide to react with the aluminium that was liberated by the photolysis of, for example, triisobutylaluminium (TIBA) on the surface [37].

In order to contribute to the understanding of how carbonaceous species are incorporated in a deposit when organometallic molecules are photolysed upon surfaces, it is helpful to study the adsorption of alkyl radicals. These radicals are produced from the precursor by a one-photon cascade process involving sequential stripping, when the light is incident upon the substrate [75].

It was observed that isobutylene adsorbs strongly on a silicon surface [37]. It has been hypothesized [38] that the adsorption of methyl radicals on Al is responsible for carbon incorporation during Al deposition in this system. Bent et al. [76] attest to the higher reactivity of the isobutyl fragments with silicon than with aluminium. They have also observed that, on an aluminium surface, the isobutylene does not adsorb strongly; therefore an explanation can be suggested for the higher carbon content in metal deposition with TMA as a precursor gas, based upon an increasing methyl radical chemisorption tendency on the metal deposit [77].

In order to avoid this high nucleation chemical barrier generating poor-quality deposits, the substrate surface can be pretreated in various ways, prior to subsequent photodeposition:

(a) by changing the physical nature of the surface (surface roughness, plasmon resonance enhancement);

(b) by changing the chemical nature of the surface to lower the chemical free energy barrier for metal nucleation;

(c) by prenucleating the surface with the same or an alternative metal, making use of the autocatalysis effect.

5.3.3.3 Competition between surface contamination and metal nucleation

The variety of contaminants in different systems makes a general discussion about this competition difficult [26b]. However, as an example, a typical scanning electron micrograph is presented in Fig. 5.6(b). The deposit has a volcano-like centre surrounded at millimetre level by four zones forming an extended circular reaction region. This is a typical example of laser-assisted (visible, CW 514 nm) aluminium nucleation from TMA pyrolysis subsequent to UV lamp (254 nm) prenucleation. In the extended circular reaction zone, two tendencies can be observed. While the Al nucleation is piling up toward the centre, the carbonaceous species show a clear out-diffusion tendency towards the circular boundary. Thus two separate zones, showing discs with different number densities of carbonaceous species, are displayed. Next, a mixed deposition zone makes the transition to the Al layer deposition. These zones are created in the absence of hydrogen flow. Having a higher sticking coefficient compared with the Al, the carbonaceous fragments match the Al atoms onto the substrate and therefore deposit as a thin layer. The displayed disc shape is appropriate when the adsorption energy is high. It is believed that, once produced, all carbonaceous fragments remain chemisorbed on the substrate. The exact chemical nature of this layer is unknown. The resultant layer is sandwiched between the substrate and the Al deposit, diminishing the Al adhesion. Hydrogen flow has been found to eliminate this effect.

5.3.3.4   The multiple roles of hydrogen

It was first believed that the role of molecular hydrogen was to protect the silicon surface and the nascent Al from oxidation. The resistance of such a Si substrate to oxidation is most probably due to the strongly chemisorbed hydrogen. The hydrogen chemisorption removes states from the top of the valence band (E_v) of Si(001). The new states are found at about 5 and 10 eV below the Fermi level for the polyhydride form and at 5 and 8 eV for the monohydride form. This adsorption is accompanied by changes of the electrical properties of the surface. When the Si(001) surface coverage approaches saturation, the initial upward band bending changes its sign due to the donor (hydrogen)–acceptor (p-Si) charge transfer. This agrees well with the passivation of a silicon surface by etching in a conventional mixture of $HF/H_2O_2/H_2O$. This procedure results in a hydrophobic surface, which displays a much slower oxidation rate. Recent publications [78] show that this is true only in a buffered basic aqueous solution or in a modified nonaqueous mixture. The silicon substrate at temperatures below $300°C$ has chemisorbed hydrogen in the polyhydride forms. The origin of the inertness of the polyhydride termination lies in the relative bond strengths concerned: the Si—H bond (323 kJ mol^{-1}) is stronger than the Si—Si bond (222 kJ mol^{-1}). Thus mono- or polyhydrides are expected to be inert to oxidation. This suggestion is indeed supported by results in the literature [79].

It is believed that the rate-determining step in the complex mechanism of metal (Al) photodeposition involves the photon-induced desorption of hydrogen from the polyhydride surface layer. The available literature data [77] support the suggestion that metallization occurs via chemisorption of a TMA dimer. We may assume that the photon energy density at 257 nm controls the hydrogen content of the surface hydride, and consequently the reactive sticking coefficient of the TMA at the surface; it apparently also controls the hydrogen content in the bulk aluminium [5], where AlH_2 may serve as an intermediate. We hope to provide more direct evidence concerning the scavenger role of hydrogen in the near future.

5.4 SPATIAL LOCALIZATION OF CLUSTERS

5.4.1 Steady state photonucleation

The main concerns here are metallization processes in which steady state photonucleation is the dominant rate-controlling process. Once the energy barrier has been overcome the shape that such a deposit acquires, in the course of its formation, is highly sensitive to deposition conditions (beam shape, power density, gas precursor partial pressure, laser beam dwell time and nucleation threshold). Figure 5.9 clearly shows that, even with a static laser beam, the shape constantly reflects the deposition mechanisms, as does the surface morphology, under steady state conditions [80]. For instance, using a CW laser pyrolysis approach yields a peculiar spike deposit of volcano-like shape, while photolysis with UV light generates a smooth deposit (Fig. 5.9). This smooth deposit, which is believed to reflect a steady state deposition rate, is mainly determined by the time elapsed from the photonucleation of the deposit. Therefore, when the photonucleation rate is low—much lower than the subsequent deposition rate—poor surface morphology results. This is then displayed as an increasing surface roughness in the final deposit. Furthermore, differences in the rate of photonucleation across the substrate surface are amplified into deposit thickness variations by the rapid rate of deposition.

5.4.2 Transient photonucleation

When the saturation ratio of metal generation in the gas phase is suddenly increased to a value at which homogeneous photonucleation occurs [9], a transient period exists during which the cluster concentration increases to its steady state value. Consequently, the photonucleation rate, defined as the flux of clusters past a certain critical size, also increases to its steady state value, corresponding to the new gas saturation ratio. When non-stationary effects connected with photonucleation parameters cannot be ignored, the mean time of expectation is no longer a meaningful characteristic feature of the process; this does effect, for example, the induction time for the onset of photonucleation, subsequent to the prenucleation time. However, practical interest in optimizing the duration of the transient period of photonucleation is found in experiments that utilize sudden changes [81] in saturation ratio, such as with the use of excimer lasers.

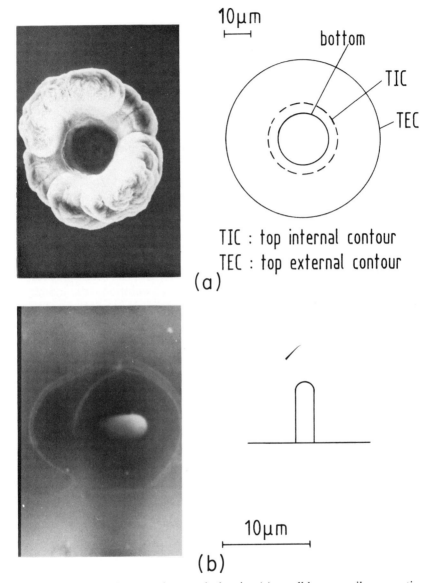

10μm

bottom

TIC

TEC

TIC : top internal contour
TEC : top external contour

(a)

10μm

(b)

Fig. 5.9 Scanning electron micrograph showing (a) a well known spike presenting a "crater" like shape characteristic of laser-assisted pyrolysis (514 nm) subsequent to UV prenucleation. (b) Smooth Al spike CW laser-induced by TMA photolysis (PAP) under steady state conditions (257 nm).

Under real photonucleation conditions, when pre-existing or intentionally prenucleated particles are frequently present, these tend to attract generated metallic atoms and clusters, thereby depressing the rate of new cluster formation. Consequently, there is an inherent difficulty, in certain metallic systems, in controlling surface roughness by using light sources that provoke a dramatic variation in the supersaturation regime.

5.5 KINETIC MODELS OF PHOTONUCLEATION

5.5.1 Kinetics of photonucleation and metallization processes

Photonucleation, and subsequent metallization, usually take place under conditions far from thermodynamic equilibrium [9, 10]. This is even more true for UV laser-induced metal deposition at interfaces: the effective pressure in the photoreaction volume, near or at the substrate surface, exceeds the condensate equilibrium pressure, at the substrate temperature, by many orders of magnitude. At these very high supersaturations, when the size of the critical nucleus is near atomic, the classical approach to nucleation is still justified [44, 48]. Under these conditions, the joining of product atoms into nuclei is thermodynamically irreversible, so kinetic factors play a substantial role. These factors are related to the dependence of the degree and nature of the photonucleation and metallization upon the localized supersaturation, the substrate temperature, and the quantity and chemical nature of the adsorbed photo by-products on the substrate or embedded in the deposit.

During heterogeneous photolytic nucleation, many new intermediate states may be formed that contribute to the localization of the photolysis [68], the photonucleation and the photodeposition process. A general scheme involving a four-phase process for heterogeneous photolytic nucleation, based on ten sequences, is shown in Figs 5.10(a–d). The localization is represented by the vertical distance from the surface versus the photon direct impingement site. In the case of a focused laser beam, the extent of the laterally localized reaction zone is on a micron level, and the effects have been discussed above (see Fig. 5.2).

Phase number one

This starts with precursor arrival into the photoreaction zone and is dominated by precursor molecule photolysis. Consider the arrival of precursor molecules (sequence 1) containing the metal deposit atom (M). The precursor molecules may become chemisorbed on photon-induced preferential (host) sites or desorbed following direct photon impingement. Surface diffusion leads the molecule from an accommodation to a precursor state (sequence 2). At a high fluence of direct photon impingement, desorption is activated from the precursor state.

Phase number two

This is dominated by competition between metal- and carbon-containing fragments (C) for nucleation onto the preferential site. The photolytic process (sequence 3 in phase number one) decomposes the chemisorbed precursor molecule, onto the semiconductor surface. An intermediate radical, which contains the deposit metal atom (M) and another carbonaceous fragment radical, is obtained under direct UV photon impingement. The UV photo-induced desorption rate of the carbonaceous fragment is rather low. Sometimes the photolytic process continues to decompose the resulting intermediates by direct collision with impinging photons. Finally, this cascade-type photoreaction yields the metal deposit atom, and fragments of different chemical nature.

The key question what can now be addressed is, whether all the photo-produced entities are accommodated onto the same type of preferential (host) site, on the semiconductor surface. The answer is not straightforward. The creation of acceptor-type centres is likely to be slower than donor-type centre creation, as judged by simple electrical transport in semiconductors. Hence, in a p-type substrate, the metal atom provided by the attached acceptor organometallic molecule benefits from a clear advantage (at the onset of nucleation). However, the birth of metastable heterogeneous clusters, metallic or carbonaceous, happens concurrently. Heterogeneous critical clusters of size i^* (where i^* is the number of fragments or metal atoms in the critical cluster) are obtained by migration, growth and coalescence of the metastable clusters, independent of their chemical nature (sequence 4). But this similarity between the behaviour towards the clusters stops there. Competition arises, in the early stages of clustering, for the majority of metals, due to kinetic factors [82]. Due to the precursor stoichiometry, the global rate of carbonaceous fragment production is several times higher than the metal production rate (e.g. for TMA it is $3:1$). The metallic cluster requires a smaller specific interface (surface site) with the substrate. Thus, at the onset of nucleation, a blocking effect of the early stages of metal clustering (sequences 4 and 5) is provoked by the carbonaceous species.

Phase number three

This is also dominated by precursor photolysis. But it exhibits an important difference to phase number one. In this case, the precursor molecule is

(a)

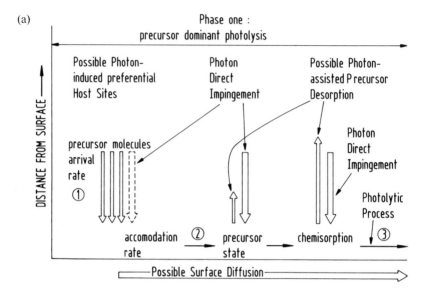

(b)

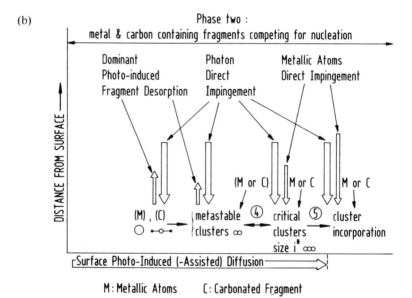

M: Metallic Atoms   C: Carbonated Fragment

Fig. 5.10 Schematic representation of a four-phase process for heterogeneous photolytic nucleation of metal: (a) phase number one—precursor photolysis; (b) phase number two—fragments competing on substrate for nucleation; (c) phase

(c)

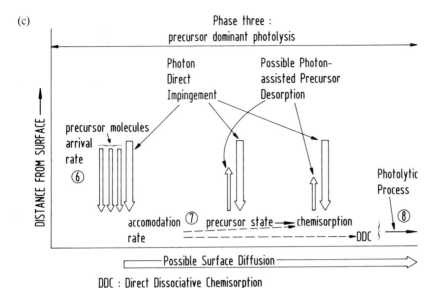

Phase three :
precursor dominant photolysis

DDC : Direct Dissociative Chemisorption

(d)

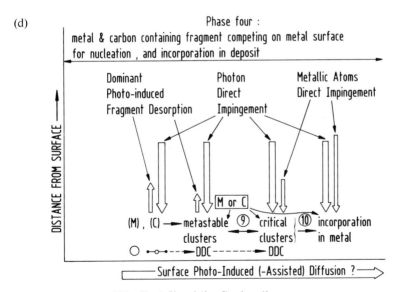

Phase four :
metal & carbon containing fragment competing on metal surface
for nucleation , and incorporation in deposit

DDC : Direct Dissociative Chemisorption
M : Metallic Atoms   C : Carbonated Fragment

number three—precursor photolysis from direct dissociative chemisorption; (d) phase number four—fragments competing on metal for nucleation and incorporation.

accommodated directly in the chemisorbed [83] state (sequence 7). We shall then expect less photo-assisted precursor desorption to occur under photon beam impingement [29]. However, this tendency may be reversed, as in the case of a direct dissociative chemisorption. The reversal occurs under direct impingement of a UV pulsed laser, at a wavelength and a corresponding fluence beam thresholds. The process starts with UV laser-induced surface charge separation. The results are specific to the photo-ion produced from the dissociatively chemisorbed layer [84]. The repulsive potential of a surface-bound ion from the surface layer is cancelled out by the adjacent created negative layer in a p-type bulk. The interaction with the surface layer is therefore weakened, permitting photo-ion emission and neutral desorption. Therefore this may represent a limiting condition for obtaining a heterogeneous nucleation on a semiconductor substrate (Si, Ge and GaAs). There is no photo-ion emission from metal surfaces. Very few experiments have been done on the selected depletion induced by UV photons [85].

Phase number four

The dominant features here are the competition between metal and carbonaceous fragments for metal nucleation and the incorporation onto the metal surface. These features mainly concern heterogeneous photolysis. Before the analysis of the limitations imposing such a regime, we must answer the question of the eventual contribution from the homogeneous gas phase. Such a contribution is certainly possible for an enhanced plasmon field, in the resonance case, when the contribution is coming from the gas phase near to the metal surface. It can be noted that metallic nanostructures can further increase this effect. In the case of an amorphous deposit, the chemical potential at any point on the surface is assumed to be a function of local morphology. Atoms or clusters sitting in a local depression can be considered to be of low chemical potential. Conversely, on a local protuberance, they are of high chemical potential. These variations in local chemical potential cause a net flow of atoms towards depressions. Finally, it is possible that this may drive the deposit towards a smooth surface [52], as is revealed in Fig. 5.9 (lower view).

Figure 5.11 is shown to exemplify some of the above phenomena. It shows a micrograph of a small area of a thin Al deposit from TMA mixed with hydrogen using a frequency-doubled Ar ion laser (257 nm). The ripples observed in the Al deposits have a regular spacing of the order of λ and are oriented perpendicular to the field strength vector of the polarized laser beam. They appear in the early stages due to electromagnetic field interference of the incoming beam with surface plasma waves having the same period as the observed ripples. The ripple structures

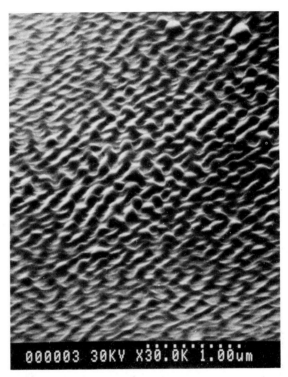

Fig. 5.11 Scanning electron micrograph of a small area of Al deposit from $Al_2(CH_3)_6$ (precursor) and H_2 (reactive carrier gas) using a polarized frequency-doubled Ar ion laser (257 nm). The laser electric field vector is oriented perpendicular to the ripples.

revealed in Fig. 5.11 are very common in photodeposition studies [26b, 66b]. If the deposit grows with a rippled structure, it can be assumed that the deposition process is reproducing the intensity variations of the light on the surface, which can be accomplished only if deposition occurs by photodissociation of surface-adsorbed species.

5.5.2 Metal direct condensation

What is the role played by surface diffusion? In which sequences is this effect less important? We are concerned with sequences 2, 3 and 4 from the suggested scheme (Fig. 5.10). The theory proposed by Eyring for rate processes [86] is useful for obtaining the adsorption flux n_a and the desorption flux n_d for the gas precursor onto the substrate with a given energetic barrier, following the normal direction to the surface. Therefore, to a first approximation, we can assume that the photoprocesses occur without interference from the presence of surface structural defects competing with preferential sites of nucleation created at ACCs by incident UV irradiation.

In the case of very high rates of metal atom incorporation onto already existing metal layers, the arrival of atoms can be assumed to be directly to the incorporating site. This approach is necessary in particular when analysing condensation on a substrate that is so cold that the desorption flux of the condensing species is many orders of magnitude less than the incident one [48]. Surface diffusion does not play a significant role. This is an important feature distinguishing an athermal-type process from other photoprocesses. Due to the atomic surface roughness, every adsorption site will show a deep potential well, ensuring direct condensation. Parity between the incorporation and the arrival rate of metal atoms can therefore be assumed. Thus a threshold temperature for direct condensation and the early stages of formation of an amorphous deposit can be determined. Knowledge of such a threshold temperature may help us to improve the lateral localization of the photonucleation process. It is suggested that the Kashchiev kinetic model [44, 87] can be useful for calculating this temperature threshold. This model is based on atomistic nucleation kinetics and emphasizes the link between the roughness, the growth rate and the temperature. The surface roughness can be defined from the average height at total surface coverage. It is found that the rate of deposition increases the surface roughness in a given time t. Hence it is conceivable that the surface roughness decreases while the interaction with the substrate increases, which is the general tendency observed for low-temperature metal deposition.

In summary, photonucleation and metallization may consist of a number of consecutive sequences (Figs 5.1 and 5.2) with competing and inhibiting steps. The suitable sequences can be juxtaposed in time (Figs 5.10a–d). These sequences are

(1) arrival of photoreactants into the photoreaction volume;

(2) adsorption of the reactant(s) onto the substrate; precursor accommodation onto the substrate;

(3) photon activation of the precursor near or at the substrate surface; transport of photoproducts to the substrate; condensation, desorption, diffusion or concurrent reactions of the photoproducts on the surface; re-evaporation or desorption (of reaction photoproducts) from the substrate;

(4) formation of any metastable clusters onto the photo-induced preferential site for nucleation; concurrent formation of any critical cluster size onto the preferential site for nucleation; transport of reaction

photo by-products out of the photoreaction volume; photo-assisted surface diffusion of clusters and cluster incorporation;

(5) cluster incorporation onto the substrate or onto other adsorbed phases; alternative heterogeneous path in photolysis onto the substrate;

(6) consecutive steps of precursor accommodation directly in the chemisorption state;

(7) precursor photolytic process in the chemisorption state;

(8) photonucleation of metastable clusters onto the metal surface; concurrent formation of stable clusters;

(9) metal cluster incorporation onto the metal.

5.5.3 Application: photonucleation of metals from metal alkyls

The basic features of laser-induced or laser-assisted metal deposition are discussed in several articles [22, 26b]. Metal alkyls are frequently employed: they are liquids with high vapour pressures at or near room temperature and have rather high decomposition temperatures. This allows photo-lytic processing to be carried out over a wide range of temperatures without complications from pyrolysis. They are among the most investigated metal precursors for laser deposition in micro-electronics [1]. Much can be learned about photonucleation and metallization mechanisms from studies involving metal alkyls. In the majority of these studies, $Al_2(CH_3)_6$ (TMA) has been used as an Al precursor [52]. However, TMA has been shown to produce carbon contamination of deposited film [23, 77]. It is believed that production and nucleation of adsorbed CH_3 radicals may account for the carbon contamination of the aluminium. A hydrogen flow was added to the photolytic process [5, 38] in further experiments. It appears that almost pure Al was then achieved. Although it is clear that these experiments give indirect evidence for a major role for H_2 as a scavenger, specific steps involving AlH_2 were postulated by Bourée et al. [5] but have yet to be experimentally confirmed. To study this, the photo-chemical mechanism and features in both the gas phase and on the surface must be followed. Detailed studies by the photolysis of surface-adsorbed Al alkyls have been conducted recently [52, 82], but have not involved flowing hydrogen. Adsorbed TMA on silicon was found to exist in a dimerized form. It has been found that alkyl radicals leave the surface much hotter than the substrate temperature during photolysis [83]. In one study [88, 89], time-of-flight measurements helped in examining the desorbing products of laser excimer photolysis of TMA, triethylaluminium (TEA) and triisobutylaluminium (TIBA) on quartz, clean silicon and the native oxide of silicon. Al, AlH and AlCH$_3$ [90, 91] were detected after laser photolysis at 193 and 248 nm. A larger AlCH$_3$/Al ratio was obtained at 248 nm for a TMA for a TMA photolysis process involving the adsorbed TMA [91, 92]. These results appear to conflict with those obtained by Higashi and Rothberg [93, 94] for TMA on oxide surfaces (248 nm), where no such products were detected at this wavelength [94]. Carbon content increases linearly with laser power at 193 nm, suggesting a multiphoton photolysis of TMA. The Al—CH$_3$ bond energy is not yet sufficiently well established in order to help explain these results. Recently, atomic hydrogen has been successfully used for decomposing TMA [95]. This work concluded that the main role of the hydrogen in an atomic state was to eliminate the carbon incorporation. This suggests that C elimination may also be explained by the role of atomic H in the case of photolysis with added molecular hydrogen. The mechanism of direct photolysis of TMA to CH$_3$ is still uncertain [1]. The calculations performed by Higashi suggested the creation of a product with little energy available for translation. This product could be in an excited metastable bound state. The result is in conflict with a more recent study [75] claiming the presence of hot methyl radicals. The key problem of carbon contamination in the resultant metal deposit is one of the main attractions of studies involving completely new gas phase precursors [1, 96, 97]. Studies concerning new precursors for metals are outlined in Chapters 6 and 16.

5.6 IMPORTANT FUTURE ISSUES

5.6.1 Weak-adhesion systems

Some results have been obtained on weak-adhesion systems showing characteristic island formation on the surface nanomorphology. The nucleation has been shown [89] to be proportional to the arrival rate R^2 or R^1. Hence the smallest stable nucleus is considered to be respectively either a pair of atoms or a single atom. The eventual explanation is based on the pair binding energy ε_2 having a higher value than the adatom adsorption energy ε_a and the surface diffusion energy ε_d ($\varepsilon_2 > \varepsilon_a > \varepsilon_d$). Additionally, it would be instructive to consider here the application of a mass-spectrometry—molecular-beam type experiment to measure these parameters directly during the metal photonucleation process. Such a direct test appears necessary since in the early stages of the process there are discrepancies with theory. The possible effect of photon impingement on migration, coalescence or desorption should be considered. Further studies of the surface potential may evaluate the contribution from surface states for preferential sites for photonucleation on a flat single-crystal semiconductor.

In the case of amorphous substrates, such as SiO or native oxides, there is no impediment to photonucleation on defect sites. Additionally, there is evidence that atomic diffusion into the surface may occur. This stresses the need for a full atomic-scale picture of the process: the total amount of material in the photodeposit, the amount of metal that arrives and also the amount desorbed.

5.6.2 Strong-adhesion systems

Most metal-on-metal systems fall into this category, whereas some others proceed by a mixed mode of adhesion. The reasons for this difference requires further studies. On semiconductor surfaces, layer deposition can be disturbed or even blocked by competition with small amounts of photo by-products acting as system self-contaminant agents. These by-products can result either from dissociative chemisorption or from photodissociation processes. Therefore the use of scavenger gases may improve metal nucleation continuity. All of these obstacles to metallization processing are suitable aims for further experimental studies, to observe the nanostructure behaviour and adatom concentrations under photon impingement, using combined spectroscopic–microscopic–crystallographic and work function *in situ* techniques.

5.6.3 Mixed-adhesion systems

These systems appear to be more common than originally perceived, evidently appearing in several metal/semiconductor groups.

More fundamental information has to be obtained about the photonucleation and metallization processes. It is not known what is the reason for, and what determines, such a type of structure. What is the role of the first adsorbed layer during irradiation over this layer? Such questions can only be answered by physicochemical and crystallographical probes of the system, both at the nano- and micrometer level, at the interface and around the island. These topics are outlined in Chapter 17.

5.6.4 Summary

This discussion has touched on a limited number of photoprocesses concerning photonucleation of metals onto semiconductors. In designing a photodeposition process, photonucleation is indicated as a key rate-controlling step.

The relationship between photon irradiation and enhanced nucleus formation onto photo-induced preferential sites has been discussed. We have suggested how the initial excited surface evolves, to yield such preferential sites, at a low observed nucleation threshold. At higher power densities, a limit is found relative to the precursor depletion by stimulated desorption from surfaces.

Spatial localization, provided by direct photolysis, is a powerful feature of photo-induced nucleation in the photolytic adlayer-dominated process, which aims to achieve a more selective heterogeneous nucleation. The selective use of layer localization, through induced preferential sites for metal photonucleation, may lead to practical applications in nanofabrication technology beyond the optical limit of the resolution at the working wavelength.

Metal photodeposition at low temperatures can be attained by exploiting favourable electronic excitation of the organometallic precursor. Concomitantly, by exciting the vibrational modes, photo by-products can be removed. Alternatively, by using scavengers, the initiation of competing channels can be a useful way to eliminate these photo by-products and obtain a pure metal deposit. This may also be effective in increasing the deposit adhesion.

Thus, using simple kinetic models, we have shown how various observations can be understood qualitatively, which may help to master the advantages of photonucleation in metal photoprocessing. To take full advantage of this kinetic control of photonucleation, it is essential that research on better precursors is carried out in several areas, starting from surface chemistry. Within this framework, information concerning the mechanisms involved with surface-adsorbed species, and thus on the nature of the rate-determining steps, is of particular interest both for microelectronic applications and chemical and physical knowledge.

ACKNOWLEDGEMENTS

This work has been partially supported by EEC Science Project SCI 0201.C and by a contract from CNET—Meylan for one of the authors (J.E.B.). The authors would like to thank Drs R. B. Jackman and I. W. Boyd for very helpful comments and suggestions during the preparation of this manuscript. They would also like to acknowledge Drs C Licoppe, B. Leon-Fong, W. C. Sinke, J. P. van Maaren and S. J. C. Irvine for fruitful discussions.

REFERENCES

[1] I. P. Herman, *Chem. Rev.* **89**, 1323 (1989).
[2] J. Haigh and K. Durose, *Prog. Crystal Growth and Charact.* **19**, 149 (1989).
[3] D. Braichotte, C. Garrido and H. van den Bergh, *Appl. Surf. Sci.* **46**, 9 (1990).
[4] R. L. Jackson, G. W. Tyndall and S. D. Sather, *Appl. Surf. Sci.* **36**, 119 (1989).
[5] (a) J. E. Bourée and J. Flicstein, *NATO ASI Series B* **198**, 33 (1989).
    (b) J. E. Bourée, J. Flicstein, J. F. Bresse and J. F. Rommeluère, *Mater. Res. Soc. Symp. Proc.* **129**, 25 (1989).
[6] R. R. Krchnavek, H. H. Gilgen, J. C. Chen, P. S. Shaw, T. J. Licata and R. M. Osgood, *J. Vac. Sci. Technol.* **B5**, 20 (1987).
[7] (a) G. Auvert, *NATO ASI Series B* **207**, 227 (1989).
    (b) G. Auvert, *French Patent* 88 15 435. (An industrial version is being sold under CNET licence by BERTIN, France.)
[8] (a) Y. I. Nissim, J. M. Moison, F. Houzay, F. Lebland, C. Licoppe and M. Bensoussan, *Appl. Surf. Sci.* **46**, 175 (1990).
    (b) The reactor is designed after *French Patents* 86 14 896 and 87 10 290. (An industrial version is being sold under CNET licence by JIPELEC, Grenoble.)
[9] S. Toschev, in *Crystal Growth: An Introduction* (ed. P. Hartman), p. 1 (North-Holland, Amsterdam, 1973).
[10] B. K. Chakraverty, in *Crystal Growth: An Introduction* (ed. P. Hartman), p. 50 (North-Holland, Amsterdam, 1973).
[11] D. E. Kotecki and P. Herman, *J. Appl. Phys.* **64**, 4920 (1988).
[12] Kratos Analytical Instruments, USA.
[13] E. van Loenen, Symposium on Surface Processing and Laser Assisted Chemistry of 1990 E-MRS Spring Conference (Strasbourg, France 29 May–1 June).
[14] P. W. Atkins, *Physical Chemistry*, 4th edn, p. 490 (Oxford University Press, 1990).
[15] G. Herzberg, *Spectra of Diatomic Molecules* (Van Nostrand, New York, 1950).
[16] C. Kittel, *Introduction to Solid State Physics*, 6th edn (Wiley, New York, 1986).
[17] I. Tamm, *Z. Phys.* **76**, 849 (1932).
[18] T. J. Chuang, *Surf. Sci. Rep.* **3**, 1 (1983).
[19] R. P. Feynman, *Quantum Electrodynamics* (Benjamin, New York, 1961).
[20] M. E. Coltrin, R. J. Kee and J. A. Miller, *J. Electrochem. Soc.* **131** 425 (1984).
[21] M. E. Coltrin, R. J. Kee and J. A. Miller, *J. Electrochem. Soc.* **133**, 1206 (1986).
[22] D. Bauerle, *Chemical Processing with Lasers* (Springer-Verlag, Berlin, 1986).
[23] (a) J. E. Bourée, J. Flicstein and Y. I. Nissim, *Mater. Res. Soc. Symp. Proc.* **75**, 129 (1987).
    (b) G. S. Higashi, G. E. Blonder and C. G. Fleming, *Mater. Res. Soc. Symp. Proc.* **75**, 117 (1987).
    (c) J. Y. Tsao and D. J. Ehrlich, *J. Cryst. Growth* **68**, 176 (1984).
[24] D. J. Ehrlich, R. M. Osgood and J. F. Deutsch, *Appl. Phys. Lett.* **38**, 946 (1981).
[25] A. F. Bernhard, B. M. McWilliams, F. Mitlitsky and J. C. Whitehead, *Mater. Res. Soc. Symp.* **75**, 633 (1987).

[26] (a) F. A. Houle, R. J. Wilson and T. H. Baum, *J. Vac. Sci. Technol.* **A4**, 2452 (1986).
      (b) F. A. Houle, *Appl. Phys.* **A41**, 315 (1986).
[27] D. J. Ehrlich, R. M. Osgood and T. F. Deutsch, *J. Vac. Sci. Technol.* **21**, 23 (1982).
[28] (a) C. J. Chen and R. M. Osgood, *Appl. Phys.* **A31**, 171 (1983).
      (b) C. J. Chen and R. M. Osgood, *J. Chem. Phys.* **81**, 327 (1984).
[29] (a) M. L. Knotek and P. J. Feibelman, *Phys. Rev. Lett.* **40**, 964 (1979).
      (b) M. L. Knotek and P. J. Feibelman, *Surf. Sci.* **90**, 78 (1979).
[30] (a) S. T. Ceyer, J. D. Beckerle, M. B. Lee, S. L. Tang, Q. Y. Yang and M. A. Hines, *J. Vac. Sci. Technol.* **A5**, 501 (1987).
      (b) S. T. Ceyer, *Ann. Rev. Phys. Chem.* **39**, 490 (1988).
[31] S. T. Ceyer, D. J. Gladstone, M. McGonigal and M. T. Schulberg, in *Physical Methods of Chemistry* (ed. J. F. Hamilton and R. C. Baetzold) (Wiley, New York, 1988).
[32] S. L. Tang, J. D. Beckerle, M. B. Lee and S. T. Ceyer, *J. Chem. Phys.* **84**, 6488 (1986).
[33] G. A. Samorjai, *Chemistry in Two Dimensions* (Cornell University Press, Ithaca, 1981).
[34] S. L. Tang, M. B. Lee, Q. Y. Yang, J. D. Beckerle and S. T. Ceyer, *J. Chem. Phys.* **84**, 1876 (1986).
[35] W. van Willigen, *Phys. Lett.* **28A**, 80 (1968).
[36] Th. Wolkenstein, *Théorie Électronique de la catalyse sur les semiconducteurs* (Masson, Paris, 1961).
[37] D. A. Mantell, *J. Vac. Sci. Technol.* **A7**, 630 (1989).
[38] J. Flicstein, J. E. Bourée, J. F. Bresse and A. M. Pougnet, *Mater. Res. Soc. Symp. Proc.* **101**, 48 (1988).
[39] N. A. Gaidai, L. Babernich and L. Guczi, *Kinet. Catal.* **15**, 868 (1974).
[40] A. Frenet and G. Lienard, *Catal. Rev. Sci. Engng.* **10**, 37 (1974).
[41] F. C. Schouten, O. L. J. Gijzeman and G. A. Bootsma, *Bull. Soc. Chim. Belg.* **88**, 541 (1979).
[42] F. Gutman and L. E. Lyons, *Organic Semiconductors* (Wiley, New York, 1967).
[43] A. A. Chernov and L. I. Trusov, *Kristallografiya* **14**, 218 (1969) [*Sov. Phys. Crystallogr.* **14**, 172 (1969)].
[44] (a) B. Lewis and J. C. Anderson, *Nucleation and Growth of Thin Films*, p. 108 (Academic Press, New York, 1978).
      (b) *Ibid.*, p. 115.
[45] E. Bauer, *Z. Kristallogr.* **110**, 378 (1958).
[46] J. A. Venables, *Thin Solid Films* **32**, 135 (1976).
[47] D. Walton, *J. Chem. Phys.* **37**, 2182 (1962).
[48] (a) S. Stoyanov, *Thin Solid Films* **18**, 91 (1973).
      (b) S. S. Stoyanov, *Current Topics in Materials Science*, Vol. 3, p. 421 (North-Holland, Amsterdam, 1979).
[49] G. Ertl, on *Molecular Processes on Solid Surfaces* (ed. E. Drauglis, R. D. Gretz and R. I. Jaffee), p. 147 (McGraw-Hill, New York, 1969).
[50] G. Lelay and R. Kern, *J. Cryst. Growth* **44**, 197 (1978).
[51] J. Woltersdorf, *Thin Solid Films* **32**, 277 (1976).
[52] J. Flicstein and J. Bourée, *Appl. Surf. Sci.* **36**, 443 (1989).
[53] C. J. Chen and R. M. Osgood, *Phys. Rev. Lett.* **50**, 1705 (1983).
[54] Yumin Gao, CNET, Personal communication (May 1988).

[55] G. S. Higashi, G. E. Blonder, C. G. Fleming, V. R. McCrary and V. M. Donnelly, *J. Vac. Sci. Technol.* **B5**, 1441 (1987).

[56] I. P. Herman, F. Magnotta and D. E. Kotecki, *J. Vac. Sci. Technol.* **A4**, 659 (1986).

[57] Mitsuo Kawasaki, Yoshihiko Tsujimura and Hiroshi Hada, *Phys. Rev. Lett.* **57**, 2796 (1986).

[58] D. J. Ehrlich and R. M. Osgood, *Thin Solid Films* **90**, 287 (1982).

[59] J. Y. Tsao, H. J. Zeiger and D. J. Ehrlich, *Surf. Sci.* **160**, 419 (1985).

[60] H. T. Zeiger, J. Y. Tsao and D. J. Ehrlich, *J. Vac. Sci. Technol.* **B3**, 1436 (1985).

[61] T. H. Wood, J. C. White and B. A. Thacker, *Appl. Phys. Lett.* **42**, 408 (1983).

[62] (a) G. Auvert, D. Bensahel, A. Georges, V. T. Nguyen, P. Hénoc, F. Morin and P. Croissard, *Appl. Phys. Lett.* **38**, 613 (1981).
(b) D. Tonneau, G. Auvert and Y. Pauleau, *Eur. Mater. Res. Soc. Symp. Proc.*, Energy Beam-Solid Interactions and Transient Thermal Processing (eds V. T. Nguyen and A. G. Cullis), p. 125 (les Editions de Physique, 1985).

[63] F. Petzold, K. Piglmayer, W. Krauter and D. Bauerle, *Appl. Phys.* **A35**, 155 (1984).

[64] D. Bauerle, *Laser Processing and Diagnostics* (ed. D. Bauerle), p. 166 (Springer-Verlag, Berlin, 1984).

[65] A. Nitzan and L. E. Brus, *J. Chem. Phys.* **74**, 5321 (1981).

[66] (a) A. Nitzan and L. E. Brus, *J. Chem. Phys.* **75**, 2205 (1981).
(b) S. R. J. Brueck and D. J. Ehrlich, *Phys. Rev. Lett.* **48**, 1678 (1982).

[67] D. J. Ehrlich and R. M. Osgood, *Chem. Phys. Lett.* **79**, 381 (1981).

[68] P. T. Leung and T. F. George, *J. Chem. Phys.* **85**, 4729 (1986).

[69] J. Y. Tsao and D. J. Ehrlich, *Appl. Phys. Lett.* **45**, 617 (1984).

[70] C. J. Chen, *J. Vac. Sci. Technol.* **A5**, 3386 (1987).

[71] J. W. Westwater, *Am. Sci.* **47**, 427 (1959).

[72] J. E. Bourée, Y. I. Nissim, J. Flicstein and C. Licoppe, Unpublished results.

[73] F. M. d'Heurle, *J. Vac. Sci. Technol.* **A7**, 1467 (1989).

[74] (a) M. Wautelet, *Appl. Phys.* **A50**, 131 (1990).
(b) A. Mesarwi and A. Ignatiev, *J. Vac. Sci. Technol.* **A7**, 1754 (1989).
(c) C. F. Yu, M. T. Schmidt, D. V. Podlesnik, E. S. Yang and R. M. Osgood, *J. Vac. Sci. Technol.* **A6**, 754 (1988).

[75] (a) D. Eres, T. Motooka, S. Gorbatkin, D. Lubben and J. E. Greene, *J. Vac. Sci. Technol.* **B5**, 848 (1987).
(b) D. Luben, T. Motooka, J. F. Wendelken and J. E. Greene, *Phys. Rev.* **B39**, 5245 (1989).

[76] (a) B. E. Bent, R. G. Nuzzo and L. H. Dubois, *Mater. Res. Soc. Symp. Proc.* **101**, 177 (1988).
(b) B. E. Bent, R. G. Nuzzo and L. H. Dubois, *J. Vac. Sci. Technol.* **A6**, 1920 (1988).
(c) B. E. Bent, R. G. Nuzzo and L. H. Dubois, *J. Am. Chem. Soc.* **111**, 1634 (1988).

[77] G. S. Higashi, *Appl. Surf. Sci.* **43**, 6 (1989).

[78] G. S. Higashi, Y. I. Chabal, G. W. Trucks and Krishnan Raghavachari, *Appl. Phys. Lett.* **56**, 656 (1990).

[79] D. Graf, M. Grudner and R. Schultz, *J. Vac. Sci. Technol.* **A7**, 809 (1989).

[80] J. E. Bourée and J. Flicstein, Unpublished results.

[81] G. Shi, J. H. Seinfeld and K. Okuyama, *Phys. Rev.* **A41**, 2101 (1990).

[82] T. E. Orlowski and D. A. Mantell, *J. Vac. Sci. Technol.* **A7**, 2598 (1989).
[83] D. Lubben, T. Motooka, J. E. Greene, J. F. Wendelken, J. E. Sundgren and
      W. R. Salanek, *Mater. Res. Soc. Symp. Proc.* **101**, 151 (1988).
[84] (a) T. A. Carlson and M. O. Krause, *J. Chem. Phys.* **56**, 3206 (1972).
      (b) W. Brenig and D. Menzel (eds), *Desorption Induced by Electronic
      Transitions: DIET II* (Springer-Verlag, Berlin, 1985).
[85] F. Träger and G. zu Putlitz (eds), *Metal Clusters* (Springer-Verlag, Berlin,
      1986).
[86] S. Glasstone, K. J. Leidler and H. Eyring, *The Theory of Rate Processes*
      (McGraw-Hill, New York, 1941).
[87] (a) D. Kashchiev, *J. Cryst. Growth* **40**, 29 (1977).
      (b) D. Kashchiev, *Thin Solid Films* **55**, 399 (1978).
      (c) D. Kashchiev, *J. Cryst. Growth* **67**, 559 (1984).
[88] Y. Zhang and M. Stuke, *Japan. J. Appl. Phys.* **27**, L1349 (1988).
[89] Th. Beuermann and M. Stuke, *Chemtronics* **4**, 189 (1989).
[90] F. Lee, T. R. Gow, R. Lin, A. L. Backman, D. Lubben and R. I. Masel,
      Mater. Res. Soc. Symp. Proc., in press (1991).
[91] T. Motooka, S. Gorbatkin, D. Lubben and J. E. Greene, *J. Appl. Phys.* **58**,
      4397 (1985).
[92] T. E. Orlowski and D. A. Mantell, *Mater. Res. Soc. Symp. Proc.* **131**, 369
      (1989).
[93] G. S. Higashi and L. J. Rothberg, *Appl. Phys. Lett.* **47**, 1288 (1985).
[94] G. S. Higashi, *J. Chem. Phys.* **88**, 422 (1988).
[95] G. V. Jagannathan, M. L. Andrews and A. T. Habig, *Appl. Phys. Lett.* **56**,
      2019 (1990).
[96] (a) T. Cacouris, G. Scelsi, P. Shaw, R. Scarmozzino and R. M. Osgood,
      Appl. Phys. Lett. **52**, 1865 (1988).
      (b) M. Hanabusa, A. Oikawa and P. Y. Kai, *J. Appl. Phys.* **66**, 3268 (1989).
[97] D. A. Tossel, C. J. Brierley, A. T. S. Wee, A. J. Murrell, N. K. Singh and
      J. S. Foord, *Proceedings of 5th UK Workshop on Photochemical Process-
      ing—An International Meeting, 28–30 March 1990, University College
      London.*
      J. S. Foord, A. J. Murrell, D. O'Hare, N. K. Singh, A. T. S. Wee and
      T. J. Whitaker, *Chemtronics* **4**, 262 (1989).

6 Considerations of the Microscopic Basis for Photon-Enhanced Chemical Beam Epitaxy

J. S. FOORD

Physical Chemistry Laboratory, University of Oxford, UK

6.1 INTRODUCTION

The growth of III–V semiconductor epilayers has been dominated by the techniques of molecular beam epitaxy (MBE) and metal organic vapour phase epitaxy (MOVPE) for a decade or more [1–4]. MBE apparently offers the ultimate in control at the atomic level, with the structure being built up atom-by-atom from the desired elemental sources. Compositional changes in the growing structure can be introduced by the instantaneous operation of shutters in front of the molecular beam sources, and *in situ* monitoring of the growth process is possible using the vast range of surface-sensitive techniques that operate under ultrahigh vacuum. Nevertheless, while MBE does possess these very significant strengths, it also suffers from important weaknesses. The Knudsen sources tend to evolve contaminants, which produce defects in the growing layer. Uniformity of growth across a large area is difficult to achieve. The need to open up the system to replenish the Knudsen ovens can result in a large "downtime" in the running of the apparatus. In growth on patterned structures, selective area growth resulting from specific chemical interactions between the surface and the impinging beams is not generally observed. The technique is inconvenient for the large-scale production of advanced materials. Work on MOVPE in recent years has therefore aimed at evolving a growth technique that does not suffer from such shortcomings and that is suitable for producing materials on a commercial scale. While dramatic progress has been made, the nature of the MOVPE experiment does itself impose new limitations. In principle, it should be possible to achieve sharper interfaces

Photochemical Processing of Electronic Materials
ISBN 0-12-121740-X

with molecular beam techniques, since flux switching is more readily achieved than in MOVPE. It is difficult to achieve very high growth uniformities with MOVPE without highly complex reactor designs; the potential for selective area growth is reduced since gas phase processes rather than purely surface chemical reactions tend to play an important role.

In the light of these considerations, interest in so-called metal organic molecular beam epitaxy (MOMBE), or chemical beam epitaxy (CBE) as it is also known, has now begun to grow [5–8]. CBE is carried out in the same type of ultrahigh vacuum environment as MBE, but it employs molecular beams comprising the reactants that are more commonly employed in MOVPE. This approach possesses many of the advantages of MBE and MOVPE without the disadvantages. High-purity chemicals can be used, and the source contamination and "source spitting" that arises in connection with the MBE Knudsen sources is thus avoided; at the same time, the vacuum system does not need to be opened to refill the evaporation cells. These are all important advantages of CBE in comparison with MBE. More rapid flux switching than in MOVPE is possible, and *in situ* monitoring of the growth process is also easier than in MOVPE. These are important advantages of CBE in comparison with MOVPE. The flux uniformities across large areas and the possibilities for selective area growth in CBE are superior both to MBE and MOVPE.

CBE is nevertheless still in the early stages of development; the bright picture presented in the previous paragraph arises in part because CBE has not yet been developed to the point where all the disadvantages can be properly appreciated! Areas where problems at present occur include the incorporation of the volatilizing ligands from the precursors into the final material and the reproducible control of the growth process. Photoenhanced CBE, that is CBE where light is used to stimulate the chemical reactions involved, is even more in a stage of infancy. Empirical studies will no doubt always play an important role in addressing the problems arising. However, it is now widely believed that an essential step in the overall development of these growth techniques is to establish an understanding of the basic chemistry involved in the growth processes. This is because such an understanding should be able to identify the problematic areas where completely new approaches—as opposed, for example, to the refinement of growth parameters—are going to be required, as well as indicating the nature of the solutions needed. The aim of the present chapter is thus to consider the nature of the basic chemistry underlying CBE and photo-enhanced CBE.

The following section assesses our current state of knowledge of a representative CBE growth system in order to consider the thermal chemistry that is likely to be involved in conventional thermal CBE. The subsequent section then deals with the ways in which photons can be used to drive reactions at surfaces. The final section then considers the likely future development of photo-enhanced CBE, the subject of particular interest in the context of the current volume, in the light of these basic considerations.

6.2 REACTION MECHANISMS FOR THERMAL CBE

Although a wide range of materials have been grown by CBE, the most detailed knowledge that exists concerns the growth of GaAs using triethylgallium and cracked arsine beams, and it is therefore worthwhile to consider this system in some detail. Early studies indicated that the growth rate was insensitive to arsine overpressure, suggesting that arsenic is not involved in the rate-limiting step for the overall reaction [9]. This idea has been further developed from RHEED oscillation data, which have come from studies of the effects of interrupting the arsine flow on the RHEED oscillations obtained [10]. The crucial experiment involves switching off the arsine flux for a time equivalent to the growth of a limited number of monolayers of GaAs and then switching this flux back on while at the same time stopping the triethylgallium (TEG) flow. Data from such experiments are illustrated in Fig. 6.1; the interesting point to note is that RHEED oscillations indicating growth equivalent to that which would be expected in the interrupted period are observed when the arsine is readmitted to the chamber. This implies that TEG conversion to Ga occurs independently of arsenic and that the overall growth process thus involves two distinct steps

$$TEG \ (g) \rightarrow Ga \ (ads) \tag{6.1}$$

$$Ga \ (ads) + As_2 \ (g) \rightarrow GaAs \ (s) \tag{6.2}$$

Step 2 is similar to the reaction in conventional MBE (using As_2), and it

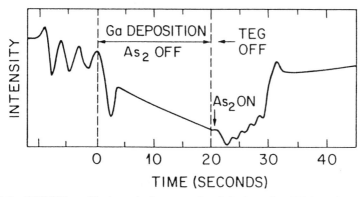

Fig. 6.1 RHEED oscillations during growth of GaAs using TEG and As_2 [10].

therefore follows that the rate-limiting step 1 is responsible for the unique characteristics encountered in CBE. This first step has thus been the subject of intense study.

Surface chemical measurements [11] using X-ray photo-electron spectroscopy and electron energy loss spectroscopy show conclusively that TEG exists in a molecular form on GaAs at low temperatures but that dissociation to partially alkylated species and adsorbed ethyl radicals occurs as the temperature is raised. Temperature-programmed desorption studies can be used to observe the species that desorb from the adsorbed Ga adlayer, and illustrative spectra are presented in Figs 6.2 and 6.3. The spectra in Fig. 6.2 are recorded with the mass spectrometer tuned to the Ga species indicated. The desorption peaks at 180 and 220 K correspond to the desorption of physisorbed and chemisorbed TEG respectively. The desorp-

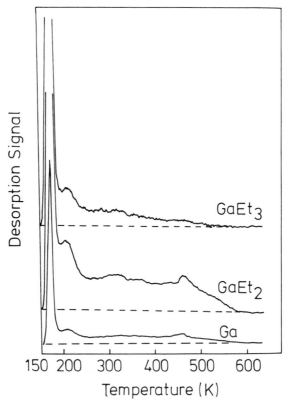

Fig. 6.2   Thermal desorption spectra following adsorption of TEG on GaAs (100) (4×1) at 100 K, monitoring species indicated.

tion of TEG at higher temperatures corresponds to the associative desorption from the partially dissociated adlayer, whereas above 400 K it is clear that DEG (diethylgallium) desorbs. In Fig. 6.3 the dissociation and desorption characteristics of the ethyl radicals formed on the surface are illustrated; the major products are ethene and hydrogen although some ethane is also detected.

The observations discussed above are obtained under static reaction conditions. Measurements can also be made under dynamic growth conditions using the technique of modulated molecular beam scattering. Early results obtained using this approach are presented in Figs 6.4 and 6.5. The observations presented were recorded *in situ* during growth and refer to the intensity of species desorbing from the surface versus surface temperature. As regards Ga species, at low temperatures both TEG and DEG desorption

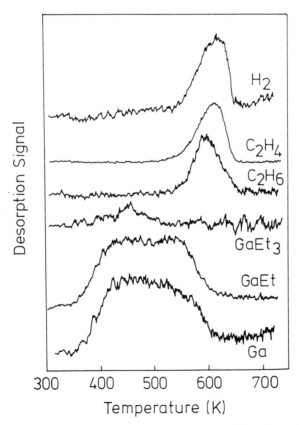

Fig. 6.3 Thermal desorption spectra as for Fig. 6.2 following adsorption at 300 K.

are detected. The TEG signal disappears rapidly as the temperature is raised, but the DEG intensity remains significant up to about 550 K, when it also drops—although it increases again at higher temperatures. The organics evolved from the surface are found to be a mixture of ethene and ethyl radicals. The temperature of 550 K corresponds to the point at which significant growth starts to take place.

The observations presented indicate the occurrence of the following sequence of reactions:

$$\text{TEG (g)} \rightarrow \text{DEG (g)} \tag{6.3}$$

$$\text{TEG (g)} \rightarrow \text{TEG (ads)} \tag{6.4}$$

$$\text{TEG (ads)} \rightarrow \text{TEG (g)} \tag{6.5}$$

$$\text{TEG (ads)} \rightarrow \text{DEG (ads)} + \text{Et (ads)} \tag{6.6}$$

$$\text{DEG (ads)} + \text{Et (ads)} \rightarrow \text{TEG (g)} \tag{6.7}$$

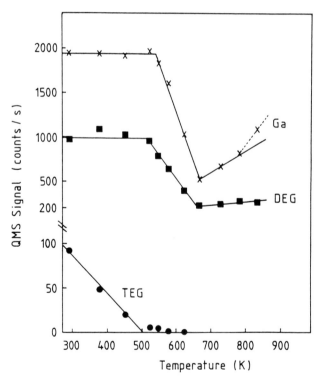

Fig. 6.4 Modulated beam scattered intensities for species indicated during growth of GaAs from TEG and As_2.

$$DEG \ (ads) \rightarrow DEG \ (g) \qquad\qquad (6.8)$$

$$DEG \ (ads) \rightarrow Ga \ (ads) + 2Et \ (ads) \qquad\qquad (6.9)$$

$$Et \ (ads) \rightarrow C_2H_4 \ (g) + H_2 \ (g) \qquad\qquad (6.10)$$

$$Et \ (ads) \rightarrow Et \ (g) \qquad\qquad (6.11)$$

The sequence of reactions (6.3)–(6.11) breaks down reaction (6.1) into its elementary steps. Reaction (6.3) refers to the reactive scattering of TEG in the form of DEG, which is observed in the growth system at low temperatures. In this regime the surface is saturated with chemisorbed species, and it is likely that this scattering occurs within a second layer. The reaction is responsible for the observation that most of the scattered flux detected at low temperatures is DEG. Above 550 K, the surface sites begin to depopulate and growth sets in arising from the reactions (6.4)–(6.11), which take place on the GaAs surface.

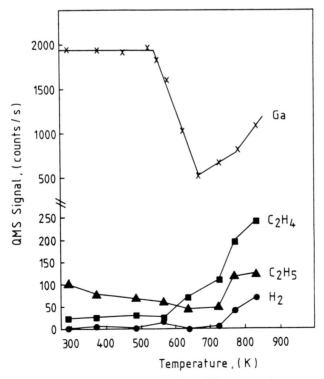

Fig. 6.5 As Fig. 6.4, but for different species.

The growth rate is determined by differing factors in differing temperature regimes. Below approximately 600 K, it is the rate of production of DEG on the surface that is rate-determining, and a rapid increase in growth rate with temperature is observed. At higher temperatures, essentially all the incoming flux is converted to DEG. However, a significant fraction of the DEG begins to desorb via reaction (6.8) rather than converting to Ga atoms. This results in a decline in growth rate at higher temperatures. This fits in with the observations presented in Fig. 6.6, recorded during the growth of GaAs.

A number of conclusions emerge from this benchmark system that probably have a general validity in related growth systems and are worth emphasizing since they are relevant to the development of photon-enhanced MOMBE processes. First, although it was convenient to break up the overall growth process into two stages (reactions (6.1) and (6.2)), the surface science studies [11] would predict that reaction (6.1) should proceed faster at low temperatures than the overall growth rate suggests; similarly, more detailed growth studies indicate that As overpressure does inhibit the growth at low temperatures, in contrast with the assumption made above, where we assumed that it had no effect [12]. The conclusion here is that the growth rate is very much controlled at low temperatures by the availability of surface sites, which can be blocked by adsorbed As species. Secondly, the surface lifetimes of all the species formed by the

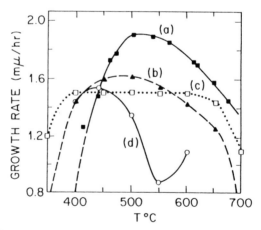

Fig. 6.6 Partial growth rate of Ga versus temperature using TEG as the Group III source. Curves (a) and (b) refer to the growth of GaAs at different TEG fluxes, while curves (c) and (d) denote the growth of AlGaAs and InGaAs respectively under the same conditions as curve (b). From M. B. Panish, in *Mechanisms of Reactions of Organometallic Compounds with Surfaces* (ed. D. J. Cole-Hamilton and J. O. Williams) (Plenum, New York, 1989), p 274.

cracking of TEG are such that the surface coverages are likely to be negligible at the temperatures that are typically used to carry out growth. Thirdly, the decomposition of metal alkyls releases organic groups, which can migrate over the surface to other sites. Finally, the surface-induced decomposition of metal alkyls displays a rate that depends quite sensitively on branching ratios between competing cracking and desorption channels of the intermediates involved. The importance of this observation is very nicely displayed in Fig. 6.6, which essentially compares the Ga incorporation efficiencies as a function of temperature measured during the growth of GaAs, InGaAs and AlGaAs. The striking feature is that In in the host matrix dramatically reduces the incorporation of Ga, whereas Al has the reverse effect. Evidence now indicates that this effect arises since the branching ratio of reactions (6.8) and (6.9) above depend quite sensitively on the nature of the surface on which they occur; reaction (6.8) is enhanced in the presence of In and blocked in the presence of Al.

6.3 PHOTODESORPTION AND PHOTOREACTION OF MOLECULES AT SURFACES

The gas density in a CBE reactor is too low to employ gas phase photolysis using conventional effusive sources. While this could be avoided, for example by photolysis of pulsed supersonic expansions of the reactants involved, the most obvious way to employ light is by irradiation of the surface itself, and the photon-induced chemistry occurring is then limited to the adsorbed phase. It is therefore crucial to understand the basic nature of the photon-induced processes that can take place at surfaces, and a number of distinct excitation channels are known to be involved.

6.3.1 Substrate heating

Large temperature rises can take place at the interface if intense light sources such as lasers are employed and the radiation is absorbed within the bulk substrate employed. This effect can be used to drive chemical reactions, and techniques such as laser chemical vapour deposition are based on this approach (see, e.g. [13]). The characteristics of this mechanism are a strong nonlinear dependence of the chemistry induced on the laser powers employed and an insensitivity to the electronic properties and optical absorption spectra of the reactants used. The effect is useful since it provides a means to achieve patterned growth by spatial definition of the light source and also provides for temperature–time profiles that could not be obtained by other means. Many examples of LCVD exist in the literature [13], and the approach can be similarly applied in CBE.

The chemistry induced is very similar to that obtained by substrate heating if modest laser-induced heating rates are obtained. However, it is interesting to note that different reaction channels can be accessed at high laser powers [14]. This is because the rate of a chemical reaction is determined by the product of the "pre-exponential factor" and the "Arrhenius factor". The latter factor is frequently the more important during the use of conventional heating sources. However, at very high heating rates, the surface phase is raised to a high temperature at which all Arrhenius factors tend to unity before reaction can take place. Under these circumstances, it is the pre-exponential factor or the entropy of activation that dominates. In terms of surface reactions, this appears to lead to the favouring of desorption over the fragmentation of an adsorbed species.

6.3.2 Photochemical reactions driven by excitation of the adsorbate

Photochemical reactions can be driven directly by photo absorption in the adsorbed layer if light of the appropriate energy range is employed and the reaction can take place on a timescale that is short in comparison with the relaxation of the excited state back to the ground state. The characteristic that is normally used to identify this mechanism is a wavelength dependence that matches the absorption spectrum of the gas phase reactants involved, although this criterion neglects the frequency shifts arising from chemisorption or the formation of new surface species.

Numerous examples of such processes are quoted in the literature, including the types of organometallic compound of interest in CBE [15–19]. Here the photochemical reaction seems to bring about the stepwise removal of ligands, resulting in the formation of new chemical entities that bond more strongly than the physisorbed or weakly chemisorbed species representing the molecular precursor. The overall effect is thus to overcome the thermal activation barrier to dissociation. There is now firm evidence that the ligands released can desorb directly into the gas phase or become trapped on the surface in a photochemically inert form [19]. It also appears to be the case that the photochemical intermediates produced become more progressively inert photochemically as photodissociation proceeds [20]. The origin of this effect is not firmly established, but it could well arise from the fact that these intermediates bond more strongly to the surface and therefore undergo very rapid relaxation from the excited state by interaction with the underlying solid before photodissociation can take place.

6.3.3 Photochemical reactions driven by photogenerated carriers

This mechanism rests on the creation of photogenerated carriers within the bulk substrate and the stimulation of reactions as they diffuse to the surface by interaction with the adsorbate complex. The key observation that is normally used to establish this mechanism is a wavelength dependence that matches the absorption spectrum of the substrate, and particularly a bandgap threshold effect. In order to observe reactions arising through this mechanism, diffusion of charge carriers to the surface must be rapid in comparison with their recombination, and a "favourable interaction" with the adsorbate is required in the surface region. Little is known concerning what constitutes a "favourable interaction" at present.

The mechanism has been clearly demonstrated with regard to photon-induced etching and photodesorption from semiconducting materials [21, 22]. With regard to metals, the faster relaxation rates are expected to diminish the effect involved—and thus, unsurprisingly, a demonstration of the mechanism here is much less clear-cut for these materials. Although the experiments carried out to date have tended to concentrate on generation of photocarriers, it should also be pointed out that a range of other substrate-induced photo-excitation effects could contribute, depending on the wavelength used. For example, while the mechanism discussed above refers to the creation of charge carriers within the energy range $E_f < E < E_{vac}$, the use of shorter wavelengths can generate photo-emitted electrons (i.e. $E > E_{vac}$). The significance of this is that organometallic compounds can be dissociated efficiently in a process known as "dissociative electron capture"; and this could therefore provide an important mechanism for bringing about the dissociation of adsorbed species at the appropriate wavelengths.

6.4 POSSIBILITIES FOR PHOTO-CBE

The thermal chemistry involved in CBE has been discussed in Section 6.2, and in Section 6.3 the ways in which light can promote reactions at surfaces has been outlined. This now provides a basis for considering the possibilities and constraints imposed by photo-assisted CBE. First, it is worthwhile to consider what one might wish to achieve by using a photo-assisted approach. We have seen in Section 6.2 that CBE processes are strongly temperature-dependent, and one obvious role for photo-assisted CBE is the lowering of growth temperatures if the precursors employed have a higher decomposition temperature than is otherwise required by the growth process. We have also seen in Section 6.2 that the surface decomposition of organometallics can follow several competing routes, which may strongly affect the growth rate; a second role for photo-assisted CBE could therefore be to favour the total decomposition route that would improve the reproducibility and reduce the substrate sensitivity, which can be unwelcome in particular applications. Carbon contamination from the volatilizing ligands is also a problem encountered in CBE that photo-assisted approaches could possibly help. Finally, selective area growth using a spatial controlled light source represents another important goal.

The most certain way to produce a photo-assisted effect is via substrate heating, since optical absorption and thermal transport properties of materials are known or can be conveniently measured, and it is a straightforward matter to select an appropriate light source in order to achieve the desired temperature rises. This approach is suitable for achieving selective area growth, as has already been done in CBE [23], but of course it is unlikely to avoid the other problems encountered in thermal CBE since the reaction route is essentially thermal. Reactions stimulated photochemically by creation of excitations within the substrate could avoid these limitations in principle, but at present the occurrence or otherwise of such effects is difficult to predict since rather little is known concerning the dynamics of the interaction of the excitation with the adsorbate complex. Evidence of the operation of this mechanism exists with regard to organometallic compounds, but it is at present difficult to assess how widespread this effect will be in semiconductor growth systems.

The most obvious way of avoiding the limitations imposed by thermal reactions is to directly excite the adsorbate in the mechanism discussed in Section 6.3.2, since the optical absorption spectrum of the precursors employed can be measured and an appropriate wavelength source employed to match the absorption frequency. However, it is important to bear in mind that this mechanism is distinct from the mechanisms previously discussed in that the primary coupling of radiation does not involve the substrate. This is significant since the magnitude of the effect induced is likely to vary monotonically with adsorbate *coverage* rather than simply the incident flux. The two are of course linked by the surface lifetime, and it follows that at short surface lifetimes this mechanism will start to play an insignificant role in comparison with the other two. Most organometallic precursors dissociate very readily to form strongly bound adsorbed layers of partially dissociated species above room temperature. Hence the surface lifetime of the molecular species at such temperatures is likely to be too short to couple effectively to the radiation. Coupling to dissociation products can take place, but even then this will not guarantee success since available evidence suggests that the rapid quenching of the excited states of such species by their interaction with the substrate may render them photochemically inert. At normal growth temperatures, it is also the case that the concentration of all adsorbed species is vanishingly small.

Although photo-CBE is still in the early stages of development, the principles discussed above are already beginning to manifest themselves in published data. McCaulley *et al.* [24] have demonstrated that the photon-enhanced growth of GaAs using TEG is thermal at practical laser fluences using pulsed excimer lasers, although they report that at low fluences direct photochemical excitation of Ga-alkyl fragments occurs. Aoyagi *et al.* [25] have studied laser-assisted ALE (atomic layer epitaxy) and have observed the dissociation of Ga-alkyl fragments by interaction with photogenerated charge carriers. The differences that arise between photo-CBE and photo-CVD as a result of the principles discussed above have been highlighted by Tossell *et al.* [26], who have studied photo-assisted growth of sapphire using a range of alkoxy-acetylacetonate Al precursors by CVD and CBE. The particular interest here is that these latter precursors only adsorb very weakly in a molecular form, and no efficient route exists for their surface dissociation. In CVD the higher gas pressure establishes a significant surface population of molecular bound species, and an efficient photo-CVD process can be established as a result of gas phase or adsorbed phase photolysis of these molecular species. In contrast, no photochemical deposition could be observed in the CBE chamber at the same growth temperature, since the surface population of the molecular form (and all other surface species) is negligible and hence direct photochemical excitation could not occur. Deposition was observed at low temperatures where the molecular species exhibits a long surface lifetime, but unfortunately the film properties were unsatisfactory at such temperatures. This example highlights the difference between photo-assisted MOCVD and CBE.

6.5 CONCLUSIONS

CBE is a relatively new growth technique that at this stage appears to offer significant advantages over MBE and MOVPE. Nevertheless, the approach also possesses its own disadvantages and it will be some years before it will be possible to assess which growth approach will have the dominant commercial role in the area of III–V materials. Photo-assisted chemistry is already important in MOVPE, where empirical work has succeeded in using photons to reduce growth temperatures, obtain sharper interfaces and bring about patterned growth.

As this chapter has shown, it is possible to obtain a molecular level understanding of CBE (in marked contrast with MOVPE), and it is therefore tempting to assume in the light of the experience gained from MOVPE that it will be possible to systematically exploit photon-assisted chemistry to overcome problems in thermal CBE as their molecular origins are recognized. However, this chapter has also demonstrated that there are underlying problems with photo-CBE that do not arise in MOVPE, where gas phase photolysis can occur and where adsorbed phase photolysis can also take place by direct coupling to intact adsorbed precursor molecules (which are physisorbed at much higher concentrations than in CBE due to higher partial pressures). It is thus unsafe to assume that the success that has been demonstrated in photo-MOVPE will also be observed in conventional photo-CBE.

REFERENCES

[1] E. H. C. Parker (ed.), *The Technology and Physics of Molecular Beam Epitaxy* (Plenum Press, New York, 1985).
[2] M. A. Herman and H. Sitter, *Molecular Beam Epitaxy: Fundamentals and Current Status* (Springer-Verlag, Berlin, 1988).
[3] P. D. Dapkus, *Ann. Rev. Mater. Sci.* **12**, 243 (1982).
[4] G. B. Stringfellow, *Organometallic Vapour Phase Epitaxy: Fundamentals and Current Status* (Academic Press, New York, 1989).
[5] W. T. Tsang, *J. Cryst. Growth* **81**, 261 (1987).
[6] W. T. Tsang, *IEEE Trans. Circuits and Devices* **4**, 18 (1988).
[7] M. B. Panish and S. Sumski, *J. Appl. Phys.* **55**, 3571 (1984).
[8] G. J. Davies and D. A. Andrews, *Chemtronics* **3**, 3 (1988).
[9] N. Putz, H. Heinecke, M. Heyen, P. Balk, M. Wayers and H. Luth, *J. Cryst. Growth* **74**, 292 (1986).
[10] T. H. Chiu, W. T. Tsang, J. E. Cunningham and A. Robertson, *J. Appl. Phys.* **62**, 2302 (1987).
[11] J. S. Foord, N. K. Singh, A. T. S. Wee, A. J. Murrell, G. J. Davies and D. Andrews, *J. Appl. Phys.* **68**, 4053 (1990).
[12] Th. Chiu, J. E. Cunningham and A. Robertson, *J. Cryst. Growth* **95**, 136 (1989).
[13] A. W. Johnson, G. L. Loper and T. W. Sigmon (eds), *Mater. Res. Soc. Symp. Proc.* **129** (1989).
[14] R. B. Hall and A. M. DeSantolo, *Surf. Sci.* **137**, 421 (1984).
[15] J. S. Foord and R. B. Jackman, *Surf. Sci.* **209**, 151 (1989).
[16] R. B. Jackman and J. S. Foord, *Surf. Sci.* **201**, 47 (1988).
[17] N. S. Gluck, Z. Ying, C. E. Batosch and W. Ho, *J. Chem. Phys.* **86**, 4957 (1987).
[18] C. E. Bartosch, N. S. Gluck, W. Ho and Z. Ying, *Phys. Rev. Lett.* **57**, 1425 (1986).
[19] C. M. Friend, J. R. Swanson and F. A. Flitsch, *Mater. Res. Soc. Symp. Proc.* **131**, 461 (1989).
[20] D. Lubben, T. Motooka, J. E. Greene, J. F. Wedelkin, J. E. Sundgren and W. R. Salaneck, *Mater. Res. Soc. Symp. Proc.* **101**, 151 (1988).
[21] F. A. Houle, *J. Chem. Phys.* **79**, 4237 (1983).
[22] H. Ebert, R. B. Jackman and J. S. Foord, *Surf. Sci.* **176**, 183 (1986).
[23] V. M. Donnelly, C. W. Tu, J. C. Beggy, V. R. McCrary, M. G. Lamont, T. D. Harris, F. A. Baiocchi and R. C. Farrow, *Appl. Phys. Lett.* **52**, 1065 (1988).

[24] J. A. McCaulley, V. R. McCrary and V. M. Donnelly, *J. Phys. Chem.* **93**, 1148 (1989).
[25] Y. Aoyagi, A. Doi, T. Meguro, S. Iwai, K. Nagata and S. Nonoyama, *Chemtronics* **4**, 117 (1989).
[26] D. A. Tossell, C. J. Brierley, A. T. S. Wee and J. S. Foord, *J. Appl. Phys.* (in press).

7   Photo-Assisted II–VI Epitaxial Growth

S. J. C. IRVINE, H. HILL, J. E. HAILS,
G. W. BLACKMORE and A. D. PITT

*Royal Signals and Radar Establishment, Malvern,
Worcestershire, UK*

7.1   INTRODUCTION

Photo-assisted epitaxial growth of II–VI semiconductors has been a particularly attractive way of reducing growth temperatures to below the normal pyrolysis thresholds, but the mechanisms have not been well understood. There is now a variety of literature on photo-assisted epitaxy of ZnSe, CdTe, HgTe and the alloy (Hg,Cd)Te. Device interest in the wider-bandgap material ZnSe is for visible electro-optic devices such as electroluminescent displays and blue-emitting diodes. Progress in the latter is restricted by the difficulty in doping this material with acceptors. Lower-temperature growth (below $350°C$) may bring about a reduction in defect concentrations sufficient to allow reproducible extrinsic doping. The situation is similar for the narrow-bandgap alloy (Hg,Cd)Te, which is used for infrared detection. For crystals grown at temperatures above $300°C$, the electrical properties are dominated by metal vacancies. Various authors have suggested that photo-assisted growth may bring about an improvement in crystalline quality in addition to the benefits of reduced growth temperatures [1–3]. It can often be difficult to establish the role of photon–surface interactions where strong photon–vapour reactions are also taking place. In this paper the relative effects of vapour and surface interactions with the photon beam will be explored. As a consequence of the photon–surface interaction, it has been shown that patterned growth can be achieved by imaging an ultraviolet laser beam onto the substrate surface [4–6]. Recent results will be presented, providing evidence for these processes and ways of enhancing patterned growth.

Photochemical Processing of Electronic Materials
ISBN 0-12-121740-X

7.2 PHOTON–VAPOUR INTERACTIONS

Photodecomposition of alkyl precursors can occur in the vapour phase as a result of photon absorption putting the molecule into an excited electronic state [7]. It has been shown by Chen and Osgood [8] that for the case of Group II alkyls excitation can take place into an antibonding excited state, resulting in the release of the Group II metal atom. However, peak absorption for Me_2Cd occurs at 220 nm and for Me_2Hg and Me_2Zn at 200 nm, requiring deep-UV emitting arc lamps or UV lasers as the photon sources. Long-wavelength absorption for Me_2Cd extends as far as 260 nm, so Hg arc lamps, KrF excimer lasers or frequency-doubled argon ion lasers can be used. The absorption cross-sections in these UV bands can be very high, up to 0.4 Å at the peak, where the absorption cross-section σ is defined by

$$\sigma = \frac{kT}{pL} \ln\left(\frac{I_0}{I}\right), \tag{7.1}$$

where T is the vapour temperature, p is the partial pressure of the organometallic, L is the path length in the vapour, I_0 is the initial intensity and I is the final intensity. For example, for a 2 cm path length cell with a partial pressure of the organometallic of 1 Torr, 48% of the UV will be absorbed for 0.1 $Å^2$ and 6% for 0.01 $Å^2$ absorption cross-sections.

Previous work has shown that high yields of Te can be gained from photodissociation of Me_2Te and Et_2Te [7]. A recent, detailed study of UV absorption spectra of organotellurium compounds has shown a series of

Table 7.1 UV Absorption cross-sections for Group II and VI organometallic compounds: σ at the peaks of the absorption curves and at some of the photon source wavelengths.

Precursor	λ_{max} (nm)	$\sigma(Å^2)$ at peak	$\sigma(Å^2)$ at source wavelength		
			257 nm	254 nm	248 nm
Me_2Te	256.3	0.121	0.118	0.1	0.404
	249.3	0.737			
	242.6	0.27			
Me_2Te_2	215.6	0.179	0.019	0.018	0.027
Et_2Te	246.9	1.07	0.091	0.147	0.985
Pr^i_2Te	249.1	0.545	0.015	0.108	0.483
Me_2Zn	200	0.4	<0.01	<0.01	<0.01
Me_2Cd	216	0.42	0.03	0.04	0.06
Me_2Hg	202	0.42	<0.01	<0.01	<0.01

electronic transitions that is more complicated than for the Group II compounds—but essentially covers the same range of wavelengths up to 260 nm [9]. Table 7.1 shows calculated absorption cross sections for Group II and Group VI organometallics for both peak values and cross-sections corresponding to wavelengths of some of the available UV sources. It should be noted that, in the case of high-pressure mercury lamps, considerable line broadening occurs, which might improve coupling into some of the organometallics. Although it is difficult to match a radiation source with a peak absorption, most of the cross-sections shown in Table 7.1 are high enough for significant absorption in a typical reactor cell.

The simplest form of photon assistance for low-temperature epitaxy is illumination parallel to the substrate, where there is no intentional illumination of the surface. This is intended to bring about photodissociation of the precursors at vapour and substrate temperatures well below the pyrolysis thresholds. This has particular advantages when using excimer lasers, which could induce high surface temperatures from the large pulse energies if the substrate surface were illuminated. Examples of this approach are those of Morris [10], who used an ArF excimer laser to grow (Hg,Cd)Te at 150°C (over 200°C below pyrolysis thresholds), Zinck et al. [11] who grew CdTe at 165°C and Shinn et al. [12], who grew ZnSe at between 200 and 475°C, all using ArF excimer lasers. Morris calculated that for the pulse energy of 348 mJ all the organometallics in the vapour stream were completely dissociated to yield Cd, Hg and Te atoms that can diffuse to the substrate surface.

Photon–vapour reactions are also important in experiments where the substrate surface is illuminated with UV radiation; normally the entire vapour stream above the substate is also illuminated. In this case it is usually more difficult to estimate the relative enhancements in growth between vapour and surface photon absorption. However, it has been shown that very high supersaturations of Cd and Te vapours arising from photodissociation of Me_2Cd and Et_2Te can lead to CdTe dust formation [13]. High concentrations of free alkyl radicals in the vapour yielded by photodissociation also appear to affect subsequent vapour reactions, and can inhibit dust formation [14]. This is an often neglected aspect of metal–organic vapour phase epitaxy (MOVPE), but possible reaction mechanisms have been discussed in detail elsewhere [14]. The phenomenon of dust formation effectively reduces growth rates at low temperature as vapour concentrations, and therefore transport rates, have to be reduced to avoid the creation of dust. However, photon–surface reactions could have an indirect influence on this process if stimulation of surface kinetics could result in the reduction of vapour concentrations close to the surface. Possible surface mechanisms will be discussed later in this chapter.

7.3 PHOTON–SURFACE INTERACTIONS

Experiments designed to compare the growth rate enhancement between vertical and horizontal illumination have been carried out by Fujita *et al.* [3] for the growth of ZnSe. A 500 W xenon arc lamp was used that gave a maximum radiant intensity on the surface of 100 mW cm^{-2}; this is not sufficient to significantly heat the surface. However, enhancement of growth rate for temperatures below 500°C was only obtained for the vertical illumination. In a previous publication by these authors it was shown that enhancement would only occur for photon energies greater than the band-gap energy of 2.5 eV [15]. This indicates that enhancement occurs as a result of photon absorption in the ZnSe surface, which is also supported by the high quantum efficiency of this reaction.

It is often difficult to deduce the effect of photon–surface interactions on layer quality or, indeed, whether the vapour photochemistry can influence layer quality other than through the effects of modified growth rate. For example, it has already been stated that the thermodynamic properties of II–VIs (i.e. point defect concentrations) are improved by reducing growth temperatures, but the kinetic rates will also be reduced, which may increase defect concentrations. It has often been a consideration of photo-enhanced growth that photon absorption at the surface may increase the kinetic rates through bond breaking, and hence improve the crystalline quality. This was suggested by Kisker *et al.* [1], who used a low-pressure arc lamp to assist growth of CdTe at 250°C. Photoluminescence spectra of the photo-assisted layers showed strong near-band-edge exciton peaks, whereas the non-photon-assisted layers showed PL that was dominated by defect-related bands [1]. It is interesting to note that, in the parallel illumination case of Shin *et al.* [12] for ZnSe growth, no improvement in PL was observed.

Photomodified growth can occur where the layer properties are changed without a change in growth rate [16]. This can be a more appropriate way of looking at the effect of photon-surface interaction on crystalline quality. An example of photomodified growth is photo-assisted molecular beam epitaxy (PA–MBE), where an argon ion laser is used to illuminate the growing surface [17]. It was shown by Bicknell *et al.* [17] that using an intensity of 150 mW cm^{-2}, which does not have a significant heating effect, can dramatically influence the conductivity of a CdTe epilayer. Indium doping without illumination does not make the layer conducting, which would be expected from substitutional doping on the Group II sites. With illumination, the layers become conducting, displaying good near-band-edge photoluminescence. The growth temperatures used for these experiments were in the range 180–250°C, which was up to 200°C below normal MBE growth temperatures for CdTe. The mechanism for activating the indium dopant is not clear, but various processes have been suggested.

Laser-assisted surface heating can be discounted because the laser power density is too low. Improved surface kinetics via photo-assisted bond breaking is unlikely because the flux of photons to the surface would be too low to have a significant effect, even if there were suitable absorption bands in the region of the laser wavelength of 514 nm. A more recent report [18] has suggested that the photons are absorbed by the CdTe layer, releasing charge near the surface that can assist the desorption of excess Te. The high absorbance of the semiconductor surface makes this a viable process for low photon fluxes, as was the argument put forward by Fujita [3] for ZnSe growth. The laser-assisted desorption of excess Te would help to maintain a stoichiometric surface, which would assist in the incorporation of indium dopant on a substitutional site. With excess Te on the surface, the In_2Te_3 phase may form, which would render the dopant inactive.

Photon–surface interaction may be important not only because of the improved epilayer quality but also because it may lead to possibilities in selected area epitaxy. If the laser beam is imaged onto the surface using a suitably patterned mask then the modulation of photon intensity on the surface should produce a modulation in growth rate or layer properties. Preliminary work by the present authors using a frequency-doubled argon ion laser has shown that CdTe growth rate enhancement does depend on the intensity of UV radiation on the surface [4]. It was shown that diffraction rings from cell wall defects reproduced a ripple pattern on the grown layer. Further experiments showed that a grid pattern could be produced from a stainless steel mesh [5]. Both mesa and bar patterns have been deposited by projecting a mask image onto the substrate using a silica lens [6].

In this chapter the mechanisms for laser-assisted selected area epitaxy will be explored further and the relative influence of laser–vapour and laser–surface interactions considered.

7.4 EXPERIMENTAL

The photo-assisted growth experiments were carried out in a horizontal MOVPE reactor with a silica window and separate purge flow to avoid reaction products depositing onto the window [19]. A schematic of the reactor cell arrangement is shown in Fig. 7.1 with the arrangement for measuring time-resolved reflectivity (TRR) [6, 11]. The precursors were electronic grade dimethyl cadmium (Me_2Cd), with either diethyl telluride (Et_2Te) or dimethyl telluride (Me_2Te). The carrier gases were hydrogen purified by Pd/Ag diffusion membranes for the nucleation layer and molecular-sieve-dried helium for the photo-assisted growth. The substrates were GaAs $(100)2^{\circ} \rightarrow (100)$ supplied by MCP Wafer Technology. Prior to growth,

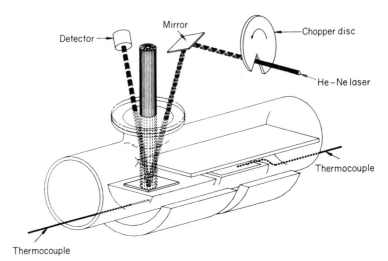

Fig. 7.1 Schematic of reactor cell, showing the set-up for TRR and UV illumination perpendicular to the substrate.

they were etched for 5 min in a $5:1:1$ solution of $H_2SO_4 : H_2O_2 : H_2O$. Photo-assisted growth was preceded by a 10 min heat clean in pure hydrogen at $350°C$ and the growth of a CdTe nucleation layer approximately 1000 Å thick, grown from the pyrolysis of Me_2Cd and Et_2Te. All layers nucleated in this way were assessed using an X-ray texture camera and found to be epitaxial [6].

The photon source was an intracavity frequency-doubled argon ion laser with a CW output of 100–150 mW at 257 nm. For the TRR experiments a 2 mW HeNe laser was used in conjunction with a 10:1 ratio beam chopper, which reduced the average intensity of radiation on the surface and assisted in background correction. It has previously been shown that a continuous HeNe beam can cause surface roughening and enhance growth rate [20]. For photopatterning experiments a silica lens and mask were inserted in the laser beam above the silica window such that the mask and substrate were a distance $2f$ either side of the lens (where f is the focal length of the biconvex lens, which was 57 mm).

7.5 MACROSCOPIC PHOTO-ENHANCED CdTe GROWTH

Before considering the variations in growth rate on a microscopic scale to bring about selected area epitaxy, the effects of photo-enhancement over a large area (>1 mm^2) will first be considered. These measurements concern the enhancement from both vapour and surface reactions, but their relative importance can be elucidated by growth rate studies. The definition of macroscopic photo-enhancement is given in Fig. 7.2. This is basically a comparison of growth rates between layers grown with the photon source on and off, and can be measured *in situ* by TRR, by pre-alignment at the centre of the UV beam where intensity at the substrate is estimated.

Figure 7.3 shows the photo-enhanced growth rates versus UV intensity for Me_2Cd with either Me_2Te or Et_2Te. The concentrations of the alkyl vapours in the two cases were similar, and it can be seen that at $250°C$ the growth rates are indistinguishable. The growth rates at $300°C$ do show a modest increase compared with rates at $250°C$, which cannot be explained by enhanced pyrolysis, and the photolytic yield should be essentially the same as at $250°C$. The broken line gives the theoretically calculated growth rate predictions, assuming vapour absorption and photo-decomposition, based on available UV absorption cross-sections (see Table 7.1), followed by diffusion of Cd and Te vapour to the surface. This is the simplest model for photo-enhanced growth and significantly does not consider growth rates limited by surface kinetics or the effect of photon–surface interactions. The experimental results differ from this prediction in two important respects: (a) the measured growth rates are substantially lower; (b) the function of growth rate with UV intensity is sublinear, with a power law ranging from $I^{0.56}$ at $300°C$ to $I^{0.75}$ at $250°C$.

The photo-enhancement factors can be simply calculated from these results by ratioing with the measured pyrolytic growth rates of 0.4 and 0.09 μm h^{-1} for Et_2Te and Me_2Te respectively at $300°C$, and 0.1 and 0.02 μm h^{-1} respectively at $250°C$. These results are shown in Fig. 7.4, and clearly indicate that larger photo-enhancement factors can be obtained

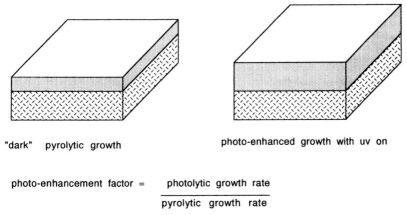

"dark" pyrolytic growth

photo-enhanced growth with uv on

photo-enhancement factor = $\dfrac{\text{photolytic growth rate}}{\text{pyrolytic growth rate}}$

Fig. 7.2 Schematic of macroscopic photo-enhancement.

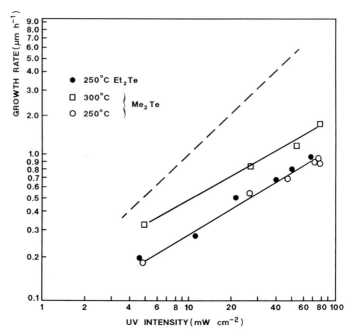

Fig. 7.3 CdTe growth rates as functions of UV intensity (at 257 nm) for alkyl partial pressures of 1.3 Torr Me₂Cd, 0.6 Torr Et₂Te and 0.7 Torr Me₂Te. The predicted transport-limited growth rate is shown as a broken line.

when using Me$_2$Te rather than Et$_2$Te. For the control of macroscopic growth using a photon beam it is desirable to grow at the lowest practicable temperature, using the most stable precursors available, so that when the beam is off no growth takes place. However, for microscopic photo-enhancement the same mechanisms may not be active, as will be discussed in Sections 7.7 and 7.8.

Growth rates have also been measured as a function of Cd/Te alkyl ratio where the Te alkyl concentration was kept constant. These experiments were designed to look for surface saturation effects that could provide data on the surface kinetic processes. The Te alkyl partial pressure was kept constant at 0.6 Torr and the alkly ratio varied between 0.8 and 4.4 for pyrolytic growth and photo-enhanced growth. The results in Fig. 7.5 show that for pyrolytic growth and low-intensity UV (31 mW cm^{-2}) the growth rates saturate with alkyl ratio, whereas, over the range of concentrations considered, the measurements made under high-intensity UV do not show saturation. These results will be considered in the light of a surface kinetic model in the next section.

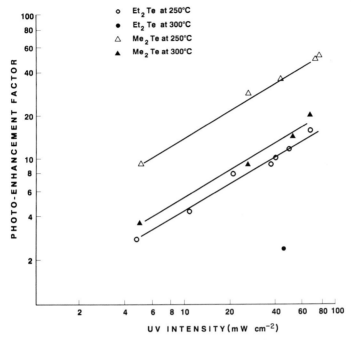

Fig. 7.4 Comparison of photo-enhancement factors for CdTe versus UV intensity.

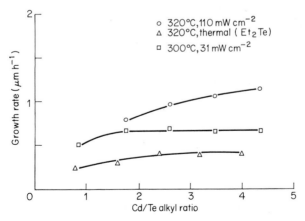

Fig. 7.5 Growth rate versus Cd/Te alkyl ratio for a Me_2Te (or Et_2Te) partial pressure of 0.7 Torr.

7.6 MODELLING OF SURFACE KINETICS

The Langmuir–Hinshelwood model has been used by previous authors to describe the surface kinetic processes that can limit growth rate in MOVPE [21]. This model can be modified to include photodecomposition of the precursors, site blocking from reaction products and laser-assisted desorption of these products.

The first step is to relate vapour concentrations of Cd and Te atoms to the inlet alkyl vapour pressures and UV intensity. By rearranging (7.1) the decrease in UV intensity over 1 cm path length can be calculated, from which the Cd and Te partial pressures can be calculated assuming one photon absorbed yields one Cd or Te atom in the vapour. This can be formally written in terms of reaction constants k_1 and k_2 for the weak absorption case:

$$p_{Cd} = \frac{k_1}{v} I_0 p_{Me_2Cd}, \tag{7.2}$$

$$p_{Te} = \frac{k_2}{v} I_0 p_{Me_2Te}, \tag{7.3}$$

where v is the flow velocity in the reactor cell and I_0 is the initial UV intensity. For independent adsorption onto the Group II and Group VI sites the Langmuir–Hinshelwood equation can be written as

$$R = \frac{k\beta_{Cd}\beta_{Te}p_{Me_2Cd}p_{Me_2Te}I_0^2}{(1 + \gamma_1 I_0 p_{Me_2Cd})(1 + \gamma_2 I_0 p_{Me_2Te})}, \tag{7.4}$$

where R is the growth rate, k is a rate constant, and β_{Cd} and β_{Te} are the ratios of rate constants for adsorption and desorption for Cd and Te respectively. The factors γ_1 and γ_2 are defined as

$$\gamma_1 = \beta_{Cd}k_1/v, \tag{7.5}$$

$$\gamma_2 = \beta_{Te}k_2/v. \tag{7.6}$$

It can be seen from (7.4) that at low intensities $R \propto I_0^2$, and at high intensities R saturates and becomes independent of I_0. Transport has not been considered but it can be seen from Fig. 7.3 that the transport-limited growth rate is proportional to I_0. At low I_0 transport will limit the growth rate, but at high I_0, where (7.4) predicts that $R \propto I_0^n$ ($n < 1$), surface kinetics will dominate. This appears to be the situation over the range of UV intensities considered in Fig. 7.3.

However, the behaviour with Cd/Te alkyl ratio cannot be explained by the simple model given in (7.4) since a saturation of growth rate with p_{Me_2Cd} requires the condition $\gamma_1 I_0 p_{Me_2Cd} > 1$. This will occur at a lower

p_{Me_2Cd} for higher values of I: the opposite behaviour to that observed in Fig. 7.5. This behaviour can be explained if a third, site-blocking, species is introduced on the surface. This could be free radicals such as CH_3 or C_2H_5 released from photodecomposition of the alkyls. For simplicity of calculation, just the radicals released from Me_2Cd will be introduced here—the qualitative results are the same as those derived from including all radicals. The partial pressure of the blocking species p_b, by analogy with (7.2) and (7.3)

$$p_b = \frac{k_3}{v} I_0 p_{Me_2Cd}. \tag{7.7}$$

Assuming adsorption onto both Cd and Te sites, (7.4) can be modified as

$$R = \frac{k\beta_{Cd}\beta_{Te}p_{Me_2Cd}p_{Et_2Te}I_0^2}{[1 + I_p p_{Me_2Cd}(\gamma_1 + \gamma_3)][1 + I_0(\gamma_2 p_{Me_2Te} + \gamma_4 p_{Me_2Cd})]}, \tag{7.8}$$

where

$$\gamma_3 = \beta_{Me}^{(Cd)}k_3/v, \tag{7.9}$$

$$\gamma_4 = \beta_{Me}^{(Te)}k_3/v. \tag{7.10}$$

$\beta_{Me}^{(Cd)}$ and $\beta_{Me}^{(Te)}$ are the ratios of adsorption to desorption of methyl radicals on Cd and Te sites respectively. If photon absorption on the surface assists desorption of the site blocking Me radicals then γ_3 becomes an inverse function of intensity. The criterion for growth rate saturation with increasing p_{Me_2Cd} is now $I_0 p_{Me_2Cd}(\gamma_1 + \gamma_3) > 1$. If at low intensities γ_3 is dominant over γ_1 (effective site blocking) and $\gamma_3 \propto I_0^{-n}$, where $n > 1$, then the criterion for growth rate saturation becomes $cp_{Me_2Cd}I_0^{1-n} > 1$, where c is a constant. This could explain the higher critical values of p_{Me_2Cd} for growth rate saturation with UV intensity observed in Fig. 7.5. An important implication of this model is the laser-assisted desorption of a site blocking species, which would be a photon–surface interaction that could bring about selected area epitaxy. This possibility will be explored further in the next section.

7.7 MICROSCOPIC PHOTO-ENHANCEMENT

The macroscopic photo-enhancement factors reported in Section 7.5 do not necessarily relate to the "light" to "dark" ratio of growth rates on neighbouring regions of the substrate, possibly only a few microns apart.

A schematic of microscopic photo-enhancement is shown in Fig. 7.6. The vapour photodecomposition reactions described in Section 7.4 could account for some of the growth rate enhancement in the macroscopic case,

photolytic yield of Cd & Te atoms in vapour

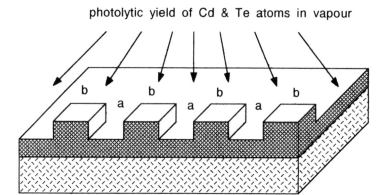

a = growth in the "dark"
b = growth in UV illumination

Fig. 7.6 Schematic of microscopic photo-enhancement.

but the mixing effect of diffusion of Te and Cd atoms through a depth of vapour 1 cm thick would be expected to create the same concentrations in the light and dark regions of the surface. In previous reports we have shown that patterned growth is possible, but measurements of the structure with profilometry have indicated that some deposition can occur due to vapour diffusion in the neighbourhood of the illuminated portions of the substrate [19].

The degree of photo-enhancement of the illuminated features has been observed to vary quite markedly with nominally the same growth conditions. It has also been observed that microscopic photo-enhancement can vary with position on the substrate. An example of this is shown in Fig. 7.7 for a 1 cm^2 pattern of mesas ranging in size from 250 μm on the outer edges to 50 μm in the centre. The micrographs show the upstream and downstream ends of the pattern, with a Dektak profilometer trace running in the upstream-to-downstream direction across the entire pattern. Due to misalignment of the central mesas with respect to the outer mesas, the Dektak peaks in the centre appear smaller than their actual height, but the dramatic decrease in height at the downstream end is genuine, with the height of the mesas going from 3 μm to approximately 0.1 μm. The Nomarski contrast micrographs show that the upstream mesas are matt but the downstream mesas have a smooth morphology. This result could be explained by a decrease in alkyl concentration going downstream due to depletion, but cannot be explained simply by a reduced supply of Cd and Te to the surface

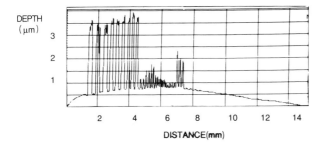

Fig. 7.7 Dektak profile and Nomarski contrast micrographs (larger mesas are on a 500 μm pitch) from each end of the pattern. The direction of flow was from left to right. The inlet alkyl concentrations were 1.3 Torr Me$_2$Cd and 0.6 Torr Et$_2$Te.

because the macroscopic growth rate measurements do not indicate a dramatic decrease in growth rate with decreasing alkyl supply concentrations. Another possibility is that the mesa height does not reflect the overall layer thickness, but the extent of vapour diffusion spread is increasing at the downstream end of the pattern.

A number of patterns grown with nominally the same inlet concentrations did not show any mesas higher than 0.2 μm above the surrounding surface. Examples are shown in Figs 7.8 and 7.9 of shallow mesa growth using Me$_2$Te and Et$_2$Te respectively. The mesas are barely discernible on the Dektak profile, although they can be clearly seen with Nomarski interference contrast. Some of the rounding in the profile, particularly at the edges, is due to substrate rounding, but in the centre it is also due to the thickness of the underlying layer following the overall intensity profile of the incident laser beam. Scanning electron micrographs (SEM) of the cleaved edges are shown in Figs 7.8 and 7.9 for both layers, taken from the centre of the pattern. They both show that the underlying CdTe layer is substantially thicker than the mesa height would suggest. In fact, from

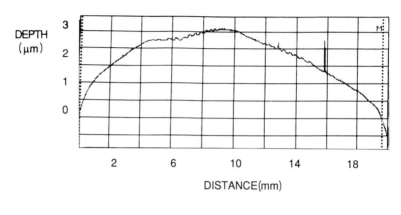

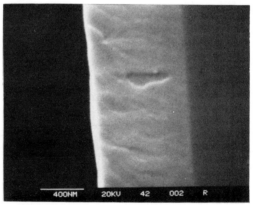

Fig. 7.8 Dektak profile of shallow patterned growth together with an SEM micrograph through a cross-section from the centre of the pattern. The growth conditions were 1.3 Torr Me_2Cd and 0.7 Torr Me_2Te.

these results it is possible to calculate the maximum microscopic photo-enhancement, factors which for Figs 7.8 and 7.9 are 1.11 and 1.07 respectively. The macroscopic photo-enhancement factors under these conditions would have been 10 for Me_2Te and 2 for Et_2Te.

To investigate the influence of inlet alkyl concentrations on mesa height, the experiments in Figs 7.8 and 7.9 were repeated, but with alkyl concentrations doubled. The Dektak profiles are shown in Figs 7.10 and 7.11 for Me_2Te and Et_2Te respectively. In Fig. 7.10 the mesa heights have increased by a factor of more than 2 to 0.6 µm in the centre. Similarly to the upstream mesas in Fig. 7.7, the surfaces of these mesas were matt due to a rough surface morphology. The X-ray texture pattern shows spots for the CdTe layer lying above the spots for the GaAs substrate, as expected from

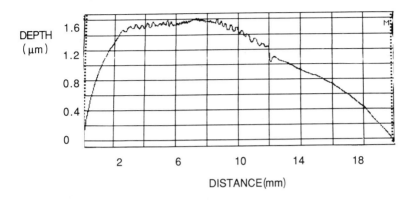

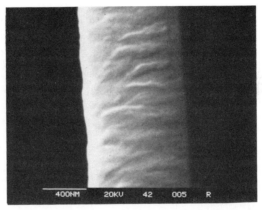

Fig. 7.9   Dektak profile of shallow patterned growth together with an SEM micrograph through a cross-section from the centre of the pattern. The growth conditions were 1.3 Torr Me$_2$Cd and 0.6 Torr Et$_2$Te.

the 14% lattice mismatch. There is still an underlying bowing of the deposit, indicating that vapour diffusion spread (deposition in the dark) is still significant.

This contrasts with the profile for the layer grown with Et$_2$Te and Me$_2$Cd shown in Fig. 7.11. The Dektak profile has now been taken diagonally to include all the features, but otherwise it is the same pattern as shown in Fig. 7.10. The profile is now dominated by distinct mesas rising to 5 μm above the surrounding layer, with a relatively flat background, indicating little diffusion spread of Cd and Te. The X-ray texture pattern again shows that the layer is epitaxial, with sharper CdTe spots than in Fig. 7.10. This can be attributed to the thicker layer where diffraction can occur away from the highly dislocated CdTe/GaAs interface.

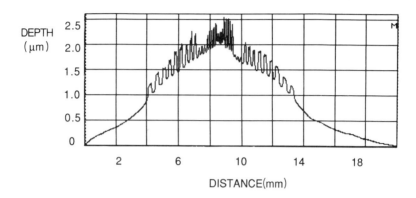

Fig. 7.10 Dektak profile with X-ray texture pattern, showing single-crystal spots for CdTe lying above equivalent reflection spots for the GaAs substrate. The alkyl concentrations were doubled compared with the pattern shown in Fig. 7.8.

It is worth noting that the corresponding macroscopic photo-enhanced growth rate for conditions comparable to those used in Fig. 7.11 was 2.5 μm h^{-1}, so the microscopic growth rate has more than doubled. Rough estimates of the microscopic photo-enhancement factors are 1.7 for Fig. 7.10 and 10 for Fig. 7.11. The latter is more than expected from the macroscopic photo-enhancement, and indeed the relative numbers for Et$_2$Te and Me$_2$Te reverse the situation, comparing microscopic and macroscopic photo-enhancement.

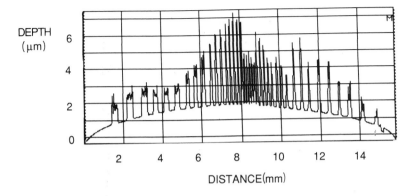

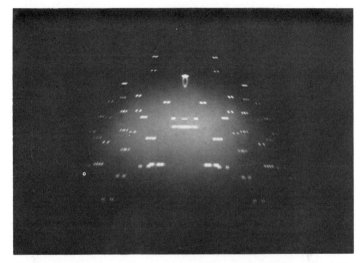

Fig. 7.11 Dektak profile with X-ray texture pattern of layer p/462. The alkyl concentrations were doubled compared with the pattern shown in Fig. 7.9.

SEM micrographs of the surface of the layer in Fig. 7.11 are shown in Fig. 7.12, where (a) shows a low-magnification image with the full range of mesa sizes in the field of view. Some diffraction effects can be seen, but mostly they are well defined and distinct mesas. The high-magnification micrograph (b) is of one of the central mesas, and shows that the pattern has become very rough and faceted, although it is still epitaxial. This surface morphology is not observed in the case of shallow mesas, and may be related to the mechanisms of microscopic photo-enhancement.

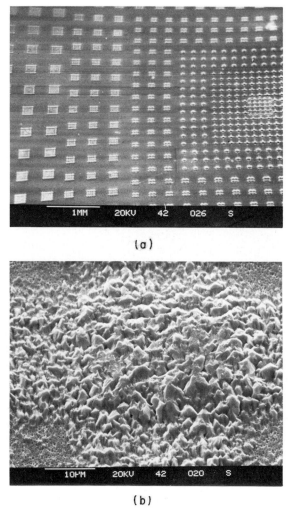

(a)

(b)

Fig. 7.12 SEM micrographs of layer p/462 showing the full range of mesa sizes from 250 μm down to 50 μm in (a) and a high-magnification micrograph of a 50 μm mesa in (b).

7.8 SUPERENHANCEMENT

A distinction needs to be made between the shallow mesas observed in Figs 7.8 and 7.9 and the much higher mesas shown in Figs 7.10 and 7.11. The mechanisms for laser–surface interactions discussed in Section 7.6, whereby adsorbed organic radicals block growth sites and are desorbed by photon absorption in the surface, is possibly only responsible for the shallow microscopic photoenhancement observed below critical alkly concentrations. Large microscopic photo-enhancement appears to occur when the surface is roughened and therefore the nature of the surface is critically changed. This situation is shown schematically in Fig. 7.13. The proposed model for super-enhancement is as follows:

(1) the high supersaturation of Cd(v) and Te(v) over the surface induces facet growth where the surface is illuminated;

(2) the increased step edge density increases the growth rate due to the higher density of nucleation sites;

(3) the growth rate is now sufficiently high that significant diffusion gradients are set up in the vapour, and the concentrations of Cd(v) and Te(v) near the surface decrease;

(4) this will decrease growth rates in the kinetically limited "dark" areas surrounding the mesas, which means that pyrolysis of the organometallics is now not a factor in the microscopic photo-enhancement;

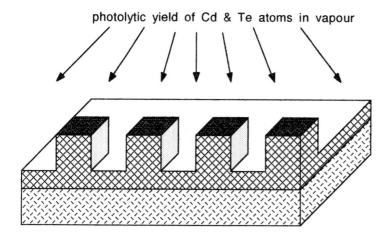

1.   high super saturation in illuminated areas causes facetting.
2.   increased step edge density increases growth rate.
3.   local concentrations of Cd & Te decrease.
4.   "dark" growth rate decreases.
5.   microscopic photo-enhancement increases.

Fig. 7.13   Schematic of super-enhanced patterned growth.

(5) microscopic photo-enhancement increases dramatically once the surface starts to roughen.

Step (1) is clearly crucial to the whole process, and it needs to be established that, in favourable environments, the photon–surface interaction can induce a roughening transition. The growth enhancement and roughening induced by the monitoring HeNe laser beam reported in an earlier publication [20] is indicative of such a process. As in the case of ZnSe

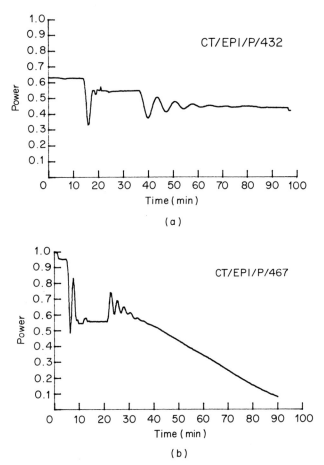

(a)

(b)

Fig. 7.14 Comparison of TRR traces between normal concentrations of Me_2Cd and Me_2Te corresponding to conditions for the layer in Fig. 7.8 (a) and double these concentrations in (b), showing a decrease in reflectivity due to a roughening transition.

growth, reported by Fujita *et al.* [15], bandgap radiation can induce growth rate enhancement without interaction with the precursor vapour. In this case a continuous HeNe beam is also sufficient to exceed critical conditions, giving rise to a roughening mode of growth.

A comparison is made in Fig. 7.14 between TRR traces for CdTe growth at 300°C with normal alkyl concentrations and doubled alkyl concentrations. With normal concentrations, the signal oscillates around a horizontal line, indicating smooth nonfaceted growth. For the increased concentrations, not only does the growth rate increase (smaller period of oscillation), but it also oscillates about a decreasing slope, indicative of faceted growth. These experiments demonstrate that this roughening transition can be controlled by alkyl concentration as well as by photon intensity.

7.9 CONCLUSIONS

Recent results on photo-assisted epitaxy of II–VI compounds have been reviewed. Processes can be broadly categorized into vapour or surface interactions, and some selection can be achieved by either parallel illumination or vertical illumination, depending on what is desired from the photostimulation. An important mechanism that is now emerging is the absorption of radiation by the semiconductor surface, with subsequent modification of the surface chemistry by charge transfer.

Recent results on photo-enhancement of CdTe growth have been presented and analysed in terms of vapour transport and surface kinetic models. Over the range of parameters considered, the growth is limited by surface kinetics and can be described by a Langmuir–Hinshelwood model with photodesorption of a site blocking species, possibly organic radicals.

Microscopic photo-enhancement, necessary for selected area epitaxy, relies on photon–surface interactions. It appears that the enhancement can be quite small due to vapour diffusion spread of Cd(v) and Te(v) unless a roughening transition occurs. The roughening transition can result in super-enhancement, with growth rates actually in excess of the macroscopic growth rates and higher photo-enhancement factors. For example, at 300°C growing from Me_2Cd and Et_2Te, the macroscopic photo-enhancement was 2, but the microscopic value was approximately 10.

More work needs to be carried out to establish these mechanisms and discover optimum conditions for selected area epitaxy. In particular, crucial areas of mechanistic study are photo-induced charge transfer and photo-induced roughening transitions.

ACKNOWLEDGEMENTS

The authors wish to thank Mrs J. Clements for technical support with substrate preparation and layer growth. Mrs O. D. Dosser is also thanked for the scanning electron microscopy.

REFERENCES

[1] D. W. Kisker and R. D. Feldman, *J. Cryst. Growth* **72**, 102 (1985).
[2] S. Haq, P. S. Dobson and S. J. C. Irvine, in *Photon, Beam and Plasma Enhanced Processing* (eds A. Golanski, V. T. Nguyen and E. F. Krimmel), Vol. 15, p. 199 (1987)
[3] Sz. Fujita, F. Y. Takeuchi and Sg. Fujita, *Japan J. Appl. Phys* **27**, L2019 (1988).
[4] S. J. C. Irvine, H. Hill, O. D. Dosser, J. E. Hails, J. B. Mullin, D. V. Shenai-Khatkhate and D. Cole-Hamilton, *Mater. Lett.* **7**, 25 (1988).
[5] S. J. C. Irvine, H. Hill, J. E. Hails, G. W. Blackmore and J. B. Mullin, *Mater. Res. Soc. Symp. Proc.* **183**, 129 (1989).
[6] S. J. C. Irvine, H. Hill, G. T. Brown, S. J. Barnett, J. E. Hails, O. D. Dosser and J. B. Mullin, *J. Vac. Sci. Technol.* **B7**, 1191 (1989).
[7] S. J. C. Irvine, J. B. Mullin, D. J. Robbins and J. L. Glasper, *J. Electrochem. Soc.* **132**, 968 (1985).
[8] C. J. Chen and R. M. Osgood, *J. Chem. Phys.* **81**, 327 (1984).
[9] J. E. Hails, S. J. C. Irvine, J. B. Mullin, D. V. Shenai-Khatkhate and D. Cole-Hamilton, *Mater. Res. Soc. Symp. Proc.* **75**, 131 (1989).
[10] B. J. Morris, *Appl. Phys. Lett.* **48**, 867 (1986).
[11] J. J. Zinck, P. D. Brewer, J. E. Jensen, G. L. Olson and L. W. Tutt, *Appl. Phys. Lett.* **52**, 1434 (1988).
[12] G. B. Shinn, P. M. Gillespie, W. L. Wilson and W. M. Duncan, *Appl. Phys. Lett.* **54**, 2440 (1989).
[13] S. J. C. Irvine, J. B. Mullin, H. Hill, G. T. Brown and S. J. Barnett, *J. Cryst. Growth* **86**, 188 (1988).
[14] S. J. C. Irvine and J. B. Mullin, *J. Cryst. Growth* **79**, 371 (1986).
[15] Sz. Fujita, A. Tanabe, T. Sakamoto, M. Isemura and Sg. Fujita, *J. Cryst. Growth* **93**, 750 (1988).
[16] S. J. C. Irvine, in *Laser Microfabrication, Thin Film Processes and Lithography* (ed. D. J. Ehrlich and J. Y. Tsao), Chap. 9 (Academic Press, New York, 1989).
[17] R. N. Bicknell, N. C. Giles and J. F. Schetzina, *Appl. Phys. Lett.* **49**, 1095 (1986).
[18] N. C. Giles, R. L. Harper, J. W. Han and J. F. Schetzina, *Mater. Res. Soc. Symp. Proc.* **161**, (1990).
[19] S. J. C. Irvine, H. Hill, J. E. Hails, J. B. Mullin, S. J. Barnett, G. W. Blackmore and O. D. Dosser, *J. Vac. Sci. Technol.* **A8**, 1059 (1990).
[20] S. J. C. Irvine, H. Hill, J. E. Hails, A. D. Pitt and J. B. Mullin, *Mater. Res. Soc. Symp. Proc.* **158**, 357 (1990).
[21] I. Bhat and S. K. Ghandhi, *J. Electrochem. Soc: Solid State Sci. Technol.* **131**, 1923 (1984).

8 IR and UV Laser-Induced Deposition of Hydrogenated Amorphous Silicon

PETER HESS

Institut für Physikalische Chemie, Universität Heidelberg, Germany

8.1 INTRODUCTION

8.1.1 General remarks

Laser-induced chemical vapour deposition (laser CVD) is a rapidly expanding field. This is mainly due to its unique possibilities in materials processing and microelectronics fabrication. There are already several books available treating the subject in the form of monographs, with emphasis on the fundamental understanding of the deposition and etching processes [1, 2], or in the form of review articles written by experts, which also include a discussion of the advances in microfabrication and electronics [3, 4].

Although laser chemical processing technology with potential applications in the semiconductor industry is growing rapidly, there is no satisfactory understanding of the fundamental chemical processes involved. The reason for this is the complexity of laser-driven chemical processes. This is also true for the laser deposition of hydrogenated amorphous silicon (a-Si : H) considered in detail in this chapter. This material has many applications, such as photovoltaic solar cells, thin-film transistors for liquid crystal displays, photoreceptors for electrophotography and laser printing, and image sensors. These technological applications are some of the driving forces behind the efforts to come to a better understanding of the spectroscopy and primary reactions of precursors, the reaction dynamics in the gas phase and at the surface, and the crucial solidification processes leading to the formation of a-Si : H films.

Photochemical Processing of Electronic Materials
ISBN 0-12-121740-X

Laser CVD of a-Si : H was reported for the first time in 1982 [5]. Compared with the conventional methods used for the deposition of this material, such as plasma CVD, HOMOCVD and reactive sputtering, this technique exhibits some unique features. Laser CVD allows spatial control of the chemical processing above the surface, avoiding wall effects. The laser method also makes possible detailed control of the deposition conditions and variables, giving new insight into the chemical mechanisms of the formation of solid networks.

All aspects of the chemical studies on a-Si : H, such as preparation and structure, optical properties, electronic and transport properties, and device applications, are considered in the four-volume treatise edited in 1984 by Pankove [6]. A review concentrating on mechanistic studies of chemical vapour deposition by thermal and plasma-assisted processes has also been published [7], while a short discussion of IR and UV laser CVD of a-Si : H can be found in [8].

8.1.2   Scope of this review

Two completely different arrangements of the IR or UV laser beam used as the excitation source and the substrate surface are possible: one where the beam travels parallel to the surface without touching it and one where the laser beam hits the surface. Detailed characterization of the deposition of a-Si : H films has been achieved for the parallel configuration, shown in Fig. 8.1. Here results will be presented mainly for this configuration, where the growth process can be controlled by the photon flux, with negligible background growth at substrate temperatures below 400°C, even for Si_2H_6 as the precursor.

In the parallel illumination configuration the laser radiation directly

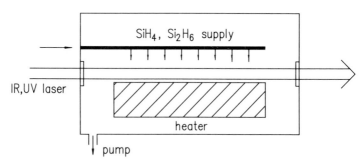

Fig. 8.1   Schematic of the experimental set-up used for a-Si : H deposition employing parallel illumination, SiH_4 or Si_2H_6 as precursors and CO_2 or ArF lasers as light sources.

excites or heats only the gas above the surface, and therefore the gas and surface temperatures may be controlled more or less independently. Thus processes occurring in the gas phase (e.g. the primary and secondary chemical reactions) and those taking place at the surface determining the film structure and composition may be synergistic. The gas phase temperature must be high enough to induce chemical processing, whereas the substrate temperature is of crucial importance for the film properties, and therefore should be optimized according to the surface chemistry yielding the best solid material. Besides photo-excitation and resulting reactions in the laser beam, transport of heat and of activated species and reaction products from the localized volume of the light beam to the surface must be considered in a theoretical description. Modelling of the deposition process is difficult, because the processes occurring in a CVD reactor with continuous flow of precursor molecules may be far from equilibrium.

Apart from the deposition geometry, the laser wavelength and intensity provide efficient control of the energetics and kinetics of the deposition process. Deposition experiments for a-Si:H have been performed using SiH_4 as the precursor and a CW CO_2 laser [5, 9–20] or pulsed CO_2 laser [21], but pulsed ArF excimer laser radiation at 193 nm has also been employed to excite Si_2H_6 [22–25] or to produce Si_2H_6 from SiH_4 [26, 27]. An important feature of laser radiation is the high photon flux available in the IR and UV spectral regions, allowing efficient heating or photodissociation of the precursor molecules. Direct bond breaking with UV lasers, multiphoton excitation with high-power IR and UV lasers and spatial control of the heated zone induce low-temperature processing with minimal heating of the whole deposition system. This low-temperature capability of laser processing is important from the point of view of energy consumption, and is essential for the processing of fragile substrates and device structures or temperature-sensitive materials.

The results obtained by direct illumination of the surface at various angles will be discussed in a separate section after detailed discussion of the parallel irradiation experiments. A mechanistic understanding of the processes induced by the laser radiation when the beam hits the surface seems to be more difficult. Despite the fact that silicon deposition was first studied for normal incidence, the data available on film properties is very limited. The available results indicate, however, that direct illumination is a promising technique for depositing high-quality a-Si:H films.

Photo-induced pyrolytic or photolytic reactions of precursors can lead to epitaxial growth at reduced temperatures if a suitable single crystal is selected as the substrate. This has been demonstrated for the growth of germanium on GaAs (001) by photodissociating GeH_4 with an ArF laser in parallel geometry [28]. However, deposition at low substrate temperatures on readily available substrates such as glasses normally leads to a

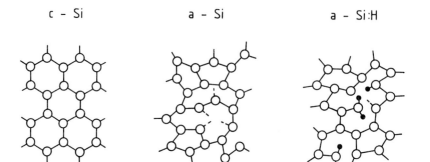

Fig. 8.2 Two-dimensional illustration of structural differences between c-Si, a-Si and a-Si : H.

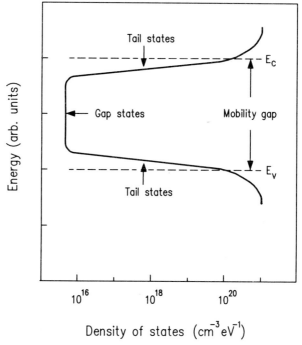

Fig. 8.3 Schematic of the energy dependence of the density of states of an amorphous semiconductor in the bandgap region.

metastable amorphous network with a higher Gibbs free energy. The properties of this amorphous material may vary considerably with the preparation conditions. As shown schematically in Fig. 8.2, amorphous silicon (a-Si) can be characterized by small distortions of the bond length and larger distortions of the bond angle in the Si network and by under- and/or overcoordination of Si atoms ("dangling" and "floating" bonds). At the low surface temperatures employed during deposition, hydrogen is incorporated into the network, saturating dangling bonds as shown schematically in Fig. 8.2. This reduces the density of defects near the middle of the bandgap of a-Si:H to such a low level that the material possesses electronic properties needed for device applications. Distortions in the Si network, especially in the bond angle, are believed to be responsible for the tail states at the bandgap edges (see Fig. 8.3). These defect states constitute traps for the charge carriers, and thus lower their mobilities. The energy between E_c and E_v (see Fig. 8.3) defines the mobility gap of the amorphous semiconductor separating nonlocalized and localized states. The optical and electrical properties observed for a-Si:H films deposited by IR and UV laser CVD will be reviewed.

8.2 LASER EXCITATION AND CHEMISTRY

8.2.1 Spectroscopy

The spectroscopy of laser excitation and the initial steps in laser deposition of a-Si:H have been reviewed recently [29, 30]. Here only the spectroscopic features important for deposition from SiH_4 and Si_2H_6 with a CO_2 laser or with an ArF laser will be considered.

In the IR experiments the CW CO_2 laser is tuned to the P(20) line at 944.195 cm^{-1}, which is near the vibrational–rotational line at 944.213 cm^{-1} of the ν_4 mode of SiH_4. Figure 8.4 is a schematic representation of the IR and UV bands of SiH_4 and the laser excitation employed. For deposition, a SiH_4 pressure of several torr is usually applied, as discussed in detail later. In this pressure region SiH_4 dissociation is a collision-assisted process [31]. In the 150 J cm^{-2} fluence range studied, collisionless multiphoton dissociation could not be achieved for SiH_4 [31]. Therefore the decomposition of SiH_4 in a 50 W CW CO_2 laser beam should be considered as a thermal pyrolytic process under the deposition conditions specified as pointed out in [15, 16]. This implies that the lowest reaction channel

$$SiH_4 \rightarrow SiH_2(\tilde{X}\ ^1A_1) + H_2 \qquad (8.1)$$

with an activation energy $E_A = 249$ kJ mol^{-1} [32] and a reaction enthalpy

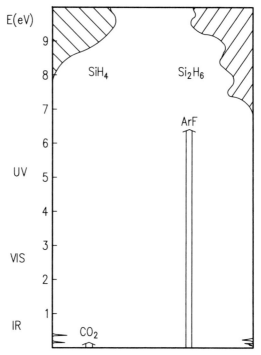

Fig. 8.4   Schematic of the IR and UV bands of SiH_4 and Si_2H_6 and the CO_2 and ArF laser excitation in the IR and UV respectively.

between 210 and 250 kJ mol^{-1} [33, 34] should be the dominant initiation reaction. In fact, this channel seems to be well separated by a lower activation and reaction energy from the electronically excited $SiH_2(A\ ^3B_1)$ species and from the $SiH_3(\tilde{X}\ ^2A_1)$ radical produced by Si—H bond rupture [29].

For Si_2H_6 irradiated with an ArF laser at 193 nm (see Fig. 8.4) a much larger number of reaction channels must be considered, since pyrolytic and photolytic pathways may contribute. The reaction with the lowest activation energy (242 kJ mol^{-1}) and a similar reaction enthalpy is

$$Si_2H_6 \rightarrow SiH_4 + SiH_2 \qquad (8.2)$$

To eliminate a hydrogen molecule from one of the silicon atoms, a somewhat higher activation energy of 256 kJ mol^{-1} is needed. However, subsequent conversion to the symmetric isomer over a barrier of about

12 kJ mol^{-1} leads to the lowest-energy product of disilane dissociation:

$$Si_2H_6 \rightarrow H_3SiSiH + H_2 \qquad (8.3)$$

$$H_3SiSiH \rightarrow H_2SiSiH_2$$

Besides these thermally favoured reactions, a series of other channels is energetically accessible with one 193 nm photon. This includes the production of radicals such as

$$Si_2H_6 \rightarrow 2SiH_3 \qquad (8.4)$$

$$Si_2H_6 \rightarrow Si_2H_5 + H \qquad (8.5)$$

with reaction enthalpies of 320 kJ mol^{-1} and 371 kJ mol^{-1} respectively [29]. It is interesting to note that even electronically excited SiH could be detected during ArF laser irradiation [22]. Thus the radicals SiH, SiH_2 and SiH_3 are produced and may contribute to the deposition process in the case of UV laser processing.

8.2.2   Gas phase chemistry

As shown before, the channels with the lowest activation energies lead to the formation of the SiH_2 radical in the case of IR and UV laser processing. This radical has been considered as one of the species mainly responsible for film growth at the surface during glow discharge [35] and laser deposition [15]. The recent direct redetermination of the rate constants for the reaction of SiH_2 with hydrogen [36] and with silane and higher silanes resulted in a nearly gas-kinetic efficiency for insertion of this radical into the silicon–hydrogen bond [37]. Therefore, during diffusion from the centre of the parallel laser beam to the substrate surface, SiH_2 reacts effectively with SiH_4, Si_2H_6, etc., to form higher-order silicon hydrides. For the CO_2 laser process the generation of higher silanes has been proved by *in situ* mass spectrometric analysis up to hexasilane [38]. The SiH_2 insertion reaction leads to polysilanes with increasingly large molecular weights, and finally to powder formation if the pressure is so high that the conditions for homogeneous nucleation are approached:

$$SiH_2 + SiH_4 + M \quad \rightarrow Si_2H_6 + M \qquad (8.6)$$

$$SiH_2 + Si_2H_6 + M \quad \rightarrow Si_3H_8 + M \qquad (8.7)$$

$$\vdots \qquad\qquad \vdots$$

$$SiH_2 + Si_{n-1}H_{2n} + M \rightarrow Si_nH_{2n+2} + M \qquad (8.8)$$

$$\rightarrow (SiH_2)_n$$

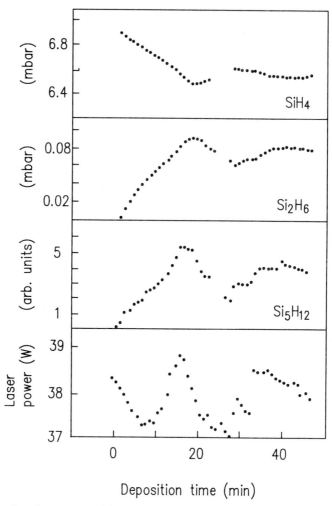

Fig. 8.5 Gas phase composition, represented by silane, disilane and pentasilane partial pressures, and laser power versus deposition time for SiH_4 as source gas and irradiation with a CW CO_2 laser [38].

Thus, in a film deposition experiment, homogeneous nucleation must be avoided because under those conditions the films are porous and only weakly bonded to the substrate. On the other hand, the highest deposition rates are obtained for pressures just below the critical pressure for homogeneous nucleation, which may be different in the IR and UV laser deposition experiments. The SiH_2 mechanism can be considered as the low-energy pathway in gas phase chemistry, contributing in IR and UV laser processing to a variable extent to film growth irrespective of the precursor. Figure 8.5 shows the results of an *in situ* mass spectrometric analysis of the gas phase during irradiation of SiH_4 with a CW CO_2 laser. It is interesting to note the sensitivity of the concentration of higher silanes to small changes in the laser power under steady state conditions.

As mentioned above, primary reactions to many species are energetically feasible, especially in the UV excitation process. The role played by these species in the gas phase and at the surface is not clear at present. No detailed *in situ* information is available; however, a-Si:H films deposited by the IR and UV methods exhibit distinct differences, which may be due to a wavelength-dependent chemistry involved. For example, in addition to the short-lived intermediate radical SiH_2 driving the gas phase chemistry, other active species may be produced in the gas phase, which reach the surface and interfere with the formation of the network. As discussed later in more detail, it has been suggested that the SiH_3 radical may play such a role because it possesses a longer lifetime. Another important effect of reactive species in the gas phase could be the restriction of growth of higher silanes in that phase.

8.2.3 Surface chemistry

The main network-forming chemical process occurring at the surface is three-dimensional bond formation between silicon atoms by the breaking of silicon–hydrogen bonds under hydrogen desorption. The surface chemistry is governed by the active gas mixture produced by laser irradiation for laser-sustained growth and the surface temperature, which strongly influences the nature of the surface. The surface temperature can be measured *in situ* with a precision of ± 1 K by employing a Ni sensor attached to the substrate surface [18]. Unfortunately, the *in situ* analysis of molecular structures and surface reactions by infrared spectroscopy was limited until recently to high-surface-area materials. Thus it has not been possible so far to monitor *in situ* the kinetics of monohydride and dihydride formation at the surface or the desorption of hydrogen due to dehydrogenation during film growth induced by laser irradiation.

However, several studies have recently been published of relevant processes on well defined surfaces such as single crystals or crystalline silicon samples with high surface area. Measurements of the reactive sticking coefficient of SiH_4 on the Si (111)–(7 × 7) surface at low hydrogen coverage indicate a value of about 10^{-5}, with essentially no dependence on the surface temperature [39]. Disilane and trisilane are roughly 10^4 times more reactive on this surface between 100 and 500°C. On a hydrogen-covered Si (111)–(7 × 7) surface the latter molecules are 10^3 times more reactive than

SiH_4 [40]. If the growing film behaves in a similar way to that found in these model studies, we have to assume that, of the stable species reaching the surface, the higher silanes are mainly responsible for film growth. Such a deposition mechanism implies a higher growth rate when starting with Si_2H_6 as the precursor, consistent with experimental observations [22]. Figure 8.6 presents the correlation detected between the growth rate measured *in situ* employing a quartz crystal microbalance and the gas phase concentration of, for example, trisilane observed by *in situ* mass spectrometry. These results were measured for CO_2 laser deposition using SiH_4 as precursor.

In order to understand the growth kinetics at the atomic level, the mechanism of reactive sticking of disilane or higher silanes as the first step to solidification must be considered. Recently, the mechanism of Si_2H_6 surface decomposition on Si (111) was studied by modulated molecular beam spectrometry [41]. According to this study, disilane decomposes, producing SiH_4. This indicates that hydrogen removal from the surface may be a slow process restricted to secondary arrangements at the surface. Figure 8.7 illustrates these channels of surface processing.

Hydrogen desorption kinetics from monohydride and dihydride species on a high-surface-area porous silicon sample was studied by Fourier transform infrared spectroscopy [42]. Annealing studies revealed that hydrogen from SiH_2 species desorbed between 640 and 700 K, with an activation barrier of 180 kJ mol^{-1}, whereas hydrogen from SiH species desorbed between 720 and 800 K, with an activation barrier of 272 kJ mol^{-1}. These results prove that the SiH group is more stable than the SiH_2 group at the surface. The same may be true for SiH_2 groups located at inner surfaces such as sponge-like structures or voids and between the rods in a columnar structure. According to the work discussed and other thermal desorption studies, we can assume that SiH_3 groups possess the lowest stability at an equilibrated surface. With increasing substrate temperature first the SiH_3-, then the SiH_2- and finally the SiH groups disappear from the surface, leading to bare Si sites near 500°C. Reactive species hitting the surface may disturb the equilibrium distribution. They may preferably attack the weaker SiH bonds, reducing the number of SiH_3 and SiH_2 groups. This can be considered as one of the important differences between IR and UV laser processing.

All three hydride species (SiH_x, $x = 1-3$) have been detected recently in real time in a flow reactor employing infrared reflection absorption spectroscopy [43]. Film deposition was carried out in these experiments by metastable argon-induced CVD. It was possible to detect *in situ* a change in the chemical bonding situation with increasing film thickness. Our current understanding of interfaces is rudimentary at best, despite their enormous technological importance. Therefore central themes of research

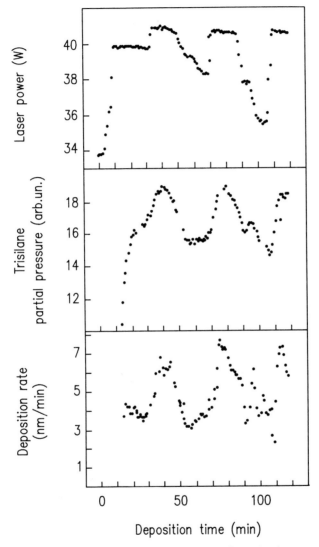

Fig. 8.6 Correlation between growth rate measured *in situ* by a quartz crystal microbalance and the partial pressure of higher silanes shown for trisilane [44].

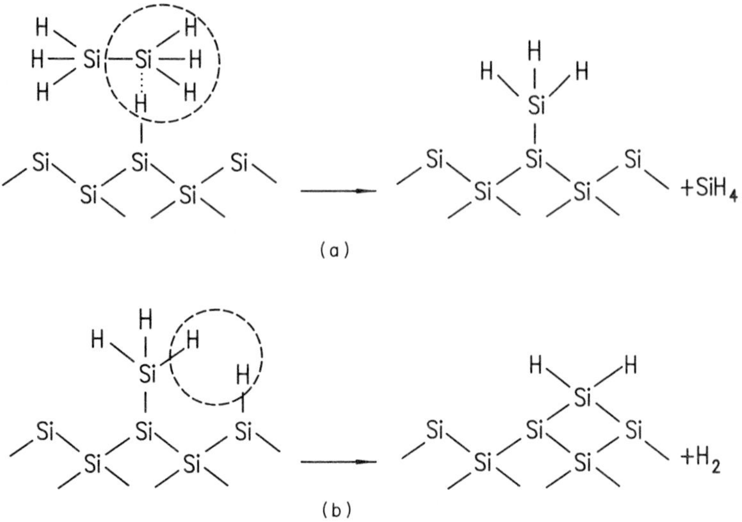

Fig. 8.7 (a) Reactive sticking of disilane on a hydrogen-covered surface, resulting in silane emission. (b) Hydrogen removal by bonding rearrangement necessary for cross-linking of the silicon network.

in this field will be the investigation of the interrelation between interfacial properties and the underlying structure and the direct observation of the growth process by *in situ* techniques.

8.3 KINETICS AND GROWTH

8.3.1 Deposition rates

The parallel SiH_4–CW CO_2 laser deposition process has been studied for growth rates between 0.2 and 5 Å s^{-1} [15, 16]. These deposition rates were determined from the deposition time and the thickness of the film measured mechanically by a stylus-type profilometer after deposition. For *in situ* analysis of film deposition rates under varying conditions a quartz crystal microbalance [44] or an optical reflectivity technique has been employed, using a HeNe laser beam at 633 nm and a photodiode to detect the interference oscillations [15, 23]

The deposition rates observed for the parallel Si_2H_6–ArF laser deposition process varied between 0.3 and 55 Å s^{-1} [22]. It is important to note that these are average growth rates obtained for excitation with 15 ns laser pulses. With pulse energies $\geqslant 100$ mJ and repetition rates of 100 Hz, comparable laser powers were applied in the pulsed UV and CW IR experiments. Depending on the time needed by the active species to reach the surface, a much higher peak deposition rate may be possible in the pulsed experiments, where light and dark periods alternate and the time needed by the active species to diffuse to the surface seems to be smaller than the time between two subsequent laser pulses [25].

8.3.2 Rate-limiting processes

To set up a realistic model for the laser CVD process, it is necessary to understand the gas phase reactions, the transport processes (e.g. the diffusion of active species to the surface) and the surface chemistry. Despite enormous efforts in recent years, this goal has not yet been achieved for any deposition system.

For the parallel SiH_4–CW CO_2 laser deposition process a simple model has been developed to describe the kinetics of film growth. In this case the rate is extremely sensitive to the SiH_4 pressure and the laser power. The model considers the effect of these parameters on the gas temperature and assumes that this temperature controls the SiH_4 decomposition rate, which is the main rate-limiting process [15]. An Arrhenius plot of the measured growth rates versus the reciprocal peak gas temperature yields an activation energy of 192 kJ mol^{-1}, which is in reasonable agreement with the value of 249 kJ mol^{-1} found for the three-centre elimination of H_2 from SiH_4 [32]. Unfortunately, the peak gas temperature could not be measured, but was estimated on the basis of a steady state energy balance model, where the energy absorbed by the gas is set equal to the energy lost due to heat conduction.

Similar results were reported in [16]. Figure 8.8 shows Arrhenius plots of the measured growth rates versus reciprocal gas temperature estimated with the energy balance model and reciprocal surface temperature measured with the Ni sensor. The first plot yields an activation energy of 200 kJ mol^{-1}, in good agreement with the former value. Correlation of the growth rate with the surface temperature leads to a much smaller activation energy.

For the Si_2H_6–ArF laser deposition process no simple kinetic model could be developed. It is generally assumed that the film precursors are formed by photolysis, and that Si_2H_6 pyrolysis should be negligible [22, 25]. If the surface processes are not rate-limiting, this would imply that diffusion of active species to the surface plays an important role. Under typical deposition conditions, the time estimated for diffusion is less than or comparable to the time between two subsequent laser pulses. This points to a periodic variation of growth with time, and to peak deposition rates exceeding the mean values discussed before.

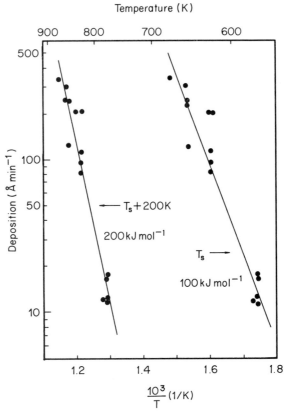

Fig. 8.8 Deposition rates versus reciprocal peak gas temperature (estimated) and substrate temperature (measured). Laser power 35 W, silane pressure 7 Torr, gas flow rate 4 sccm.

8.3.3 Modelling

The processes occurring in a laser CVD reactor employing a parallel configuration have been considered in terms of finite element analysis of a two-dimensional form of the temperature and species fields [45]. Such a description gives a fairly complete analysis of the transport processes (e.g. gas flow, heat and species transport). At the low pressures used for the laser CVD process, transport is diffusion-controlled, resulting in a spatial nonuniformity of the deposited film. Convection becomes important at higher pressures. Horizontal orientation of the substrate with respect to gravity is optimal because of the stagnation-like flow. This has been observed in recent experiments studying different laser beam−substrate positions with respect to gravity [20]. For the gas phase chemistry a simplified version of the SiH_4 decomposition model was assumed. The surface processes were modelled by selecting appropriate sticking coefficients for the active species and by neglecting hydrogen evolution. These drastic simplifications of the gas phase and surface processes limit the usefulness of this model to predicting film properties such as film composition and binding configurations. The main assumption of this modelling study is the existence of two temperatures, causing temperature gradients between the laser beam and the surface.

For pulsed laser processing an analytical model has recently been suggested, describing diffusion-limited growth including depletion effects and relaxation of the system between pulses [46]. A reasonable description of the experimental results has been obtained with this model for the deposition of chromium films from $Cr(CO)_6$ employing a KrF laser in perpendicular configuration. This indicates that at least mass transport can be modelled successfully in suitable pulsed processing systems. The main problem remaining is the incorporation of a realistic chemistry into the model describing gas phase and surface reactions.

8.4 MATERIAL PROPERTIES

8.4.1 Chemical composition

At the relatively low surface temperatures employed during laser deposition (200–400°C), a silicon–hydrogen alloy is grown in a one-step process. Figure 8.9 illustrates the chemistry of this binary system. It shows the transition from silicon hydrides to a polymeric chain at a composition with a silicon-to-hydrogen ratio of 1:2 and finally to a three-dimensional network, which can be formed if the hydrogen concentration falls below 50%. However, this scheme only describes stoichiometry and cannot be used to determine chemical structures in an alloy containing a certain amount of hydrogen. For example, in a film with 20% hydrogen we may find not only SiH groups but also SiH_2 groups, and furthermore these groups may not be homogeneously distributed in the chemical network. It turns out that it is difficult to obtain a high-quality film with a hydrogen content above 25%, and therefore such films are not important for electronic device applications.

Hydrogen incorporation into the film changes with the surface temperature. As can be seen from Fig. 8.10, a different dependence is observed for the parallel SiH_4–CW CO_2 laser and the Si_2H_6–ArF laser deposition processes. This indicates characteristic differences in the chemical mechanism.

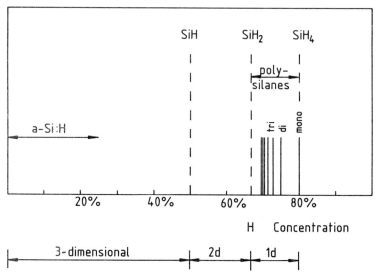

Fig. 8.9 Chemistry in the silicon–hydrogen system. The possible dimensionality of bonding is shown and the region of useful a-Si : H films indicated.

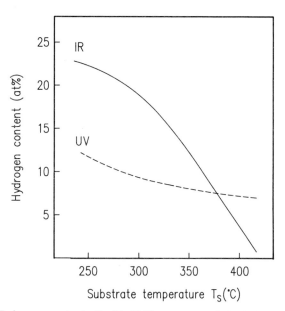

Fig. 8.10 Hydrogen content of a-Si : H films versus substrate temperature for IR laser deposition [16] and UV laser deposition [22] in the parallel configuration.

As yet, however, it is not clear why the variation of the hydrogen concentration is much smaller for the UV process. A major contribution to this effect could be due, for example, to dehydrogenation of the surface by reactive species generated by UV photons.

8.4.2   Structure

Under the preparation conditions employed for IR and UV laser deposition, an amorphous hydrogenated silicon network is prepared. Such an amorphous material can be characterized by bond angle distortions and bond length distortions, under- and/or overcoordination of silicon atoms in the network ("dangling" and "floating" bonds), topological structures (distribution of rings with five, six or seven members), and the morphology of the film (voids, columnar structure). This gives rise to the formation of a large variety of network structures, and in fact, it can be assumed that the properties of the material depend strongly on the conditions of preparation.

The energy spectrum near the edges of the bandgap ("tail state distribution") depends primarily on the structural disorder of the silicon network, such as the presence of bond-angle distortions. To date, no information is available concerning these network distortion effects for laser-deposited material. Under- and/or overcoordination of silicon atoms are responsible for the defect states in the middle of the bandgap ("gap states"). Hydrogen is able to passivate these gap states caused by unsaturated bonds. Films of a-Si : H have been prepared by the IR and UV laser deposition methods where the density of these defects could be reduced from about $10^{19}\,\mathrm{cm}^{-3}\,\mathrm{eV}^{-1}$ to about $10^{16}\,\mathrm{cm}^{-3}\,\mathrm{eV}^{-1}$ [15, 25]. This is the defect density found in good single-crystal semiconductors due to impurities. No results have been reported on the topology and ring structures in laser-deposited a-Si : H films.

The morphology was studied by TEM [17] and by small-angle X-ray scattering (SAXS) [44]. At least the films grown at low surface temperatures and higher growth rates possess a heterogeneous morphology. Columnar or sponge-like structures have been detected in the IR laser-deposited material. The islands consist of a network with higher density than the regions in between. These latter regions may contain a high concentration of hydrogen (e.g. SiH_2 groups and/or polysilanes). Figure 8.11 shows FTIR spectra of films deposited with the IR laser at different substrate temperatures. With decreasing temperature, the characteristic bands of the SiH_2 group (2090 cm^{-1}) and the $(SiH_2)_n$ group (890 cm^{-1} and 845 cm^{-1}) appear in the IR spectrum, indicating a heterogeneous composition of the film. Similar IR spectra were found for UV laser-deposited

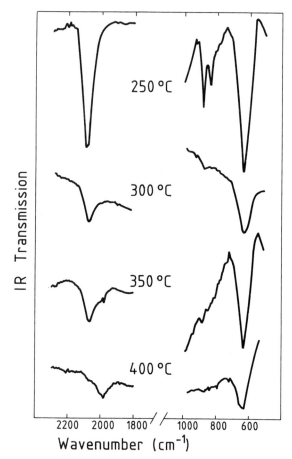

Fig. 8.11 FTIR transmission spectra for a-Si:H films grown at substrate temperatures of 250, 300, 350 and 400°C [17].

films [22]. The main difference seems to be a shift by about 50 K to lower surface temperatures. This is consistent with the fact that UV films possess a lower hydrogen content in this temperature region, and the formation of SiH_2 and $(SiH_2)_n$ groups depends on the hydrogen concentration. It has been suggested that the observation of the stretching band at 2090 cm^{-1} always indicates film inhomogeneities like microvoids [47].

8.4.3 Optical properties

Compared with crystalline silicon, which has a bandgap of 1.1 eV, the optical gap of a-Si:H depends on the hydrogen content and is considerably larger (e.g. about 1.7 eV for 10% hydrogen in the film). Due to structural disorder, the optical absorption is also considerably higher in the visible part of the spectrum (red, 1.7 eV; blue, 3.0 eV). Therefore the material is ideally suited for photosensitive devices applied in the visible part of the spectrum to transform photon energy.

By changing the hydrogen concentration in a-Si:H, the optical gap can be varied within a certain range. As can be seen from Fig. 8.12, this range is larger for IR laser-deposited material, consistent with the larger variation of the hydrogen content. Unfortunately, it does not seem to be possible to obtain high-quality material over a very large range of hydrogen concentrations. At low hydrogen concentrations, the density of gap states increases because the saturation of dangling bonds is no longer complete. For extensive hydrogen incorporation the heterogeneity of the amorphous network becomes substantial, and this leads to a deterioration in the optical and electrical properties of the material. Detrimental effects caused by extensive alloying have also been observed in other chemical systems. For example, incorporation of germanium into the network (a-SiGe:H) lowers the bandgap, whereas alloying with carbon (a-SiC:H) or nitrogen (a-SiN:H) widens the bandgap. Thus these alloys are of great interest for bandgap engineering. However, material with good properties can be obtained only within a limited concentration range. Incorporation of more than about 10–20% Ge, C or N leads to an increase in the density of gap states and a decrease in the efficiency of doping. In the case of the intrinsic material

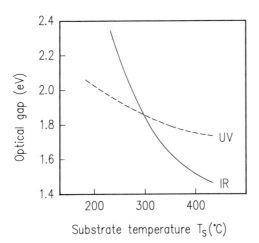

Fig. 8.12 Optical gap of a-Si:H films versus substrate temperature for IR [16] and UV [22] laser deposition.

a-Si : H the best film properties were achieved for a hydrogen content of about 10%, yielding a bandgap of 1.7–1.8 eV.

8.4.4 Electrical conductivity

Photoconductivities and dark conductivities have been extensively measured for a-Si : H films deposited by the parallel SiH_4–CW CO_2 laser and the Si_2H_6–ArF laser techniques. Very good agreement between different groups was obtained for the dependence of the conductivities on surface temperature in the case of IR laser deposition [15, 17]. On the other hand, the results reported by two groups on the conductivities of UV laser-deposited films deviate considerably [22, 25]. The photo- and dark conductivities obtained in [22] increase with the surface temperature in a way similar to that found in the IR experiments (see Fig. 8.13), whereas in [25] a report is given of a maximum of the photoconductivity near 250°C and of the dark conductivity near 300°C. The reason for these deviations in the temperature dependence is not clear.

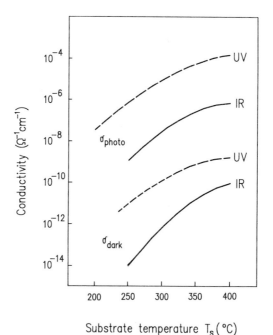

Substrate temperature T_s (°C)

Fig. 8.13 Photoconductivities and dark conductivities versus substrate temperature for IR [17] and UV [22] laser deposition at 1 Å s^{-1}.

The structure of a film and its properties such as electrical conductivity depend critically on the growth rate. Thus films grown at comparable rates should be compared in order to judge different deposition techniques. This is done in Fig. 8.13, where the photo- and dark conductivities are shown of films that were deposited by the IR and UV laser-deposition methods at a mean rate of about 1 Å s^{-1}. A similar temperature dependence is observed. However, the UV laser-deposited films possess considerably higher photo- and dark conductivities, indicating a superior film quality. A photoconductivity of about 10^{-4} Ω^{-1} cm^{-1} and a photo-to-dark conductivity ratio around 10^5 are among the best values obtained for this material. It is interesting to note that the conductivities of films grown with a mean rate of 4 Å s^{-1} in the UV experiment agree quite well with the conductivities shown in Fig. 8.13 for the IR experiment employing a lower growth rate of 1 Å s^{-1}. This clearly demonstrates the pronounced dependence of the film properties on the dynamics of the deposition process. Figure 8.14 illustrates this important point, showing as an example the dependence of the photoconductivity on the growth rate for films deposited from SiH$_4$ using a CO$_2$ laser. Extrapolation of the measured dependences to lower

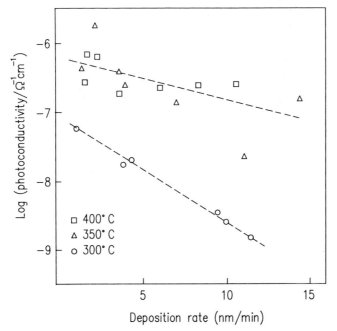

Fig. 8.14 Photoconductivity versus growth rate for substrate temperatures of 300, 350 and 400 [38].

growth rates leads to the conclusion that good values of the photoconductivity in the $10^{-4}\,\Omega^{-1}\,cm^{-1}$ range could be obtained at sufficiently low deposition rates.

8.5 DEPOSITION BY DIRECT SUBSTRATE ILLUMINATION

8.5.1 CO_2 laser CVD in perpendicular configuration

One of the first reports on laser CVD, published as early as 1978, treated silicon deposition from SiH_4 using a CW CO_2 laser for direct substrate illumination [48]. However, the deposit was not very well characterized in this work, and therefore it is not clear whether the films were polycrystalline or amorphous. That the deposition of hydrogen-containing amorphous films is possible for normal incidence with a CW CO_2 laser was shown in [49] for SiH_4 and in [50] for Si_2H_6 as the donor gas. In the latter experiment the deposition mechanism was thermal decomposition of Si_2H_6 at the laser-heated surface. Even for a relatively high surface temperature of $480°C$, effective hydrogen incorporation was observed mainly in the form of SiH bonding. At these high temperatures, large deposition rates above $150\,\text{Å}\,s^{-1}$ could be achieved. Unfortunately, the optical and electrical film properties were only partially characterized.

At substrate temperatures above $650°C$, it was possible to carry out low-temperature epitaxial growth employing a CO_2 laser and SiH_4 as the deposition gas [51]. The single-crystalline silicon was grown on a (100) oriented p-type Si wafer with a rate of $15\,\text{Å}\,s^{-1}$, employing perpendicular illumination.

8.5.2 Excimer laser CVD by direct surface irradiation

The formation of single-crystalline silicon films was observed at even lower temperatures (around $450°C$) by splitting an ArF laser beam for simultaneous parallel and perpendicular illumination [26]. Parallel irradiation resulted in the generation of amorphous films. With the additional perpendicular beam, polycrystalline and single-crystalline material could be grown at very low temperatures.

The deposition of a-Si:H films using pulsed excimer laser radiation and Si_2H_6 as the precursor has been studied for normal incidence and various other angles, as shown in Fig. 8.15. A large angle of $81°$ between laser beam and surface normal was chosen in [52], where mixtures of silanes served as donor gases and ArF and F_2 lasers were employed as light sources. At surface temperatures near $200°C$, the films contained $5-10\%$

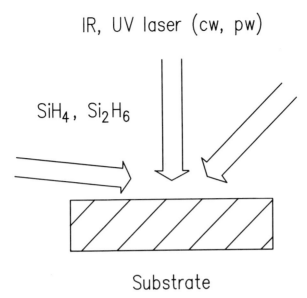

IR, UV laser (cw, pw)

SiH$_4$, Si$_2$H$_6$

Substrate

Fig. 8.15 Schematic of the different laser beam—surface configurations used for direct substrate illumination.

hydrogen, and relatively low photoconductivities (10^{-7}–10^{-8} Ω^{-1} cm^{-1}) and dark conductivities (5×10^{-9}–5×10^{-12} Ω^{-1} cm^{-1}) were measured. The deposition rate was varied between 1 and 10 Å s^{-1} in these experiments.

An angle of incidence of 30° was selected in [53] for the deposition from Si$_2$H$_6$ at surface temperatures between 100 and 350°C using an ArF laser. The large value and extensive variation of the optical gap (1.8–2.7 eV) indicate strong hydrogen incorporation under these conditions. For the photoconductivity the highest value of 2×10^{-6} Ω^{-1} cm^{-1} was found near 250°C, whereas the dark conductivity varied between 10^{-10} and 10^{-11} Ω^{-1} cm^{-1} in the whole temperature range studied.

The most complete study of a-Si:H deposition using an ArF laser in normal incidence and Si$_2$H$_6$ as the precursor was reported in [54]. The substrate temperature was varied between 35 and 350°C. In addition to the electrical film properties, optical parameters were evaluated from the optical absorption spectra showing Tauc, Urbach and Urbach shoulder regions. Good material properties were obtained for a substrate temperature of 280°C. The density of gap states was 3×10^{16} cm^{-3}, the photoconductivity 1.25×10^{-5} Ω^{-1} cm^{-1} and the photo-to-dark conductivity ratio 5×10^{5} at this deposition temperature. An optical bandgap of 1.82 eV and a refractive index of 3.2 were observed. The Urbach edge slope parameter had a constant value of 0.3 eV, whereas the slope parameter of the shoulder region increased from 0.1 to 1.0 eV in the temperature range investigated. At 260°C the two slope parameters possessed identical values, indicating a transition from insulator-like to semiconductor-like behaviour.

8.6 OPTIMIZED MATERIAL FOR APPLICATIONS

8.6.1 Structure and stability

From the detailed discussion given above, it can be concluded that the best film properties can be expected from chemically and structurally homogeneous material. Besides the bending mode peak at 635 cm^{-1}, the IR spectrum of such a film should show only the band of the SiH stretching mode at 2000 cm^{-1}. This is an empirical formulation of a structure–property relationship extracted from many experiments. Below a certain surface temperature, which depends on the deposition method employed, the hydrogen concentration in the film becomes so high that SiH$_2$ groups are incorporated in the material. The connection between structural inhomogeneities and the occurrence of SiH$_2$ groups and polysilane structures is not understood and must be studied in more detail.

Besides possible detrimental effects of SiH$_2$ groups on the structure of the network, these species possess a lower stability than SiH groups, as shown at least for the surface species in [42]. Thus these groups also limit the thermal stability of the material. They may also contribute to the Staebler–Wronski effect insofar as weak hydrogen sites that are clustered in microvoids and inner surfaces are involved. Staebler and Wronski [55] observed a decrease in the photo- and dark conductivities of plasma-deposited a-Si:H films caused by long exposure to light, which is reversible by annealing.

This suggestion is supported by recent annealing experiments performed with a Q-switched Nd:YAG laser on a-Si:H films prepared by glow discharge deposition [56]. For a-Si:H samples deposited at higher growth rates the absorption intensity of the 2090 cm^{-1} band due to SiH$_2$ bonds decreased appreciably, whereas the intensity of the SiH band at 2000 cm^{-1} showed only a small change. On the other hand, both bands decreased upon annealing of samples grown at a lower deposition rate, which are expected to have a more homogeneous structure.

8.6.2 Electronic properties

In the amorphous material the periodic arrangement of constituents in the network prevailing in crystalline material is heavily distorted. These imperfections and structural disorder reduce the mean free path and the mobility of the charge carriers. If the density of gap states is below 10^{16} cm^{-3} eV^{-1}, the tail states at the bandgap edges constitute the predominant traps for the charge carriers. These states are responsible for the transition from extended or nonlocalized to localized states, and limit the movement of carriers across the material when an electric field is applied and carriers are generated by absorption of photons. To come to a better understanding of the relationship between structural disorder and carrier movement, it is important to measure charge carrier mobilities and lifetimes in the material. This has not been done extensively for laser-deposited material.

As can be seen from Fig. 8.3, the density of states decreases not to zero at the band edges E_c and E_v ("forbidden gap"), as in the ideal cases, but reduces more or less slowly to the density of gap states. Therefore the distribution of tail states near the band edge should be investigated in more detail. This can be done by determining the Urbach energy from the slope of the exponential absorption curve. The absorption spectrum can be obtained by the constant-photocurrent method (CPM) [57]. Only for the case of direct laser illumination of the surface is information on slope parameters available, as discussed above.

8.6.3   Other methods

The deposition of optimized material of device quality can be achieved by several methods. Results obtained by plasma CVD will be discussed here for comparison with laser methods. Photo-CVD of a-Si:H using lamps is reviewed by Milne and Robertson in Chapter 9.

Plasma CVD is normally used to manufacture a-Si:H films for device applications. It is one of the most complex deposition processes. Nevertheless, a-Si:H films with properties that are state of the art were obtained by plasma CVD several years ago. Typical material constants for device-grade glow discharge films are: spin density $< 10^{16}$ cm^{-3}, photoconductivity $10^{-3}\,\Omega^{-1}$ cm^{-1}, dark conductivity $3 \times 10^{-9}\,\Omega^{-1}$ cm^{-1}, optical bandgap 1.7–1.8 eV, conductivity activation energy 0.76 eV, electron mobility 0.5–1.0 cm^2 V^{-1} s^{-1}, hole mobility 10^{-3}–5×10^{-3} cm^2 V^{-1} s^{-1}, carrier diffusion length > 1 μm, refractive index 3.43 and density 2.2 g cm^{-3} [58].

A related technique developed recently is remote plasma-enhanced CVD (PERCVD), which has been applied to the growth of intrinsic and doped a-Si:H films [59]. The films were grown with a relatively low rate of 0.13 Å s^{-1}, and monohydride groups dominated for surface temperatures above 100°C. The defect state concentrations were about 10^{16} cm^{-3} at a surface temperature of $100^{\circ}C$ and 5×10^{15} cm^{-3} at $235^{\circ}C$. The quantum efficiency–mobility–lifetime product was $> 10^{-5}$ cm^2 V^{-1} at $235^{\circ}C$.

The examples given here and in Chapter 9 clearly show that, with photo-CVD including lasers as light sources, a-Si:H films of high quality have been deposited. Unfortunately, film characterization was not complete in most cases.

8.7 CONCLUSIONS

8.7.1 Scenarios for network formation

The different deposition methods discussed in this review provide characteristic scenarios for the growth of a-Si:H networks. It is interesting to compare these scenarios, despite the fact that our mechanistic understanding of the details, especially at the surface, is very limited.

The softest method seems to be the decomposition of silanes with a CW CO_2 laser (photon energy approximately 0.1 eV) via the lowest-energy channels. Here the SiH_2 radical and higher silanes formed by secondary reactions of this radical play the dominant role. The hydrogen content of the film increases rapidly with decreasing surface temperature, and this is connected with an increasing inhomogeneity of the amorphous network and degrading of film properties for device applications.

Processing with an ArF laser (photon energy 6.4 eV) opens several high-energy channels in the chemistry of silanes. Thus species such as SiH and SiH_3 may be generated. However, they have to diffuse to the surface in the case of a parallel laser beam–surface configuration. Different active species may be responsible for the reduction of hydrogen incorporation into the film, shifting the occurrence of SiH_2 peaks in the IR spectrum to smaller surface temperatures [22]. This leads to improved film properties.

With the photo-CVD technique, channels with even higher energy are accessible in the silane chemistry, for example by employing a Xe lamp (photon energy 8.4 eV). Using direct illumination of the surface, these photons produce the highly active precursors directly above the surface. Under these conditions, the incorporation of hydrogen and the formation of SiH_2 groups in the network may be further reduced. More details on UV lamp deposition using high photon energies for dissociation can be found in Chapter 9.

Well characterized material with very good film properties has been deposited by the remote plasma-enhanced CVD method [59]. Here helium is excited in an RF plasma separated from the substrate, and excited helium atoms control the chemistry (He* metastable energy approximately 20 eV). It has been suggested by the authors that the main effect may be the forma-

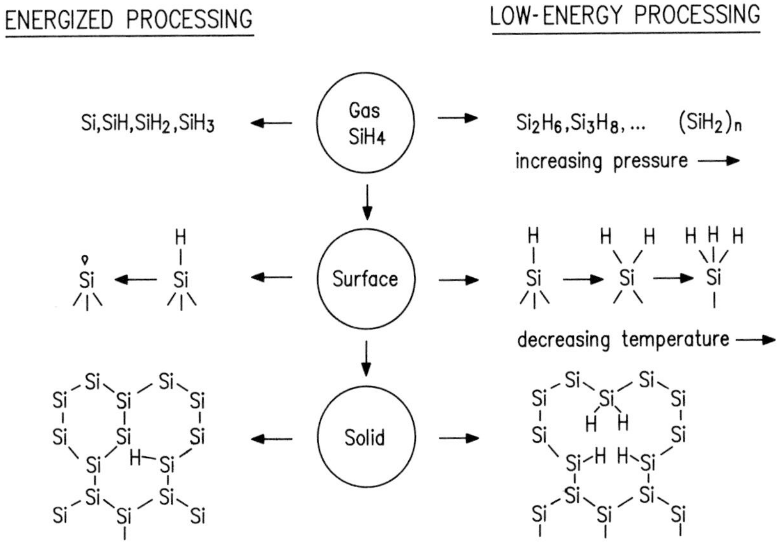

ENERGIZED PROCESSING   LOW-ENERGY PROCESSING

Fig. 8.16 Illustration of important steps in energized and low-energy processing of a-Si : H films.

tion of SiH_3 radicals and the restriction of polysilane formation. However, a dominant role of ions cannot be excluded. In these films the formation of SiH_2 groups starts around $100°C$ [59].

This discussion clearly shows that it may be necessary to avoid charged particles such as ions and electrons. However, species with high energy or chemical reactivity are needed to restrict hydrogen incorporation into the film and to grow a homogeneous network at low substrate temperatures. Thus it is not the softest methods but rather the deposition techniques using energetic particles in a controlled way that give the best films.

Figure 8.16 gives a comparison between the main aspects of energized and low-energy processing. The production of radicals (e.g. SiH_2) in the gas phase starts a series of thermally activated reactions leading to higher silanes and powder formation if the pressure is too high. These secondary reaction processes can only be restricted when active species are present, modifying the nature and number of hydrogen bonds.

8.7.2   Role of laser methods

The growth of an amorphous network by chemical vapour deposition is an extremely complicated process. Therefore it is not surprising that our present understanding of the critical processes governing gas phase and surface chemistry (e.g. hydrogen desorption and incorporation) is not satisfactory. Considerable improvements have been made by empirical optimization of device fabrication. However, trial and error may not be the best procedure to develop techniques for the production of high-quality materials.

Laser methods offer unique possibilities for performing CVD experiments under well defined conditions and for *in situ* analysis of the complicated processes occurring in such an experiment. They can provide the specific data needed to develop a chemical mechanism for the deposition process and to come to a better understanding of the fundamental processes involved. The goal is the development of realistic models describing all relevant processes taking place in the CVD reactor, including transport processes and gas phase and surface reactions. Such an approach provides a broader view of the problems, and therefore may lead to new ways for growing tailored amorphous networks with specific properties. The results already obtained by laser methods clearly show that these techniques will play a major role in the development of this field.

ACKNOWLEDGEMENTS

Financial support of our work on silicon deposition by the German Ministry of Research and Technology (BMFT) under Contract 13 N 5363/8 and by the Fonds der Chemischen Industrie is gratefully acknowledged.

REFERENCES

[1] D. Bäuerle, *Chemical Processing with Lasers* (Springer-Verlag, Berlin, 1986).
[2] I. W. Boyd, *Laser Processing of Thin Films and Microstructures* (Springer-Verlag, Berlin, 1987).
[3] K. G. Ibbs and R. M. Osgood, *Laser Chemical Processing for Microelectronics* (Cambridge University Press, 1989).
[4] D. J. Ehrlich and J. Y. Tsao, *Laser Microfabrication* (Academic Press, New York, 1989).
[5] R. Bilenchi, I. Gioninoni and M. Musci, *J. Appl. Phys.* **53**, 6479 (1982).
[6] J. I. Pankove (ed.), *Semiconductors and Semimetals*, Vol. 21: *Hydrogenated Amorphous Silicon*, Parts A–D (Academic Press, London, 1984).
[7] J. M. Jasinski, B. S. Meyerson and B. A. Scott, *Ann. Rev. Phys. Chem.* **38**, 109 (1987).
[8] P. Hess, *Spectrochim. Acta* **46A**, 489 (1990).
[9] R. Bilenchi, I. Gioninoni and M. Musci, *Mater. Res. Soc. Symp. Proc.* **17**, 199 (1983).
[10] R. Bilenchi, M. Musci and R. Murri, *Proc. SPIE* **459**, 61 (1984).
[11] I. Gianinoni and M. Musci, *J. Non-Cryst. Solids* **77/78**, 743 (1985).

[12] M. Meunier, T. R. Gattuso, D. Adler and J. S. Haggerty, *Appl. Phys. Lett.* **43**, 273 (1983).
[13] T. R. Gattuso, M. Meunier, D. Adler and J. S. Haggerty, *Mater. Res. Soc. Symp. Proc.* **17**, 215 (1983).
[14] J. H. Flint, M. Meunier, D. Adler and J. S. Haggerty, *Proc. SPIE* **459**, 66 (1984).
[15] M. Meunier, J. H. Flint, J. S. Haggerty and D. Adler, *J. Appl. Phys.* **62**, 2812 and 2822 (1987).
[16] D. Metzger, K. Hesch and P. Hess, *Appl. Phys.* **A45**, 345 (1988).
[17] K. Hesch, P. Hess, H. Oetzmann and C. Schmidt, *Mater. Res. Soc. Symp. Proc.* **131**, 495 (1989).
[18] K. Hesch, P. Hess, H. Oetzmann and C. Schmidt, *Appl. Surf. Sci.* **36**, 81 (1989).
[19] P. González, M. D. Fernández, B. Leon and M. Pérez-Amor, in *Proceedings of the 9th EC Photovoltaic Solar Energy Conference*, p. 1017 (Kluwer, Dordrecht, 1989).
[20] E. Golusda, R. Lange, G. Mollekopf and H. Stafast, *Statusreport Photovoltaik, Projektträger Biologie, Energie, Ökologie (BEO), Forschungszentrum Jülich* (1990).
[21] Y. Pauleau, D. Tonneau and F. Auvert, in *Laser Processing and Diagnostics* (ed. D. Bäuerle), p. 215 (Springer-Verlag, Berlin, 1984).
[22] A. Yamada, M. Konagai and K. Takahashi, *Japan. J. Appl. Phys.* **24**, 1586 (1985).
[23] D. Eres, D. H. Lowndes, D. B. Geohegan and D. N. Mashburn, *Mater. Res. Soc. Symp. Proc.* **101**, 355 (1988).
[24] D. Eres, D. B. Geohegan, D. H. Lowndes and D. N. Mashburn, *Appl. Surf. Sci.* **36**, 70 (1989).
[25] T. R. Dietrich, S. Chiussi, H. Stafast and F. J. Comes, *Appl. Phys.* **A48**, 405 (1989).
[26] M. Murahara and K. Toyoda, in *Laser Processing and Diagnostics* (ed. D. Bäuerle), p. 252 (Springer-Verlag, Berlin, 1984).
[27] T. Taguchi, M. Morikawa, Y. Hiratsuka and K. Toyoda, *Appl. Phys. Lett.* **49**, 971 (1986).
[28] V. Tavitian, C. J. Kiely, D. B. Geohegan and J. G. Eden, *Appl. Phys. Lett.* **52**, 1710 (1988).
[29] H. Stafast, *Appl. Phys.* **A45**, 93 (1988).
[30] I. P Herman, *Chem. Rev.* **89**, 1323 (1989).
[31] T. F. Deutsch, *J. Chem. Phys.* **70**, 1187 (1979).
[32] C. G. Newman, H. E. O'Neal, M. A. Ring, F. Leska and N. Shipley, *Int. J. Chem. Kinet.* **11**, 1167 (1979).
[33] P. Ho, M. E. Coltrin, J. S. Binkley and C. F. Melius, *J. Phys. Chem.* **89**, 4647 (1985).
[34] S. K. Shin and J. L. Beauchamp, *J. Phys. Chem.* **90**, 1507 (1986).
[35] F. J. Kampas and R. W. Griffith, *Appl. Phys. Lett.* **39**, 407 (1981).
[36] G. Inoue and M. Suzuki, *Chem. Phys. Lett.* **122**, 361 (1985).
[37] J. M. Jasinski and J. O. Chu, *J. Chem. Phys.* **88**, 1678 (1988).
[38] K. Hesch, P. Hess, H. Oetzmann and C. Schmidt, *Appl. Surf. Sci.* **46**, 233 (1990).
[39] S. M. Gates, C. M. Greenlief, D. B. Beach and R. R. Kunz, *Chem. Phys. Lett.* **154**, 505 (1989).

[40] S. M. Gates, B. A. Scott, D. B. Beach, R. Imbihl and J. E. Demuth, *J. Vac. Sci. Technol.* **A5**, 628 (1987).
[41] S. K. Kulkarni, S. M. Gates, C. M. Greelief and H. H. Sawin, *J. Vac. Sci. Technol.* **A8**, 2956 (1990).
[42] P. Gupta, V. L. Colvin and S. M. George, *Phys. Rev.* **B37**, 8234 (1988).
[43] Y. Toyoshima, K. Arai, A. Matsuda and K. Tanaka, *Appl. Phys. Lett.* **56**, 1540 (1990).
[44] K. Hesch and P. Hess, To be published.
[45] S. Patnaik and R. A. Brown, *J. Electrochem. Soc.* **135**, 697 (1988).
[46] L. Konstantinov, R. Nowak and P. Hess, *Appl. Surf. Sci.* **46**, 102 (1990).
[47] W. Beyer and H. Wagner, *J. Non-Cryst. Solids* **59/60**, 161 (1983).
[48] C. P. Christensen and K. M. Lakin, *Appl. Phys. Lett.* **32**, 254 (1978).
[49] M. Hanabusa, S. Moriyama and H. Kikuchi, *Thin Solid Films* **107**, 227 (1983).
[50] T. Iwanaga and M. Hanabusa, *Japan J. Appl. Phys.* **23**, L473 (1984).
[51] T. Meguro, Y. Ishihara, T. Itoh and H. Tashiro, *Japan J. Appl. Phys.* **25**, 524 (1986).
[52] Y. Toyoshima, K. Kumata, U. Itoh and A. Matsuda, *Appl. Phys. Lett.* **51**, 1925 (1987).
[53] A. Yoshikawa and S. Yamaga, *Japan J. Appl. Phys.* **23**, L91 (1984).
[54] H. Zarnani, H. Demiryont and G. J. Collins, *J. Appl. Phys.* **60**, 2523 (1986).
[55] D. L. Staebler and C. R. Wronski, *J. Appl. Phys.* **51**, 3262 (1980).
[56] B. Zhong, Z. Lu and R. Tang, *J. Non-Cryst. Solids* **112**, 85 (1989).
[57] M. Vaněček, J. Kočka, J. Stuchlik, Z. Kožišek, O. Stika and A. Třiska, *Solar Energy Mater.* **8**, 411 (1982).
[58] M. Hirose, in [6], Part A, p. 9.
[59] G. N. Parsons, D. V. Tsu and G. Lucovsky, *J. Vac. Sci. Technol.* **A6**, 1912 (1988).

9 UV Lamp Deposition of a-Si:H and Related Compounds

W. I. MILNE and P. A. ROBERTSON*

Cambridge University Engineering Department, UK

9.1 INTRODUCTION

Over the past 20 years, hydrogenated amorphous silicon (a-Si:H) has emerged to become an economically viable semiconducting material for electronic applications. Its main use has been in the manufacture of terrestrial solar cells [1]; however, more recently it has become increasingly important in large-area active matrix addressed liquid crystal displays, where it is used to produce the switch that must be included in each picture element. This switch usually, but not always, takes the form of a thin-film transistor (TFT) [2, 3].

Good quality a-Si:H is conventionally produced using a plasma enhanced chemical vapour deposition (PECVD) method. However, it is felt by some workers that the bombardment of the growing films by energetic species in the PECVD process could cause damage to the films. This will obviously limit the film quality and adversely affect the properties of interfaces required in the above devices.

Numerous techniques including remote plasma CVD, hydrogen radical CVD [4] and various biased and gridded PECVD methods have been used in an attempt to minimize the potential damage, but the most effective method is to eliminate the plasma entirely by using a photo-CVD technique. In this the source gases are dissociated by means of UV light. The photon energies required to dissociate the reactant gases are insufficient to cause significant ionization, and this method therefore has the potential to produce material with a lower defect state density, and will give cleaner interfaces.

Using excimer and CO_2 lasers, good-quality a-Si:H has been produced

---

* Permanent address: P. A. Technology, Melbourn, Hertfordshire, UK.

Photochemical Processing of Electronic Materials
ISBN 0-12-121740-X

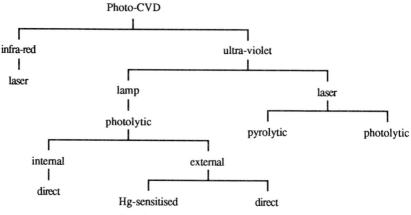

Fig. 9.1 Photo-CVD techniques.

from monosilane and disilane (see Chapter 8), but the bulk of the published work on the photo-CVD of a-Si : H and related compounds has involved the decomposition of silanes using incoherent UV light sources. Such a process can occur either by direct absorption of photons in the vacuum region, or by the absorption of photons by other atoms, which then react with the silane molecules, i.e. by sensitized reactions. The various possible methods are illustrated in Fig. 9.1.

This chapter will review the different UV lamp photo-CVD techniques that have been utilized for the deposition of a-Si : H and related compounds.

9.2 MERCURY-SENSITIZED DEPOSITION

The most common a-Si : H photo-CVD method utilizes a low-pressure mercury discharge lamp with mercury sensitization [5], where a small amount of mercury vapour is admitted to the reaction chamber with the source gases. A schematic of such a system is shown in Fig. 9.2. The UV light (185 and 254 nm) is transmitted into the reaction vessel through a fused silica window (transparent to $\lambda > 160$ nm) and couples into the mercury-saturated reaction gas, leading to deposition of a-Si : H on heated substrates.

Using monosilane as the reactant gas, several possible reaction paths have been proposed to explain the primary processes that take place leading to the deposition of a-Si : H:

$$Hg^0(^1S_0) + h\nu(185 \text{ nm}) \rightarrow Hg^*(^1P_1) \qquad (9.1)$$

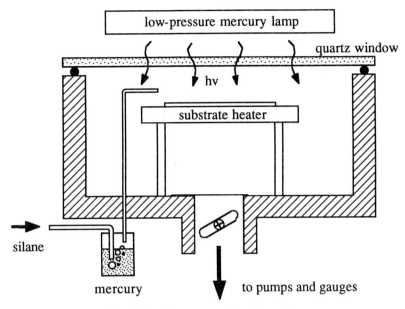

Fig. 9.2 Mercury-sensitized reactor.

$$Hg^0(^1S_0) + h\nu(254 \text{ nm}) \rightarrow Hg^*(^3P_1) \qquad (9.2)$$

$$Hg^*(^{1,3}P_1) + SiH_4 \rightarrow Si + 2H_2 + Hg^0(^1S_0) \qquad (9.3)$$

Reactions (9.1) and (9.2) represent the absorption of the UV light by the ground state mercury atoms admitted to the reaction chamber. These are excited by the photons into energetic states, which transfer their energy to the silane molecules (reaction (9.3)). This takes place through several steps:

$$Hg^* + SiH_4 \rightarrow SiH_2 + 2H + Hg^0 \qquad (9.4)$$

$$Hg^* + SiH_4 \rightarrow SiH_3 + H + Hg^0 \qquad (9.5)$$

$$H + SiH_4 \rightarrow SiH_3 + H_2 \qquad (9.6)$$

$$Hg^* + H_2 \rightarrow 2H + Hg^0 \qquad (9.7)$$

$$\rightarrow H + HgH \qquad (9.8)$$

$$SiH_2 + SiH_4 \rightarrow Si_2H_6 \qquad (9.9)$$

$$SiH_2 + \text{surface} \rightarrow \text{a-Si:H (solid)} \qquad (9.10)$$

$$SiH_3 + \text{surface} \rightarrow \text{a-Si:H (solid)} \qquad (9.11)$$

Reactions (9.4) and (9.5) represent quenching of the energetic mercury atoms by the silane molecules, resulting in the formation of silyl radicals. The radicals either form solid a-Si : H from surface reactions (9.10) and (9.11) or form higher silanes from gas phase reactions such as (9.9). Quantum yields exceeding unity for silicon atoms incorporated into the film may be accounted for by reactions (9.6) and (9.7). The role of hydrogen radicals is also very important in determining the deposition of a-Si : H films, and reaction (9.6) is known to occur very rapidly.

The rates of quenching in reactions (9.4), (9.5), and (9.7) are determined by the quenching cross-section σ^2 of the molecules for Hg^* atoms. The approximate values of the quenching cross-section for $Hg^*(^3p_1)$ atoms for relevant gas molecules are shown in Table 9.1 [6–8].

Table 9.1 Quenching cross-sections of reaction gas molecules [45].

	σ^2 (Å^2)
H_2	10
SiH_4	26
Si_2H_6	80
N_2O	21
O_2	17
NH_3	4.2

Hence the deposition rate obtained from disilane may be several times that of monosilane for otherwise similar conditions. It has also been reported [9] that the use of $Hg^*(^1p_1)$ atoms, from the mercury 185 nm line, results in a much increased (> 100 times) photon efficiency compared with that obtained from $Hg^*(^3p_1)$ atoms with the 254 nm line.

Films of SiO_x and SiN_x have also been deposited using mercury-sensitized photolytic reactions [10]. In these cases a mixture of SiH_4 and O_2, N_2O or NH_3 is used, resulting in the decomposition of each reactant gas, with a rate dependent on the quenching cross-section and partial pressure. Reactions (9.12) to (9.14) represent the decomposition of N_2O, O_2 and NH_3 respectively, producing the radicals responsible for film deposition:

$$N_2O + Hg^* \rightarrow N_2 + O(^3p) + Hg^0 \qquad (9.12)$$

$$O_2 + Hg^* \rightarrow 2O(^3p) + Hg^0 \qquad (9.13)$$

$$NH_3 + Hg^* \rightarrow NH_2 + H + Hg^0 \qquad (9.14)$$

9.3 DIRECT PHOTO-CVD

Direct photo-CVD systems are very similar in construction to those used for sensitized depositions. A system designed in our laboratory is shown schematically in Fig. 9.3. To ensure the direct photodeposition of a-Si : H from monosilane, it can be seen from Table 9.2 that the minimum activation energy for dissociation is 2.43 eV. Hence at least this amount of

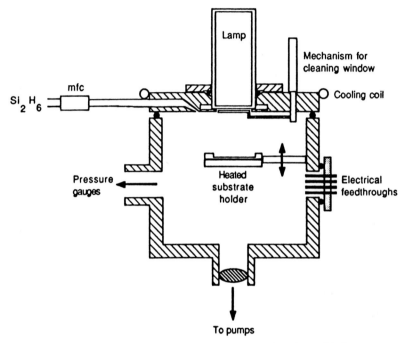

Fig. 9.3 Schematic of photo-CVD reactor chamber with lamp.

Table 9.2 Primary gas phase reactions of SiH₄.

	ΔH (eV)	E_A (eV)
(1) $SiH_4 \rightarrow SiH_3^+ + H + e^-$	12.3	12.3
(2) $SiH_4 \rightarrow SiH_2^+ + H_2 + e^-$	11.4	11.9
(3) $SiH_4 \rightarrow SiH_2 + 2H$	6.72	6.74
(4) $SiH_4 \rightarrow SiH + H_2 + H$	5.68	—
(5) $SiH_4 \rightarrow Si + 2H_2$	4.20	—
(6) $SiH_4 \rightarrow SiH_3 + H$	3.86	3.86
(7) $SiH_4 \rightarrow SiH_2 + 2H$	2.08	2.43

energy must be transferred to each molecule to effect its decomposition. If this energy transfer is to be accomplished directly then the optical absorption coefficient at the exciting wavelength must be high enough to ensure adequate energy coupling in the deposition system.

As shown in Fig. 9.4, the direct photo-enhanced decomposition of a-Si : H is only possible using photon wavelengths below 160 nm, where the gas begins to absorb significantly. The low-pressure mercury lamps used in the sensitized deposition usually have a high electrical to optical power conversion efficiency, but unfortunately their shortest emission wavelength is 185 nm. As is obvious from Fig. 9.4, these lamps cannot be used for the direct photolysis of SiH_4, but have been successfully used for the deposition of a-Si : H from Si_2H_6 and Si_3H_8. At present, however, these gases are very much more expensive than SiH_4 (e.g. £1300 kg^{-1} for SiH_4 compared with £20 000 kg^{-1} for Si_2H_6 in 1990).

Gas discharge sources that emit in the hard-UV region are listed in Table 9.3, together with an approximate typical conversion efficiency and emitted photon wavelength range. The conversion efficiency of these short-wavelength VUV sources is very low compared with the highly efficient low-pressure mercury discharge lamp. The lamp window may also be a source of attenuation to the VUV. There are very few window materials transparent to photons in the far-UV that are also stable under intense UV exposure and/or exposure to atmospheric moisture (see Table 9.4).

Thus few deposition systems have been designed for direct photolysis of SiH_4 because of the difficulty of producing a high-output far-UV lamp, but

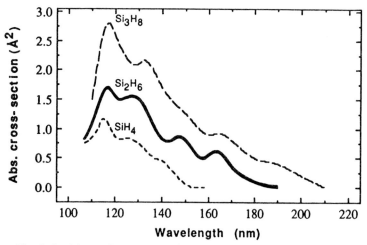

Fig. 9.4 Absorption cross-sections for SiH_4, Si_2H_6 and Si_3H_8.

Table 9.3 Vacuum-ultraviolet sources.

Discharge medium	Wavelength (nm)	Efficiency (%)
Mercury (low-pressure)	185, 254	50
Hydrogen, deuterium	100–300	0.33, 0.05
Nitrogen	120–320	0.08
Argon	105–135	≈ 0.01
Krypton	125–170	≈ 0.01
Xenon	150–180	≈ 0.01

Table 9.4 Properties of ultraviolet window materials.

Material	Short-wavelength cut-off (nm)	Comments
LiF	105	Expensive ($£10$ cm^{-2}), poor moisture and and UV stability
MgF$_2$	115	Very expensive ($£100$ cm^{-2}), good posture and UV stability
Saphire	142	Very expensive, stable
Fused silica, "suprasil"	160	Inexpensive ($£1$ cm^{-2}), opaque to extreme vacuum UV, stable
Quartz	180	Inexpensive ($£1$ cm^{-2}), opaque to extreme vacuum UV, stable
"Pyrex" glass	320	Inexpensive, stable, transmits near-UV only

Perkins *et al.* [13] have studied the photolytic decomposition of SiH$_4$ (and Si$_2$H$_6$) using a xenon microwave discharge lamp and a LiF window.

The primary photochemical dissociation reactions of SiH$_4$ at 147 nm [11] are reported to be

$$SiH_4 + h\nu \rightarrow SiH_2 + 2H \qquad (9.15)$$

$$SiH_4 + h\nu \rightarrow SiH_3 + H \qquad (9.16)$$

Other primary reactions forming excited luminescent products have also been reported at shorter wavelengths, but the total quantum yield of these products is only a few percent [12]:

$$SiH_4 + H \rightarrow SiH_3 + H_2 \qquad (9.17)$$

The secondary reaction (9.17) involving the formation of atomic hydrogen by reactions (9.15) and (9.16) may account for the dissociation of over four silane molecules for each photon absorbed. This reaction is known to occur very rapidly [13].

In this instance the reaction chain leading to the deposition of a-Si : H is very complex and will vary with experimental conditions, such as excitation wavelength, gas pressure and temperature. However, it is very likely that the principal silyl radical leading to the formation of a-Si : H in a photo-CVD system at 147 nm is SiH_3 due to the rapid depletion of SiH_2 radicals by the following secondary gas phase reactions [12]:

$$SiH_2 + SiH_4 \rightarrow Si_2H_6 \tag{9.18}$$

$$SiH_2 + SiH_4 \rightarrow SiH_3SiH + H_2 \tag{9.19}$$

$$SiH_2 + H_2 \rightarrow SiH_4 \tag{9.20}$$

Reactions (9.21) to (9.24) represent possible secondary gas phase reactions of SiH_3 radicals to form higher silanes; however, these inter-radical reactions are much less probable than the SiH_2 reactions (9.18) and (9.19), which involve molecular silane:

$$SiH_3 + SiH_3 \rightarrow Si_2H_6 \tag{9.21}$$

$$SiH_3 + SiH_3 \rightarrow SiH_2 + SiH_4 \tag{9.22}$$

$$SiH_3 + SiH_3 \rightarrow SiH_3SiH + H_2 \tag{9.23}$$

$$SiH_3SiH + SiH_4 \rightarrow Si_3H_8 \tag{9.24}$$

There are no significant reactions between SiH_3 radicals and SiH_4 or H_2.

The surface reactions that lead to polymerization of the solid a-Si : H are still under debate. It has been suggested [14] that hydrogen bonded to a surface silicon atom may be etched by a monoradical (e.g. SiH_3), to leave a dangling bond. Alternatively, a diradical such as SiH_2 may displace the hydrogen atom and bond to the silicon. These steps are illustrated in Fig. 9.5.

These reactions are thermoneutral and occur with a very low activation energy, <0.2 eV. Following step (a) in Fig. 9.5, adjacent dangling bonds

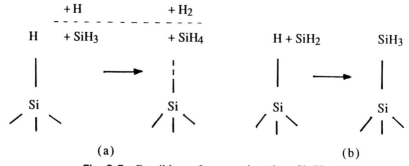

Fig. 9.5 Possible surface reactions in a-Si : H.

cross-link the structure or capture other radicals to continue the growth process. Alternatively, following step (b), hydrogen may be scavenged by radicals or eliminated from energetically activated pendant groups. The resulting dangling bonds may then cross-link or bond to another radical. In practice a combination of these schemes is likely to occur, with the kinetics of hydrogen elimination and cross-linking depending on the nature and energies of the reacting species. These in turn will depend on the deposition parameters such as substrate temperature, gas pressure and excitation source.

The direct photo-enhanced CVD of SiO_x and SiN_x thin films has been achieved from reaction gas mixtures of SiH_4 or Si_2H_6 and O_2, N_2O or NH_3. Ultraviolet sources employed include low-pressure mercury [10, 15, 16] and deuterium [17] lamps.

The primary photolytic reaction in the production of nitride films is the photolysis of NH_3 represented by the reaction

$$NH_3 + h\nu \rightarrow NH_2 + H \qquad (9.25)$$

Both fragmentation radicals may react with silane and silyl radicals to form nitride groups or other silyl radicals, but neither species reacts readily with ammonia [6].

The primary step in the production of SiO_2 films is the generation of oxygen radicals from the photolysis of the oxidant gas (usually N_2O or O_2). Oxygen is not particularly suitable for this application, because of its ability to react spontaneously with molecular silane, leading to porosity in the deposited film. This is usually reflected in the high etch rate of films deposited using oxygen. Nitrous oxide N_2O is a better choice, as it is relatively unreactive and has a higher optical cross-section at 185 nm than oxygen: i.e. 1.5×10^{-3} Å^2 compared with 4×10^{-4} Å^2 [8]. Another possible gas that has been investigated is NO_2 [18], which has an absorption cross-section of 0.02 Å^2 at 185 nm and does not react spontaneously with silane.

The primary photolytic reactions at 185 nm for these gases are [16]

$$O_2 + h\nu \rightarrow 2O(^3P) \qquad (9.26)$$

$$N_2O + h\nu \rightarrow N_2 + O(^1D) \qquad (9.27)$$

$$NO_2 + h\nu \rightarrow NO + O(^3P, {}^1D) \qquad (9.28)$$

The secondary reactions leading to the formation of the solid SiO_x are very complex, including those involving oxygen and silyl radicals with both molecular and radical species. A branching chain reaction has been proposed [19] that is initiated by oxygen radicals:

$$SiH_4 + O \rightarrow SiH_2 + H_2O \qquad (9.29)$$

$$SiH_4 + O \rightarrow SiH_3 + OH \qquad (9.30)$$

and these are followed by branching reactions:

$$SiH_2 + O_2 \rightarrow SiH_2O + O \qquad (9.31)$$

$$SiH_2 + O \rightarrow SiH_2O \qquad (9.32)$$

$$OH + SiH_4 \rightarrow SiH_3 + H_2O \qquad (9.33)$$

$$SiH_3 + O \rightarrow SiH_2O + H \qquad (9.34)$$

$$SiH_2O + O_2 \rightarrow SiH_2O_2 + O \qquad (9.35)$$

and by the regenerating terminating reaction

$$SiH_2O_2 + O_2 \rightarrow SiO_2 + H_2O + O \qquad (9.36)$$

SiH_2O is a silanane, which is very likely to undergo rapid secondary reactions [20]. If an SiH_xO_y radical is absorbed into the growing film surface, the hydrogen may be incorporated into the film or may combine with oxygen and be driven off as water at high temperatures. Deposited SiO_x films usually contain several atomic percent of hydrogen.

As well as the abovementioned problem regarding the choice of suitable VUV lamps and UV-transparent window material, the other major problem that besets photo-CVD systems is the deposition of material onto the inner surface of the window itself. This greatly attenuates the flux of VUV photons, and hence the deposition rate will decrease continuously. There have been several schemes proposed to reduce this effect, such as the use of a layer of perfluoroether oil (Fomblin), which reduces the sticking coefficient of the radicals from 0.5 for the a-Si : H surface to 0.001 for the oil surface [21]. The other methods include directing an inert gas (He) stream [22] or a reactive etching XeF_2 jet [23] at the window, using a wiper blade system [24], or even the insertion of a movable flexible polymer curtain running between the window and reaction region [25].

These antifouling systems have all been applied to mercury lamp systems at 185 and 254 nm for sensitized reactions and the direct photolysis of Si_2H_6. However, at the shorter wavelengths required for direct SiH_4 photolysis, the use of Fomblin is unlikely to be suitable due to attenuation of the high-energy photons by the oil or polymer film. We have observed a 40% decrease in transmission through an MgF_2 window in the 115–160 nm region when it has a thin layer of Fomblin applied to its surface. Also, dissociation products from the oil have been detected in our mass spectrometer data. None of these products were detected when using radiation of wavelength > 160 nm. Thus the use of Fomblin in a direct photolysis system that utilizes photon wavelengths < 160 nm will almost certainly result in contamination of the deposited film by dissociation products. None of the other methods has been shown to be entirely satisfactory either, and therefore we have developed a windowless deposition system for both a-Si:H and insulator deposition [26].

9.4   DIRECT PHOTO-CVD USING AN INTERNAL LAMP

With such a system, it is possible to use monosilane as the silicon source gas and nitrous oxide as the oxidant. As shown in Fig. 9.6, the photon source is an internal discharge lamp operating with either hydrogen or nitrogen for the deposition of a-Si:H or SiO_x respectively. A detailed cross section of the lamp is shown in Fig. 9.7. The discharge lamp itself consists of two concentric closed-end metal cylinders, which form the anode and cathode, and a tungsten filament, which acts as an electron source that localizes and helps to strike and maintain the discharge. This is maintained between two molybdenum plates containing small apertures. The gas flow for the lamp is supplied via the inner cathode cylinder and escapes into the reaction vessel through the apertures. This arrangement also prevents the electrodes becoming coated with the film material.

By employing such a deposition technique, we may use monosilane as the

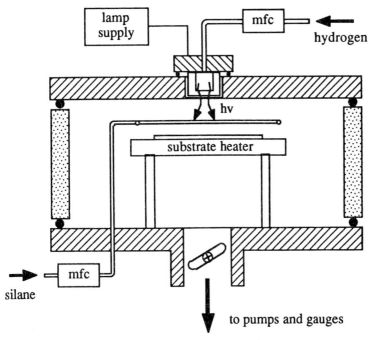

Fig. 9.6   Schematic of internal lamp system.

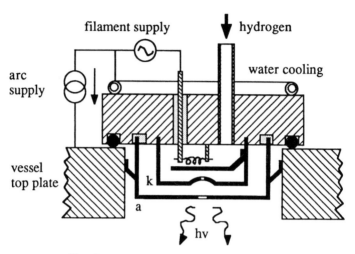

Fig. 9.7   Cross-section of discharge lamp.

reaction gas without the necessity of using mercury sensitization, and we
have also eliminated the window clouding problem. However, it is possible
that the internal lamp system could be operating with contributions from
primary reactions other than just direct photolysis. There are several possible contributions to the deposition process, the most probable mechanisms
being

    (a) plasma dissociation—where the reaction gases may diffuse into the
        lamp region and undergo dissociation by electron and/or ion
        bombardment;

    (b) radical reactions—where radicals produced in the lamp arc region
        may react directly with the silane (cf. hydrogen radical enhanced
        CVD);

    (c) pyrolysis—where the thermal dissociation of the reaction gases may
        occur by contact with hot surfaces;

    (d) direct photolysis—where dissociation of the reactive gases occurs by
        absorption of the UV light from the discharge lamp.

We use optical emission spectroscopy (OES) and mass spectroscopy to
attempt to distinguish between the above reactions. The data obtained
from these techniques are complementary in that OES provides information
on short-lifetime energetic species in the lamp plasma region and mass
spectroscopy provides information on stable species from the reaction
region. Using OES, we sampled the arc region and found no evidence of

any Si emission either from the atomic lines or from the SiH* radical lines. This implies that there is very little silane actually reaching the arc discharge region. This has been confirmed by mass spectroscopy readings, which indicate that <0.5 at% of the gas in the arc region is silane, i.e. the contribution from plasma dissociation is small. To test for the contribution from radical reactions, where for example atomic hydrogen produced in the arc region could react directly with the silane, we have placed a molybdenum shield directly in front of the lamp aperture. This has the effect of preventing the photons from entering the reaction chamber while still allowing the passage of energetic molecules and radicals from the lamp region to the reaction region. Thus any dissociation products detected without the shield in place are due to both photon and radical effects, whereas any products detected in the presence of the shield are due to radical reactions alone. We estimate from detailed investigation of the mass spectra that the primary reaction of hydrogen accounts for approximately 30% of the total silane decomposition.

The experiments have been described in more detail elsewhere [26] but we simply point out here that they show that about 70% of the gas decomposition is due to direct photolysis, and the remainder is thought to be due to hydrogen radical effects. For the deposition of a-Si : H the primary reactions in our system are therefore thought to be

$$\text{SiH}_4 + h\nu \rightarrow \text{SiH}_2 + 2\text{H} \qquad (9.37)$$

$$\text{SiH}_4 + h\nu \rightarrow \text{SiH}_3 + \text{H} \qquad (9.38)$$

$$\text{H} + \text{SiH}_4 \rightarrow \text{SiH}_3 + \text{H}_2 \qquad (9.39)$$

Reaction (9.39) may be important where the lamp arc is also a source of atomic hydrogen. Furthermore, the following reactions may also play a part:

$$\text{H} + h\nu(121.6 \text{ nm}) \rightarrow \text{H}^*(^2\text{P}) \qquad (9.40)$$

$$\text{H}^*(^2\text{P}) + \text{H}_2 \rightarrow 3\text{H} \qquad (9.41)$$

and any reactions of H$^*(^2$P) with SiH$_4$, etc.

For the deposition of oxide films from silane and an oxidant gas (such as nitrous oxide), the silane need not be dissociated. Instead, as mentioned above, the process relies mainly on the reaction of molecular silane with oxygen radicals, derived from the photolysis of the oxidant gas, which usually forms by far the major fraction of the reaction gases. The simplified reaction scheme is as follows:

$$\text{N}_2\text{O} + h\nu \rightarrow \text{N}_2 + \text{O} \qquad (9.42)$$

$$\text{SiH}_4 + 2\text{O} \rightarrow \text{SiO}_2 + 2\text{H}_2 \qquad (9.43)$$

Although, in our system, photolysis by the ultraviolet light from the lamp is thought to be the primary method of dissociating the nitrous oxide, contributions may also occur from a number of other reactions, as detailed above.

For the deposition of oxides we use a nitrogen discharge lamp. Optical emission spectroscopy was used to examine the spectral output from the nitrogen arc in the lamp, both with and without the reaction gas present. The optical spectra obtained were identical to within experimental determination. In particular, there was no evidence of the $O(^1S)-O(^1D)$ transition at 557.7 nm nor from the NO β bands over the range 300–500 nm. It is concluded that there is no significant contribution to decomposition of the nitrous oxide by plasma dissociation in the lamp.

Quadrupole mass spectrometry using a commercial Spectramass VISA and an in-house designed sampling and lock-in detection system was also employed to analyse the gas composition in the lamp arc and within the reaction chamber. It was found that, under normal deposition conditions, the lamp arc region comprises approximately 4% nitrous oxide, of which 1% is dissociated to produce oxygen, nitrogen and nitric oxide. The net balanced equation for the nitrous oxide dissociation derived from the mass spectrometry data is

$$N_2O \rightarrow 0.46NO + 0.27O_2 + 0.77N_2 \qquad (9.44)$$

and compares very closely with the net photolysis reaction of nitrous oxide with 185 nm radiation reported by Greiner [27]:

$$N_2O \rightarrow 0.50NO + 0.25O_2 + 0.75N_2 \qquad (9.45)$$

The quantum yield observed for the nitrous oxide decomposition is approximately 20 molecules per photon (compared with a value of 1.7 reported by other researchers [28]). It is therefore concluded that there is a significant contribution to dissociation from the reactions of radicals produced in the discharge region. Although ground state nitrogen atoms are known to be unreactive towards nitrous oxide [29], and the product of a reaction of metastable $N(^2D, {}^2P)$ with nitrous oxide is not observed, we propose the following regenerative scheme producing atomic oxygen:

$$N_2O + h\nu \rightarrow N_2 + O^* \qquad (9.46)$$

$$O^* + N_2O \rightarrow N_2 + O_2 \qquad (9.47a)$$

$$\rightarrow 2NO \qquad (9.47b)$$

$$N + NO \rightarrow N_2 + O \qquad (9.48)$$

This regenerative scheme can explain the observed dissociation rate, and requires the lamp to dissociate approximately 1% of the nitrogen passing through it. These schemes have been confirmed by the use of isotopic $^{30}N_2$ and shown to be consistent with observed mass spectrometry data.

We therefore conclude that the primary reactions are photo-initiated, but the radicals created in the lamp play an important role in determining the rate of secondary reactions. The roles of pyrolytic and plasma dissociation of the reactant gases has been determined to be negligible.

9.5   MATERIAL PROPERTIES

9.5.1   a-Si:H

Table 9.5 summarizes typical UV lamp techniques for the deposition of undoped and doped a-Si:H. As can be seen, all of these techniques appear

Table 9.5   Dark and light conductivities for doped and undoped a-Si:H prepared by photo-CVD.

σ_D $(\Omega\,cm)^{-1}$ (dark conductivity)	σ_{ph} $(\Omega\,cm)^{-1}$ (AM1 photoconductivity)	Comments	Ref.
10^{-10}	10^{-5}	Si$_2$H$_6$, 0.1 Å s^{-1}, 184.9 m Si$_3$H$_8$, 0.5 Å s^{-1}, no Hg	[22]
8×10^{-7}	5.3×10^{-4}	SiH$_4$, 35 Å min^{-1}, Hg	[46]
$10^{-11} - 10^{-10}$	$10^{-5} - 10^{-4}$	Si$_2$H$_6$, 2 Å s^{-1}, 254 nm, Hg	[47]
$10^{-9} - 10^{-7}$	10^{-4}	SiH$_4$, 400 Å min^{-1}, 184.9 nm, Hg	[48]
4×10^{-9}	6×10^{-4}	Si$_2$H$_6$, 15 Å min^{-1}, 254 nm plus weak 185 nm, no Hg	[49]
10^{-10}	10^{-4}	SiH$_4$, H$_2$ lamp, 4 Å s^{-1}, windowless	[41]
10^{-11}	10^{-5}	Si$_2$H$_6$, Xe discharge lamp, 147 nm, 75 Å min^{-1}	[50]
3.5×10^{-9}	4×10^{-4}	Si$_2$H$_6$, He lamp, 3.5 Å s^{-1}, windowless	[51]
$\approx 10^{-9}$	$\approx 10^{-4}$	Si$_2$H$_6$, deuterium lamp, 50 Å min^{-1}	[24]
10^{-3} max.	No data	Si$_2$H$_6$/PH$_3$, 253.7 nm, Hg	[5]
5×10^{-3}	No data	Si$_2$H$_6$/B$_2$H$_6$, Si$_2$H$_6$/PH$_3$, } 3.2 Å m^{-1} no Hg (254 + 185 nm)	[49]

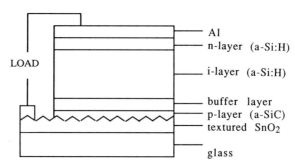

Fig. 9.8 Structure of a p–i–n type a-Si : H solar cell.

to produce device-quality a-Si : H. However, in order to deposit at reasonable rates—a requirement for commercial applications—it is necessary either to use a higher silane as reactant gas or else employ a windowless discharge lamp.

Photo-CVD was originally applied to the deposition of a-Si : H in an attempt to improve material/interface properties and thence enhance device performance. Several groups have now used the material in both solar cell [30] and TFT manufacture [31] in order to test this idea.

A schematic structure of a p–i–n type solar cell, incorporating a SiC p layer and graded band buffer layer at the p–i interface and based on photo-CVD material is illustrated in Fig. 9.8. The p layer and buffer layer were prepared by direct photo-CVD using a mercury lamp with lines at 253.7 and 184.9 nm, and the i and n layers were deposited by mercury sensitization using a 253.7 nm low-pressure mercury lamp. The highest efficiency obtained for such a structure was 11.4% for a 3 mm × 3 mm cell. This compares to the best results found on conventional PECVD-produced cells.

The other major application of a-Si : H is in the manufacture of TFTs for use as the switching elements which must be placed in each pixel in large-area active matrix addressed liquid crystal displays. Suzuki *et al.* [31] have therefore used photo-CVD produced a-Si : H in the simple TFT structure shown in Fig. 9.9. As the purpose of the study was to evaluate the performance of the a-Si : H film itself, thermally grown SiO_2 was employed at the gate insulator. The field effect mobility of 0.5 $cm^2 V^{-1} s^{-1}$ was also comparable to that obtained using conventionally grown a-Si : H as the active layer.

In order to produce a totally integrated large-area liquid crystal display, however, it is necessary that a-Si : H also be used in the manufacture of the peripheral circuits used to drive the display. These require higher speeds

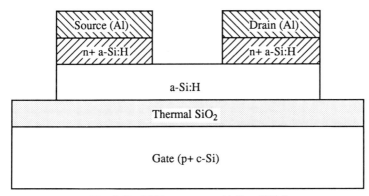

Fig. 9.9   TFT structure. (After [31].)

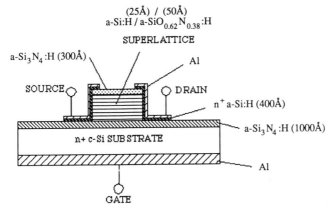

Fig. 9.10   Cross-section of an a-Si:H/a-Si$_{1-x}$N$_x$:H superlattice. (After [32].)

and current drive capabilities than those presently achievable using conventional a-Si:H devices. In an attempt to overcome this problem, the Hiroshima University Group [32] have produced a superlattice TFT. A cross-section of such a device incorporating a-Si:H/a-Si$_{1-x}$N$_x$:H layers is shown in Fig. 9.10. Using a PECVD method to produce such a structure may introduce surface defects and compositional mixing at the heterojunction interface. In contrast with this, the use of photo-CVD eliminates the possibility of ion bombardment. Also, the low deposition rate associated with photo-CVD, in this instance, is a positive advantage, since it allows close control of the approximately 50 Å layer required. A five times enhancement in field effect mobility over conventional PECVD-produced TFTs was obtained.

9.5.2 Insulators

Silicon oxide and nitride are used extensively as insulators in the fabrication of many semiconductor devices. These materials are typically deposited by CVD at temperatures up to 800°C, but, in order to scale down the new generation of VLSI devices, a substantial decrease in wafer processing temperature is required. This is also needed in opto-electronics technology, since compound semiconductors are unstable at relatively low temperatures. Finally, as hydrogen evolution from a-Si:H takes place at $T < 500°C$, thermal growth techniques cannot be used to produce the gate materials needed in TFTs. Photo-CVD has therefore been developed for the deposition of both silicon oxide and nitride. A selection of typical results is given in Table 9.6.

Initially, the photo-CVD deposition of SiO_x was carried out from SiH_4/N_2O mixtures [29] using a mercury-enhanced technique. The oxides produced showed excellent physical and electrical properties, and no mercury contamination could be detected in the films. However, suspicion that mercury contamination was possible led to the investigation of direct photo-CVD methods. Although mercury has since been detected in films produced using mercury sensitization [34] (in concentrations of 90–160 ppm), many groups have continued to use such a technique due to the higher deposition rates achievable. In fact, an Hg-sensitized photo-CVD system capable of depositing over 1 m^2 has recently been developed by Fuji Electric Co. [35].

Tarui et al. [10] used a combination of nitrogen diluted SiH_4 and O_2 to form SiO_x films using a low-pressure mercury lamp that had an output of 3.4 mW cm^{-2} at 185 nm and 30 mW cm^{-2} at 254 nm. In spite of the very low intensity, the deposition process was found to be strongly controlled by the 185 nm emission.

Okuyama et al. [36] and Mishima et al. [37] have also produced oxides using O_2 as a reactant gas. Okuyama et al. were the first to use a deuterium lamp to deposit SiO_x from nitrogen-diluted SiH_4/O_2 mixtures. On the other hand, Mishima et al. reported on the direct photo-CVD of SiO_x from helium-diluted Si_2H_6/O_2 mixtures using a 110 W mercury lamp. However, both SiH_4 and Si_2H_6 are pyrophoric, and their mixtures with O_2 are extremely reactive, which leads to gas phase reactions and subsequent degradation of film properties.

A much more interesting piece of research was carried out by Marks et al. [18], who deposited SiO_x from SiH_4/NO_2 mixtures using 106.6 m radiation from a windowless microwave excited argon discharge lamp. They contended that Hg-sensitized films were incompletely oxidized and had adhesion problems. They also claimed that their deposition method was superior because (i) VUV radiation was known to aid the complete oxidation of silicon, (ii) NO$_2$ was a stronger oxidant than N$_2$O and (iii) the 106.6 nm line could be coupled directly into SiH$_4$. However, they also found strongly substrate-dependent deposition rates, and attributed these to VUV-stimulated bond breaking in the native oxide layers on their substrates. This deposition technique is somewhat similar to the method

Table 9.6 Electrical properties of SiO$_x$ and SiN$_x$ prepared by UV-lamp photo-CVD.

SiO$_x$	$n = 1.45$–1.47, H incorp. $< 1\%$, 100 Å m^{-1} typ.	N$_2$O/SiH$_4$ windowless Ar microwave lamp at 106.6 nm, no sensitization	[18]
	$n = 1.46$, $E_{br} = 6$–8 MV cm^{-1} H incorp. 0%	Hg lamp, 185 nm, SiH$_4$/N$_2$O, no sensitization	[39]
	$n = 1.435$–1.47, H incorp. 1% (?), 9.5 Å min^{-1} (120°C) to 125 Å min^{-1} (200°C)	O$_2$/N$_2$/SiH$_4$, Hg at 185 nm and 254 nm no sensitization	[40]
	$n = 1.460$, $E_{br} = 9$–10 MV cm^{-1}, >3 Å s^{-1}	Windowless N$_2$ discharge lamp, SiH$_4$/N$_2$O, no sensitization	[41]
	$n = 1.456$, $E_{br} = 6$–7 MV cm^{-1}, H incorp. 2%, 1–2 Å s^{-1}	D$_2$ lamp, N$_2$O$_3$/Si$_2$H$_6$	[24]
SiN$_x$	$n = 2.0$, $E_{br} = 3$–5 MV cm^{-1} H incorp. $\approx 20\%$, 1.6×10^9 dyn cm^{-2} compression	Hg lamp, 185 nm, NH$_3$/Si$_2$H$_6$/N$_2$	[39]
	$n = 1.95$	Hg lamp, sensitized, NH$_3$/SiH$_4$	[42]
	$n = 1.95$–2.0, 2×10^9 dyn cm^{-2} compression (70°C) to 1.8×10^9 dyn cm^{-2} tensile (200°C)	NH$_3$/SiH$_4$, Hg lamp	[43]
	13.6 Å min^{-1}	NH$_3$/SiH$_4$, Hg lamp	[44]

pioneered in our laboratory [26], where we used a windowless nitrogen discharge lamp to decompose SiH_4/N_2O mixtures. We concluded in our windowless system that a significant fraction of the deposition was in fact due to radical-enhanced effects rather than direct photo-CVD itself. Marks *et al.* do not comment on this aspect in their paper. However, they later [38] compared the VUV (at 106.6 nm) and UV (at 300–400 nm) photo-CVD of SiH_4/N_2O mixtures, the UV (253.7 nm) photo-CVD of Si_2H_6/NO_2 mixtures and remote plasma CVD (RPCVD) of N_2O/SiH_4 mixtures. The Si_2H_6/NO_2 mixtures yielded porous and nonstoichiometric (oxygen-rich) films. The UV photo-CVD of SiH_4/NO_2 and the RPCVD of SiH_4/N_2O were found to produce higher-quality films.

Bhatnagar and Milne [24] have since extended this work to include the deposition of oxides from Si_2H_6/N_2O_3 mixtures. This resulted in films with better properties than those obtainable using other Si_2H_6/oxidant mixtures in our laboratory.

The photo-CVD of silicon nitride has also been extensively investigated. However, in this instance there is much less variety in the types of lamp employed. As shown in Table 9.6, all of these workers used a mercury discharge lamp as the UV source and NH_3 as the source of nitrogen. As mentioned previously, monosilane is not decomposed directly by the mercury lamp output. However, ammonia is very effectively photochemically decomposed by the 185 nm mercury line, and the photodissociation of NH_3 produces atomic hydrogen, which then decomposes the SiH_4. Inushima *et al.* [39] have departed slightly from the norm, and, although they still use a mercury lamp, they now employ Si_2H_6 as the source of silicon. In this instance, as well as the above reaction with atomic hydrogen, the Si_2H_6 may also be directly decomposed by the 185 nm line.

9.6 CONCLUSIONS

We have considered the various possible photo-CVD methods that can be used to deposit high-quality a-Si : H and related insulators. As device dimensions in VLSI continue to decrease, these techniques will become increasingly important for low-temperature processing in the semiconductor industry.

REFERENCES

[1] S. Nakano, Y. Kuwano and M. Onishi, *Appl. Phys.* **A41**, 267 (1986).
[2] P. G. LeComber, A. J. Snell, K. D. Mckenzie and W. E. Spear, *J. Phys. (Paris) Colloq.* **42**, C4–423 (1981).

[3] R. L. Weisfield, H. C. Tuan, L. Fennell and M. J. Thompson, *Mater. Res. Soc. Symp. Proc.* **95**, 469 (1987).

[4] S. Oda, S. Ishihara, N. Shibata, H. Sirai, A. Miyauchi, K. Fukada, A. Tanabe and I. Shimizu, *Japan. J. Appl. Phys.* **25**, L188 (1986).

[5] T. Inoue, M. Konagai and K. Takahashi, *Appl. Phys. Lett.* **43**, 774 (1983).

[6] H. Okabe, *Photochemistry of Small Molecules* (Wiley, New York, 1978).

[7] M. Nay, G. Woodall, O. Strausz and H. Gunning, *J. Am. Chem. Soc.* **87**, 179 (1965).

[8] T. Pollock, H. Sandhu, A. Jodhan and O. Strausz, *J. Am. Chem. Soc.* **95**, 1017 (1973).

[9] Y. Tarui, S. Sorimachi, K. Fujii and K. Aota, *J. Non-Cryst. Solids* **50/60**, 711 (1983).

[10] Y. Tarui, J. Hidaka and K. Aota, *Japan. J. Appl. Phys.* **23**, L827 (1984).

[11] G. G. A. Perkins, E. R. Austin and F. W. Lampe, *J. Am. Chem. Soc.* **101**, 1109 (1979).

[12] M. Suto and L. Lee, *J. Chem. Phys.* **84**, 1160 (1986).

[13] E. Austin and F. Lampe, *J. Phys. Chem.* **81**, 1134 (1977).

[14] B. Scott and J. Reimer, *J. Appl. Phys.* **54**, 6853 (1983).

[15] Y. Numasawa, K. Yamazaki and K. Hamano, *J. Electron. Mater.* **15**, 27 (1986).

[16] Y. Numasawa, K. Yamazaki and K. Hamano, *Japan. J. Appl. Phys.* **22**, L792 (1983).

[17] K. Tamagawa, T. Hayashi and S. Komiya, *Japan. J. Appl. Phys.* **25**, L728 (1986).

[18] J. Marks and R. E. Robertson, *Appl. Phys. Lett.* **52**, 810 (1988).

[19] L. Meiners, *J. Vac. Sci. Technol.* **21**, 655 (1982).

[20] T. Kudo and S. Nagase, *J. Chem. Phys.* **88**, 2833 (1984).

[21] J. Perrin, T. Broekhuizen and R. Benfehrat, in *Proceedings of the European Materials Research Society, Strasbourg* (Les Editions des Physique, Les Ullis, 1986).

[22] K. Kumata, U. Itoh, Y. Toyoshima, N. Tanaka, H. Anzai and A. Matsuda, *Appl. Phys. Lett.* **48**, 1380 (1986).

[23] A. Langford, B. Stafford and Y. Tsuo, in *Proceedings of the 19th IEEE Photovoltaic Specialists Conference, New Orleans, May 1987*, p. 573.

[24] Y. K. Bhatnagar and W. I. Milne, *Thin Solid Films* **163**, 237 (1988).

[25] R. Rocheleau, S. Jackson, S. Hedgeus and B. Baron, in *Proceedings of the Materials Research Society* (eds D. Adler, Y. Hamakawa and A. Madan), *Spring Meeting, Palo Alto, April 1986*, p. 37.

[26] W. I. Milne, F. J. Clough, S. C. Deane, S. D. Baker and P. A. Robertson, in *Beam Processing and Laser Chemistry* (ed. I. W. Boyd and E. Rimini), p. 277 (North-Holland, Amsterdam, 1989).

[27] N. Greiner, *J. Chem. Phys.* **47**, 4373 (1967).

[28] W. Groth and H. Schierholz, *Planetary and Space Sci.* **1**, 333 (1959).

[29] A. Wright and C. Winkler, *Active Nitrogen* (Academic Press, New York, 1968).

[30] W. Y. Kim, A. Shibata, Y. Kazama, M. Konagai and K. Takahashi, *Japan. J. Appl. Phys.* **28**, 311 (1989).

[31] K. Suzuki, Y. Yukawa, H. Takao, K. Kuroiwa and Y. Tarui, *Japan. J. Appl. Phys.* **25**, L811 (1986).

[32] M. Tsukude, S. Akamatsu, S. Miyazaki and M. Hirose, *Japan. J. Appl. Phys.* **26**, L111 (1987).

[33] J. W. Peters, in *Technical Digest, IEEE International Electronic Devices Meeting, New York, 1981.*
[34] K. Usami, Y. Mochizuki, T. Minagawa, A. Iada and Y. Gohshi, *Japan. J. Appl. Phys.* **25**, 1449 (1986).
[35] Fuji Electric Press Release, 1989.
[36] M. Okuyama, Y. Toyoda and Y. Hamakawa, *Japan. J. Appl. Phys.* **23**, L97 (1984).
[37] Y. Mishima, M. Hirose, Y. Osaka and Y. Ashida, *J. Appl. Phys.* **55**, 1234 (1984).
[38] J. Marks, Personal communication.
[39] T. Inushima, Personal communication.
[40] Y. Nissim, J. L. Regolinl, D. Bensahel and C. Licoppe, *Electron. Lett.* **24**, 488 (1988).
[41] P. A. Robertson and W. I. Milne, in *Proceedings of Materials Research Society Fall Meeting, Symposium B, Boston, 1986.*
[42] R. Padmanabhan and N. C. Saha, *J. Vac. Sci. Technol.* **A4**, 2226 (1988).
[43] N. Arnold, I. Schleicher and T. Grave, in *Proceedings of 14th International Symposium on GaAs and Related Compounds, Heraklion, Greece, 1987.*
[44] S. Miyazaki and M. Hirose, *Phil. Mag.* **B60**, 23 (1989).
[45] P. A. Robertson, PhD Thesis, Cambridge University (1987).
[46] T. Saitoh, S. Muramatso, T. Shimada and M. Migataka, *Appl. Phys. Lett.* **42**, 678 (1983).
[47] M. Konagai, in *Proceedings of Materials Research Society* (eds D. Adler, Y. Hamakawa and A. Madan), *Spring Meeting, Palo Alto, 1986*, p. 257.
[48] Y. Tarui, K. Aota, K. Kamisato, S. Susuki and T. Hiramoto, in *Proceedings of 16th International SSDM, Kobe, Japan, 1984*, p. 429.
[49] Y. Mishima, M. Hirose, Y. Osaka, K. Nagamine, Y. Ashida, N. Kitigawa and K. Isogaya, *Japan. J. Appl. Phys.* **22**, 146 (1983).
[50] T. Fuyuki, K.-Y. Du, S. Okamoto, S. Yasuda, T. Kimoto and H. Matsunami, in *Technical Digest, International Photovoltaic Science and Engineering Conference, Tokyo, 1987*, p. 21.
[51] G. J. Collins, Personal communication.

10 UV Photo-Assisted Formation of Silicon Dioxide

ERIC FOGARASSY

Laboratoire PHASE, Centre de Recherches Nucléaires, Strasbourg, France

10.1 INTRODUCTION

One of the most important steps in the fabrication of electronic devices is the formation of good quality SiO_2 films. From the point of view of device performances, it is highly desirable to form thin-film oxides rapidly at lower substrate temperatures or with a significant reduction in high-temperature processing times. This is the reason why, in recent years, various photo-assisted methods have been proposed for growing and depositing silicon dioxide using both incoherent (lamps) and coherent (CW and pulsed lasers) light sources [1, 2]. In addition, lasers offer the advantage over lamps of spatially confining the chemical and physical reactions in very specific areas of the substrate surface to form localized oxides by direct writing. This appears to be of great interest for reducing the device geometry in the fabrication of microcircuits, in particular for very large-scale integration (VLSI) applications.

Following the intensity and wavelength of the incident radiation, the nature and pressure of the reactant gases, and the configuration of the experimental set-up, various photothermal or/and photochemical reactions can be induced both in the gas phase, at the gas/solid interface and in the irradiated solid.

The photoformation of SiO_2 thin films, using ultraviolet (UV) photons of high energy (4–7 eV), able to initiate nonthermal molecular photodissociation processes in various oxidizing atmospheres, presents some unique advantages compared with visible and infrared wavelengths, which will be detailed in this review.

Several of the basic photochemical reactions that can take place with UV light, both in the gas phase and at surfaces of silicon or growing oxide, are not yet well understood. Among the unsolved questions, one of the most

Photochemical Processing of Electronic Materials
ISBN 0-12-121740-X

important is related to the influence of highly reactive photogenerated species, such as atomic oxygen (neutral and ionized) and ozone, on the oxide formation.

Laser light may also change the properties of irradiated surfaces in different ways, including lattice phonon excitations, electron–hole pair generation ($> 10^{21}$ cm^{-3} with pulsed lasers) and electron emission, that are able to strongly modify surface adsorption of oxidizing species and their transport into the solid.

Finally, from the point of view of device performance, it also seems important to take into account the role of photon-induced defects in the growing oxide, especially when using deep-UV light.

10.2 PROCESSING MODES AND BASIC MECHANISMS

UV photoformation of silicon dioxide can be achieved at low temperature ($<500°$C), in various oxidizing atmospheres, by different methods, which

Table 10.1 UV photoformation of SiO_2 thin films.

Processing mode	Oxidizing atmosphere	UV
Photon-induced oxidation of Si	AIR [3, 4] O_2 [3–10]	Pulsed excimer laser (193, 248 nm) [4, 6, 7, 9, 12] (308 nm) [5]
	N_2O [11]	Xe/Hg lamp (250–350 nm) [9]
	$O_2 + NF_3$ [12]	Low-pressure Hg lamp (185, 254 nm) [10, 11]
Photochemical vapour deposition of SiO_2	$SiH_4 + N_2O + Hg$ [13] $SiH_4 + N_2O$ [14–17] $SiH_4 + O_2$ [18–20] $Si_2H_6 + O_2$ [21, 22] $Si_3H_8 + O_2$ [23] TEOS $+ O_2$ [24]	Low-pressure Hg lamp (185, 254 nm) [13–15, 18, 19, 21] D_2, H_2 lamps (<160 nm) [22, 23] Windowless N_2 lamp (<160 nm) [17] Xe lamp (>180 nm) [22] Pulsed excimer lasers (193, 248 nm) [16, 20, 24]
Laser ablation deposition of SiO_2 from SiO and Si targets	O_2 [25, 26]	Pulsed excimer lasers (193, 248 nm) [25, 26]

are briefly summarized in Table 10.1. Three basic mechanisms can be recognized behind these different processing modes, which are schematically represented in Fig. 10.1.

In *photon-induced incorporation* (Fig. 10.1a), the light source impinges upon an absorbing substrate, heating it locally over the beam area, possibly up to its melting point. Oxygen molecules, in the vicinity of this localized region, undergo photothermal (pyrolysis) or photochemical (photolysis) decomposition, depending on the wavelength, leading to incorporation and diffusion of oxygen atoms into the substrate, which is subsequently oxidized [27]. Several studies demonstrated the possibility of using pulsed lasers of different UV wavelengths and arc lamps, to induce direct oxidation of silicon (Table 10.1).

(a) PHOTON-INDUCED INCORPORATION

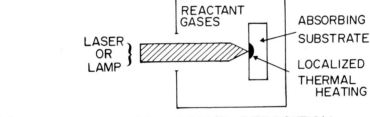

(b) PHOTOCHEMICAL VAPOUR DEPOSITION

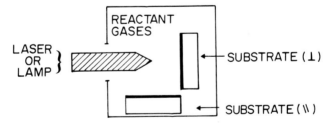

(c) LASER ABLATION DEPOSITION

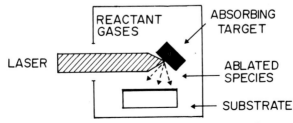

Fig. 10.1 Schematic representation of the three basic mechanisms.

In the *photochemical vapour deposition process* (photo-CVD, Fig. 10.1b) the light source can directly dissociate the gaseous species if the photon energy is above the binding energy of the reactant molecules. Products of this photodissociation then react chemically with other species present in the gas phase, diffuse and finally condense upon the substrate surface. When the substrate is mounted in the chamber parallel to the photon beam, the film deposition can be achieved by a purely photolytic process. By contrast, when the incident light interacts with the substrate (perpendicular configuration), the situation is more complex since the deposition can also be due partly to thermal decomposition of the reactant gases in the vicinity of the irradiated surface. As shown in Table 10.1, SiO_2 thin-film deposition by photo-CVD has been performed, in parallel ($\parallel$) and perpendicular ($\perp$) configurations, using various reactant gas mixtures and different UV light sources (lamps and pulsed lasers) providing high-energy photons (>5 eV).

In the *laser ablation deposition process* (Fig. 10.1c), a high flux of photons is directed onto a solid target. The laser removes atoms from the target, which are transported under vacuum or in controlled atmosphere (reactive laser ablation) toward a substrate, where they are collected [23]. The composition of the deposit is strongly related to the chemical reactions that take place in the gas phase and at the surface of the growing layer between the emitted species (excited and ionized) and gaseous environment, as confirmed for SiO_2 thin films deposited by reactive UV laser ablation from Si and SiO targets in an O_2 atmosphere (Table 10.1).

In this chapter, the most significant results related to the use of deep-UV light ($h\nu > 5$ eV) sources to form silicon dioxide, for each one of the three processing modes described above will be reviewed and discussed.

10.2.1 Photon-induced oxidation of Si

The major difficulty in the interpretation of photonic oxidation processes is the accurate evaluation of the thermal contribution resulting from the interaction between the incoming light and the solid. Surface heating during laser irradiation is difficult or impossible to measure experimentally, especially for pulsed lasers. Consequently, the thermal contribution of pulsed lasers, such as excimer lasers, in irradiated Si has to be evaluated numerically [28] by solving the one-dimensional heat flow equation

$$\left(C_s\rho \, \frac{\partial T}{\partial t} \right)(x, t) = \frac{\partial}{\partial x}\left[k\left(\frac{\partial T}{\partial x}\right)(x, t) \right] + S(x, t), \qquad (10.1)$$

where C_s is the specific heat, ρ the density and k the thermal conductivity of the sample. The temperature T and heat generation S are functions of space and time. The heat generation

$$S(x, t) = I(t)(1 - R)\alpha\, e^{-\alpha x} \tag{10.2}$$

depends on the optical parameters of the solid: the optical absorption α and the surface reflectivity R. $I(t)$ is the time-dependent power density of the laser beam. The accuracy of the modelling is therefore strongly related to how precisely we know the experimental values of these different parameters, and their dependence on temperature and wavelength.

The optical parameters (α and R), which depend upon wavelength, are specially important, since they control the amount of incident photon energy participating in the heating process (through (10.2)). We show in

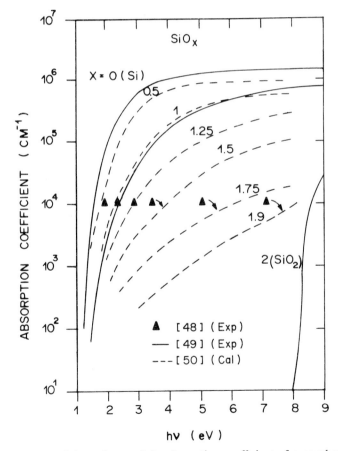

Fig. 10.2 Spectral dependence of the absorption coefficient of amorphous Si and SiO$_x$ ($0.5 \leqslant x \leqslant 2$) [29].

Fig. 10.2 the evolution with wavelength of the absorption coefficients of Si and SiO$_x$, with x ranging between 1 and 2 [29]. These data show the strong absorption of UV light ($h\nu > 4$ eV) in both Si (absorption length $L_{abs} \approx 100$ Å) and SiO ($L_{abs} \approx 250$ Å). By contrast, these UV photons are not absorbed by a stoichiometric SiO$_2$ layer. We note also the influence of the oxide stoichiometry (SiO$_x$) on absorption coefficient. This reveals another important difficulty, namely the modification of surface optical absorption resulting from the continuous charge in the atomic composition of the growing oxide under laser treatment.

Finally, we also have to take into account, in the thermal calculations, possible changes in surface optical reflectivity resulting from the growth of a thin SiO$_2$ layer on top of a Si substrate, as shown in the simulation Fig. 10.3, performed in both the visible and UV ranges.

Thermal calculations [28] show that, under the irradiation conditions of crystalline silicon with a pulsed excimer laser, we have two main regimes: at high intensity, above the Si surface melting threshold ($E_T \approx 0.5$ J cm^{-2} with a pulsed ArF (193 nm) excimer laser of 20 ns pulse duration as deduced from the simulations of the melting front dynamics, reported in Fig. 10.4), the oxidation of silicon results from the fast diffusion (of order 10^{-4} cm^2 s^{-1}) of oxygen from air (or an O$_2$ controlled atmosphere) into

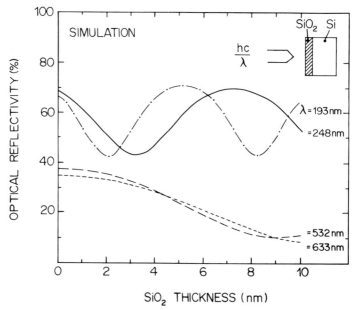

Fig. 10.3 Influence of a SiO$_2$ thin film on the optical reflectivity of Si at visible and UV wavelengths (simulation).

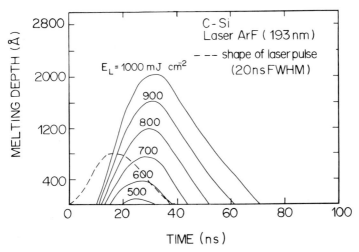

Fig. 10.4 Si melting depth as a function of time. Calculations are for a 20 ns pulsed ArF excimer laser and energy densities between 0.5 and 1 J cm^{-2} (laser pulse shape is represented by the dashed line).

liquid Si, and its trapping at the solid–liquid interface during resolidification of the molten phase. As shown in Fig. 10.4, below 1 J cm^{-2} energy density, the depth of fusion is less than 2000 Å for a total duration of the melting phase not exceeding 80 ns (similar results hold for KrF (248 nm) and XeCl (308 nm) excimer lasers of the same pulse duration) [28]. In this way, thick oxides (in excess of 1000 Å) have been grown very quickly (about 100 Å s^{-1}), using both pulsed XeCl [5] and ArF [3,4] excimer lasers working at 100 Hz repetition rate. However, the oxidation process is believed to be mainly thermal in nature (with no specific influence of UV light).

At low intensity, below the surface melting threshold, photo-oxidation of silicon, in air or in an O$_2$ atmosphere, can be achieved in the solid phase regime, using pulsed excimer lasers working both at 248 nm [6,8,9] or 193 nm [4,7,8] wavelengths [6,9]. The thickness of the grown oxides obtained at these two wavelengths for similar conditions of irradiation by different authors [4,7,8] are plotted in Fig. 10.5 as functions of the number of laser shots. After about 2×10^4 shots, and for laser energy densities ranging between 150 and 230 mJ cm^{-2}, the grown oxides reach a maximum thickness of 25–35 Å, for a total time of illumination not exceeding 0.5 ms (with a laser pulse duration of about 20 ns). The time-dependent Si surface temperature rise for irradiation with a pulsed ArF laser of 20 ns pulse duration is shown in Fig. 10.6. These data (similar for

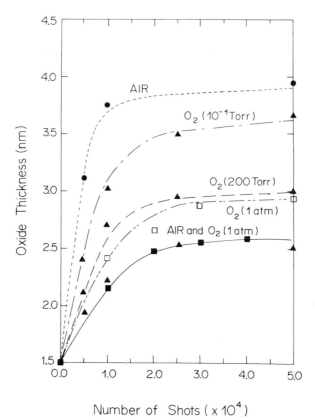

Fig. 10.5 Oxide thicknesses grown as a function of the number of ArF and KrF laser shots (deduced from ellipsometry, IR absorption and nuclear microanalysis experiments) [4, 7, 8]: •, ArF laser, $z = 15$ ns, $E = 200$ mJ cm^{-2} [7]; ▲, ArF, 20 ns, 150 mJ cm^{-2} [4]; ■, ArF, 23 ns, 230 mJ cm^{-2} [8]; □, KrF, 23 ns, 230 mJ cm^{-2} [8].

ArF and KrF lasers of same pulse duration) show that, for laser energy densities below 250 mJ cm^{-2}, the surface temperature does not exceed 500°C during a time of less than 100 ns. This produces an equivalent thermal oxidation rate ranging between 10^4 and 2×10^4 Å s^{-1}. This value is considerably higher than the thermal oxidation rates of thin-film oxides grown at high temperature in a furnace (about 4×10^{-2} Å s^{-1} at 821°C), as deduced from the experimental data of Taft [30] (Fig. 10.7). These results demonstrate the possibility of rapidly growing SiO$_2$ thin films at very low temperature by UV laser-induced oxidation. Recently, Boyd [9] observed similar behaviour when using a UV Xe/Hg arc lamp to grow, at very low temperature (about 450°C), thin-film oxides (about 30 Å). In this case, the

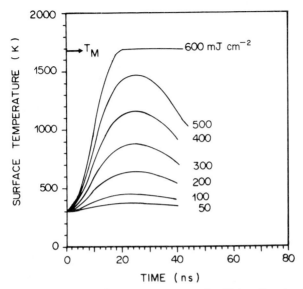

Fig. 10.6   Time-dependent surface temperature of c-Si irradiated with a 20 ns pulsed ArF laser (193 nm): influence of pulse energy density.

Si photo-oxidation rate in UV light is about five times greater than for thermally grown SiO$_2$. The same group [10] also confirmed these results by using a low-pressure Hg lamp emitting at 185 and 254 nm. Their data, shown in Fig. 10.8 for Si samples of $\langle 100 \rangle$ orientation, show that oxides up to 40 Å can be grown, through a Cabrera−Mott type mechanism [31], in 120 min at temperatures around 500°C. By contrast, thermal growth of oxide in dry O$_2$ is very slow at such temperatures, since typically only 1−2 Å of oxide is formed at 612°C after 360 min [30].

Very similar results were reported by Ishikawa *et al.* [11] when replacing molecular oxygen by nitrous oxide (N$_2$O) as the oxidizing agent to form thin-film oxides at low temperature (< 500°C) with UV light. In this case, the 185 nm wavelength radiation of a low-pressure Hg lamp is able to directly dissociate N$_2$O molecules (through reaction (10.7)), to generate oxygen atoms responsible for the oxidation process.

The nonthermal contribution of the UV light to the photo-oxidation process, especially when using high-energy photons ($h\nu \geqslant 5$ eV), is now well established [32, 33]. With UV photons of 5 and 6.4 eV provided by pulsed KrF and ArF lasers respectively, it is possible to induce internal photo-emission of electrons at the Si−SiO$_2$ interface directly from the valence band of Si into the conduction band of SiO$_2$, which needs a photon energy of 4.2 eV [34], as shown in Fig. 10.9. These photo-injected electrons are able to combine with oxygen dissolved in SiO$_2$ to form negatively charged

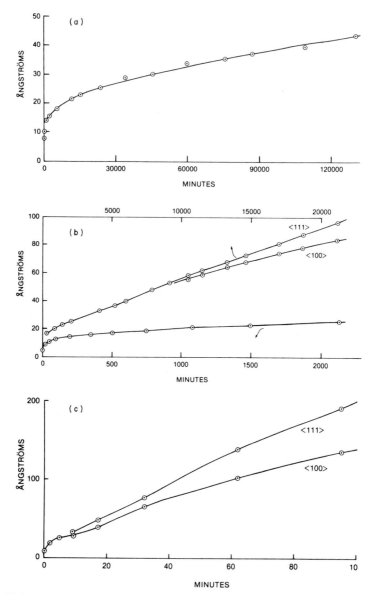

Fig. 10.7 Growth curves of dry thermal oxide on ⟨111⟩ and ⟨100⟩ silicon for three temperatures: (a) 450°C, (b) 612°C; (c) 821°C. (From Taft [30].)

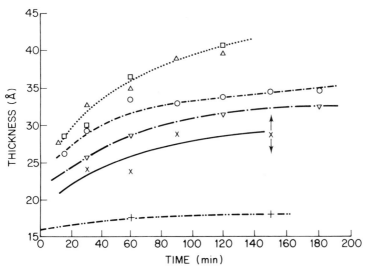

Fig. 10.8 Photo-assisted oxide growth as a function of substrate temperature and time. (From Nayar *et al.* [10].)

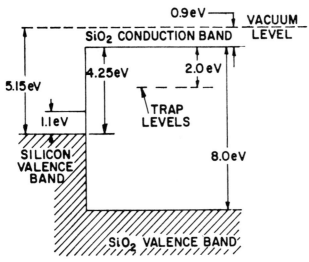

Fig. 10.9 Energy band diagram for the interface between Si and SiO₂. (From Williams [34].)

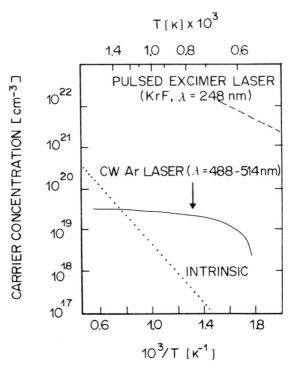

Fig. 10.10 Carrier population in Si as a function of temperature: intrinsic and after CW argon and pulsed excimer laser irradiation [6, 9].

species, which may play an important role in the oxidation process. The fast diffusion of oxygen through the SiO_2 layer towards the $Si-SiO_2$ interface may also be promoted by the photodissociation of O_2 molecules into oxygen, which begins above 5.1 eV (see Fig. 10.11). In addition, Chao et al. [35] have recently demonstrated that the generation of ozone and excited oxygen radicals by a low-pressure mercury (185, 254 nm) lamp through a three-step process (reactions (10.9), (10.12) and (10.13)) could also play an important role in SiO_2 film growth by enhancing the surface reactions, possibly through reactive chemisorption of oxygen.

Finally, fluorine is recognized as an efficient agent for stimulating SiO_2 growth. Morita et al. [12] observed a significant increase in the oxidation rate of Si in an $O_2 + NF_3$ gas mixture under ArF excimer laser irradiation. In this experiment, fluorine radicals, such as NF, NF_2 and F_2, created by the direct UV photodissociation of NF_3 molecules at 193 nm, could be mainly responsible for the oxidation rate enhancement observed.

Several authors [6, 9] have also argued that the oxidation reaction may be limited by the availability of "free" Si bonds in the lattice. In consequence, it may be expected that the creation of high photocarrier concentrations, due to the strong absorption of intense radiation from a laser beam by the solid, will appreciably increase the density of broken Si—Si covalent bonds and thus enhance the oxidation rate. As shown in Fig. 10.10, the intrinsic carrier population, which is of the order of 10^{16} cm^{-3} in Si at around 500°C, is strongly increased under irradiation with a CW Ar laser (about 10^{19} cm^{-3}) and even more so with a pulsed KrF excimer laser ($> 10^{21}$ cm^{-3}). This effect, however, needs to be investigated in more detail, to be clearly understood.

Defects could be photogenerated in the growing oxide with a pulsed UV laser by inducing various changes in the atomic arrangement of SiO₂. After multipulse (5×10^3 shots) irradiation of SiO₂ with a KrF excimer laser, operating at 300 mJ cm^{-2} pulse energy density, Devine and Fiori [36, 37] observed, by electron spin resonance, a structure characteristic of the so-called nonbridging oxygen hole centres and a resonance due to the oxygen vacancy centres (fixed oxide charges). At the same time as defects are created, film thickness measurements indicate a compaction and densification of the oxide layer resulting from surface oxygen desorption, which leads to the formation of superficial substoichiometric oxide (SiO$_x$ with $x < 2$). The SiO$_x$ layer can be subsequently photo-ablated for laser energy densities in excess of 50 mJ cm^{-2}, as a result of its strong optical absorption in the UV range (Fig. 10.2).

Optimization of the laser grown oxide quality requires good control of these different photon-induced defects. The various charges and traps associated with thin-film oxides have a strong influence on their electrical properties. Located at the Si–SiO₂ interface, interface trapped charges Q_{it} have energy states in the Si forbidden bandgap and can interact electrically with the underlying silicon. These charges are thought to result from several sources, including structural defects related to the oxidation process, impurities or bond breaking by UV light. Interface trapped charge densities of less than 10^{10} cm^{-2} eV^{-1} have already been measured for high-temperature oxidized silicon. The fixed oxide charges Q_f (usually positive) are located in the oxide, within approximately 30 Å of the Si–SiO₂ interface. Their density ranges from 10^{10} to 10^{12} cm^{-2}, depending on oxidation and annealing conditions (temperature, and dry or wet oxygen) as well as orientation. The mobile ionic charges Q_m are attributed to alkali ions in the oxide as well as to negative ions and heavy metals. The alkali ions (e.g. sodium and potassium) are mobile even at room temperature when electric fields are present. Their densities range from 10^{10} to 10^{12} cm^{-2} or higher and are related to processing materials, chemical environment or handling.

Up to now, the different research groups working on Si photo-oxidation have reported only partial and preliminary information concerning the relative concentrations of these different charges in the growth oxides,

Table 10.2 Capacitance–voltage measurements on MOS structures.

Reference	[37]	[11]	[10]	[5]
UV irradiation	KrF laser ($E = 200$ mJ cm^{-2})	Hg lamp (185, 254 nm)	Hg lamp (185, 254 nm)	XeCl laser ($E = 900$ mJ cm^{-2})
Oxidation temperature (°C)	500 (simulation)	500	500	$\geqslant 1410$, liquid Si
Oxide thickness (Å)	200	36	40	>1000
Fixed oxide charge Q_f (cm^{-2})	1.8×10^{10}	2×10^{11}	7.5×10^{11}	$(3–8) \times 10^{11}$
Interface trap density D_{it} (cm^{-2} eV^{-1})	1.6×10^{10}			5×10^{11}

which have been deduced from low- and high-frequency capacitance–voltage measurements performed on MOS structures. These are summarized in Table 10.2.

10.2.2   Photochemical vapour deposition of SiO2

Low-temperature ($<400°C$) UV photochemical vapour deposition of insulating compounds, such as SiO_2, proceeds in several steps: photolytic extraction of volatile species (atoms and/or radicals) from parent molecules that are able to diffuse and react in the gas phase or at the surface of the substrate to form the desired compound. Since this method depends upon the direct photodissociation of the reactant gases, the most critical parameter is the wavelength of the excitation light used. Photolytic deposition of SiO_2 thin films (Table 10.1) was successfully achieved using various reactant gas mixtures, SiH_4, Si_2H_6, Si_3H_8, N_2O, O_2 and TEOS (tetraethyl orthosilicate), irradiated with different UV light sources, including low- (Hg) and high- (D_2, H_2, N_2, Xe) pressure discharge lamps and pulsed ArF and KrF excimer lasers. The photochemical processes involved in these

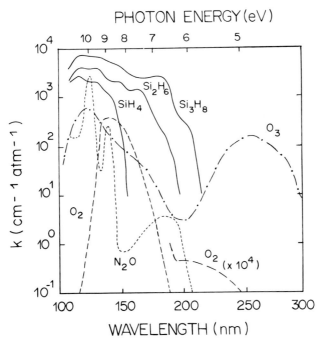

Fig. 10.11   UV absorption spectra of SiH_4, Si_2H_6, Si_3H_8, N_2O, O_2 and O_3 [38].

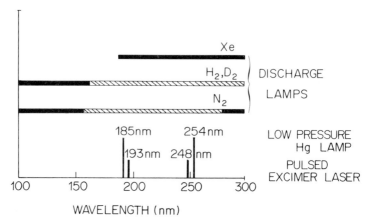

Fig. 10.12 Emission range for various UV light sources used in photo-CVD: ━━━━━, high emission intensity; ⩵⩵⩵⩵⩵, low emission intensity.

various experiments depend strongly on both the optical absorption of gaseous species (Fig. 10.11) and the incident UV radiation spectra, represented schematically in Fig. 10.12.

In the first experiment [13], a mixture of SiH_4, N_2O, and Hg atoms was photoreacted by the Hg photosensitization process. The 254 nm resonance line of a low-pressure mercury lamp was used to excite the Hg atoms via the reaction

$$Hg(^1S_0) + h\nu(254 \text{ nm}) \rightarrow Hg^*(^3P_1) \qquad (10.3)$$

The excited Hg atoms subsequently transfer their energy to N_2O and SiH_4 by collision, forming highly reactive species such as oxygen atoms through the reaction

$$Hg^* + N_2O \rightarrow Hg + N_2 + O(^3P) \qquad (10.4)$$

and intermediate radicals such as SiH_2 and SiH_3 [39–41] via

$$Hg^* + SiH_4 \rightarrow \begin{cases} Hg + SiH_3 + H & (10.5) \\ Hg + SiH_2 + H_2 & (10.6) \end{cases}$$

SiH_4, SiH_3, SiH_2, and O can react together in different manners to form the SiO_2 thin film.

In order to avoid any possible Hg contamination of the deposit, an alternative approach was proposed consisting in using the 185 nm wavelength radiation of the low-pressure mercury lamp [14] to dissociate N_2O molecules directly through the reaction:

$$N_2O + h\nu(185 \text{ nm}) \rightarrow N_2 + O(^1D) \qquad (10.7)$$

After collisional relaxation of the $O(^1D)$ to the ground state, $O(^3P)$, oxygen atoms can react with SiH_4 molecules to form SiO_2:

$$SiH_4 + O(^3P) \rightarrow SiO_2 + products \tag{10.8}$$

By replacing nitrous oxide in this experiment by oxygen gas, SiO_2 deposition with a low-pressure mercury lamp (185, 254 nm) is achieved through the direct photolysis of O_2 molecules at 185 nm and subsequent formation of oxygen atoms and ozone in the gas phase, which could play an important role in the process [18, 19]:

$$O_2 + h\nu(\lambda < 245 \text{ nm}) \rightarrow 2O(^3P) \tag{10.9}$$

$$O_2 + h\nu(130 < \lambda < 175 \text{ nm}) \rightarrow O(^3P) + O(^1D) \tag{10.10}$$

$$O_2 + h\nu(\lambda < 130 \text{ nm}) \rightarrow O(^3P) + O(^1S) \tag{10.11}$$

$$O_2 + O(^3P) \rightarrow O_3 \tag{10.12}$$

$$O_3 + h\nu(\lambda < 310 \text{ nm}) \rightarrow O_2 + O(^1D) \tag{10.13}$$

The main chemical reactions for the deposition can be expressed as

$$SiH_4 + O_3 \text{ or } O \rightarrow SiO_2 + products \tag{10.14}$$

In some experiments [21, 22] silane was replaced by disilane (Si_2H_6/O_2 mixtures), which presents a significant optical absorption below 200 nm wavelength (Fig. 10.11). Under these conditions, Si_2H_6 photodissociation with the 185 nm UV radiation from a low-pressure mercury lamp [21] or on using a D_2 discharge lamp [22] can generate radical species through the following primary reactions [42, 43]:

$$Si_2H_6 + h\nu(\lambda < 200 \text{ nm}) \rightarrow \begin{cases} SiH_2 + SiH_3 + H & (10.15) \\ SiH_3 + 2H & (10.16) \\ Si_2H_5 + H & (10.17) \end{cases}$$

which are able to participate in SiO_2 formation.

Recently, SiO_2 thin films have also been grown at very low temperature (25–390°C) from Si_3H_8 and O_2 gases by photo-CVD using double UV excitation with D_2 and Hg lamps [23]. In these experiments, Si_3H_8 molecules, which have a larger absorption coefficient in the VUV range (Fig. 10.11) than SiH_4 and Si_2H_6, are more easily photodissociated. Details of the chemical reactions induced by the strong absorption of Si_3H_8 below the 200 nm wavelength have not yet been clarified, but many kinds of reactive species, including silicon hydrides and hydrogen, must be photogenerated in a similar way as with Si_2H_6.

Boyer et al. [16] first demonstrated the use of an ArF laser to deposit SiO_2 from the photolytic dissociation of a SiH_4/N_2O mixture with a much higher efficiency and growth rate (up to 300 nm min^{-1}) than achieved using incoherent UV light (10–100 nm min^{-1}). Above 160 nm wavelength

Table 10.3 Capacitance–voltage measurements on MOS structures.

Reference	[14]	[18]	[21]	[23]	[17]	[20]	[16]	[24]	[14, 16]
UV irradiation	Hg lamp (254 nm)	Hg lamp (185, 254 nm)	Hg lamp (185, 254 nm)	D_2, Hg lamp (<185 nm)	N_2 lamp (<160 nm)	KrF laser (248 nm)	ArF laser (193 nm)	ArF laser	Thermal oxidation
Deposition temperature (°C)	200	300	350	280	300	200	400	250	1000
Post-thermal treatment (°C)	—	—	350	—	320	—	425	—	—
Oxide thickness (Å)	1000	—	—	1000	1000	—	1600	413	1000
Mobile ion charge Q_m (cm^{-2})	8×10^{10}	—	7.7×10^{10}	—	—	—	—	8×10^{10}	10^{10}
Fixed oxide charge Q_f (cm^{-2})	—	10^{11}	2.3×10^{11}	2×10^{11}	2×10^{11}	5×10^{11}	2.6×10^{11}	3×10^{11}	3×10^{10}
Interface trapped density D_{it} (cm^{-2} eV^{-1})	—	—	—	3.6×10^{10}	5×10^{11}	—	3×10^{11}	—	—

(7.8 eV), silane, which is optically transparent to UV light (Fig. 10.11), cannot be dissociated by a single-photon process. However, high-energy photons of 6.4 eV (193 nm) provided by a pulsed ArF excimer laser of high intensity (>30 mJ cm^{-2}) [40] can dissociate SiH$_4$ in the gas phase by multiphoton excitation of electronic molecular state, and the following reactions can be written [44]:

$$SiH_4 + 2h\nu(193 \text{ nm}) \rightarrow SiH_4{}^* \rightarrow \begin{cases} SiH_3 + H & (10.18) \\ SiH_2 + 2H & (10.19) \end{cases}$$

In 1989, Klumpp and Sigmund [24], using a new precursor, deposited at very low temperature ($\geqslant 200°$C) SiO$_2$ layers on silicon wafers from a mixture of TEOS and oxygen by ArF excimer laser irradiation performed in a perpendicular configuration (Fig. 10.1b). The formation of SiO$_2$ results from the photolytic dissociation of O$_2$ into atomic oxygen and ozone, designed by reactions (10.9)–(10.13), which can react with the ethyl group of the TEOS molecule, forming Si—O—Si bonds [45].

It must, however, be noted that the direct UV irradiation of the substrate during this experiment could have some significant influence on the final properties of the deposit by inducing photothermal and photochemical reactions leading to bond rearrangement and defect creation in the oxide.

The electrical properties of UV photo-CVD oxide films are summarized in Table 10.3. The values found in the literature for the mobile ion charge density, fixed oxide and interface trapped charge densities by different authors, using various methods of UV excitation, are of the same order of magnitude and appear to be characteristic of low-temperature processing ($T \leqslant 400°$C), similar to other low-temperature SiO$_2$ deposition methods.

10.2.3   Laser ablation deposition of SiO₂

For most compounds, laser ablation deposition under vacuum gives film stoichiometry close to the target [46]. In contrast, modifications in the composition of the deposit and formation of new compounds can be achieved by working in reactive atmospheres.

Very recently [25, 26], we have demonstrated the possibility of depositing SiO$_2$ thin films at ambient temperature by laser ablation from a strongly absorbing silicon monoxide (SiO) target, performed (Fig. 10.1c) under an oxygen atmosphere with high-power pulsed ArF and KrF excimer lasers. As shown in the infrared spectra of Fig. 10.13, the oxide deposited in O$_2$ is fully converted into SiO$_2$, as confirmed by the peak position at 1080 cm^{-1}. In contrast, the 1020 cm^{-1} IR absorption stretching bond, corresponding to the oxide layer deposited under vacuum, is indicative of SiO$_x$ bond structure with $1 < x < 2$.

These results are confirmed by Rutherford back-scattering spectrometric (RBS) analysis of these samples (Fig. 10.14), which shows that the laser-grown oxide under oxygen presents a nearly uniform depth distribution with a well defined atomic composition ($x \approx 2$) corresponding to the formation of a 2500 Å thick stoichiometric SiO_2 layer, in comparison with the oxide deposited under vacuum, which is nonstoichiometric ($x \approx 1.5$).

SiO_2 formation was also observed on replacing SiO by a Si target. Laser ablation in an oxygen atmosphere seems to play a key role in the formation of the SiO_2 layer, through the chemical reactions that take place in the gas phase end at the surface of the growing oxide between the excited and ionized species, ablated from the target and oxygen molecules, which could also be photodissociated at the 193 nm wavelength of the excimer laser (section 10.9).

High-frequency capacitance–voltage measurements performed on these oxides give a fixed oxide charge density $Q_f \approx 8 \times 10^{10}$ cm^{-2} after rapid thermal treatment at $600°C$ during 1 min, characteristic of low-temperature processing.

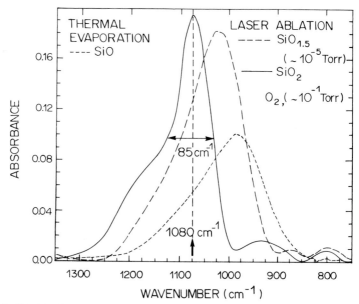

Fig. 10.13 IR absorption spectra of oxide films deposited onto silicon by thermal evaporation and laser ablation from an SiO target under vacuum and in an O_2 atmosphere.

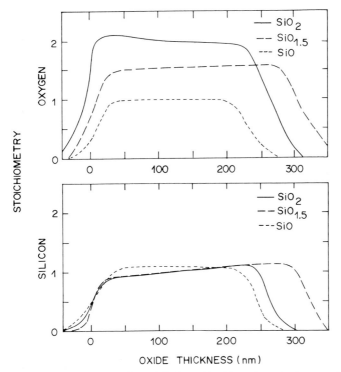

Fig. 10.14   Depth profiles of O and Si in oxide films deposited onto silicon by thermal evaporation and laser ablation from an SiO target under vacuum and in an O_2 atmosphere.

10.3   CONCLUSIONS

We have reviewed the main photochemical reactions involved in the various photo-assisted techniques developed to form SiO_2 at low temperature under different UV excitation light sources.

Molecular oxygen photodissociation with deep-UV photons ($h\nu > 5.1$ eV) and the subsequent generation of ozone in the gas phase seems to play an important role in the oxide formation, whatever the processing mode considered.

We must also take into account of possible photon-induced surface reactions when the UV light impinges directly on the growing oxide (in the perpendicular configuration), which could be the origin of the defect generation that can degrade the final quality of the dielectric films.

In addition, these photogenerated oxides generally present electrical properties comparable to those obtained using other more conventional low-temperature processing techniques.

Finally, it should be noted that photon-assisted methods are not only restricted to silicon technology, since they have also recently been applied successfully to form, at lower temperatures, various oxides and insulating layers such as nitrides on other materials, including III–V compound semiconductors and metals [47].

REFERENCES

[1] I. W. Boyd, in *Laser Processing Diagnostics* (ed. D. Baüerle), p. 274 (Springer-Verlag, Berlin, 1984).
[2] I. W. Boyd, in *Dielectric Layers in Semiconductors: Novel Technologies and Devices* (ed. G. G. Bentini, E. Fogarassy and E. Golanski), p. 171 (Les Editions de Physique, Les Ulis, 1986).
[3] E. Fogarassy, C. W. White, A. Slaoui, C. Fuchs, P. Siffert and S. J. Pennycook, *Appl. Phys. Lett.* **53**, 1720 (1988).
[4] E. Fogarassy, A. Slaoui and C. Fuchs, in *Emerging Technologies for In-Situ Processing* (ed. D. J. Ehrlich and V. T. Nguyen), p. 249 (Martinus Nijhoff, Dordrecht, 1988).
[5] T. E. Orlowski and H. Richter, *Appl. Phys. Lett.* **45**, 241 (1984).
[6] C. Fiori, *Phys. Rev. Lett.* **2**, 2077 (1984).
[7] J. Siejka, R. Srinivasan, J. Perriere, R. Braren and S. Lazare, in *Dielectric Layers in Semiconductors: Novel Technologies and Devices* (ed. G. G. Bentini, E. Fogarassy and A. Golanski), p. 213 (Les Editions de Physique, Les Ulis, 1986).
[8] V. Nayar, I. W. Boyd, F. N. Goodall and G. Arthur, *Appl. Surf. Sci.* **36**, 134 (1989).
[9] I. W. Boyd, *Mater. Res. Soc. Symp. Proc.* **129**, 421 (1989).
[10] V. Nayar, P. Patel and I. W. Boyd, *Electron. Lett.* **26**, 205 (1990).
[11] Y. Ishikawa, Y. Takagi and I. Nakamichi, *Japan. J. Appl. Phys.* **28**, L1453 (1989).
[12] M. Morita, S. Aritome, T. Tanaka and M. Hirose, *Appl. Phys. Lett.* **49**, 699 (1986).
[13] J. W. Peters, in *Technical Digest, IEDM-81*, p. 240 (IEEE, Washington, D.C., 1981).
[14] J. Y. Chen, R. C. Henderson, J. T. Hall and J. W. Peters, *J. Electrochem. Soc.* **131**, 2146 (1984).
[15] T. Inushima, N. Hirose, K. Urata, K. Ito and S. Yamazaki, *Appl. Phys.* **A47**, 229 (1988).
[16] P. K. Boyer, G. A. Roche, W. H. Ritchie and G. J. Collins, *Appl. Phys. Lett.* **40**, 716 (1982).
[17] S. D. Baker, W. I. Milne and P. A. Robertson, *Appl. Phys.* **A46**, 243 (1988).
[18] Y. Tarui, J. Hidaka and K. Aota, *Japan. J. Appl. Phys.* **23**, L827 (1984).
[19] J. Takahashi and M. Tabe, *Japan. J. Appl. Phys.* **24**, 274 (1985).
[20] S. Nishino, H. Honda and H. Matsunami, *Japan. J. Appl. Phys.* **25**, L87 (1986).

[21] Y. Mishima, M. Hirose, Y. Osaka and Y. Ashida, *J. Appl. Phys.* **55**, 1234 (1984).

[22] K. Inoue, M. Michimori, M. Okuyama and Y. Hamakawa, *Japan. J. Appl. Phys.* **26**, 805 (1987).

[23] M. Okuyama, N. Fujiki, K. Inoue and Y. Hamakawa, *Appl. Surf. Sci.* **33/34**, 427 (1988).

[24] A. Klumpp and H. Sigmund, *Appl. Surf. Sci.* **36**, 141 (1989).

[25] E. Fogarassy, C. Fuchs, A. Slaoui and J. P. Stoquert, *Appl. Phys. Lett.* **57**, 664 (1990).

[26] E. Fogarassy, A. Slaoui, C. Fuchs and J. P. Stoquert, *Appl. Surf. Sci.* **46**, 195 (1990).

[27] E. Fogarassy, in *The Physics and Technology of Amorphous SiO$_2$* (ed. R. A. B. Devine), p. 321 (Plenum Press, New York, 1988).

[28] S. De Unamuno and E. Fogarassy, *Appl. Surf. Sci.* **36**, 1 (1989).

[29] E. Fogarassy, S. De Unamuno, J. L. Regolini and C. Fuchs, *Phil. Mag.* **B55**, 253 (1987).

[30] E. A. Taft, *J. Electrochem. Soc.* **131**, 2460 (1984).

[31] N. Cabrera and N. F. Mott, *Rep. Prog. Phys.* **12**, 163 (1949).

[32] S. A. Schafer and S. A. Lyon, *J. Vac. Sci. Technol.* **21**, 422 (1982).

[33] E. M. Young and W. A. Tiller, *Appl. Phys. Lett.* **50**, 46 and 80 (1987).

[34] R. Williams, *Phys. Rev.* **140**, A569 (1965).

[35] S. C. Chao, R. Pitchai and Y. H. Lee, *J. Electrochem. Soc.* **136**, 2751 (1989).

[36] R. A. B. Devine and C. Fiori, *Mater. Res. Soc. Symp. Proc.* **60**, 303 (1986).

[37] C. Fiori and R. A. B. Devine, *Le Vide, Les Couches Minces* **40** (227), 327 (1985).

[38] J. C. Calvert and J. N. Pitts (eds), *Photochemistry* (Wiley, New York, 1966).

[39] E. R. Austin and F. W. Lampe, *J. Phys. Chem.* **80**, 2811 (1976).

[40] C. Fuchs and E. Fogarassy, *Mater. Res. Soc. Symp. Proc.* **75**, 195 (1987).

[41] J. Perrin and T. Broekhuisen, *Mater. Res. Soc. Symp. Proc.* **75**, 201 (1987).

[42] G. G. A. Perkins and F. W. Lampe, *J. Am. Chem. Soc.* **102**, 3764 (1980).

[43] E. Boch, C. Fuchs, E. Fogarassy and P. Siffert, *Mater. Res. Soc. Symp. Proc.* **129**, 195 (1989).

[44] G. G. A. Perkins, E. R. Austin and F. W. Lampe, *J. Am. Chem. Soc.* **101**, 1109 (1979).

[45] H. Sigmund, A. Klumpp and G. Springholz, *Mater. Res. Soc. Symp. Proc.* **158**, 190 (1989).

[46] H. Sankur and J. T. Cheung, *Appl. Phys.* **A47**, 271 (1988).

[47] I. W. Boyd, *Laser Processing of Thin Films and Microstructures* (Springer-Verlag, Berlin, 1987).

[48] E. Holzenkampfer, F. W. Richter, J. Stuke and U. Voget-Grote, *J. Non-Crystalline Solids* **32**, 327 (1979).

[49] H. R. Philipp, *J. Phys. Chem. Solids* **32**, 1935 (1971).

[50] G. Zuther, *Phys. Stat. Sol. (a)* **59**, K109 (1980).

11 Gas Immersion Laser Doping (GILD) in Silicon

FRANÇOIS FOULON

Department of Electrical Engineering, Imperial College of Science, Technology and Medicine, London, UK

11.1 INTRODUCTION

Doping of semiconductors is an important step in micro-electronic device fabrication. Precise control of the final dopant profile, with good uniformity and reproducibility, is required. Traditionally, this stage is performed by one of two techniques: solid state diffusion or ion implantation.

In the former technique, diffused layers are typically formed in a two-step process. In the first, or predeposition, step, impurity atoms supplied by a gaseous, liquid or solid source are introduced into the semiconductor wafer heated to high temperature (typically $950°C$). In the second step, the so-called drive-in diffusion step, impurities are diffused and electrically activated to provide a suitable carrier concentration distribution without any more impurities being added to the semiconductor. This second step is also achieved at high temperature (typically $1100°C$). However, the solid state diffusion technique impose some severe limitations on the design flexibility of devices. For example, the total number of diffusions in any given region is usually limited to two, and the circuit designer has little control over the shape of the impurity profile resulting from diffusion, which generally approximates to a Gaussian or complementary error function.

Some of the limitations of solid state diffusion have been overcome by ion implantation. In this technique, ions emitted by a gaseous source are filtered by a mass separator and accelerated in an electric field to energies between 3 and 500 keV. To achieve spatially uniform doping, the beam is swept across the surface. The depth of penetration of the atoms and the density of the implanted ions can be precisely controlled by the ion energy and the beam current respectively, and result in a Gaussian distribution of dopant to some distance from the surface of the semiconductor. Multiple implantation allows summation of several of these profiles, and hence flat

Photochemical Processing of Electronic Materials
ISBN 0-12-121740-X

doping profiles can be achieved. Unfortunately, ion implantation introduces point defects, which induce crystal damage and generally the formation of an amorphous layer. Moreover, the implanted impurities are not
electrically activated since they are not primarily in substitutional positions
in the lattice. Thus high-temperature annealing ($T > 600°C$) is needed to
recover the crystallinity and to achieve electrical activation of the implanted
atoms.

So, in these two doping techniques, the complete wafer, and not only the
area to be doped, suffers from high-temperature processing steps, which
can induce undesirable effects such as a decrease of the minority carrier
diffusion length in the bulk or dopant atom redistribution in any previous
prepared underlayers. The former effect is of particular importance for
solar cell fabrication, while the second is a big technological problem with
the advent of very large-scale integration.

The decrease of the lateral device dimensions in the CMOS microfabrication process requires the formation of a heavily doped box profile and
shallow junctions (typically $C = 10^{20}$ at cm^{-3}, $X_j = 100$ nm) for the source
and drain contact junctions in order to avoid the short-channel effect of
MOS transistors. With the doping techniques previously described, it is
difficult to form such junctions. Solid state diffusion methods do not form
box profile junctions, and are more adapted to achieving deeper junctions
($X_j > 0.5$ μm). In the same way, it is difficult to obtain such characteristics
by ion implantation followed by thermal annealing, particularly in the case
of boron doping into silicon because of the inherent boron channeling
tail [1, 2]. To overcome this problem, various techniques have been developed to reduce the implantation depth, such as silicon pre-amorphization
by silicon or germanium implantation [3–5], implantation of BF$_2$ molecules rather than ^{11}B [6, 7] and implantation through an SiO$_2$ overlayer
[7]. In addition, rapid thermal annealing has been used to limit the dopant
redistribution during the necessary high-temperature processing using these
methods [7]. Thus, by a combination of all the above techniques, boron-
doped junctions as shallow as 150 nm have been obtained [7]. However,
this form of processing strongly increases the number of fabrication steps
and results in anomalous profiles in which high-concentration defects are
believed to be present at the metallurgical junction [8, 9].

In this context, a large number of studies have been carried out to investigate laser-assisted doping of semiconductors. A pulsed laser can deliver
heat only to a thin surface layer, since the incident light is absorbed in the
near-surface region (typically 10 nm for UV light) [10], avoiding high-
temperature processing of all the wafer. Historically, the laser found its
first role as a simple substitute for thermal annealing, complementary to
dopant incorporation by ion implantation. More recently, the development

of techniques in which the laser is used to achieve both dopant incorporation from a solid, liquid or gaseous source and redistribution and electrical activation of impurities have been the focus of considerable activity. The gas immersion laser doping (GILD) technique has been shown to be particularly attractive, since it allows the formation of heavily doped shallow junctions in a one-step process with good control of the profile shape.

This doping process offers additional advantages directly linked to the properties of laser light. The directionality and coherence of laser radiation permit strong spatial localization of heat and/or chemical treatments of semiconductors, with a resolution theoretically limited by diffraction to the laser wavelength. These properties open the way to projection patterned processes that would allow direct doping without the requirement for resist patterning. The directionality of the laser light allows *in situ* processing of material placed in a controlled ambient, limiting pollution problems. Finally, the monochromaticity and the high energy density of laser light as well as the large choice in laser wavelength from infrared to far-ultraviolet (CO_2 laser 10 600 nm, F_2 excimer laser 157 nm) allows access to a large number of laser-induced physical and chemical processes on various materials, with a high degree of process control.

As a result, laser-induced processing of silicon can actually cover all of the necessary device fabrication steps, including material annealing, recrystallization, ablation, deposition and etching, as well as surface modifications like doping or oxidation [11−17]. Consequently, one can hope to gather together all these laser processes in order to achieve the fabrication of a complete integrated circuit with a unification of the processing techniques. In this way, McWilliams [18] has reported the fabrication of NMOS transistors and simple logical functions (NOR, NAND and NOT gates) with a resolution of about 1 μm by pyrolytic laser direct writing. Finally, laser processing is not limited to planar device fabrication, but allows three-dimensional fabrication as well. All these facts explain why laser processing of semiconductors has received a great deal of attention during the past 15 years.

This chapter concentrates on gas immersion laser doping of silicon with pulsed lasers. The second section describes the different laser-assisted doping techniques that led to the emergence of the GILD technique at the beginning of the last decade. The third section describes the principle of the GILD technique, which involves three basic mechanisms: (i) laser−silicon interaction; (ii) laser−gas interaction; and (iii) redistribution of the impurities. In the fourth section, experimental results on the GILD technique are reported and discussed from a technological point of view. In the fifth section, these results are analysed from a fundamental point of view in order to summarize the different mechanisms controlling the doping process. In conclusion, the expected applications and the problems encountered in the GILD technique are discussed.

11.2 LASER-ASSISTED DOPING TECHNIQUES

11.2.1 Laser annealing

The suitability of laser irradiation for annealing ion-implanted silicon samples was shown for the first time in 1975 by Soviet scientists [19]. In their experiments, boron, phosphorus and antimony ion-implanted silicon samples were annealed either by Nd : YAG pulsed laser irradiation or thermal annealing. Comparison between the electrical characteristics obtained after these two kinds of treatment demonstrated the capability of laser irradiation to anneal a thin surface layer without any modification of the bulk properties.

In principle, laser annealing is done by irradiating (generally in an air ambient) the sample surface with the focused laser beam. The incident light, strongly absorbed in a thin surface layer (of $10–10^3$ nm depth, depending on the laser wavelength), induces an increase in the surface temperature. If the laser fluence is higher than the melting threshold, irradiation results in rapid surface melting and epitaxial recrystallization from the underlying substrate in a very brief time (typically 100 ns), since the duration of the illuminating pulse itself is very short. This mechanism is responsible for the annealing of the implantation defects and for the electrical activation of the dopant impurities.

Since 1975, a large number of studies have been carried out to investigate laser annealing of ion-implanted (B, P, As, Sb, ...) silicon samples using various lasers [11, 20–27]. Experimental results show that, during this process, redistribution of impurities takes place by liquid phase diffusion in the molten layer under nonequilibrium conditions because of the high recrystallization velocities (typically 2–6 m s^{-1}). For this reason, the segregation coefficient at the liquid–solid interface and the solubility limit are both increased in comparison with their values under equilibrium conditions [22, 25–27]. This phenomenon allows the formation of junctions with high surface concentrations, for example up to 5×10^{21} at cm^{-3} [28] in the case of phosphorus, which is well above its solubility limit under equilibrium conditions (1.5×10^{21} at cm^{-3} [29]). Samples exhibit good crystallinity and are largely free from extended defects like dislocations or stacking faults [30].

Diodes and solar cells have been processed on laser-annealed samples [24, 31–33]. Leakage and recombination currents of these devices are relatively high, providing evidence that electrical defects are induced by the laser treatment at the sample surface. Another source of electrical defects is constituted by the defects of the implantation tail if the melt depth is not sufficient.

In order to eliminate the latter kind of defects resulting from ion implantation and to simplify the doping process, recent studies have addressed the possibility of using laser heating to induce dopant incorporation from a range of sources.

11.2.2   Laser-induced doping from solid dopant sources

In 1968, Fairfield [34] reported for the first time the formation of np junctions by irradiating silicon samples covered by a thin film of deposited phosphorus with a ruby laser. Two years later, Harper [35] processed n^+p junctions by Nd : YAG laser-induced diffusion from an evaporated aluminium layer. However, it was only at the end of the 1970s that laser-induced doping from solid sources came under more intensive study [27, 31, 36–39]. Incorporation and diffusion of various impurities (deposited on the silicon surface by different methods, such as evaporation, spin-on or painting) has since been achieved using various pulsed lasers.

Laser irradiation, carried out at energy densities above the melting threshold, induces incorporation and redistribution in the melted surface layer of the impurities supplied by the deposited dopant source. As in laser annealing, impurity redistribution takes place by liquid phase diffusion under nonequilibrium conditions. Thus junctions with solubility limits much higher than those obtained under equilibrium conditions can be reached.

In spite of the fact that laser-induced doping from deposited solid sources avoids the creation of defects induced by ion implantation, this technique is still a two-step method, with (i) dopant film deposition and (ii) laser-induced redistribution of the dopant atoms. Therefore one-step laser doping processes have been investigated.

11.2.3   Laser-induced doping from liquid dopant sources

In 1981, Stuck [40] proposed a one-step laser doping process involving a liquid dopant source. With this technique, the silicon sample to be doped is placed in a doped organic liquid. It is then irradiated through the liquid with a pulsed ruby laser, which induces surface melting, allowing impurity redistribution in the sample surface. As with previous laser-assisted doping

Table 11.1. Gas immersion laser doping (GILD) in silicon.

Dopant gas	Laser[a] (λ (nm), τ (ns))	Absorbant gas	Doping mechanisms:[b] $I \propto t_d^{\gamma}$	Dopant sources	Ref.
$B(CH_3)_3$	ArF (193, 7)	Yes			[42]
$B(C_2H_5)_3$	ArF (193, 7)	—			[43, 44]
B_2H_6	Alexendrite (730, 180)	No			[45]
	XeCl (308, 27)	No	Adsorbed layer: $\gamma \approx 1$	Adsorbed layer: pyrolysis	[46–50]
BCl_3	ArF (193, 7)	Yes	Gas ambient: $\gamma \approx 0.6$	Adsorbed layer	[51, 52]
	ArF (193, 21)	Yes	Adsorbed layer: $\gamma \approx 0.5$ Gas ambient: $\gamma \approx 0.75$	Chemisorbed + physisorbed molecules: pyrolysis + photolysis	[53–55]
	KrF (248, 34)	No	Adsorbed layer: $\gamma \approx 1$ Gas ambient: $\gamma \approx 1$	Chemisorbed + physisorbed molecules: pyrolysis	[56]
	XeCl (308, 55)	No	Adsorbed layer: $\gamma \approx 1$ Gas ambient: $\gamma \approx 1$	Chemisorbed + physisorbed molecules: pyrolysis	[54]
	XeF (351, 7)	No		Adsorbed layer: pyrolysis	[51, 52]

	Laser^a (λ, τ)	Incorporated dose^b	γ	Process	Ref.
BF₃	ArF (193, 17)	No	Absorbed layer: $\gamma \approx 1$ Gas ambient: $\gamma \approx 1$	Adsorbed layer: pyrolysis	[57]
	ArF (193, 21)	No	Adsorbed layer: $\gamma \approx 1$ Gas ambient: $\gamma \approx 1$	Chemisorbed + impinging molecules: pyrolysis	[55, 58]
	XeCl (308, 27)	No			[59]
PH₃	Alexendrite (730, 200)	No		Pyrolysis	[60]
PCl₃	ArF (193, 7)	—	Gas ambient: $\gamma \approx 1$	Adsorbed layer	[51, 52]
	XeCl (308, 55)	No	Gas ambient: $\gamma \approx 1$	Chemisorbed + physisorbed molecules: pyrolysis	[61–64]
POCl₃	ArF (193, 17)	Yes		Impinging molecules: photolysis	[65]
PF₅	ArF (193, 21)	No	Adsorbed layer: $\gamma \approx 0.5$ Gas ambient: $\gamma \approx 0.75$	Chemisorbed + physisorbed molecules: pyrolysis	[66, 67]
AsH₃	Alexendrite (730, 180)	No			[45]
	XeCl (308, 27)	No	Gas ambient: $\gamma \approx 1$	Adsorbed layer: pyrolysis	[47–50]
Al(CH₃)₃	ArF (193, 7)	Yes			[42]

^{a} λ and τ are respectively the laser wavelength and pulse duration.
^{b} Incorporated dose I as a function of doping time t_d, deduced from experimental measurements.

techniques, shallow and heavily doped junctions can be obtained. However, due to the high dopant concentration in the liquid at the liquid–semiconductor interface, control of the junction surface concentration is difficult.

11.2.4 Laser-induced doping from gaseous sources: the GILD technique

The first results on gas immersion laser doping of silicon and gallium arsenide placed under an arsenic or antimony atmosphere were reported by Solomon in 1968 [41]. However, until the commercialization of the excimer laser and the development of the laser photochemical processing of semiconductors at the beginning of the 1980s, no additional work was carried out in this area.

In 1980, Deutsch [42] reported on gas immersion laser doping of silicon under an organometallic atmospheric ($B(CH_3)_3$ and $Al(CH_3)_3$ dopant gas) with an ArF excimer laser. The dopant atoms supplied at the sample surface by the photochemical decomposition of the organometallic molecules constitute the dopant source for impurity incorporation and redistribution during sample surface melting.

Since then, gas immersion laser doping of silicon has been performed using various dopant gases and pulsed lasers in order to form n- or p-type doped layers. Experimental conditions and results for the gas immersion laser doping in silicon are reported in Table 11.1.

11.3 THE MECHANISMS OF THE GILD TECHNIQUE

11.3.1 Experimental principle of the GILD technique

The silicon sample to be doped is enclosed in a stainless steel cell under an appropriate dopant gas atmosphere. The sample surface is then irradiated through the gas with a focused pulsed laser beam. At each pulse, the laser irradiation is used both to melt a controlled thickness of the silicon surface, by the laser–silicon interaction, and to dissociate by pyrolysis and/or photolysis the dopant molecules, by the laser–gas interaction. The resulting dopant atoms are then partly incorporated into the molten layer, where they are redistributed by liquid phase diffusion before epitaxial recrystallization of the sample surface, as shown in Fig. 11.1. As can be seen, the GILD technique is dependent on three basic mechanisms: (i) the laser–silicon interaction; (ii) the laser–gas interaction; and (iii) the incorporation and diffusion in the liquid phase of the impurities, which are discussed below.

11.3.2 Laser–silicon interaction

The mechanism of energy transfer from incident photons to the lattice is now well known [68–71]. It occurs via three steps: (i) carrier creation in the sample surface by photon absorption; (ii) carrier thermalization and recombination; and (iii) energy transfer from the carrier to the lattice. In

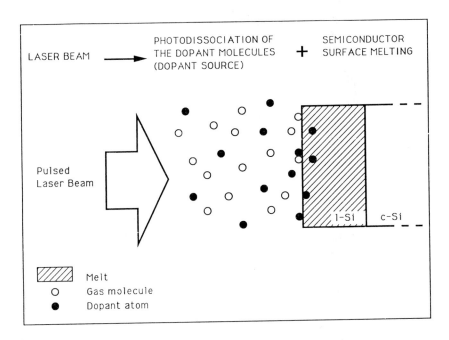

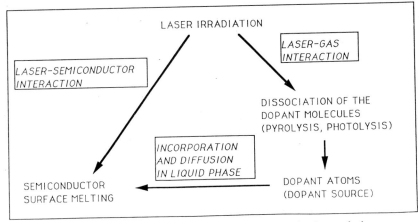

Fig. 11.1 Principle of the gas immersion laser doping technique.

the case of high-fluence laser irradiation, it has been shown that the dominant effects take place in about 10^{-11} s, which is much less than the laser pulse duration (typically $\tau = 10^{-8}$ s). Thus the energy transfer can be assumed to be instantaneous and a thermal description of the laser–silicon interaction can be used, involving the heat conduction equation.

As the optical and thermodynamic parameters involved in this equation depend on the silicon temperature, it cannot be solved analytically, and it is therefore numerically solved using computers. A large number of simulation programs have been developed for the laser-induced melting of silicon [11, 48, 63, 72–76]. These programs yield important information for the GILD process.

11.3.2.1   Melting threshold

The melting threshold E_s is the minimum laser energy density needed to melt the silicon surface. It depends on the laser wavelength and on the shape and duration of the laser pulse. It can be experimentally determined by following the silicon surface reflectivity during laser irradiation: time-resolved reflectivity (TRR) measurement [11, 48, 50, 76]. As the laser doping process is achieved by the redistribution of dopant atoms in the liquid phase, the laser energy density supplied for the GILD process must be higher than this threshold.

11.3.2.2   Melt depth and melting time

The simulation programs also give the maximum depth reached by the solid–liquid interface during the melting process, which is generally called the melt depth X_m, and the time during which the silicon surface is molten, i.e. the melting time t_m. We shall see in Section 11.3.5 that the value of X_m controls the layer in which the dopant atoms are redistributed and t_m controls the time available for these atoms to be incorporated and diffused.

Melt depth and melting time are generally plotted as functions of the laser energy density E_L, as shown in Figs 11.2(a, b) for the case of irradiation with three excimer lasers, working at different wavelengths and having different pulse durations [74, 76, 77]. We can observe for each laser increases in the melt depth and melting time as the laser energy supply increases. The melting thresholds can also be deduced from these curves, as shown in Fig. 11.2(a).

11.3.2.3   Influence of silicon parameter variation on simulation results

Simulation programs are generally developed using the thermodynamic and

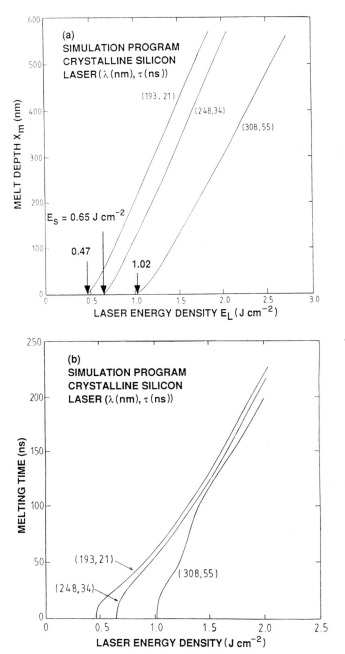

Fig. 11.2 Evolution of the melt depth X_m and melting time t_m as functions of the laser energy density. Calculations are shown for three excimer lasers: ArF ($\lambda = 193$ nm, $\tau = 21$ ns), KrF (248, 34) and XeCl (308, 55) [77].

optical parameters of weakly doped crystalline silicon. However, the GILD process in silicon induces an increase in the doping level and can also lead to the formation of a thin film of adsorbed compounds or to the degradation of the processed surface layer. All these phenomena can result in modification of the melt depth and melting time. Such an effect has been observed to lead to a decrease of $150°C$ in the melting temperature of heavily doped silicon (3.5×10^{21} As cm^{-3}) [78]. This can induce an increase of about 20% in the melt depth with an ArF excimer laser ($\lambda = 193$ nm, $\tau = 21$ ns, $E_L = 0.75$ J cm^{-2}) as predicted by the simulation program developed by De Unamuno [77].

Furthermore, Fogarassy [79] has observed that an increase in doping level can lead to an important decrease in both solid and liquid reflectivities of silicon. An increase in the boron concentration to 10^{20} B cm^{-2} has been found to reduce the surface reflectivity by about 10%, which results in an increase of 50% in the melt depth. The influence of such effects therefore has to be considered in the case of GILD processing.

11.3.3   Laser–gas interaction

When the dopant gas is introduced into the experimental chamber, adsorption of molecules occurs at the sample surface. Therefore two kinds of laser–gas interaction have to be considered: those taking place in the gas phase and those taking place in the adsorbed phase. In addition, dissociation of the dopant molecules can result either from a pyrolytic or a photolytic process. This gives rise to four different ways for the formation of dopant atoms, which are schematically described in Fig. 11.3.

11.3.3.1   Interaction in the gas phase

As shown in Fig. 11.3, dopant atoms can be supplied either by photolysis or pyrolysis of molecules in the gas phase.

In the former mechanism, laser light absorbed by the molecule breaks a chemical bond, inducing the formation of dopant species, which can diffuse up to the melted silicon surface, where they are incorporated. Such a photolytic process occurs only if the photon energy is sufficient to bring the molecule into a dissociative state. The dissociation energy of the molecule, E_d, can be deduced, for example, from the relation between E_d and the enthalpy variation associated with the equivalent chemical process of decomposition [80–82]. This can then be compared with the photon energy. Such a simple calculation leads to the prediction that photolysis of boron trichloride gas ($E_d = 5.5$ eV) could occur using 193 nm light (ArF

1. PHOTOLYSIS OF THE ADSORBED MOLECULES

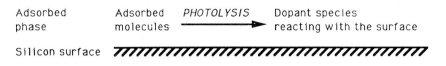

2. PYROLYSIS OF THE ADSORBED MOLECULES

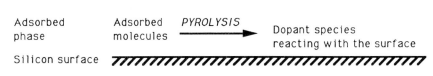

3. PHOTOLYSIS OF THE MOLECULES IN GAS PHASE

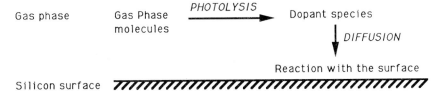

4. PYROLYSIS OF THE MOLECULES IN GAS PHASE

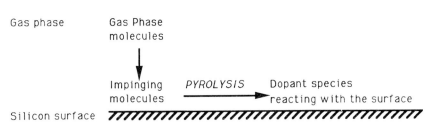

Fig. 11.3 Dopant atom supplies in the GILD process.

laser, $E_{photon} = 6.4$ eV), but boron trifluoride ($E_d = 12.1$ eV) would only be decomposed by a pyrolytic mechanism at this wavelength.

In fact, experimental measurements of the gas optical transmission at the laser wavelength, giving the gas absorption cross-section, are necessary to determine the yield from photolytic decomposition. Moreover, these measurements must be performed at the high laser energy densities used for the GILD technique. Indeed, for high fluences, multiphoton excitation of the molecules can take place in addition to a single-photon process, as reported in the case of boron trichloride gas [83]. Precise knowledge of the gas absorption cross-section is also essential for the GILD process in order to calculate the accurate laser fluence irradiating the sample surface through the gas.

Pyrolytic decomposition of gas molecules during laser irradiation can take place when gas molecules strike the hot molten surface. Heat transfer from the silicon surface to the molecule results in its decomposition by a thermochemical reaction. The resulting dopant species located at the silicon surface are then incorporated and redistributed in the molten surface region.

The number of dopant atoms supplied by these two mechanisms is roughly proportional to the number of dopant species impinging on the surface, and then is assumed to increase linearly with the gas pressure.

11.3.3.2 Interaction in the adsorbed phase

Two types of adsorption are generally distinguished, chemisorption and physisorption, corresponding respectively to molecules strongly (0.5–5 eV) and weakly (0.05–0.5 eV) bound to the surface. This difference in surface bond strength results from the nature of the bonds concerned (van der Waals and chemical respectively). While the population of a physisorbed layer can often be simply controlled by variation of surface temperature or chamber pressure, chemisorbed species are more difficult to control. The species present in this case does not necessarily conform to a simple molecular bound form of the precursor gas, but may be a highly reactive or dissociated state. Understanding the influence of the laser on this type of phase is clearly much more complex.

As in the gas phase, molecule decomposition can take place either by a pyrolytic or a photolytic mechanism. For photolytic decomposition, a small shift to a higher wavelength in the absorption spectrum (from 215 to 225 nm, for example, in the case of $Cd(CH_3)_3$ gas) and an extension of the absorption band are generally observed for the chemisorbed molecules. In general, only an extension of the absorption band is observed for physisorbed molecules [84–86].

Whatever the decomposition mechanisms, the quantity of dopant atoms supplied by the adsorbed phase is expected to be proportional to the quantity of adsorbed molecules and thus to be linked to the adsorption isotherm of the dopant gas on the silicon surface.

11.3.4   Impurity redistribution in the liquid phase

Incorporation and diffusion of dopant atoms during laser-induced melting processes have been largely studied for other laser-assisted doping techniques [11, 22, 25, 26, 87–90]. It has been shown that dopant redistribution can be predicted by solving the simple one-dimensional Fick diffusion equation:

$$\frac{\partial C(x,t)}{\partial t} = D_\ell \frac{\partial^2 C(x,t)}{\partial x^2}, \tag{11.1}$$

where $C(x,t)$ is the impurity concentration, taking into account the time dependence of the liquid–solid interface position during the melting process. As pointed out above, dopant redistribution in laser-assisted doping techniques has particular characteristics because it is not an equilibrium process.

11.3.4.1   Diffusion coefficient

Diffusion of impurities in the liquid phase is much faster than in the solid phase. Diffusion coefficients D_ℓ in the liquid phase are typically in the range of 10^{-4} cm^2 s^{-1} as against 10^{-11} cm^2 s^{-1} for D_s at high temperature in the solid phase (see Table 11.2). As a result, impurity diffusion in the solid is negligible during the short doping time (typically $t_d = 1$ μs), and thus the junction depth X_j is limited by the melt depth.

11.3.4.2   Segregation coefficient

Nonequilibrium segregation of impurities at the solid–liquid interface results in higher values of the segregation coefficient K in comparison with its value under equilibrium condition (see Table 11.2). This trend results in a diminution of impurity segregation at the solid–liquid interface, which allows the formation of a uniform dopant concentration along the molten layer and thus the formation of boxlike profile junctions.

11.3.4.3   Solubility limit

Values of the solubility limit C_{max} under equilibrium and nonequilibrium conditions are reported in Table 11.2. They show that higher solubility

Table 11.2. Values of the diffusion coefficient of boron, phosphorus and arsenic in the solid phase at high temperature (1200°C) D_s, and in the liquid phase, D_ℓ, in silicon. The segregation coefficient K and solubility limit C_{max} of these impurities under equilibrium (subscript zero) and nonequilibrium conditions are also shown.

Impurity	Boron	Phosphorus	Arsenic
D_s (cm^2 s^{-1}) [91]	1.3×10^{-12}	1.7×10^{-12}	1.3×10^{-13}
D_ℓ (cm^2 s^{-1})	2.4×10^{-4} [92]	5.1×10^{-4} [92]	3.3×10^{-4} [92]
	3.3×10^{-4} [93]	2.7×10^{-4} [93]	2.5×10^{-4} [48]
K_0 [29]	0.8	0.35	0.3
K	0.96 [87]	0.85 [87]	~1 [48, 94, 95]
	0.9–1 [86]	0.9–1 [86]	
$C_{max,0}$ (cm^{-3}) [29]	6×10^{20}	1.5×10^{21}	1.5×10^{21}
C_{max} (cm^{-3})	2×10^{21} [88]	5×10^{21} [94, 96]	6×10^{21} [92]

limit values are obtained with laser-induced doping techniques, allowing the formation of heavily doped junctions. The maximum substitutional solubility of impurity is limited by different mechanisms. For boron doping in silicon, for example, it has been shown, in the case of laser annealing of ion-implanted samples, that limitation occurs due to lattice strain when the local boron concentration exceeds about 4 at% [97, 98].

11.3.5 Doping parameters

From the discussion above, one can determine the main doping parameters in the GILD process. The choices of dopant gas and laser wavelength determine the type of mechanism leading to the decomposition of the molecule and the nature of the dopant source. The laser energy density controls the melt depth and hence limits the junction depth. The gas pressure, which determines the quantity of adsorbed molecules and the flux of molecules striking the molten surface, fixes the dopant atom supply. Finally, the number of laser pulses controls the incorporated dopant dose and the final profile shape.

11.4 TECHNOLOGICAL ASPECTS

11.4.1 Role of surface cleaning

Before gas immersion laser doping, silicon samples are usually cleaned in a chemical solution in order to remove native oxide, and are then immediately placed in the doping chamber [48, 49, 53–55, 58, 64–67]. The importance of this stage has been stressed by comparing the evolution of the surface sheet resistance of both cleaned and uncleaned samples as a function of the number of laser pulses for GILD processing under a BCl_3 atmosphere with an ArF excimer laser [53].

The experimental results show that, during the first pulse, the doping process is completely inhibited by the presence of native oxide at the sample surface. This can be easily understood by calculating the boron diffusion length L_d through a silicon oxide layer during the surface melting time, typically $t_m = 100$ ns. At the silicon melting temperature ($T = 1680$ K), the boron diffusion coefficient in silicon dioxide is of order 10^{-15} cm^2 s^{-1} [91], resulting in a diffusion length of about 10^{-4} nm, which is much smaller than the native oxide thickness, which is of order 2 nm [99].

After repeated pulses, the sheet resistance of uncleaned samples decreases, but is always higher than that of cleaned ones. This is attributed to the degradation of the native oxide by the breaking of the Si—O bond by the laser irradiation at 193 nm [100]. The incorporation of dopant atoms becomes easier, and thus the surface sheet resistance is reduced.

11.4.2 Influence of doping parameters

11.4.2.1 Laser energy density

The influence of laser energy density on the doping process has been studied by following the evolution of the surface sheet resistance when E_L is increased [47, 52, 57, 58, 67]. Figure 11.4 shows a typical evolution of the sheet resistance as a function of E_L obtained for the case of doping in 50 Torr of BF_3 after five pulses with an ArF excimer laser [58]. For laser energy densities below 0.5 J cm^{-2}, the sheet resistance is constant and has the substrate value (about 1.8×10^4 $\Omega/\square$), providing evidence that the doping process does not occur. For values above 0.5 J cm^{-2}, the sheet resistance decreases sharply with laser energy density, due to the incorporation and electrical activation of impurities in the sample surface. This doping threshold is directly correlated with the melting threshold of silicon, which is about 0.47 J cm^{-2} for the laser used in this experiment. As the laser fluence increases above the melting threshold, the melting time becomes longer and the melt depth becomes deeper (Section 11.3.2.2). Consequently, a greater number of dopant atoms are incorporated and redistributed into the molten silicon, resulting in lower sheet resistance values.

The influence of the laser energy density on the profile shape has also been reported by other authors [47–50, 63, 64]. Figure 11.5 shows the impurity concentration profiles of boron-doped silicon samples irradiated at two different laser energy densities for the same gas pressure and number

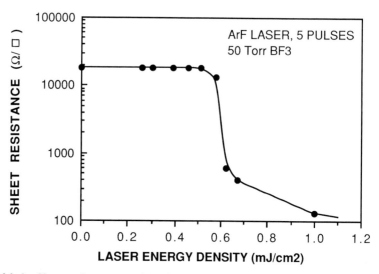

Fig. 11.4 Sheet resistance as a function of the laser energy density after five pulses in 50 Torr BF$_3$ with an ArF excimer laser (193 nm, 21 ns).

of pulses. The calculated melt depths [77] for these two energy densities are also shown. One can observe an increase in junction depth and dopant incorporated dose, which corresponds to the integral of the concentration profile, as functions of the laser energy density. This is in agreement with the sheet resistance measurements. These data also show that the junction depths are in good correlation with the predicted melt depths. This behaviour confirms that the dopant redistribution is limited to the melted layer as a result of the large difference between the diffusion coefficients in the liquid and solid phases. Thus accurate *in situ* control of GILD-processed junction depths can be achieved by measuring the surface melting time during the doping process (TRR measurement) and using the simulation program calculations giving the melt depth as a function of the melting time. This technique, used by Carey [48, 50], allows the formation of heavily doped, box profile, ultrashallow (down to 50 nm) junctions with a good control and reproducibility.

11.4.3.2 Number of laser pulses

For laser energy densities above the doping threshold, repetitive irradiation under a gas atmosphere induces a decrease in surface sheet resistance [45, 48–54, 57, 58, 65, 67]. Figure 11.6 shows a typical dependence of sheet resistance on number of pulses in the case of doping at 1 J cm^{-2} with an

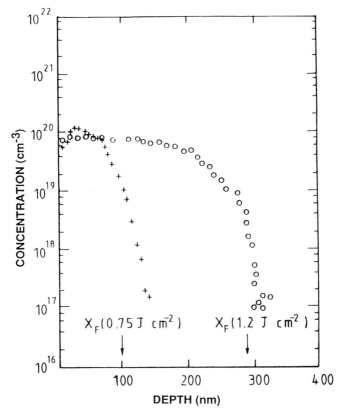

Fig. 11.5 Boron concentration profiles from SIMS measurements for samples doped in 5 Torr BCl_3 after five pulses at 0.75 ($+$) and 1.2 ($\circ$) J cm^{-2} with an ArF excimer laser (193 nm, 21 ns).

ArF excimer laser under different conditions of BF_3 gas pressure [58]. Curve (a) was obtained using the adsorbed layer of BF_3 molecules as the dopant source. This was achieved by irradiating the sample in the doping chamber, which was initially filled with 50 Torr BF_3 gas and then pumped down to 10^{-2} Torr, making the quantity of molecules from the gas phase striking the molten surface negligible. Curves (b), (c) and (d) show the results for doping under 1, 10 and 50 Torr BF_3 respectively. These curves are quite similar. After the first pulse, the sheet resistance decreases strongly from the substrate value (1.8×10^4 $\Omega/\square$) and continues to decrease with the number of pulses to reach values as low as 10 $\Omega/\square$ after 50 or 200 pulses, depending on the BF_3 gas pressure. The decrease in sheet resistance with repetitive irradiation is explained by the incorporation into silicon and the subsequent redistribution and activation of an increasing

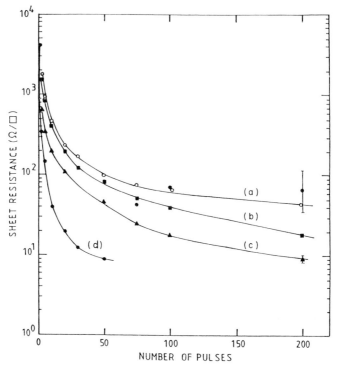

Fig. 11.6 Sheet resistance as a function of the number of laser pulses after doping at 1 (b, ■), 10 (c, ▲) and 50 (d, ●) Torr BF$_3$ ambients and using the adsorbed layer (a, ○) with an ArF excimer laser (193 nm, 21 ns) at $E_L = 1.5$ J cm^{-2}.

quantity of dopant atoms supplied by the adsorbed layer and/or by the gas phase.

Evolution of dopant concentration profiles with number of pulses is also commonly reported. Figure 11.7 shows the SIMS profiles of boron-doped silicon samples after different numbers of laser pulses. The surface concentration and the junction depth increase with number of pulses as larger quantities of dopant atoms are incorporated and diffused in the molten surface after each pulse. The shape of the profile is explained by the very fast diffusion of the dopant atoms in the molten silicon. Indeed, a boron diffusion coefficient of about 3×10^{-4} cm^{-2} s^{-1} has been determined from these profiles, in good correlation with the boron diffusion coefficient in the liquid phase reported in the literature (see Table 11.2). Due to this fast diffusion, square-shaped profiles are obtained within a few pulses (about 20). The junction depth after 20 pulses, $X_j = 220$ nm, is in good agreement with the calculated melt depth, which is about 200 nm. However, an

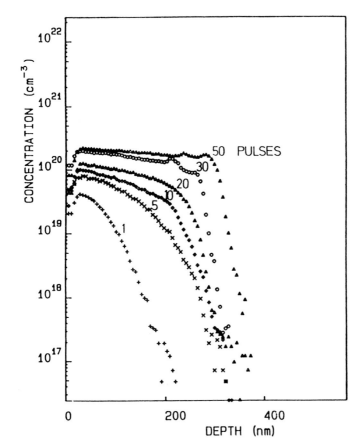

Fig. 11.7   Boron concentration profiles from SIMS measurements for samples doped in 5 Torr BCl$_3$ at 1.2 J cm^{-2} with an ArF excimer laser (193 nm, 21 ns) after 1 (+), 5 (×), 10 (◇), 20 (△), 30 (○) and 50 (▲) pulses.

increase in junction depth beyond the predicted melt depth with the repeated irradiation has often been reported both for boron and phosphorus gas immersion laser doping in silicon [53–55, 57, 63, 65]. This effect can be observed in Fig. 11.7, where the junction depth after 50 pulses, $X_j \approx 300$ nm, is 50% larger than the predicted melt depth. This phenomenon is generally attributed to the increase in melt depth resulting from the decrease on surface reflectivity and from some variations in the thermodynamic parameters of heavily doped silicon, as discussed in Section 11.3.2.3. However, as will be discussed below, it seems that the junction depth increase cannot be simply attributed to the increase in dopant surface concentration.

11.4.3.3 Gas pressure

Studies of the influence of gas pressure on the GILD process have generally shown that an increase of gas pressure results in an increase in the number of incorporated atoms, which in turn induces a diminution of the surface sheet resistance. This behaviour can be observed in Fig. 11.6. For a given number of pulses, the sheet resistance decreases as the gas pressure is increased. This trend is related to the increase in the number of molecules striking the molten surface, and the number of molecule adsorbed on the silicon surface between pulses, with gas pressure [51–53, 58, 64, 66, 67].

However, Kato [57] has not observed a substantial increase in incorporated dose on doping with an ArF excimer laser under 50 Torr BF_3 instead of using the adsorbed layer of BF_3. From this result, he concluded that the dopant atoms were mainly supplied by the adsorbed phase and that the gas phase contribution is negligible even at 50 Torr. However, our results [58] on ArF excimer laser doping with BF_3 are fundamentally different, showing that a large quantity of dopant atoms is supplied by the gas phase when the BF_3 gas pressure is increased (see Fig. 11.6). The reason for this discrepancy is not known.

The influence of gas pressure on dopant profiles is shown in Fig. 11.8. SIMS profiles were obtained at different gas pressures in the range $10^{-2}–10^2$ Torr for the same number of pulses and laser energy density. Boxlike profiles with a depth of about 250 nm are obtained whatever the gas pressure, but the surface concentration increases strongly from 4×10^{19} to 10^{21} B cm^{-3} with gas pressure. The surface concentration at 100 Torr is well above the equilibrium solubility of boron in silicon (Table 11.2), providing evidence that the GILD process is a nonequilibrium one. However, it can be seen that the strong increase in surface concentration does not result in a significant increase in junction depth and thus in melt depth. This suggests that the mechanisms leading to increased melt depth is complex and needs more investigation. A decrease in surface reflectivity resulting from the formation of a thin layer of photoproducts with repetitive irradiation, for example, could explain the decrease in melt depth.

11.4.4 Structural characterization of GILD-processed silicon

11.4.4.1 Surface status

The surface status of GILD-processed silicon under halogenated dopant gases (BCl_3, BF_3, PF_5, PCl_3) has been monitored using scanning electron microscopy (SEM) and profilometer techniques [58, 67]. These studies have focused particularly upon the possibility of surface etching following repetitive irradiation of the sample surface under chlorine and fluorine

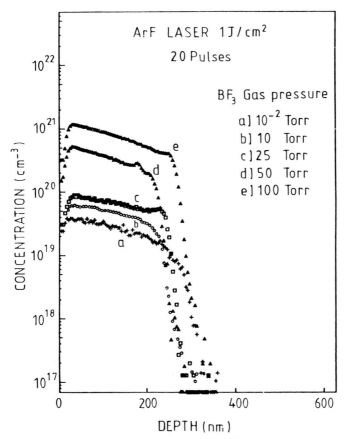

Fig. 11.8   Boron concentration profiles from SIMS measurements for samples doped after 20 pulses at 1.2 J cm^{-2} with an ArF excimer laser (193 nm, 21 ns) for different gas pressures (in Torr): (a) 10^{-2}; (b) 10; (c) 25; (d) 50; (e) 100.

compounds, which can result from decomposition of the dopant parent gas.

Surface degradation in the form of surface etching or roughness has not been observed within the detection limit of these techniques in most cases.

In contrast, strong surface degradation has been observed for heavily boron-doped samples with concentrations close to the nonequilibrium solubility limit (about 2×10^{21} B cm^{-2}) after doping in a BF$_3$ atmosphere [58]. Cracks approximately 1 µm wide and deep are present in the near-surface region of the sample. These cracks result from the lattice strain mechanisms, which has been shown to limit the dopant activation of heavily boron-implanted laser-annealed silicon samples (Section 11.2.3).

11.4.4.2 Undesired impurity incorporation

When the GILD process is performed using a halogen-containing gas, it is of importance to know if chlorine or fluorine are incorporated in the melted layer during the doping process. For this purpose, SIMS and RBS measurements have been performed for doping in BCl_3, BF_3, PF_5 [56, 58, 67] and PCl_3 [62] atmospheres. No significant incorporation of chlorine or fluorine has been detected even after a large number of pulses with high gas pressures (up to 200 pulses under 100 Torr PF_5, for example), within the detection limits of these techniques.

11.4.4.3 Lattice defects

The crystalline quality of GILD-processed samples has been investigated by RBS channelling measurements [45, 47, 56, 62]. Recrystallized layers exhibit excellent crystalline quality (minimum yield $X_{min} = 3-5\%$) within the accuracy of this measurement technique, providing evidence that doped layers are free of extended defects. These results can be related to those obtained for other laser-assisted doping techniques (Section 11.2).

Recently, Nipoti [101] has analysed lattice structures resulting from the laser melting process using X-ray diffraction, complemented by chemical (SIMS) and electrical (automatic resistivity–Hall mobility stripping: ARHS) profile measurements. The results confirm that extended defects with strong displacement fields do not form during the laser melting process. However, analysis of the strain profiles obtained for both GILD-processed samples and samples irradiated under high vacuum shows that the laser melting process is able by itself to generate a rather high density of vacancy-type melt-related point defects. It was also observed that the defect distribution exceeds the melt depth, suggesting that the lattice is perturbed also in the region immediately below the melted layer.

11.4.5 Electrical characterization

11.4.5.1 Electrical activation

The proportion of dopant atoms (B, P and As) electrically active in substitutional sites has been determined by different techniques, by comparing the electrical (ARHS) and chemical (SIMS) concentration profiles [56, 101] or by comparing RBS profiles under channelling and nonchannelling conditions [45, 47, 62]. Results show that almost all the dopant atoms ($>90\%$) are in substitutional positions after the GILD process, even for heavily doped samples above the equilibrium solubility limit and below the nonequilibrium one.

11.4.5.2   Carrier mobility

Measurements of the profiles of mobility variation relative to the best values published in the literature [102] have been performed by the ARHS technique for both n and p laser-doped silicon [54, 56, 101]. The results show that carrier mobilities are noticeably poorer than the reference value ($\Delta\mu/\mu_{\text{best value}} \approx -30\%$), in all cases studied. This behaviour is consistent with the presence of defects, which act as scattering centres in the laser-doped layer. It has been shown that a rapid thermal annealing ($T = 1000°C$, $t = 10$ s) is able to reduce the point defects resulting from the laser-doping process, and then improves the mobility values without affecting the dopant profile [54].

11.4.5.3   Electrical characteristics

Fabrication of diodes, MOSFETs and solar cells by the GILD technique has been reported in the literature. Electrical characteristics of these devices are discussed below.

(i) Diodes

Capacity versus reverse-bias voltage measurements ($C-V$) have been performed on p^+n and n^+p diodes processed by the GILD technique [43–47, 49, 53, 54, 58]. Linear trends of $1/C^2$ versus V are generally observed, confirming that the electrical junctions are abrupt, but in the case of doping under BF_3 and PF_5 atmospheres deviation from a straight line at low reverse bias is observed [56, 58]. This behaviour is attributed to the presence of highly electrically active defect concentrations induced by the laser thermal treatment in an extended region (up to 1 μm) deeper than the emitter region. Built-in potentials deduced from $C-V$ measurements are typically about 0.65–0.9 V.

Forward and reverse current–voltage ($I-V$) characterstics have shown that laser-doped diodes generally exhibit relatively high ideality factors $n = 1.2-1.9$ and high leakage current values, typically a few tenths of a microamp per square centimetre at -1 V. These results confirm that electrical defects are present in the laser-doped layer. This is corroborated by recent deep-level transient spectroscopy measurements performed on Schottky diodes made on ArF excimer laser-irradiated silicon samples [103, 104].

However, it has been shown that these microscopic electrical defects can be removed by furnace annealing ($T > 650°C$) [104] or rapid thermal annealing ($T = 950°C$, $t = 10$ s) [45–47, 54]. Post-annealed diodes exhibit ideality factors close to unity and leakage currents as low as 47 nA cm^{-2}

at -10 V [46]. Moreover, experiments performed to determine the source of the excess leakage currents after GILD processing have shown that optimizing the number of laser pulses and laser energy density allows the formation of junctions with good electrical characteristics without a post-doping annealing step [59]. Thus reverse leakage currents of less than 10 nA cm^{-2} at -5 V have been reported for p^{+}n diodes that had not received high-temperature post-doping treatment ($T < 400°$C).

(ii) MOSFETs

Fabrication of submicrometre p- and n-channel MOSFETs by GILD processing have been reported by Carey [49, 59]. Figure 11.9 shows a cross-section (not to scale) of the NMOS device structure during the GILD process achieved in a dilute AsH$_3$ ambient with an XeCl excimer laser. In this process, the dopant atoms are supplied by the AsH$_3$ molecules adsorbed at the silicon surface, and the GILD process is used both to form the p^{+} source/drain contacts and to dope the polysilicon gate. The field oxide and the gate act as a mask for the doping process. The GILD process results in shallow junctions ($X_j \approx 110$ nm) with surface concentrations of about 10^{20} at cm^{-3}. In order to compare the GILD process with a standard

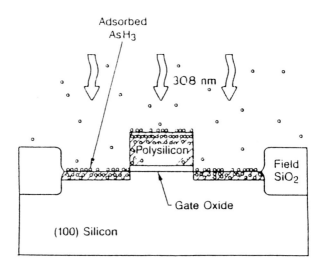

Fig. 11.9 Fabrication of a N MOSFET by the GILD technique. The cross-section of the MOS device shows the adsorption of the dopant species (in this case AsH$_3$) onto the polygate and source–drain regions. The laser irradiation results in dopant incorporation and redistribution in the regions not covered by the field oxide [49]

fabrication process, a set of samples have been doped in parallel by the ion implantation technique, followed by thermal annealing.

Electrical measurements performed on GILD-processed MOS devices show excellent drain characteristics even without thermal annealing, down to channel lengths of 0.8 μm, which is the shortest gate length defined in the process. Transconductances of ion-implanted and laser-processed MOS show comparable values, ranging from 15 to 20 μs for the implanted devices and from 12 to 33 μs (depending on the laser energy density) for the laser-doped devices. In addition, a study of the evolution of threshold voltage versus gate length, from 20 to 0.8 μm, has been carried out for both GILD-processed and implanted MOS to compare the short-channel behaviour. Results show that the laser-processed devices are at least as good as the standard implanted one. Moreover, it has been shown that GILD-processed devices have virtually no source–drain encroachment under the gate (less than 0.05 μm) due to the very short diffusion time.

(iii) Solar cells

The main advantage of the GILD process for solar cell formation is its ability to leave the substrate at room temperature during the doping process, avoiding degradation of the minority carrier diffusion length of the substrate. Both p^+n and n^+p cells have been fabricated using the GILD technique [42, 60, 61]. The best results report efficiences of about 9.6% without an antireflecting coating and without optimization of the solar cell processing.

11.4.6   GILD process modelling program

Simulation programs that model the GILD process have been developed by Carey [47, 48, 50] and Bentini [62, 63]. They include simulations of both the laser-induced melting process and the dopant redistribution in the melted layer. The melting process simulation was discussed above (Section 11.3.2). It combines the laser parameters of pulse shape and duration, the energy density and the number of pulses with the silicon thermodynamic and optical parameters to calculate the melt depth and the melting time. For the dopant redistribution, the simulation program numerically or semi-analytically resolves the one-dimensional diffusion equation (11.1), taking into account the motion of the liquid–solid interface during the melting process. Two kinds of dopant incorporation models are considered: an adsorption model and an impingement model.

The first assumes that the dopant source is a layer of adsorbed atoms

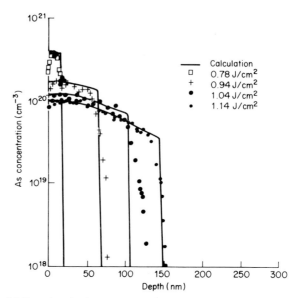

Fig. 11.10 RBS and calculated profiles of GILD-processed samples under an AsH₃ atmosphere (10% AsH₃, 90% He; 500 Torr) after 10 pulses with an XeCl excimer laser (308 nm, 27 ns) at different laser energy densities (in J cm⁻²) [48]: □, 0.78; +, 0.94; •, 1.04; ·, 1.14; ——, calculation.

that forms on the surface prior to its melting. The initial boundary condition at the gas–silicon interface is then

$$C(x = 0, t = 0) = N_{ad}, \tag{11.2}$$

where N_{ad} is the quantity of adsorbed material.

The second model considers a flux of gas molecules Φ_i crossing the liquid surface. The boundary condition becomes

$$\Phi_i = -D_\ell \left.\frac{\partial C(x, t)}{\partial x}\right|_{x=0} = \chi N_i, \tag{11.3}$$

where N_i is the number of molecules striking the surface per unit time and area, and χ is the sticking coefficient, which represents the fraction of molecules efficiently incorporated into the melt.

The results of the melting program simulations are then combined with the diffusion process parameters (Section 11.3.4) and the doping parameters (gas pressure, sticking coefficient and adsorbed dose) to predict the final dopant profile and the surface sheet resistance of the doped region.

Thus, by an *in situ* determination of the pulse shape and duration by TRR measurements and by using the values of the sticking coefficient and adsorbed dose deduced from previous experiments, an accurate simulation of the GILD process can be achieved. Figure 11.10 shows the RBS and calculated laser-doped profiles for doping under AsH_3 gas with an XeCl laser at different laser energy densities. Experimental and calculated profiles are in good agreement.

11.5 DOPING MECHANISMS

11.5.1 Impurity diffusion

Studies have been performed in order to determine the mechanisms that control the incorporation and redistribution of dopant atoms in the melted layer during the GILD process [52, 54–58, 64, 66]. For this purpose, evolution of the surface sheet resistance, or the related dopant incorporated dose I, as a function of the total surface melting time, which is proportional to the number of pulses, is generally recorded. This evolution is expressed as $I \propto t_d^\gamma$, where t_d is the doping time, approximated by the total surface melting time. The values of γ are deduced from experimental data. We have reported in Table 11.1 the values of γ for various doping conditions given in the literature. Three kinds of behaviour are observed: (i) a square-root dependence of incorporated dose on doping time ($\gamma = 0.5$); (ii) a t_d^γ dependence with $0.5 < \gamma < 1$; and (iii) a linear increase of incorporated dose with doping time ($\gamma = 1$). These results have been interpreted by reference to the diffusion theory.

Deutsch [52] and Kato [57] have assumed that the doping process is limited by diffusion of the incorporated atoms in the molten silicon. In this case, the incorporated dose is expected to be proportional to the diffusion length in the latter, $L = (D_\ell t_d)^{0.5}$, so that it is expected to vary as $t_d^{0.5}$. For doping under BCl_3 and PCl_3 atmospheres with an ArF excimer laser, Deutsch found that the incorporated dose varies as $t_d^{0.6}$, so he concludes that the doping process is mainly diffusion-limited. In contrast, Kato has observed a linear increase of incorporated dose with doping time when doping is performed under a BF_3 atmosphere with an ArF excimer laser. He suggests that, in this case, the doping process is not limited by diffusion of impurities, but rather is controlled by an external rate limitation—which means by the dopant supply.

Experimental results on the evolution of the incorporated dose and impurity concentration profiles with diffusion time have also been interpreted by solving the one-dimensional diffusion equation (11.1) [54, 56, 58, 64, 66]. The solution is obtained by making the approximation that the diffusion takes place in a semi-infinite material ($C(x = \infty, t) = 0$)

and using boundary conditions at the silicon interface suggested by experimental results.

The evolution of the theoretical incorporated dose as a function of t_d, given by

$$I(t_d) = \int F \, dt = D_\ell \int \left. \frac{\partial C(x, t)}{\partial x} \right|_{(0,t)} dt, \qquad (11.4)$$

where F is the flux of dopant atoms crossing the silicon surface, is then compared with the experimental evolution $I \propto t_d^\gamma$. These studies have shown that the doping process can be limited by three mechanisms, each associated with a specific variation of the incorporated dose with doping time. This is summarized in Table 11.3. On going from the first to the third case, there is an increase in the influence of external rates on the diffusion mechanism. Case (i) corresponds to a doping process only limited by the diffusion mechanism; then $\gamma = 0.5$ as assumed by Deutsch [52]. The dopant source supplied by photodecomposition of the dopant molecules can be approximated as infinite. In case (ii), this dopant source is still infinite, but the doping process is limited by the rate of transport of the dopant atoms through the gas–silicon interface. The incorporated dose varies as t_d^γ with $0.5 < \gamma < 1$. Finally, in case (iii), the doping process is limited by the dopant supply, which is now finite. The incorporated dose is then directly proportional to the diffusion time.

Using these results, one can analyse the experimental results reported in Table 11.1. When the doping process is performed with a dopant gas that is not absorbent at the laser wavelength used, it seems to be limited mainly by the dopant supply, since the dopant atoms are only supplied by pyrolytic decomposition of the dopant molecules at the gas–silicon interface. Alternatively, doping performed in an absorbing gas that then supplies a larger quantity of dopant due to the photolytic contribution results in an infinite dopant source, and the doping mechanism is then limited by mechanism (i) or (ii). This is illustrated by the case of doping under a BCl_3 atmosphere at 193, 248 and 308 nm. For the two latter wavelengths, the gas is transparent. The dopant atoms are supplied only by pyrolytic decomposition of the parent molecules. The doping process is then limited by the dopant supply since $\gamma = 1$. However, at 193 nm, the gas is absorbent, and the additional quantity of dopant atoms supplied by photolytic decomposition of the parent molecules results in a modification of the limiting mechanism. The dopant source is infinite, and the doping process is limited either by diffusion or the rate of transport of dopant atoms through the gas–silicon interface, depending on the gas pressure conditions.

Obviously, the doping process is also dependent on the yield of the decomposition process and on the interaction between the dopant species and the silicon surface. Thus, in spite of the fact that PF_5 gas is transparent

Table 11.3. The three mechanisms that limit the doping process.

Experimentally observed boundary conditions at the gas–silicon interface	Theoretical expression[a] for incorporated dose I as a function of doping time t_d	Dopant source	Limiting mechanism
(i) Surface concentration constant and equal to its equilibrium concentration in silicon, C_0: $C(x=0, t_d) = C_0$	$I = 2C_0(D_\ell t_d)^{0.5}$, $\gamma = 0.5$	Infinite	Diffusion of incorporated impurities in the melted layer
(ii) Surface concentration increases with t_d up to C_0: $-\dfrac{\partial C(x, t_d)}{\partial x}\bigg|_{(0,t)} = h[C_0 - C(0, t_d)]$	$I \propto C_0 t_d^{\gamma}$, $0.5 < \gamma < 1$	Infinite	Rate of transport of dopant atoms through the gas–silicon interface
(iii) Constant flux F_0 of dopant crossing the melted silicon surface: $-\dfrac{\partial C(x, t_d)}{\partial x}\bigg|_{(0,t)} = \dfrac{F_0}{D_\ell}$	$I = F_0 t_d$, $\gamma = 1$	Finite	Dopant supply (rate of formation of dopant atoms)

[a] Obtained by solving the diffusion equation (11.1) with the boundary conditions of the first column.

at 193 nm, it gives rise to an infinite dopant source whatever the gas pressure, suggesting a higher efficiency of the pyrolytic decomposition of parent molecule and/or dopant species reactions with the silicon surface.

11.5.2 Dopant sources

As discussed in Section 11.3.3, the dopant atoms can be supplied by two kinds of dopant sources: the adsorbed layer, which includes both chemisorbed and physisorbed layers, and the gas phase. The quantity of dopant atoms supplied by the first is expected to be related to the gas isotherm, since it is expected to increase linearly with gas pressure in the second case. Thus, in order to determine the source of dopant atoms, the evolution of the incorporated dose as a function of gas pressure has been measured experimentally for different dopant-gas/laser combinations [52, 54, 58, 64, 66]. Let us consider two typical cases observed when doping is performed with BF_3 [58] and BCl_3 [54] dopant gases.

(i) BF_3 gas/ArF excimer laser

Figure 11.11 [58] shows the dependence of the incorporated dose I, on BF_3

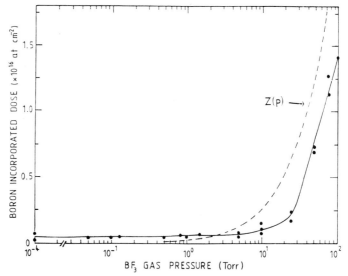

Fig. 11.11 Incorporated boron dose after 20 pulses at 0.75 J cm^{-2} with an ArF excimer laser (193 nm, 21 ns) as a function of the gas pressure ($\bullet$ are experimental data points). The number of molecules striking the melted surface during the 20 pulses is also shown (———).

pressure p for silicon samples irradiated at a fixed fluence after 20 pulses. A first set of samples was irradiated using this procedure: the reaction chamber was pumped down to 10^{-2} Torr, and the required pressures were then obtained by introducing a controlled amount of BF_3 gas. In the second set, this procedure was reversed so that the pressures were decreasing towards 10^{-2} Torr. Both procedures gave rise to the same relationship between I and p. The quantity of molecules striking the melted silicon surface during the 20 laser pulses is also reported as a function of the gas pressure (the dashed line). It corresponds to the maximum quantity of dopant atoms that the gas phase could supply. Two different pressure ranges can be distinguished.

At pressures lower than about 5 Torr, incorporation is constant and higher than the gas contribution. This provides evidence for the presence of an adsorbed layer supplying the dopant atoms. Moreover, the constant incorporation even when the gas pressure is decreased to 10^{-2} Torr shows that this layer is strongly bound to the silicon surface. Thus dopant atoms are expected to be supplied by a chemisorbed layer of BF_3 molecules. This layer has been evaluated as at least about 2×10^{13} BF_3 molecules cm^{-2}. This value is lower than the calculated surface density for a monolayer of BF_3, which is 5.8×10^{14} molecules cm^{-2}.

For the higher pressure range, the increase in incorporated dose as a function of gas pressure seems to follow a linear law. In fact, changing the pressure by one order of magnitude (from 10 to 100 Torr) leads to an incorporated dose variation after 20 pulses from about 1.8×10^{15} to 1.4×10^{16} at cm^{-2}. This trend and the reversibility of the process, evidenced by the same behaviour of I versus p for the two sets of samples, suggest that pyrolysis of molecules striking the hot surface during irradiation is the dominant supply process. Deviation of the experimental data from the theoretical curve can be explained by the fact that the sticking coefficient of BF_3 molecules is not unity, in addition to some losses of dopant species due to laser-stimulated desorption [105].

(ii) BCl₃ dopant gas/ ArF and XeCl excimer lasers

Figure 11.12 shows the dependence of the incorporated dose on BCl_3 gas pressure after irradiation at fixed fluences and for 30 pulses at 308 nm (XeCl excimer laser), where only pyrolytic gas decomposition occurs. Also shown are data for 193 nm irradiation (ArF excimer laser), where both pyrolytic and photolytic gas decomposition take place. The number of molecules striking the molten surface during the doping process as a function of the gas pressure is shown by the dot-dashed and dashed lines for the two lasers. The curves $I = f(p)$ exhibit the same behaviour at 193 and 308 nm, and two pressure ranges can also be distinguished.

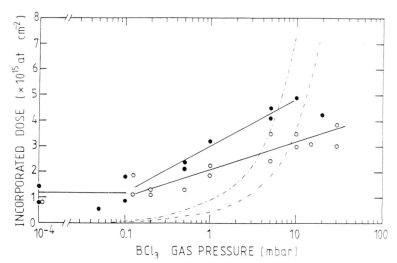

Fig. 11.12 Incorporated boron dose after 30 pulses at 1.2 (●) and 1.5 (○) J cm^{-2}, with ArF (193 nm, 21 ns) and XeCl (308 nm, 55 ns) excimer lasers respectively, as a function of gas pressure. The number of molecules striking the melted surface during the 30 pulses is also shown for the two lasers (— — —, ArF; — · — ·, XeCl).

At gas pressures lower than about 0.1 Torr, for both wavelengths, the incorporated dose is constant in the pressure range 10^{-1}–10^{-4} Torr and is always well above the quantity of striking molecules. As for the previous case, this provides evidence that the dopant atoms are supplied by a chemisorbed layer of BCl$_3$ molecules.

In the higher-pressure range, the increase in I as a function of p seems to follow a logarithmic law. The deviation of the experimental data from the dot-dashed and dashed curves shows that the striking molecules give a negligible dopant supply. This trend and the reversibility of the process suggest that the decomposition of one or more weakly bound physisorbed layers supplies the additional dopant atoms. These results are consistent with the conclusion of Deutsch regarding doping with the same dopant gas and an XeF excimer laser (351 nm).

However, comparison of the incorporated doses per unit time at 193 and 308 nm has shown that the incorporated dose at 193 nm is increased by a factor of about 2 compared with those at 308 nm, due to the additional dopant atoms supplied by the photolytic decomposition of the adsorbed molecules.

We have summarized in Table 11.1 the experimental results concerning the dopant sources reported in the literature for different dopant-gas/laser-wavelength combinations. The nature of the dopant source is dependent on the dopant gas. For almost all the gases, the dopant atoms are supplied by pyrolytic and eventually photolytic decomposition of the adsorbed molecules. In fact, dopant species striking the melted surface give rise to a negligible dopant atom supply, except for $POCl_3$ and BF_3 gases in the high-pressure range.

However, the nature of the dopant source seems to be independent of laser wavelength and, more precisely, of the decomposition mechanism. Indeed, the adsorbed layer of BCl_3 still remains the dominant dopant atom supply even when photolytic decomposition of the molecules in the gas phase occurs (193 nm). In this case, only photolytic decomposition of adsorbed molecules gives an important dopant supply.

11.6   CONCLUSIONS

We have seen that the gas immersion laser doping technique yields good control of the dopant concentration profile. The laser energy density, the gas pressure and the number of pulses control respectively the junction depth, the doping level and the profile shape. Thus heavily doped box-profile ultrashallow junctions can be processed. Moreover, the development of simulation programs that model the surface melting and the dopant redistribution in the molten layer allow *in situ* control of junction characteristics. Laser-doped junctions exhibit good structural characteristics, but microscopic electrical defects, which need an annealing step to be removed, have generally been shown to degrade their electrical characteristics. However, optimization of the doping process (number of pulses and laser energy density) allows the formation of junctions with electrical characteristics at least as good as those of junctions formed by ion implantation followed by rapid thermal annealing. This result and the fact that the GILD process allows the formation of MOSFETs having heavily doped ultrashallow contacts with virtually no source–drain encroachment under the gate makes this technique a good candidate for ULSI device fabrication, which requires restrictive characteristics difficult to achieve by the ion-implantation technique.

However, two problems encountered in our discussion seem to need more investigation. The first is the increase in melted depth with repetitive irradiation that has been observed many times in GILD processing in silicon. The second concerns the exact origin of the laser-induced electrical defects, which seem to be dependent on the number of pulses and on the laser energy density. Additional studies would be useful to give better understanding of these two points.

ACKNOWLEDGEMENT

This chapter is based on work carried out at Laboratoire PHASE, Centre de Recherches Nucléaires, Strasbourg, France, with A. Slaoui and E. Fogarassy.

REFERENCES

[1]   M. Morgan (ed.), *Channeling* (Wiley, New York, 1973).
[2]   T. M. Liu and W. G. Oldham, *IEEE Electron Device Lett.* **4**, 59 (1983).
[3]   R. G. Wilson, *J. Appl. Phys.* **54**, 6879 (1983).
[4]   A. C. Ajmera, G. A. Rozgonyi, *Appl. Phys. Lett.* **49**, 1269 (1986).
[5]   M. C. Ozturk and J. J. Wortman, *Appl. Phys. Lett.* **52**, 963 (1988).
[6]   T. E. Seidel, in *VLSI Technology* (ed. S. M. Sze), p. 253 (McGraw-Hill, New York, 1983).
[7]   T. E. Seidel, *IEEE Electron. Device Lett.* **4**, 353 (1983).
[8]   R. Kalish, T. O. Sedgwich, S. Mader and S. Shatus, *Appl. Phys. Lett.* **44**, 107 (1984).
[9]   T. Sands, J. Washburn, R. Gronsky, W. Maszara, D. K. Sadana and G. A. Rozgonyi, *Appl. Phys. Lett.* **45**, 982 (1984).
[10] M. Von Allmen, in *Laser and Electron Beam Processing of Materials* (ed. C. W. White and P. S. Peercy), p. 3 (Academic Press, New York, 1980).
[11] R. F. Wood, C. W. White and R. T. Young (eds), *Semiconductors and Semimetals*, Vol. 23: *Pulsed Laser Processing of Semiconductors* (Academic Press, New York, 1984).
[12] D. Bauerle (ed.), *Chemical Processing with Lasers* (Springer-Verlag, Berlin, 1986).
[13] I. W. Boyd (ed.), *Laser Processing of Thin Films and Microstructures* (Springer-Verlag, Berlin, 1987).
[14] D. J. Ehrlich and J. Y. Tsao, *J. Vac. Sci. Technol.* **B1**, 969 (1983).
[15] Y. Rytz-Froidevaux, R. P. Salathe and H. H. Gilgen, *Appl. Phys.* **A37**, 121 (1985).
[16] D. Bauerle, *Appl. Phys.* **B46**, 261 (1988).
[17] M. Rothschild and D. J. Ehrlich, *J. Vac. Sci. Technol.* **B6**, 1 (1988).
[18] B. M. McWilliams, I. P. Herman, F. Mitlitsky, R. A. Hyde and L. L. Wood, *Appl. Phys. Lett.* **43**, 946 (1983).
[19] E. I. Shtyrkov, I. B. Khaibullin, M. M. Zaripov and M. F. Galyautudinov, *Sov. Phys. Semicond.* **9**, 1309 (1975).
[20] I. B. Khaibullin, E. I. Shtyrkov, M. M. Zaripov, M. F. Galyautudinov and G. G. Zakirov, *Sov. Phys. Semicond.* **11**, 190 (1977).
[21] J. Krynicki, J. Suski. S. Ugniewski, R. Grotzshel, R. Klabes, U. Kreissig and J. Rudiger, *Phys. Lett.* **A61**, 181 (1977).
[22] P. Baeri, J. M. Poate, S. U. Campisano, G. Foti, E. Rimini and A. G. Cullis, *Appl. Phys. Lett.* **37**, 912 (1980).
[23] J. Narayan, J. Fletcher, C. W. White and W. H. Christie, *J. Appl. Phys.* **52**, 7121 (1981).
[24] R. T. Young, J. Narayan, W. H. Christie, G. A. van der Leeden, J. I. Levatter and L. J. Cheng, *Solid State Technol.* **26**, 183 (1983).

[25] C. W. White, *J. Phys.* (*Paris*) **44**, Colloq. C5-145 (1983).

[26] S. U. Campisano, *Appl. Phys.* **A30**, 195 (1983).

[27] E. P. Fogarassy, D. H. Lowndes, J. Narayan and C. W. White, *J. Appl. Phys.* **58**, 2167 (1985).

[28] M. Finetti, P. Negrini, S. Solmi and D. Nobili, *J. Electrochem. Soc.* **128**, 1313 (1981).

[29] F. Trumbore, *Bell Syst. Tech. J.* **39**, 205 (1960).

[30] J. Narayan, R. T. Young and C. W. White, *J. Appl. Phys.* **49**, 3912 (1978).

[31] J. C. Muller, E. Fogarassy, D. Salles, R. Stuck and P. Siffert, *IEEE Trans. Electron Devices* **27**, 815 (1980).

[32] A. Greenwald, S. J. Hogan, C. J. Keavney and W. E. Wieler, in *Proceedings of IEEE Photovoltaic Specialists Conference, Las Vegas, 1985*, p. 804 (IEEE, New York, NY, 1985).

[33] R. F. Wood, R. D. Westbrook and G. E. Jellison, *IEEE Electron Device Lett.* **8**, 249 (1987).

[34] J. M. Fairfield and G. H. Schwuttke, *Solid State Electron.* **11**, 1175 (1978).

[35] F. E. Harper and M. I. Cohen, *Solid State Electron.* **13**, 1103 (1970).

[36] J. Narayan, R. T. Young and R. F. Wood, *Appl. Phys. Lett.* **33**, 338 (1978).

[37] K. Affolter, W. Luthy and M. von Allmen, *Appl. Phys. Lett.* **33**, 185 (1978).

[38] E. Fogarassy, R. Stuck, J. J. Grob and P. Siffert, *J. Appl. Phys.* **52**, 1076 (1981).

[39] T. Shameshima and S. Usui, *J. Appl. Phys.* **62**, 711 (1987).

[40] R. Stuck, E. Fogarassy, J. C. Muller, M. Hodeau, A. Wattiaux and P. Siffert, *Appl. Phys. Lett.* **38**, 715 (1981).

[41] R. Solomon and L. F. Mueller, *US Patent* 3 364 087 (1968).

[42] T. F. Deutsch, D. J. Ehrlich, R. M. Osgood and Z. L. Liou, *Appl. Phys. Lett.* **36**, 847 (1980).

[43] K. G. Ibbs and M. L. Lloyd, *Mater. Res. Soc. Symp. Proc.* **17**, 243 (1983).

[44] M. L. Lloyd and K. G. Ibbs, *Mater. Res. Soc. Symp. Proc.* **29**, 35 (1984).

[45] T. W. Sigmon, P. G. Carey. R. L. Press, T. S. Falhen and R. J. Pressley, *Mater. Res. Soc. Symp. Proc.* **23**, 247 (1984).

[46] P. G. Carey, T. W. Sigmon, R. L. Press and T. S. Falhen, *IEEE Electron Devices Lett.* **6**, 291 (1985).

[47] T. W. Sigmon, *Mater. Res. Soc. Symp. Proc.* **75**, 619 (1987).

[48] E. Landi, P. G. Carey and T. W. Sigmon, *IEEE Trans. Computer Aided Design* **7**, 205 (1988).

[49] P. G. Carey, K. Bezjian, T. W. Sigmon, P. Gildea and T. J. Magee, *IEEE Electron Devices Lett.* **7**, 440 (1986).

[50] P. G. Carey and T. W. Sigmon, *Appl. Surf. Sci.* **43**, 325 (1989).

[51] T. F. Deutsch, D. J. Ehrlich, D. D. Rathman, D. J. Silversmith and R. M. Osgood, *Appl. Phys. Lett.* **39**, 825 (1981).

[52] T. F. Deutsch, *Mater. Res. Soc. Symp. Proc.* **17**, 225 (1983).

[53] F. Foulon, A. Slaoui, E. Fogarassy, R. Stuck, C. Fuchs and P. Siffert, *Appl. Surf. Sci.* **36**, 384 (1989).

[54] A. Slaoui, F. Foulon, M. Bianconi, L. Correra, R. Nipoti, R. Stuck, S. Unamuno, E. Fogarassy and S. Nicoletti, *Mater. Res. Soc. Symp. Proc.* **129**, 591 (1989).

[55] A. Slaoui, F. Foulon and P. Siffert, *Appl. Phys.* **A50**, 479 (1990).

[56] F. Foulon and A. Slaoui, Unpublished work.

[57] S. Kato, T. Nagahoni and S. Matsumoto, *J. Appl. Phys.* **62**, 3656 (1987).

[58] F. Foulon, A. Slaoui and P. Siffert, *Appl. Surf. Sci.* **43**, 333 (1989).

[59] P. G. Carey, K. H. Weiner and T. W. Sigmon, *IEEE Electron Devices Lett.* **9**, 542 (1988).

[60] G. B. Turner, D. Tarrant and G. Pollock, *Appl. Phys. Lett.* **39**, 967 (1981).

[61] G. G. Bentini, M. Bianconi, L. Correra, R. Lotti and S. Summonte, in *Proceedings of 7th European Photovoltaic Specialists Conference, 1986* (eds A. Gretzberger, W. Palz and Willeke), p. 1044 (D. Reidel Publishing Company, 1987).

[62] G. G. Bentini, M. Bianconi, L. Correra, R. Nipoti, S. Summonte, C. Cohen and J. Siejka, in *Proceedings of European Materials Research Society Meeting, June 1987*, Vol. 15, p. 273 (Les Editions de Physique, Les Ullis, 1987).

[63] G. G. Bentini, M. Bianconi and S. Summonte, *Appl. Phys.* **A45**, 317 (1988).

[64] L. Correra, G. G. Bentini, M. Bianconi, R. Nipoti and D. A. Patti, *Appl. Surf. Sci.* **36**, 394 (1989).

[65] S. Kato, H. Saeki, J. Wada and S. Matsumoto, *J. Electrochem. Soc.: Solid State Science and Technol.* **135**, 1030 (1988).

[66] A. Slaoui, F. Foulon, E. Fogarassy and P. Siffert, *Mater. Res. Soc. Symp. Proc.* **158**, 237 (1990).

[67] A. Slaoui, F. Foulon and P. Siffert, *J. Appl. Phys.* **67**, 6197 (1990).

[68] E. J. Yoffa, *Phys. Rev.* **B21**, 2415 (1980).

[69] E. J. Yoffa, *Appl. Phys. Lett.* **36**, 37 (1980).

[70] M. von Allmen, in *Laser Annealing of Semiconductors* (ed. J. M. Poate), p. 43 (Academic Press, New York, 1982).

[71] B. R. Appleton and G. K. Celler (eds), *Lasers and Electron Beam Interactions with Solids* (North-Holland, Amsterdam, 1982).

[72] D. H. Lowndes, R. F. Wood and J. Narayan, *Phys. Rev. Lett.* **52**, 561 (1984).

[73] R. F. Wood and G. A. Geist, *Phys. Rev.* **B34**, 2606 (1986).

[74] S. Unamuno, M. Toulemonde and P. Siffert, in *Laser Processing and Diagnostic* (ed. D. Bauerle), p. 35 (Springer-Verlag, Berlin, 1984).

[75] S. Unamuno and E. Fogarassy, *Appl. Surf. Sci.* **36**, 1 (1989).

[76] F. Foulon, E. Fogarassy, A. Slaoui, C. Fuchs, S. de Unamuno and P. Siffert, *Appl. Phys.* **A45**, 361 (1988).

[77] S. de Unamuno, Personal communication.

[78] P. S. Peercy and M. O. Thompson, *Mater. Res. Soc. Symp. Proc.* **35**, 53 and 169 (1985).

[79] E. Fogarassy, C. Fuchs, S. de Unamuno and P. Siffert, *Appl. Surf. Sci.* **43**, 316 (1989).

[80] S. W. Benson, *J. Chem. Educ.* **42**, 502 (1965).

[81] J. A. Kerr, *Chem. Rev.* **66**, 465 (1966).

[82] Y. D. Orlov, Y. K. Knobel and Y. A. Lebedev, *Russ. J. Phys. Chem.* **61**, 1676 (1987).

[83] A. Slaoui, F. Foulon, C. Fuchs, E. Fogarassy and P. Siffert, *Appl. Phys.* **A50**, 317 (1990).

[84] Y. Rytz-Froidevaux, R. P. Salathe, H. H. Gilgen and H. P. Weber, *Appl. Phys.* **A27**, 133 (1982).

[85] C. J. Chen and R. M. Osgood, *Appl. Phys.* **A31**, 171 (1983).

[86] C. J. Chen, *J. Vac. Sci. Technol.* **A5**, 3386 (1987).

[87] R. F. Wood, J. R. Kirkpatrick and G. E. Giles, *Phys. Rev.* **B23**, 5555 (1981).

[88] R. F. Wood, *Phys. Rev.* **B25**, 2786 (1982).

[89] C. W. White, D. M. Zehner, J. Narayan, O. W. Holland and B. R. Appleton, *Mater. Res. Soc. Symp. Proc.* **13**, 287 (1983).

[90] E. Fogarassy, R. Stuck, J. J. Grob and P. Siffert, in *Laser Annealing of Semiconductors* (ed. J. M. Poate), p. 121 (Academic Press, New York, 1982).

[91] W. E. Beadle, J. C. C. Tsai and R. D. Plummer (eds), *Quick Reference Manual for Silicon Integrated Circuit Technology* (Wiley-Interscience, New York, 1985).

[92] H. Kodera, *Japan J. Appl. Phys.* **2**, 212 (1963).

[93] Y. M. Shashkov and V. M. Gurevich, *Russ. J. Phys. Chem.* **42**, 1082 (1968).

[94] C. W. White, S. R. Wilson, B. R. Appleton and F. W. Young, *J. Appl. Phys.* **51**, 738 (1980).

[95] C. W. White, B. R. Appleton, B. Stritzker, D. M. Zehner and S. R. Wilson, *Mater. Res. Soc. Symp. Proc.* **1**, 31 (1981).

[96] M. Tamura, N. Natsuaki and T. Tokuyama, in *Laser and Electron Beam Processing of Materials* (ed. C. W. White and P. S. Peercy), p. 247 (Academic Press, New York, 1980).

[97] R. F. Wood, B. R. Appleton, B. Stritzker, D. M. Zehner and S. R. Wilson, in *Laser and Electron Beam Solid Interactions and Material Processing* (ed. J. F. Gibbons, L. D. Hess and T. W. Sigmon), p. 109 (North-Holland, Amsterdam, 1981).

[98] B. C. Larson, C. W. White and B. R. Appleton, *Appl. Phys. Lett.* **32**, 801 (1978).

[99] I. W. Boyd and J. I. B. Wilson, *J. Appl. Phys.* **53**, 4166 (1982).

[100] D. A. Mantell and T. E. Orlowski, *Mater. Res. Soc. Symp. Proc.* **74**, 141 (1987).

[101] R. Nipoti, M. Bianconi, R. Fabbri and M. Servidori, *Appl. Surf. Sci.* **43**, 321 (1989).

[102] G. Masetti, M. Severi and S. Solmi, *IEEE Trans. Electron Devices* **30**, 764 (1983).

[103] B. Hartiti, A. Slaoui, J. C. Muller and P. Siffert, *J. Appl. Phys.* **66**, 3934 (1989).

[104] B. Hartiti, A. Slaoui, J. C. Muller and P. Siffert, *Mater. Sci. Engng* **B4**, 257 (1989).

[105] T. J. Chuang, *Surf. Sci.* **A8**, 763 (1986).

12 Photochemical Etching of III–V Semiconductors

RICHARD B. JACKMAN

*Department of Electronic and Electrical Engineering,
University College London, UK*

12.1 INTRODUCTION

Semiconductor wafers must be cleaned prior to the growth of additional semiconducting layers and the subsequent steps in device formation. This prevents the incorporation of contamination, which will lead to poor device performance or even device failure. Cleaning is often achieved by removing a thin layer of the semiconductor through chemical corrosion. In addition, the formation of a given micro- or opto-electronic device geometry relies upon the selective removal of various regions of the semiconductor to yield the surface features that are required. Metallization is then carried out and dielectric material deposited to realize the circuit being created. Thus "etching" of the semiconductor material, or other regions of a circuit that is being formed, is a vital processing step.

12.1.1 Etching: available methods and limitations

Conventional approaches to etching, as utilized in current VLSI technology, falls into two categories. The first involves a purely chemical reaction between a liquid (usually an acid) and the solid semiconductor concerned. Transport of the semiconductor atoms away from the surface occurs through the dissolution of the resultant compounds into the liquid [1].

The second involves the interaction of the semiconductor with an energetic gas phase, and is known generally as "dry etching". An important form of dry etching involves the formation of energetic species, often ionic in nature, from a precursor gas in a plasma discharge. Once formed, these

Photochemical Processing of Electronic Materials
ISBN 0-12-121740-X

species impinge on the semiconductor, and a combination of physical sputtering and chemical reaction (with the formation of volatile compounds containing elements from the semiconductor) leads to the etching step. To create the chemical component of this etching action, the precursor is often a halogen or halogen-containing species [2].

The processes described above have certain limitations. For example, wet etching requires physical masking to define the required feature, and can often lead to poor definition of the shape required through undercutting of this mask [3]. Plasma etching also requires physical masking, but can yield highly anistropic features through the directed nature of the impinging particles [4]. However, the incorporation of defects into the semiconductor through physical bombardment by energetic particles can limit device performance, especially when using III–V materials [5]. Furthermore, several, poorly understood, reaction mechanisms are believed to operate sequentially, leading to the etching process [6]. This leads to poor process reproducibility; routine deployment in different laboratories without prior empirical evaluation is difficult.

The ability to incorporate all growth and processing steps into a single ultraclean vacuum chamber would allow the realization of devices of unique purity [7]. Such an *in situ* processing concept requires compatibility between all of the steps concerned, and would certainly preclude large-scale use of "wet" chemistry (including current photolithographic masking methods) or high gas pressures of reactive species. This is considered in more detail in Section 12.3.

12.1.2   Etching for devices in III–Vs: an example

Figure 12.1 shows a typical device structure for a high-electron-mobility transistor (HEMT) fabricated in GaAs [8]. To form such a device, the GaAs substrate must be cleaned and very thin epitaxial layers of GaAs and AlGaAs deposited by a technique such as molecular beam epitaxy (MBE). The formation of the recessed and non-recessed gates then requires a highly anisotropic etch process to yield straight-walled features with submicron resolution. Extremely precise control over etch depth is also required; simply "turning off" the etching process to a precision of fractions of a micron is difficult. Ideally, an etch that stops automatically when meeting a new GaAs or AlGaAs layer would be utilized.

Some success in achieving well defined features and selectivity in etching GaAs with an automatic stop at an AlGaAs layer has been demonstrated using plasmas derived from CF_2Cl_2 [9]. In this case the formation of a tenacious AlF_3 species is believed to "quench" the etching step when an AlGaAs layer is reached [10]. A limitation to this approach, however, is

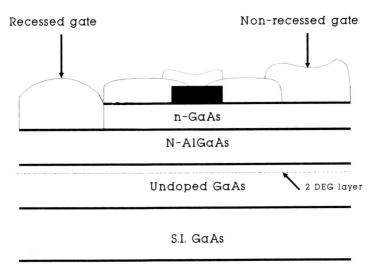

Fig. 12.1 A typical high-electron-mobility transistor (HEMT) structure, revealing the need for anisotropic etching to form the gate features.

the radiation damage that this form of etching often provokes in the AlGaAs channel under the gate. The use of energetic particles can lead to lattice damage at a much deeper level than the penetration depth of the initial ion. This is believed to be caused by collisional cascades and a "thermal spike" mechanism [11]; deep-level traps can result. Very low defect and trap concentrations are required in this AlGaAs layer ($< 10^{15}$ cm^{-3}) in order to minimize threshold variations and instabilities in the current–voltage relationship, as well as device sensitivity to light and temperature changes.

12.1.3 A role for photochemical etching?

Recent advances in dry etching have shown the potential of using an ion *beam* source for the reactive species, as opposed to direct use of a plasma, to stimulate the etching reaction [12]. This approach, often referred to as reactive ion beam etching (RIBE), may allow more careful control over the reaction environment. An alternative method involves the introduction of the energetic particle and the reaction species separately, and is often known as chemically assisted ion beam etching (CAIBE). Here an inert gas ion beam (such as argon) is used to irradiate the material to be etched, while a suitable gas is introduced directly to the surface. Beams may be readily incorporated into vacuum chambers, but may still present problems of physical damage to the material being processed. Replacing this ion beam with a photon beam that stimulates a (dry) photochemical reaction between a gas and the semiconductor may yield the advantages that have been discussed in earlier chapters for other processing and growth steps. Thus the use of photons from lamp or laser sources, if carefully controlled, may allow us to realize a patterned etch (through the use of direct "writing" with a highly focused beam or by projecting prepatterned light) that is non-physical in origin, in a single step. Such an approach gives rise to the potential for extremely high spatial resolution, particularly if short-wavelength light, such as UV, is used. This has been discussed in Chapters 1 and 3. Furthermore, if a single photochemical reaction pathway dominates the reaction kinetics then considerable scope for producing a highly reproducible and selective etch may exist. Table 12.1 summarizes some of the

Table 12.1 A brief comparison of plasma and dry laser methods for semiconductor chemical etching.

	Plasma etching	Laser etching
Energy source	Plasma discharge and e field	Photon flux
Chemical species	Various, many halogenated compounds	Halogens and halogenated compounds
Etching mechanism	Chemical and physical through reactive ion and radical bombardment, sputtering	Chemical through photolysis, carrier excitation or thermal activation
Phase in which reactive species is created	Gaseous	Gaseous or adsorbed
Reaction temperature	Generally around room temperature	Can be high locally to the reaction zone
Method for localization	Resist masking	Can be maskless; use of focused or patterned beam
Anistropy	Excellent with mask	Relies upon confinement of gaseous reaction or surface reaction without mask
Resolution	$< 1\ \mu m$	$< 1\ \mu m$
Etch rate	Often slow: $0.06-80\ \mu m\ min^{-1}$ GaAs	Can be very high: $0.1-2000\ \mu m\ min^{-1}$ GaAs
Damage	Can be very high	Extremely low
Likely applications	Widespread	More specialized circuit repair; high-value low-volume fabrication

differences between laser-promoted photochemical etching and plasma etching.

It is the way in which we can attempt to promote photochemical etching that is the topic of this chapter. The fundamental reactions that must be driven are considered with particular reference to incorporation of the photochemical etching reaction onto an ultrahigh-vacuum system for *in situ* etching of III–V materials following growth. The rate-determining steps in the etching reactions concerned are first considered, the fundamental processes that occur are then addressed and the limitations that these naturally impose on the approach discussed. Examples of photochemical etching processes that have been successfully realized on III–Vs are then presented.

12.2 PHOTOCHEMICAL ETCHING: THE BASICS

During photochemical etching, a suitable gas phase species is introduced near to the surface of a semiconductor. A photon beam is either projected onto a given area of the solid or is projected parallel to the solid. A reaction is stimulated, giving rise to volatile reaction products containing elements of the semiconductor material. A number of reaction systems have been studied involving light in the wavelength range $0.19–11$ μm and differing precursor gases. An early review of work in this area is that by Chuang [13], while more recent progress is included in an excellent review by Haigh and Ayllet [14]. Gases utilized have included halogens [15], alkyl halides [16] and inorganic halides [17]. Etching of a wide range of semiconductors [18], metals [19], insulators [20] and polymers [21] has been attempted. Photochemical etching of III–V semiconductors is often more complex than etching of a material such as silicon. The number of detailed microscopic level studies on III–Vs is small. For this reason, examples from silicon etching processes will be included in this chapter where this will help the reader to consider the mechanisms that pertain. Examples of III–V semiconductor photochemical etching are more explicitly considered later in the chapter.

12.2.1 Reaction mechanisms

Light may promote several reaction pathways that lead to surface etching. If the flux of the incident light is sufficiently high then direct photon sputtering or ablation of the material may occur. This form of material removal is not chemical in origin and is discussed in some detail in Chapter 16. Surfaces of III–V materials etched in this manner are often rough and highly damaged. At all fluxes lower than this threshold, chemically assisted etching can be promoted. The light can effectively enhance the formation of etch products through one—or a mixture—of three routes.

(i)   The light may influence the precursor species while it is still in the gas phase. Under these conditions, a more reactive species may be produced that spontaneously forms volatile products when collisions with the surface subsequently occur.

(ii)   In any real system the precursor gas will adsorb on the surface to some extent. The influence of the light on this state may lead to reaction and yield the etching step required.

(iii) The light may exert little influence on the precursor in either of these phases, but may be primarily absorbed by the solid semiconductor itself. This can result in the formation of electron–hole pairs and/or lattice heating. These effects may then enhance the reaction probability of impinging species.

These three limiting cases are represented in Fig. 12.2(a–c). In addition, there is the extent to which etching will occur in the "dark" to be considered, as pictured in Fig. 12.2(d). Any significant etching reaction that occurs spontaneously between the precursor gas and the solid will clearly limit our ability to localize the modification to the beam irradiated area.

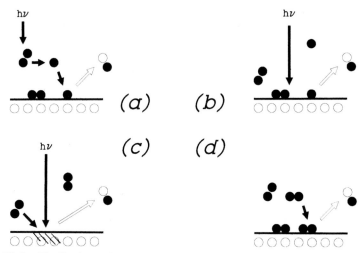

Fig. 12.2 Idealized reaction pathways for photo-assisted chemical etching. Open circles signify surface atoms, while filled circles represent the etching gas: (a) light-induced gas phase reaction; (b) light-induced adsorbed phase reaction; (c) absorption of light by the solid and subsequent reaction of impinging species; (d) spontaneous reaction between the gas and solid in the absence of light.

The simplest chemistry leading to etching of, say, silicon involves the formation of silicon halides by a reaction with a halogen gas. The use of chlorine for silicon etching in the presence of various wavelengths of light is in fact one of the most studied photochemical etching systems [22–32]. Silicon tetrachloride $SiCl_4$ can be produced as a stable compound and is readily vaporized (under standard conditions, it is a liquid with b.p. $57°C$) [33]. Other chlorides $SiCl_x$ ($x = 1–3$) have also been observed as reaction products during photo-assisted etching in this system. The use of short pulses of UV light to assist chlorine etching of silicon and copper is considered in detail in Chapter 13.

12.3 *IN SITU* PROCESSING

Before addressing the etching mechanisms that we may favour in detail, it is important to consider the environmental restrictions imposed on photochemical etching if it is to be applied as an *in situ* technique following growth.

Methods associated with silicon VLSI for growth and subsequent device fabrication are largely inappropriate for III–V materials. In recent years molecular beam epitaxy (MBE), metal–organic MBE (MOMBE), metal–organic chemical vapour deposition (MOCVD) and photo-assisted methods of deposition have proved powerful methods for the formation of epitaxial III–V semiconductor layers (see [36] and Chapter 6). Virtually atomic-scale control over the properties of the growing layers (thickness and composition) can be achieved.

Many of these growth techniques are carried out under ultrahigh-vacuum conditions. Full integration of subsequent processing with these methods relies upon the use of a common environment. Thus any etching step must not only achieve the low damage and high spatial resolution (promised by photo methods), but must also be compatible with a UHV system. If such compatibility can be realized then small numbers of ultraclean devices may be fabricated completely without exposing the device to atmospheric contamination at any stage of its growth or processing. This unique achievement is the goal of *in situ* growth and processing. The UHV environment would also allow the operation of a number of surface-sensitive chemical and structural probes, as well as accurate gas analysis, for careful process control.

Having become committed to a UHV system for growth, the additional expense of complete fabrication *in situ* may not be prohibitive for high-value devices that are required in small numbers. Alternatively, prototyping or customization may be envisaged without the expense of the construction of masks that may be required for only a short time. A

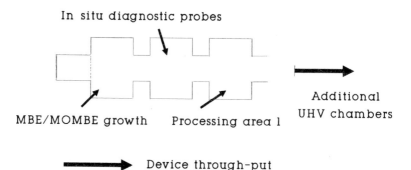

Fig. 12.3 *In situ* growth and processing concept. Semiconductor layers are grown and transferred, without atmospheric exposure, into neighbouring UHV chambers for characterization and processing into device structures.

multichamber system may be envisaged, with the growing device travelling between the different regions. Figure 12.3 shows a schematic representation of this approach.

UHV equipment is not only expensive to purchase, maintain and staff, but can also readily suffer a catastrophic loss of performance if exposed to high pressures of a reactive gas. Thus, at all stages of operation, gas pressures must be carefully controlled. Furthermore, if a system is to be operated with various reactive gases, care is required to prevent a "memory" effect occurring, through a persistent contaminant in the chamber. With these constraints in mind, the fundamental photo-assisted mechanisms that can be promoted will be re-examined.

12.4 REACTION MECHANISMS REVISITED

At an atomic scale, the interaction of the etching precursor and the light to be used with a (clean) semiconductor surface can give rise to a whole host of reactions. These are schematically represented in Fig. 12.4 and include gas phase diffusion and reaction, adsorption, surface diffusion and reaction, surface-to-bulk transport, and desorption. Light may influence one or more of these processes. Under *all* reaction conditions, any etching that occurs will involve the formation of an adsorbed state that allows surface bound species to react and desorb as volatile products. However, under certain conditions, the reaction kinetics can be dominated by gas phase processes. This case is discussed further below.

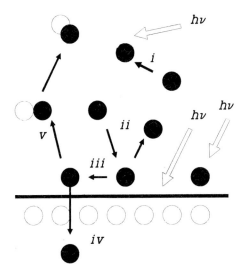

Fig. 12.4 Schematic representation of the fate of a gaseous species near a surface. Reactions may include (i) gas phase diffusion and reaction; (ii) adsorption; (iii) surface diffusion and reaction; (iv) surface to bulk transport and (v) desorption. The light may influence any of these steps.

12.4.1 Gas phase processes

As light passes through a precursor gas, photochemical dissociation may occur, giving rise to reactive radicals. If these species collide with the surface and lead to products containing elements of the semiconductor that have a very short surface lifetime (i.e. that spontaneously desorb at the process temperature) then the reaction is essentially dominated by the gas phase species. In this case the rate of collision of these species with the surface will control the reaction rate. The flux of such a reactive gas phase on the surface relies upon

(i) the cross-section for photodissociation of the molecule concerned at the wavelength used;

(ii) the mean free path between collisions and subsequent homogeneous destruction of the radical at the pressure used;

Experimentally, etch rates achieved with a laser source *only* creating excitation in the gas phase are always lower than those with at least a contribution from excitation in the solid (of whatever form). For example, in the Si–Cl$_2$ system, light from an Ar$^+$ ion laser is capable of dissociating the chlorine and can also be absorbed by the Si. By using the laser beam in a configuration parallel or perpendicular to the semiconductor surface, the relative contribution of these mechanisms can be studied. Ehrlich *et al.* [22] found that absorption by the solid gave rise to significantly higher rates of Si loss from the surface. Kullmer and Bauerle [35] have observed the need for substrate excitation or heating for efficient etching in this chemical system with other photon wavelengths.

In addition, the actual reaction zone may extend outside the beam-irradiated area. Let us consider an etching reaction on silicon that involves a 1 μm wide track being etched across a 1 cm length of wafer. Assuming a Si—Si bond length of 2.35 Å [33], the surface of this region will consist of around 5.5×10^{12} Si atoms. It is widely reported that the primary etch products of such a reaction would be $SiCl_2$ and $SiCl_4$. Assuming an average requirement of three chlorine atoms per desorbing Si atom, and that the reaction probability upon collision for a given chlorine atom is unity, 1.65×10^{13} Cl atoms would be required to etch a single monolayer of the Si. For a 0.1 μm deep etch (which is rather shallow) around 800 layers like this must be removed. To achieve this over a timescale of, say, 10 min would therefore require a flux of 2.1×10^{13} molecules on this area per second. This assumes that the reaction does not experience any inhibition through mass transport, i.e. that products are immediately desorbed and removed from the reaction region. According to gas kinetic theory (see, e.g. [36]),

$$\text{flux (molecules m}^{-2}\text{ s}^{-1}) = \frac{P}{(2\pi MkT)^{1/2}}, \qquad (12.1)$$

where P is the pressure (in Pa) of a gas of molecular mass M. Thus an equivalent pressure of approximately 5 Torr is required. Of course, in practice the light will not convert every Cl_2 species in its path into Cl radicals, and radicals far from the surface will be lost before reaction can occur. It is, however, clear that very high pressures of chlorine (compared with the UHV background) would be required for *in situ* processing in this manner.

The mean free path of a species may be calculated from

$$\lambda = \frac{1}{2^{1/2}\sigma} \frac{kT}{p}, \qquad (12.2)$$

where σ is the collision cross-section for the gas concerned. On this basis, the mean free path between collisions for a gaseous species at this pressure (at 300 K) is about 5 μm. Thus spatial localization at submicron levels is likely to be severely hindered by etching outside the beam irradiated area. Indeed, experimental observations directly support this belief. Chlorine etching of SiO_2 by Chuang [37] with an Ar^+ ion laser focused to 5 μm gave rise to an etched feature of a width greater than 50 μm.

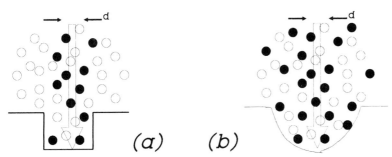

Fig. 12.5 The addition of radical "scavengers" to a gas phase etching process can lead to an anisotropic etch as displayed in (a) as opposed to the broader, isotropic, etch pictured in (b) obtained in the absence of such scavengers. A laser beam of diameter d is shown, with open circles representing the etching gas and filled circles photo-activated species. (After Chuang [13].)

This loss of spatial resolution is also likely to yield a process that is isotropic. Chuang [13] has discussed the addition of radical "scavengers" to reduce the spread of reactive species outside the beam-irradiated area and enhance the anisotropic nature of the etching reaction. In the case of chlorine etching through the formation of Cl radicals, it was suggested that the addition of ethane leads to C_2H_5Cl and HCl in a single collision. This would tend to confine the reaction region by reducing the mean free path of the Cl radicals. Of course, the addition of hydrocarbons of this nature may lead to considerable contamination of the surface by residual carbon (or carbon-containing) species. This approach can also be applied to other chemical systems. For example, gas phase excitation of SF_6 by radiation from a CO_2 laser is known to give rise to a fluorine-based silicon etch through vibrational excitation of the SF_6 molecule [38]. Collisional deactivation leads to very short mean free paths for the excited species at pressures in the region of 100 Torr. Confinement of the reaction region may again be envisaged, and this is pictured in Fig. 12.5.

The gas pressures that are described above are not easily integrated with UHV processing. Even if this is attempted, considerable reaction outside of the beam, which does *not* give rise to etching, may occur at these pressures. This may degrade the rest of the device being fabricated. This is discussed further below.

12.4.2 Spontaneous surface chemical reactions

The topic of adsorption of a gaseous species onto a solid has been introduced in Chapter 1 and is covered in detail in a number of textbooks

(a)

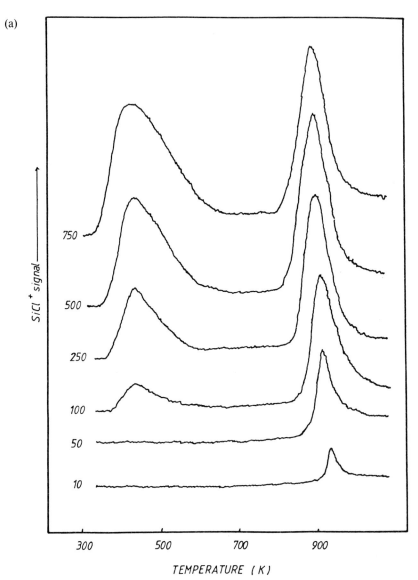

SiCl⁺ signal →

750

500

250

100

50

10

300   500   700   900

TEMPERATURE (K)

Fig. 12.6 Thermal desorption spectra (TDS) taken following the adsorption of Cl_2 onto Si(100). (a) Data collected for $SiCl^+$ species, plotted as a function of increasing exposure of the Si to the chlorine gas. Two peaks are evident, labelled α and β. (b) Uptake curves for the adsorption of chlorine onto Si(100) based upon the two peaks evident in (a). (After [25].)

(e.g. [39]). Of interest here is the tendency of halogen, or halogen-containing, molecules to dissociatively chemisorb when in contact with a number of clean semiconductor surfaces.

The adsorption of Cl_2, Br_2 and I_2 has been studied on clean Si(100) by a range of surface-sensitive techniques under UHV conditions [25, 40–43]. One of these—thermal desorption spectroscopy (TDS)—involves the application of a linear temperature ramp to the sample following an adsorption experiment. Any surface phases formed are desorbed at a given temperature (which is related to the strength of the surface bond concerned) into an immediately adjacent mass spectrometer [44]. This form of analysis has revealed that initially chlorine reacts with a high sticking probability (0.7) [25]. A strongly bound surface-adsorbed phase is formed, which saturates at a monolayer coverage, with a heat of adsorption of $235 \ kJ \ mol^{-1}$. Interestingly, further exposure of this surface to chlorine leads to a continued uptake, although into a more weakly bound phase (heat of adsorption $115 \ kJ \ mol^{-1}$). Spectra revealing this trend are reproduced in Fig. 12.6(a). The intensity of one of the fragments ($SiCl^+$) detected in the mass spectrometer is plotted against the surface temperature

(b)

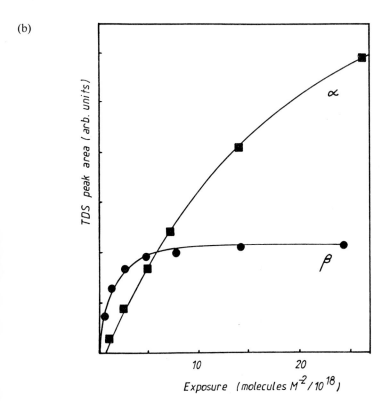

of the silicon. Two peaks are clearly evident, marked α and β, distinguishing the two states discussed above. Spectra are plotted as a function of increasing exposure of the silicon to chlorine. The upper curves represent the desorbed products from a surface that has been exposed to around 7.5×10^{18} molecules m^{-2}. A complete analysis of the desorbing products reveals that the α state is predominantly SiCl$_4$, while the β state gives rise to a mixture of SiCl$_2$ and SiCl$_4$. Figure 12.6(b) shows the relative uptake of chlorine into these states calculated from these peaks. The β state is believed to represent the presence of a simple strongly bound monolayer of chlorine; the α state represents the thickening of this phase into a multi-layer "corrosion" type layer containing Si and Cl species. Temperatures above 500 K allow desorption of this state; further exposure to Cl$_2$ at this temperature will therefore lead to effective thermal etching. Interestingly, this same study identified a very small amount of spontaneous etching (in the dark) during adsorption at temperatures as low as 100 K. It is thought that this is due to a relaxation process that accompanies adsorption.

This form of analysis is very useful in identifying the products expected during thermally driven chlorine etching of silicon. However, these results also clearly indicate that adsorption, or inclusion, of chlorine occurs at levels significantly above a monolayer at room temperature at relatively low exposures of chlorine gas. Thus any etching process where a *clean* surface exists that is subsequently exposed to chlorine at the pressures discussed in Section 12.4.1 may be expected to have a reacted surface layer present on all regions outside the beam-irradiated area. This type of behaviour is also expected on InP and GaAs in the presence of chlorine [45, 46]. Such forms of contamination may be very undesirable during the *in situ* processing of devices; surface states may be introduced and the electronic properties of the interface that is subsequently fabricated adjusted.

12.4.3 Solid phase etching processes

An obvious way to alleviate the isotropic nature of the gas phase processes discussed in Section 12.4.1 is to provide the energy for enhancing the etching reaction to the surface region of the solid alone. In this manner only molecules hitting this excited region will proceed to react. Provided that the spontaneous reactions that occur between the precursor gas and the semiconductor outside this irradiated area are negligible, other regions will remain unaffected. That this is not the case for pure halogen gases on clean surfaces has been described above. Thus alternative etching gases are desirable for III–V *in situ* photochemical etching.

Halogen-containing molecules such as alkyl halides (including Freons) or inorganic halides may be suitable for this role. Many of these species are widely used in plasma-assisted etching reactions. In addition to the necessary physical properties (a gas or liquid with a high vapour pressure for ease of beaming onto the semiconductor concerned), such species must be readily excited to yield the "active" halogen component in the beam-irradiated area. For example, in the case of simple alkyl halides, thermal degradation should appear readily [33], and the presence of strong UV absorption bands has been known for some years [47].

12.4.3.1 Photothermal etching reactions

Photon irradiation of solids readily leads to heating, as has been discussed in Chapter 1. Most wavelengths (other than infrared) give rise to significant electron excitation rather than direct lattice phonon excitation. However, this excitation quickly relaxes, giving rise to lattice heating. Calculations reveal that at visible wavelengths a considerable amount of the absorbed energy would be converted to heat after only 10^{-11} s through electron–phonon scattering (see [48] and Chapter 1). Adsorbed phases present, or freshly impinging molecules, may then react with the underlying solid. Assuming any gas phase that is present to be transparent to the wavelength of light utilized, the resolution in this mode of operation will be controlled by the extent of the region that is heated sufficiently to exceed any activation barriers for the region concerned. There are two distinct regimes that must be considered.

(i) Lamp or CW laser radiation

Under this form of radiation, thermal modelling predicts that a focused beam will produce a heating effect over a surface radius of around three times larger than the beam radius [49, 50]. A CW laser beam is expected to induce a Gaussian temperature distribution around a central peak temperature increase [51, 52]. The peak temperature value will be influenced by the photon flux present and the thermal conductivity of the semiconductor concerned. If the reaction to be provoked between an impinging particle and this hot region has an activation barrier that is a strong function of temperature, etching may be confined to a region even less than the width of the incident beam. For UV and visible lasers this can be in the region of fractions of a micron.

An example of this form of etching comes from the work of Takai et al. [54, 55], who exposed GaAs surfaces to Cl_2 gas at low pressures in the presence of visible radiation from an argon ion laser. They believed substrate heating to be the dominant effect of the beam, the extent of gas phase photolysis of the Cl_2 being small at the pressures used (<0.1 Torr). The beam was focused and then scanned across the GaAs. The resultant etched lines reflected the thermal profile expected from the beam and deepened from 0.5 to 7 μm as the scan speed was reduced (leading to greater heating).

(ii) Pulsed laser irradiation

A surface under irradiation from a series of very short pulses of light (of the order of nanoseconds or less) will not exhibit equilibrium thermal chemistry. Under these conditions, a shallow region of the surface may be made hot—very quickly. Typical repetition rates are tens or hundreds of hertz; rapid cooling will occur between pulses. The temperature dependence of chemical reactions is often simply modelled by the familiar Arrhenius equation,

$$k = A \exp\left(-\frac{E_a}{RT}\right),$$
(12.3)

where k is the reaction rate, E_a is the activation energy for a given chemical route and A is the so-called pre-exponential factor, which relates to kinetic rather than thermodynamic factors. Under conditions of thermal equilibrium, the ability to overcome the required activation barrier may be rate-determining for a given chemical reaction. Under nonequilibrium conditions, the value of E_a may be surpassed rapidly compared with the kinetic factors involved. For example, in the case of thermally induced chlorine etching of silicon, the formation of $SiCl_4$ requires that the constituent elements *find* each other as well as having adequate energy to react when they do so. Under very rapid heating, the activation barrier for this reaction may be rapidly exceeded, but the surface may be cool again before sufficient migration has occurred for the species to collide with each other. In such a case high laser powers or increased pulse duration times may be required, which lead to short-duration surface melting before the surface mobility is high enough to allow an etching reaction to proceed.

It is clear that these conditions may give rise to significantly modified etching behaviour in comparison with the case of thermal equilibrium. This is considered in more detail in Chapter 13.

(iii) Suitable precursor compounds

This section has highlighted some of the restrictions that must be placed on the photochemical etching precursor. In the case of photothermal etching the precursor gas must not strongly absorb the light used. This may be achieved by reducing the interaction of the gas phase with the light in comparison with the interaction of the gas phase with the solid—for example by cooling the semiconductor to allow condensation of the precursor gas from a rather disperse gas phase. However, this may give rise to trapped

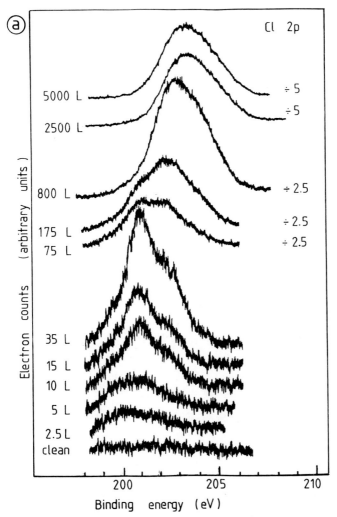

Fig. 12.7 XPS data taken following the adsorption of CCl₄ onto Si(100) at 150 K. (a) Binding energy of the Cl 2p peak as a function of increasing CCl₄ exposure, expressed in langmuirs (L): (b) Binding energy of the Cl 2s and C 1s as in (a). (c) Binding energy of the Cl 2s and C 1s for an adsorbed layer formed after exposure to 5000 L of CCl₄, plotted as a function of increasing surface temperature. (After [56].)

reaction products on the surface. An alternative approach is simply to utilize wavelengths transparent to the gas concerned. This may reduce the usefulness of the UV excimer laser wavelengths (which can be produced with excellent spatial resolution; see Chapter 4) for many halogenated precursors. In all cases it will be important to have a knowledge of the temperatures required to promote effective etching and the extent of any surface chemical disruption that is induced by these reactions.

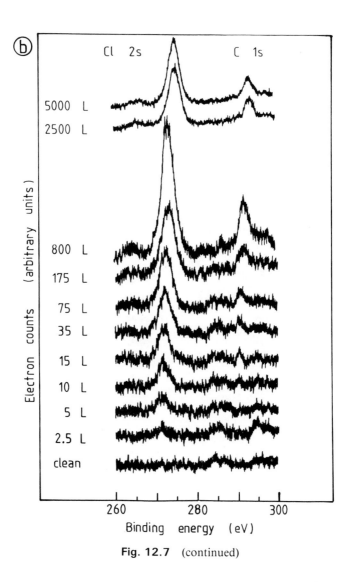

Fig. 12.7 (continued)

Recent studies of the interaction of CCl₄ with clean Si(100) surfaces [56]
act as a good example of the potential of halogen-containing molecules
(rather than halogens themselves) as precursors. Figure 12.7(a) shows
X-ray photoelectron spectroscopy data (XPS) relating to the Cl 2p transition following CCl₄ adsorption at 150 K. The spectra shown were collected *in situ* and are plotted as a function of increasing exposure of CCl₄

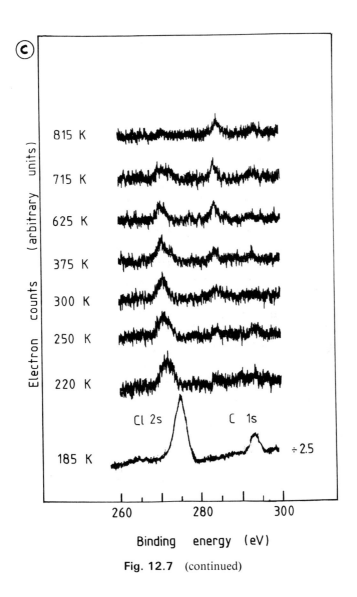

Fig. 12.7 (continued)

to the Si surface. It is immediately apparent that for all exposures above 15 L (1 L = 1 langmuir = 10^{-6} Torr. s) a shift occurs in the energy of the observed peak. This is indicative of the formation of two chemically distinct chlorine-containing adsorbed phases on the surface as the thickness of the adsorbed overlay increases. Figure 12.7(b) shows the XPS data taken during the same experiments for the C 1s and Cl 2s features; qualitatively similar behaviour is evident.

TDS analysis allows the nature of these two states to be assigned. Figure 12.8(a) shows spectra relating to the observation of Cl$^+$ desorption from the surface, again as a function of increasing exposure of the surface to CCl$_4$. Intriguingly, three peaks are evident. The high-temperature peaks are at the same temperatures as those observed for Cl$_2$ adsorption on Si (see Section 12.4.2); the low-temperature peak reveals the presence of a phase that is only weakly held to the surface. The uptake into each of these states is shown in the curves in Fig. 12.8(b), derived from the areas of each of the peaks in (a). It is apparent that the high-temperature peaks populate up to a level of around two monolayers; subsequent adsorption of CCl$_4$ gives rise to the weakly held state. A full analysis of all desorbing fragments in the TDS experiment shows the high-temperature desorbing products to be SiCl$_x$ ($x = 1 - 4$) compounds, while the low-temperature state is due to the desorption of molecular CCl$_4$.

The data from these two forms of analysis show that the carbon tetrachloride molecule dissociatively adsorbs for the first two layers on the surface. The species that exist give rise to thermal etching behaviour analogous to that of pure Cl$_2$ adsorption. Further adsorption (at low temperatures) *does not lead to any further surface reactions*, but simply gives rise to physisorbed molecules of CCl$_4$. It is important to consider the fate of the carbon in this reaction scheme. Figure 12.7(c) shows C 1s XPS data for a thick (several layers) adsorbed layer plotted as a function of increasing surface temperature. It is apparent that, while the quantity of surface carbon drops at the temperature at which the molecular layer desorbs, a carbon feature is seen to re-emerge at higher temperatures. The most likely origin of this is the incorporation of carbon fragments into the subsurface region of the Si during the dissociative chemisorption step. This carbon segregates back to the surface at high temperatures. The simplest reaction scheme to explain these observations is given in Fig. 12.9.

This analysis of the CCl$_4$/Si system shows the importance of precursor selection. The move from Cl$_2$ has allowed the degree of surface reaction that will occur outside any beam irradiated region to be limited to one or two monolayers. This is clearly very desirable. However, the thermally promoted surface phase etching reaction will lead to considerable carbon contamination. Precursors designed to avoid this are discussed in more detail in Section 12.5.3.

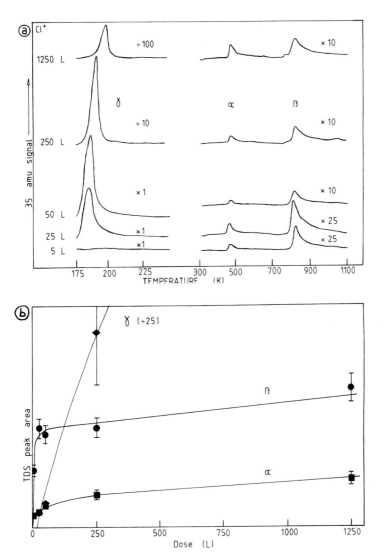

Fig. 12.8 TDS data taken following the adsorption of CCl_4 onto Si(100) at 150 K. (a) Data collected for Cl^+ species plotted as a function of increasing exposure of the Si to CCl_4 revealing three peaks, labelled α, β and γ. (b) Uptake curves for the three states observed in (a), plotted as observed peak intensity vs exposure of the Si to CCl_4. (After [56].)

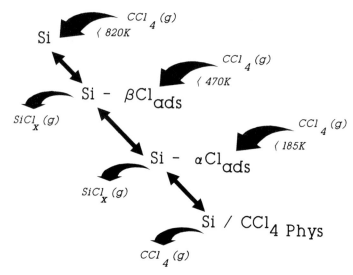

Fig. 12.9 Reaction scheme to explain the spontaneous processes that occur when CCl₄ comes into contact with Si(100), including the species thermally desorbed form each state that is formed. (After [56].)

12.4.3.2 Etching promoted by semiconductor photo-excitation

Surface irradiation by light of energy greater than the bandgap of the material concerned can give rise to excitation of electrons from the valence band to the conduction band. Holes are created in the valence band. The adsorption of strongly electronegative species, such as halogens, onto semiconductor surfaces can severely alter the electronic structure of the surface region. Such an effect can lead to a shift in the position of the energy of the valence and conduction bands (as referenced to the Fermi level) as the surface is approached. This "band bending" can lead to drift of the electron–hole pairs formed towards the surface from the bulk of the solid. In an n-type semiconductor holes are swept to the surface while the electrons move into the bulk of the solid; in a p-type semiconductor the converse is true. The appearance of either moiety at the surface of the semiconductor can lead to charge transfer between the adsorbed phase present and the surface atoms. Thus chemical reactions can be promoted that may in turn lead to etching.

This form of photochemical etching has been elegantly demonstrated and discussed by Houle [57–60]. This work involved exposing silicon to XeF₂ in the presence and absence of visible light from an Ar⁺ ion laser. This compound is known to dissociatively chemisorb on Si [61, 62], acting as a convenient source of fluorine. Spontaneous etching can occur at temperatures in the region of $320°C$ [62]. The introduction of light was found to enhance the etch rate by around four times, and the etch products were more highly coordinated (namely SiF_x, where x is seen to increase under illumination as compared with purely thermal etching products) [60]. A further example comes from chlorine chemistry. At low temperatures, the etching reaction of p-type Si with Cl radicals is negligible. It is believed that this is due to an absence of electron density, which is necessary for the formation of Cl^- ions prior to $SiCl_4$, formation [63]. However, under illumination by 308 nm light from an excimer laser, electrons can be swept to the surface, and etching at a rate of around 0.1 nm per J cm^{-2} of exposure can then be detected.

In terms of III–V photochemical etching, the ability to drive an etching reaction that displays selectivity between different-bandgap materials, or materials with differing doping levels, would appear very attractive. Progress in this area will be considered in Section 12.5.1. However, it is important to consider the potential of this approach to yield a highly localized and anisotropic etching reaction. The diffusion lengths of electrons and holes in the semiconductor concerned is likely to limit both of these properties. For example, even the minority carriers in bulk p-type GaAs (at a doping level of 8×10^{16} cm^{-3}) may be expected to have a diffusion length of around 2.8 μm at 300 K [64]. The extent of electron–hole pair excitation will clearly depend upon the intensity of the incident light; the etched profile may therefore reveal the intensity profile of the beam concerned. This may lead to features with sloped walls and further limit the anisotropic nature of the approach in a maskless *in situ* etching environment.

12.4.3.3 Adsorbed phase photolysis

The final approach to photochemical etching that may be adopted involves pure photolysis of the precursor compound (or an intermediate) while in the form of an adsorbed phase. The absorption of a photon of light can lead to electronic excitation of the absorbing species. Relaxation of these excited forms can occur through a number of pathways as illustrated in Fig. 12.10 [65]. Charge transfer may occur, leading directly to a chemical reaction between the absorbing species and a neighbouring atom. Alternatively, ionization or dissociation may give rise to a species that in turn can chemically react with its neighbours. Other methods of relaxation lead to a return to the ground state without chemical reaction of the primary species. In the case of an adsorbed layer on a semiconductor either route to "photolysis" may lead to the formation of (more) volatile products containing elements from the semiconductor, and hence etching will occur. When photolysis is limited to the adsorbed phase, the reaction will clearly be confined to where the photon beam strikes the surface.

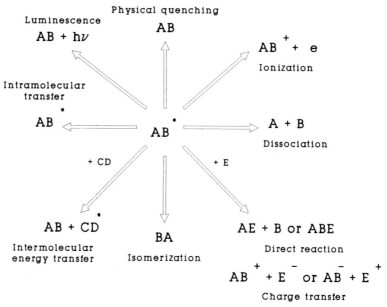

Fig. 12.10 Relaxation of an electronically excited species can occur by several routes. Some lead to the atom or molecule simply returning to its ground state, while others involve chemical changes that could lead to etching. (After Wayne [65].)

The Grotthus–Draper law of photochemistry states that "only radiation actually absorbed by the reacting system can initiate a reaction" [65]. This simple statement highlights the fact that the light being used must be matched to the absorption spectrum of species present. Adsorption of a molecule on a surface leads to some degree of charge transfer between this species and the solid; this can lead to a shift in the observed absorption spectrum as compared to its gas phase counterpart. This process could be exploited to allow photolysis to be driven in the adsorbed phase exclusively. However, these shifts are typically small compared with the absorption band widths concerned, and this route is unlikely to yield the extent of selectivity that is required. Thus the direct photolytic stimulation of an etching reaction between a precursor and the semiconductor is likely to rely upon the formation of the adsorbed phase in the presence of a disperse gas phase, as discussed in Section 12.4.3.

An alternative approach involves controlling the surface chemistry such that dissociative chemisorption occurs, which naturally gives rise to different species to those present in the gas phase. If these species were subject to photolysis, while the gas phase was transparent to the light used, surface domination of the reaction would be achieved.

A simple example of this form of reaction is highlighted by considering the $Cl_2/Si(100)$ [25] and $CCl_4/Si(100)$ systems [56]. The effect of low-intensity UV irradiation (produced from a mercury arc lamp) on the adsorbed phases formed in these systems (which have been discussed in Section 12.4.2) is illustrated in Fig. 12.11. In Fig. 12.11 thermal desorption spectra for the evolution of silicon halides are plotted for three different surface coverages of Cl_2 on Si(100), both in the presence and absence of low-intensity radiation. It is immediately apparent that the more strongly bound state is depopulated at the expense of the more weakly held one. This interconversion of the surface forms is indicative of the formation of more strongly coordinated silicon halides, which occupy more weakly bound surface sites. It has been suggested that this reaction may be promoted by the excitation of surface electron–hole pairs in the semiconductor [24], although the reaction could in principle be photolytic in origin. Intriguingly, the effect of similar light on the $CCl_4/Si(100)$ system reveals qualitatively similar behaviour for the strongly bound Cl states that are present in the first two monolayers [56]. Figure 12.12 reveals uptake curves for these states, and the low-temperature state that is present due to a molecular CCl_4 overlayer, as a function of increasing exposure to light for a fixed initial surface coverage. These are labelled α, β and γ respectively. It is clear that concurrent to interconversion of the high-temperature states (as in the $Cl_2/Si(100)$ case), loss of species from the low-temperature state is seen. Thus one mechanism for enhancing the etching reaction involves the light-stimulated reaction of adsorbed chlorine with silicon; CCl_4 simply acts as a "feedstock" of adsorbed Cl species. We have already noted the surface passivation that occurs when CCl_4 is exposed to Si in the absence of light; thus this form of photo-enhanced etching may allow the rest of the device to remain largely unperturbed by the precursor species. This passivation effect will also provoke a high degree of anisotropy in the etching reaction being driven. CCl_4 itself is unlikely to be an ideal etching precursor; the carbon contamination that may arise during etching has been shown above. However, this compound demonstrates the nature of surface reactivity that is required of the precursor ultimately chosen.

To successfully achieve photochemical etching of III–Vs in this manner will therefore require precise control over the spontaneous surface chemistry of the (more complex) precursor on the (more complex) semiconductor that gives rise to the desired intermediate. Photolytic surface conversion of this form to give rise to etching is then required. In practice, this may be difficult to achieve in isolation from photothermal effects. However, the spatial resolution that would be realized by this route is directly related to the region of the surface that is illuminated; other regions will remain passivated, and this will give rise to significant localization of the etch and, importantly, anisotropy.

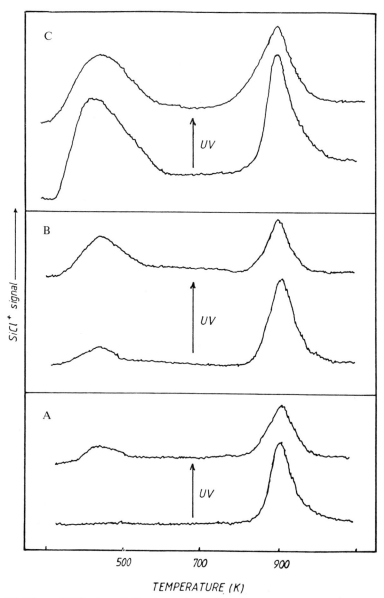

Fig. 12.11 TDS data following adsorption of Cl₂ onto Si(100) for three different coverages: (a) sub-monolayer, (b) three monolayers, (c) several multilayers of adsorbate. Spectra are shown prior to and subsequent to exposure to UV irradiation from a mercury arc lamp. (After [25].)

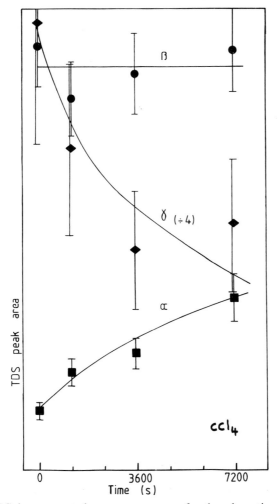

Fig. 12.12 TDS data presented as coverage curves for the adsorption of CCl₄ onto Si(100) plotted as a function of increasing exposure to UV light, again from a mercury arc lamp. The α and β peak intensities are seen to vary in accordance with Fig. 12.11, while the peak is seen to significantly decrease in intensity. (After [56].)

12.5 III–V SEMICONDUCTOR PHOTOCHEMICAL ETCHING

Now that we have surveyed the possible approaches, let us consider recent progress in III–V photochemical etching, concentrating on GaAs. The requirements for an *in situ* photochemical etching process for III–V semi-conductors will then be addressed.

12.5.1 Recent results

The first reports of localized laser assisted chemical etching of GaAs came from Ehrlich *et al.* [66]. They employed UV light (257 nm, 4 mW, frequency-doubled argon ion laser), with optics that gave a Fresnel diffraction pattern on the semiconductor surface. Methyl halides (CH_3Br, CH_3Cl, CF_3I) were used; these are known to strongly absorb light in this region of the spectrum [67]. Etching at a rate of 0.3 μm min^{-1} was achieved for GaAs(100), with a surface pattern being produced with around 1 μm feature sizes, which reflected the surface intensity of the light. Ehrlich *et al.* speculated upon a mechanism involving the gaseous formation of radicals, followed by the reaction of these with the GaAs to give rise to volatile Ga and As halides, i.e. in the case of CH_3Br,

$$CH_3Br(g) \xrightarrow{h\nu\ =\ 257\ nm} CH_3 + Br \qquad (12.4)$$

$$nBr \longrightarrow (nBr):(GaAs)_{ads} \qquad (12.5)$$

$$(nBr):(GaAs)_{ads} \longrightarrow GaBr_n\ (g) + AsBr_n\ (g) \qquad (12.6)$$

The addition of 514 nm light which cannot dissociate the CH_3Br but can heat the surface, revealed an enhancement in the etch rate of some 6.6 times for a 600°C temperature rise. At low laser fluences and gas pressures the Br radicals diffuse to the surface without significant recombination losses. Under these conditions, the rate of production of these species was suggested to be rate-determining. At high pressures and laser powers this is no longer the case. Osgood *et al.* [67] subsequently reported a GaAs etching reason with CF_3Br using UV radiation at 193 nm from an excimer laser (50 Hz, 35 mJ cm^{-2}). Rapid etching rates (up to 1 μm min^{-1}) were achieved at elevated temperatures, but the reaction rate was low at room temperature. The requirement for a photothermal and a photochemical contribution to the reaction was attributed to the involatile nature of the products. The authors speculated that these included condensed films of $Ga(CF_3)_3$ on the semiconductor surface [68].

Further studies into the reaction between CH_3Br and CF_3Br with GaAs under the influence of pulsed UV light [69] found physical ablation and laser-enhanced thermal desorption to dominate any etching at high energy densities (>35 mJ cm^{-2}, 50 Hz). These processes lead to rough surface morphologies and damaged material. At lower energy densities (around 35 mJ cm^{-2}, giving a GaAs temperature increase of around 400°C over a depth of 1 μm) the etching rate was at least 100 times higher than the ablation rate and left a less damaged surface, with etched features being anisotropic in nature. Use of the beam perpendicular and parallel to the surface again indicated a nonthermal initiation pathway for the reaction. The

Table 12.2 Physical properties of etching reaction products of GaAs with CH_3Br and CF_3Br taken from [69].

Etching gas	As products mp/bp ($^\circ$C)		Ga products mp/bp ($^\circ$C)	
CH_3Br	$As(CH_3)_3$	—/70	$Ga(CH_3)_3$	—/56
	$As(CH_3)_2Br$	—/51	$Ga(CH_3)_2Br$	—/58
	$As(CH_3)Br_2$	—/89	$Ga(CH_3)Br_2$	—/—
	$AsBr_3$	32/221	$GaBr_3$	121/278
CF_3Br	$As(CF_3)_3$	—/33	$GA(CF_3)_3$	—/—
	$As(CF_3)_2Br$	—/59	$Ga(CF_3)_2Br$	—/—
	$As(CF_3)Br_2$	—/118	$Ga(CF_3)Br_2$	—/—
	$AsBr_3$	32/221	$GaBr_3$	121/278

authors did, however, find that the reaction *rate* depended upon the thermal removal of the reaction products. The type of products detected by mass and optical spectroscopies were tabulated against their melting and/or boiling points, and this information is reproduced in Table 12.2.

The use of HBr as an etching precursor for GaAs in the presence of UV excimer light has also been reported [70, 71]. This gas was chosen since it is not expected to spontaneously react with GaAs in the temperature range 0–100°C [72]. Rapid etch rates were achieved (up to 8 μm min^{-2}) with 193 nm at 32 mJ cm^{-2}. The HBr is readily dissociated at the wavelengths employed, and the etching rate was linear with the laser power above 10 mJ cm^{-2}. Below this threshold energy, little etching occurred. The authors attributed this to the need to thermally desorb the surface bound products of the reaction, $GaBr_3$ and $AsBr_3$, with activation barriers of 56 and 44 kJ mol^{-1} respectively [73]. Interestingly, the etching reaction showed a strong dependence upon the crystallographic orientation of the GaAs:

$$GaAs\,[111]\,B > [100] > [110] > [111]\,A$$

This effect allowed anistropic etch features to be formed when a surface was masked, even though the "active" etching agent was produced in the gas phase. This type of effect has also been observed in Br_2 plasma etching of III–V semiconductors [73]. The addition of a buffer gas (Ar plus H_2 or C_2H_2) was investigated as a means of improving the spatial resolution of the process during direct laser projection patterning of the GaAs [71]. In both cases spatial resolution equivalent to the optical limit of the system (4 μm) was achieved. In the case of Ar/H_2 buffer mixtures a reduction in the mean free path of the Br radicals was the dominant effect. For C_2H_2 the formation of passivating polymer overlayers was thought to contribute to the improvement in confining the lateral spread of the etching reaction.

The presence of such films on the GaAs following etching may, however, be highly undesirable.

Hirose *et al.* have also investigated GaAs photochemical etching with UV wavelengths (excimer laser, 193 nm, 30–40 mJ per shot, 80 Hz) [74]. HCl (5% in He, 6.65×10^3 Pa) was introduced to GaAs (covered with a native oxide layer), with initial substrate temperatures in the region 20–200°C. An average etching rate of 23 nm min^{-1} was achieved. At room temperature, XPS data revealed the presence of several layers of GaCl$_3$ on top of an oxide species, while any AsCl$_3$ was believed to have spontaneously desorbed. When the GaAs substrate was heated to 200°C, no peaks due to chlorine or chloride species were observed. Under these conditions, around 10 laser shots eliminated the XPS signals due to Ga$_2$O$_3$ and As$_2$O$_3$. Yokoyama *et al.* [75] carried out an investigation into the usefulness of this reaction for device formation. Metal masks (Pt) were deposited prior to etching. Significant undercutting of the mask occurred, indicative of a gas phase process. Laser-etched material was compared with wet-etched GaAs. Photoluminescence measurements revealed much higher signals for the laser-etched surface, which decayed over a number of hours. This effect was attributed to the removal of the oxide layer by the laser process, which remained during wet etching. Schottky barrier diodes were fabricated that displayed similar I–V characteristics; the authors concluded that laser etching did not introduce any more surface defects than wet etching.

A number of articles by Ashby have discussed selectivity in photochemical etching of III–V semiconductors [76–80]. If significant thermal etching can be avoided then the etch rate for a given material may be related to its electronic properties. Since fluorine tends to lead to GaAs etching at relatively low temperatures, Ashby employed Cl radicals (created remotely from HCl) to etch GaAs, GaP and GaAsP in the presence of various wavelengths of light. For GaAs surface irradiation by light at all wavelengths greater than the bandgap gave rise to an etching reaction (Ar ion laser, 514 nm, etch rate 0.16 μm min^{-1} at 361 K, n- and p-type GaAs) [76]. The number of Ga and As atoms removed compared with the number of incident photons gave a ratio of around 1 in 10^5–10^6. The etched surface was particularly smooth, in contrast with thermally promoted laser etches [81]. The etching reaction was linearly dependent upon laser intensity, and the profile of a photo-etched hole replicated the Gaussian shape of the beam profile. The Arrhenius dependence of surface temperature observed for the thermally promoted GaAs reaction with chlorine was not observed; the reaction was attributed solely to the presence of photogenerated electron–hole pairs.

This reaction was investigated as a function of the GaAs doping level [77]. Application of a negative bias voltage across an n$^+$-type GaAs sample led to etching suppression; p$^+$-type GaAs etching remained unaffected.

This effect is believed to be associated with the "band bending" that the presence of halogens on the GaAs may cause (see Section 12.4.3.3). Application of the negative voltage raises the Fermi level relative to the surface band edge. This decreases the potential difference between the bulk band position and the surface band by an amount proportional to the applied voltage. Ultimately, flat bands or an accumulation layer may occur in the n^+-type GaAs; this suppresses hole migration to the surface (following illumination), and hence etching ceases. The "shut-off" of the etching in this manner was found to have an extremely sharp threshold. The etching rate measured by Ashby as a function of this applied voltage is reproduced in Fig. 12.13(a). Furthermore, since the degree to which band bending occurs is expected to vary with doping level (n-type GaAs has more

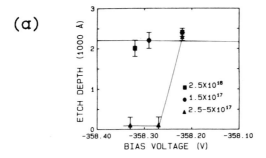

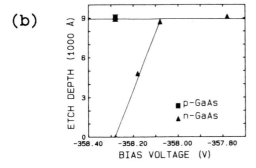

Fig. 12.13 Photo-etching of GaAs in the presence of Cl atoms promoted by carrier excitation in the solid. The application of a bias voltage allows the etching reaction to be suppressed. This effect has been exploited by Ashby to achieve impressive selectivity between n- and p-type GaAs and GaAs at different doping levels. (a) Plot of etch depth versus applied voltage for three different doping levels of n-GaAs for $10 \, W \, cm^{-2}$, 514 nm illumination. (b) Plot of etch depth versus applied voltage for n- and p-type GaAs for $1000 \, W \, cm^{-2}$, 514 nm illumination. (Reproduced from [77].)

band bending than n^+-type GaAs), the shut-off voltage for n^+-type GaAs will not prevent n-GaAs etching: 20:1 selectivity was achieved by exploiting this effect, as shown in Fig. 12.13(b).

The variation in bandgap in the $GaAs_{1-x}P_x$ system was also utilized by Ashby to generate selective etching [78–80]. The gap in GaAs (1.38 eV) and $GaAs_{0.8}P_{0.2}$ (1.58 eV) can be exceeded by 766 nm photons (from a dye laser), while the gap in $GaAs_{0.63}P_{0.67}$ cannot. Under this form of illumination, Cl atoms were found to etch the former materials but not the latter. Illumination by 514 nm light (with an energy larger than the bandgap of all three) promoted etching in all cases. Localized suppression of etching in this system was achieved by the implantation of damage by ion bombardment. The introduction of bulk defects reduces the lifetimes of the carriers produced under illumination; localized creation of defective regions by an ion beam can result in a means of selecting regions of the GaAs that will etch more slowly. Irradiation of n-type GaAs (doped to 10^{17} cm^{-3}) by a flux of 10^{14} boron ions cm^{-2} was found to almost eliminate all Cl atom etching. Subsequent thermal etching can be used to remove the damaged regions once a surface pattern has been achieved [80].

Very recently, Matz et al. [82] have investigated excimer laser-induced etching of InP with HBr, HCl and Cl_2. The results achieved exemplify the differences between gas phase and surface phase processes and are worthy of discussion here. The study involved an etching chamber attached to a UHV system allowing XPS studies to be carried out of the InP(100) surface following etching. Patterned light (193 nm, ArF laser, 150 mJ pulse, 1–80 Hz) was projected into the system through the gas precursor (local pressure 0.35–0.7 mbar). In the presence of HBr, gas phase dissociation to yield Br occurred, leading to surface adsorbed layers. Surface temperatures above 150°C lead to loss of adsorbed In halide species. Such a surface etched rapidly, but Gaussian-shaped trench profiles were achieved of around 100 μm size features. Etching outside these features was negligible. As the projected feature size was reduced, the etch rate decreased. This was attributed to relatively more halogen being lost through lateral diffusion and spatial resolution was found to suffer. Matz et al. suggested using a spontaneously reacting, nonabsorbing gas to improve upon this situation. In such a case the laser locally removes the corroded layer thermally. Studies with Cl_2 on InP revealed the spontaneous growth of an $InCl_3$ corrosion layer. At laser powers above 120 mJ cm^{-2} the authors suggest this layer plus phosphorus may be removed directly and an etch rate of around 0.2 μm min^{-1} may be achieved. The spatial resolution achieved by this photothermal approach was suggested to be limited by the thermal conductivity of the InP to around 0.5–1 μm. Thus the authors summarize that spontaneously reacting gases are more desirable for InP etching.

12.5.2   Surface chemistry and precursor selection

It is clear from the discussion above that photochemical etching of III–V semiconductors has been achieved under a range of conditions. A number of the reported examples rely upon a gas phase photolytic reaction accompanied by photothermal excitation of the solid, to encourage desorption of the etching reaction products. In these cases the gas phase component is undesirable for the *in situ* high-resolution UHV-compatible etching process that we are seeking. Etching promoted by photo-excited carrier generation in the semiconductor has shown excellent promise in terms of etch selectivity and the ability to quickly stop the etching reaction at a given layer. However, the reported reactions employed halogen and halogen radicals as precursor species, which are not ideal for the application being considered here. Furthermore, the use of this form of excitation may be inappropriate in a maskless environment when highly anisotropic etched features are required. It is apparent that a precursor showing ideal surface behaviour has not yet been identified.

To find such a precursor requires a study of the surface reactivity of a range of halogenated compounds on III–V materials. Detailed knowledge of the reactions of the halogens themselves is a prerequisite to such studies. To this end, a study of the adsorption of Cl_2 on InP(100)(4 × 2) has recently been carried out. The extent of surface perturbation that may be expected if this gas is used directly as an etchant is also demonstrated [45]. Figure 12.14 shows LEED patterns achieved for a range of exposures of InP(100)(4 × 2) to Cl_2. Analysis of these diffraction patterns reveals that the initial uptake of adsorbed Cl reverts the (4 × 2) surface structure to (4 × 1). Further exposure to around a monolayer leads to a (1 × 1) surface structure. Very high exposures (3000 L) reveal the complete loss of a LEED pattern. The nature of the surface reconstruction is known to have an effect upon the nature of any subsequently grown layers. TD spectra were also measured to enable the chemisorption behaviour that accompanies these structural changes to be evaluated. Figure 12.15 shows a series of spectra taken for increasing exposure of the InP to Cl_2, at room temperature, for three of the fragments observed in the mass spectrometer: $InCl^+$, $InCl_2^+$ and P_4^+. The presence of structure in these spectra at two different temperature regimes indicates that adsorption is a two-stage process. The higher temperature state reveals the desorption products from a two-dimensional chemisorbed monolayer, while the lower-temperature peaks arise from desorption from a corrosion phase. This latter state continues to populate without limit up to the exposure investigated (around 3000 L). Cracking pattern analysis reveals that the low-temperature In halide species is in fact $InCl_3$, other observed species arising from parent molecule fragmentation

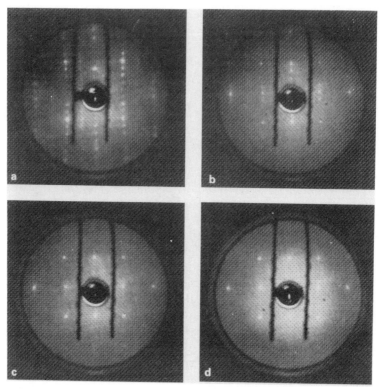

Fig. 12.14 LEED patterns from InP(100): (a) clean (4×2) reconstructed surface; (b), (c) and (d) after exposure to 2, 4 and 10 L of Cl_2. (After [45].)

in the mass spectrometer. Thus it is immediately apparent that this III–V semiconductor will show significant reactivity to Cl_2 at room temperature at the pressures discussed in Section 12.4.1, as was the case for Si.

The addition of ligands to this halogen to gain optimal precursor activity, for *in situ* maskless III–V etching, would therefore appear to require the creation of a compound that undergoes

(i)   spontaneous reaction with the surface of the III–V material, limited to monolayer thickness (and hence surface passivation);

(ii) spontaneous or light-induced surface reaction releasing the "active" halogen;

(iii) light-induced or spontaneous reaction with semiconductor;

(iv) light-induced or spontaneous desorption of (stable) reaction by-product.

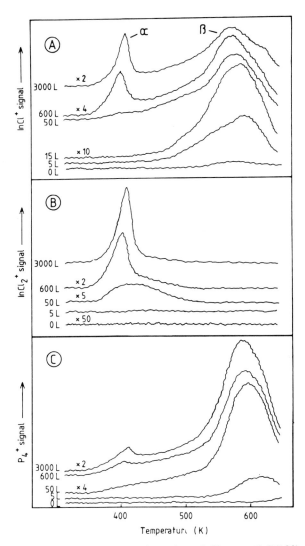

Fig. 12.15 TDS spectra following adsorption of Cl$_2$ onto InP(100) at 300 K: (a) InCl$^+$; (b) InCl$_2^+$; (c) P$_4^+$. Spectra are plotted as a function of increasing exposure to the chlorine gas, expressed in Langmuirs (L). (After [45].)

The role of the light may be to promote reactions through photolysis, pyrolysis or carrier excitation in steps (iii) and (iv). A purely photolytic process may give rise to a high degree of anistropy, as highlighted in Section 12.4.3. Ultraviolet photons can readily promote photochemistry in a range of halogenated compounds and can also be projected in submicron patterns. However, the gaseous precursor must then be inactive to this very energetic beam (typically > 6 eV), or rather disperse in the region through which the beam travels. If pulsed UV light from an excimer laser is used then the precursor gas could also be introduced in pulsed form. Supersonic gas beams, with choppers, could be used to allow the precursor to be adsorbed on to the III–V prior to the irradiation of this phase while the gas source was effectively "off". Alternatively, electronic and vibrational excitation may be readily achieved with visible or infrared wavelengths, to which these compounds may be made transparent more readily. However, under maskless conditions, limits to the spatial resolution that can be achieved with these wavelengths will become apparent. A combination effect that may be worthy of investigation involves the use of carrier excitation in the solid with a photothermal reaction. If the products of the carrier-promoted reaction still require heat to remove them (but less heat than the surface bound precursor) a spatially localized, anisotropic, etch may be achieved. The selectivity offered by the carrier generation approach could then be utilized. A photolytic reaction *in the adsorbed phase* along with a photothermal contribution to the reaction or simply a photothermal reaction, with a well defined activation energy, would also be an excellent solution for III–V etching in this environment. All available wavelengths will lead to surface heating of some degree, particularly in the presence of an adsorbed phase.

The precursor must spontaneously adsorb; studies discussed above have revealed the tendency of some halogenated compounds to do this. However, simple alkyl halides based upon a single carbon unit are likely to lead to contamination. The use of halogenated compounds containing carbon units that may lead to stable by-products that readily desorb would appear an important area for further study [83, 84].

A limitation on the usefulness of certain compounds may be the Montreal protocol, which was enhanced and ratified by 61 nations in London in June 1990. This agreement concerns certain halogenated compounds that are believed to lead to significant environmental damage. The production and use of a number of fully halogenated chlorofluorocarbons (CFCs) and bromofluorocarbons are to be phased out by the year 2000. The only exceptions allowed are certain safety applications without current alternatives. A summary of these substances is given in Table 12.3 [86].

Only halogenated compounds have been directly addressed as precursors for *in situ* (dry) photochemicals etching of III–Vs. Photo-induced oxidation followed by photothermal removal of the oxide species is an alternative approach. However, the dissimilar thermal oxidation rates of Group III and Group V elements may restrict the usefulness of this method. It is quite possible that photothermal etching could be promoted between the III–V materials and several other elements; no such studies have yet been published.

Table 12.3 Summary of the compounds to be phased out by the year 2000 under the revised Montreal protocol. Data taken from [86].

Substance		Relative ozone depleting potential
$CFCl_3$	CFC11	1.0
CF_2Cl_2	CFC12	1.0
CF_3Cl	CFC13	1.0
C_2FCl_5	CFC111	1.0
$C_2F_2Cl_4$	CFC112	1.0
$C_2F_3Cl_3$	CFC113	0.8
$C_2F_4Cl_2$	CFC114	1.0
C_2F_5Cl	CFC115	0.6
C_3FCl_7	CFC211	1.0
$C_3F_2Cl_6$	CFC212	1.0
$C_3F_3Cl_5$	CFC213	1.0
$C_3F_4Cl_4$	CFC214	1.0
$C_3F_5Cl_3$	CFC215	1.0
$C_3F_6Cl_2$	CFC216	1.0
C_3F_7Cl	CFC217	1.0
CF_2BrCl	Halon 1211	3.0
CF_3Br	Halon 1301	10.0
$C_2F_4Br_2$	Halon 2402	—
CCl_4	Carbon tetrachloride	1.1

12.6 CONCLUDING REMARKS

The process of etching has been discussed with reference to the various ways in which light can be used to promote a reaction between a precursor gas and a semiconductor. The specific use of photo-induced chemical etching of III–V compounds for an *in situ* maskless process has been addressed. While various reaction mechanisms can occur leading to etching, those based upon a rate-limiting process that occurs on the surface of the semiconductor seem favourable for this application. The direct reaction of a halogen species with the III–V may allow this condition to be fulfilled. However, the extent of perturbation beyond the beam-irradiated area may lead to an intolerable degree of material modification across the rest of the device. Precursors that display limited surface reactivity, followed by beam-promoted release of the halogen component and subsequent reaction seem preferable. In this case it is critical that the by-products of the reaction are sufficiently stable (and volatile) that their loss from the surface may be achieved. Finding compounds with appropriate surface reactivity would appear to be the primary area for research in this subject area. The choice of photon wavelength, duration of exposure (pulse length and repetition rate) and power are all fundamental in affecting the anisotropic nature of the etching process. High intensities simply give rise to ablation of the III–V, which leads to rough and damaged surfaces. Processes based upon carrier excitation in the semiconductor or photothermal effects may produce etched features that image the beam profile. However, in the latter case the activation barrier for the thermal process may allow highly spatially resolved sharp-sided features to be produced. Purely photolytic surface reactions give the most hope for a highly anisotropic process; these reactions also require, however, the most precise control of gas phase/surface phase photochemistry.

The use of particle beams for this form of etching has been mentioned (Section 12.1.3). The inherent problems of material damage induced by this form of energetic beam are very real. However, recent progress in ion beam source design may allow reasonable gas fluxes to be produced at energies as low as 10 eV. Since this represents the sputter threshold energy for GaAs, such a beam may be utilized to drive surface chemical reactions in the near absence of physical perturbation [84].

ACKNOWLEDGEMENTS

The thoughts and activities of Glenn Tyrrell, Judith Beckman and Duncan Marshall have greatly contributed to this chapter. The author is also very grateful for sponsorship from LEYBOLD Ltd in support of some of the work discussed.

REFERENCES

[1] W. Kern and G. L. Schnable, in *The Chemistry of the Semiconductor Industry* (ed. S. J. Moss and A. Ledwith), Chap. 11 (Blackie, Glasgow, 1987).
[2] D. L. Flamm and J. A. Mucha, in *The Chemistry of the Semiconductor Industry* (ed. S. J. Moss and A. Ledwith), Chap. 15 (Blackie, Glasgow, 1987).
[3] T. Tadokoro, F. Koyama and K. Iga, *Japan. J. Appl. Phys.* **27**, 389 (1988).
[4] K. P. Hilton and J. Woodward, *Vacuum* **38**, 519 (1988).
[5] K. L. Seaward, N. J. Moll and W. F. Stickle, *J. Electron. Mater.* **19**, 385 (1990).
[6] J. W. Coburn, H. F. Winters and T. J. Chuang, *J. Appl. Phys.* **48**, 3532 (1977).
[7] D. J. Ehrlich and V. T. Nguyen (eds), *Emerging Technologies for In-Situ Processing* (Martinus Nijhoff, Dordrecht, 1988).
[8] T. Mimura, S. Hiyaamizu, T. Fujii and K. Nanbu, *Japan. J. Appl. Phys.* **19**, 225 (1980).

[9] K. P. Hilton and J. Woodward, *Vacuum* **38**, 519 (1988).
[10] K. L. Seaward, N. J. Moll and W. F. Stickle, *J. Vac. Sci. Technol.* **B6**, 1645 (1988).
[11] P. Sigmund, *J. Vac. Sci. Technol.* **A7**, 585 (1989).
[12] Y. Yuba, T. Ishida, K. Gamo and S. Namba, *J. Vac. Sci. Technol.* **B6**, 253 (1988).
[13] T. J. Chuang, *Surf. Sci. Rep.* **3**, 1 (1983).
[14] J. Haigh and M. R. Aylett, *Prog. Quantum Electron.* **12**, 1 (1988).
[15] F. A. Houle, *Mater. Res. Soc. Symp. Proc.* **29**, 203 (1984).
[16] D. J. Ehrlich, R. M. Osgood and T. F. Deutsch, *Appl. Phys. Lett.* **36**, 916 (1980).
[17] G. L. Loper and M. D. Tabat, *Appl. Phys. Lett.* **46**, 654 (1985).
[18] R. J. Bienstock, *Appl. Phys. Lett.* **54**, 54 (1989).
[19] G. N. A. van Veen, T. S. Baller and J. Dieleman, *Appl. Phys.* **A47**, 183 (1988).
[20] S. Yokoyama, Y. Yamakage and M. Hirose, *Appl. Phys. Lett.* **47**, 389 (1985).
[21] R. Srinivasan and V. Mayne-Banton, *Appl. Phys. Lett.* **41**, 576 (1982).
[22] D. J. Ehrlich, R. M. Osgood and T. F. Deutsch, *Appl. Phys. Lett.* **38**, 1018 (1981).
[23] T. Arikado, M. Sekine, H. Okano and Y. Horiike, *Mater. Res. Soc. Symp. Proc.* **29**, 167 (1984).
[24] W. Sesselman and T. J. Chuang, *J. Vac. Sci. Technol.* **B3**, 1507 (1985).
[25] R. B. Jackman, H. Ebert and J. S. Foord, *Surf. Sci.* **176**, 183 (1986).
[26] T. Baller, D. J. Ostra, A. de Vries and G. N. A. van Veen, *J. Appl. Phys.* **60**, 2321 (1986).
[27] R. Kullmer and D. Bauerle, *Appl. Phys.* **A43**, 227 (1987).
[28] P. Mogyorosi, K. Piglmayer, R. Kullmer and D. Bauerle, *Appl. Phys.* **A45**, 293 (1988).
[29] R. Kullmer and D. Bauerle, *Appl. Phys.* **A47**, 377 (1988).
[30] S. Van Nyugen, S. Fridman and J. Rembetski, *Mater. Res. Soc. Symp. Proc.* **101**, 33 (1988).
[31] W. Sesselmann, E. Hudeczek and F. Bachmann, *J. Vac. Sci. Technol.* **B7**, 1284 (1989).
[32] T. S. Baller, J. Van Zwol, S. T. de Zwart, G. N. A. van Veen, H. Feil and J. Dieleman, *Mater. Res. Soc. Symp. Proc.* **129**, 299 (1989).
[33] R. C. Weast (ed.), *CRC Handbook of Chemistry and Physics*, 69th edn (CRC Press, Florida, 1988).
[34] M. J. Howes and D. V. Morgan (eds), *Gallium Arsenide: Materials, Devices and Circuits* (Wiley, Chichester, 1985).
[35] R. Kullmer and D. Bauerle, *Proc. SPIE (Trends in Quantum Electron.)*, **1033**, 232 (1988).
[36] W. Kauzmann, *Kinetic Theory of Gases* (Benjamin, New York, 1966).
[37] T. J. Chuang, *IBM J. Res. Dev.* **26**, 145 (1982).
[38] T. J. Chuang, *J. Chem. Phys.* **74**, 1453 (1981).
[39] G. Ertl and J. Kuppers, *Low Energy Electrons and Surface Chemistry* (Verlag-Chemie, 1974).
[40] P. K. Larson, N. V. Smith, M. Schluter, H. H. Farrell, K. M. Ho and M. L. Cohen, *Phys. Rev.* **B17**, 2612 (1978).
[41] J. V. Florio and W. D. Robertson, *Surf. Sci.* **18**, 398 (1969).
[42] R. J. Price, R. B. Jackman and J. S. Foord, *Appl. Surf. Sci.* **36**, 296 (1989).
[43] G. C. Tyrrell, I. W. Boyd and R. B. Jackman, *Appl. Surf. Sci.* **43**, 439 (1990).

[44] P. A. Redhead, *Vacuum* **12**, 203 (1962).
[45] A. J. Murrell, R. J. Price, R. B. Jackman and J. S. Foord, *Surf. Sci.* **227**, 197 (1990).
[46] G. C. Tyrrell, R. D. Marshall, J. Beckman and R. B. Jackman, *J. Phys. D. (Condens. mat.)* (in press) (1991).
[47] C. R. Zobel and A. B. F. Duncan, *J. Am. Chem. Soc.* **77**, 2611 (1955).
[48] W. L. Brown, in *Laser and Electron Beam Processing of Materials* (ed. C. W. White and P. S. Pearcy), p. 20 (Academic Press, New York, 1980).
[49] M. Lax, *J. Appl. Phys.* **48**, 3919 (1977).
[50] J. Mazumder and W. M. Steen, *J. Appl. Phys.* **51**, 941 (1980).
[51] J. E. Moody and R. H. Hendel, *J. Appl. Phys.* **53**, 4364 (1982).
[52] I. E. Calder and R. Sue, *J. Appl. Phys.* **53**, 7545 (1982).
[53] K. Piglmayer, J. Doppelbauer and D. Bauerle, *Mater. Res. Soc. Symp. Proc.* **29**, 42 (1984).
[54] M. Takai, J. Tokuda, H. Nakai, K. Gamo and S. Namba, *Japan. J. Appl. Phys.* **22**, L757 (1983).
[55] M. Takai, J. Tokuda, H. Nakai, K. Gamo and S. Namba, *Mater. Res. Soc. Symp. Proc.* **29**, 211 (1984).
[56] C. L. French, R. J. Price, R. B. Jackman and J. S. Foord, *Surf. Sci. (in prep.)*.
[57] F. A. Houle, *Chem. Phys. Lett.* **95**, 5 (1983).
[58] F. A. Houle, *J. Chem. Phys.* **79**, 4237 (1983).
[59] F. A. Houle, *J. Chem. Phys.* **80**, 4851 (1984).
[60] F. A. Houle, *Mater. Res. Soc. Symp. Proc.* **29**, 203 (1984).
[61] H. F. Winters and F. A. Houle, *J. Appl. Phys.* **54**, 1218 (1983).
[62] B. Roop, S. Joyce, J. C. Schultz, N. D. Shinn and J. I. Steinfeld, *Appl. Phys. Lett.* **46**, 1187 (1985).
[63] T. Arikado, M. Sekine, H. Okano and Y. Horiike, *Mater. Res. Soc. Symp. Proc.* **29**, 167 (1984).
[64] D. R. Wight and K. Duncan, *Datareview 5.13: Properties of GaAs* (Inspec, London, 1986).
[65] R. P. Wayne, *Principles and Applications of Photochemistry* (Oxford University Press, 1988).
[66] D. J. Ehrlich, R. M. Osgood and T. F. Deutsch, *Appl. Phys. Lett.* **36**, 698 (1980).
[67] R. M. Osgood, H. H. Gilgen and P. Brewer, *J. Vac. Sci. Technol.* **A2**, 504 (1984).
[68] R. M. Osgood, *J. Phys. (Paris) Colloq.* Suppl. 10 **44**, C5–133 (1983).
[69] P. Brewer, S. Hale and R. M. Osgood, *Appl. Phys. Lett.* **45**, 475 (1984).
[70] P. D. Brewer, D. McClure and R. M. Osgood, *Appl. Phys. Lett.* **47**, 310 (1985).
[71] P. D. Brewer, D. McClure and R. M. Osgood, *Appl. Phys. Lett.* **49**, 803 (1986).
[72] T. Aridado, H. Okano, M. Sekine and Y. Horiike, *Mater. Res. Soc. Symp. Proc.* **29**, 167 (1984).
[73] D. Ibbotson, D. Flamm and V. Donnelly, *J. Appl. Phys.* **54**, 5974 (1983).
[74] M. Hirose, S. Yokoyama and Y. Yamakage, *J. Vac. Sci. Technol.* **B3**, 1445 (1985).
[75] S. Yokoyama, T. Inoue and M. Hirose, in *Proceedings of International Symposium on GaAs and Related Compounds*, p. 325 (IOP, Bristol, 1985).
[76] C. I. H. Ashby, *Appl. Phys. Lett.* **45**, 892 (1984).

[77] C. I. H. Ashby, *Appl. Phys. Lett.* **46**, 752 (1985).
[78] C. I. H. Ashby and R. M. Biefeld, *Appl. Phys. Lett.* **47**, 62 (1985).
[79] C. I. H. Ashby, *J. Vac. Sci. Technol.* **A4**, 666 (1986).
[80] C. I. H. Ashby and D. R. Myers, *Solid State Technol.* **1989**, 129 (1989).
[81] N. Tsukada, S. Semura, H. Saito, S. Sugata, K. Asakawa and Y. Mita, *J. Appl. Phys.* **55**, 3417 (1984).
[82] R. Matz, J. Meiler and D. Haarer, *Mater. Res. Soc. Symp. Proc.* **158**, 307 (1990).
[83] EEC "SCIENCE" Programme in "Photo-Assisted Chemical Processing of GaAs" No. 021 (R. B. Jackman (co-ordinator), B. Leon, J. Flicstein, W. C. Sinke, I. W. Boyd and M. Green).
[84] Collaborative Research in Electronic Engineering, University College London (R. B. Jackman) and Physical Chemistry, University of Oxford (J. S. Foord).
[85] Collaborative research between Electronic Engineering, University College London (R. B. Jackman) and Atom Tech Ltd, Teddington, Middlesex, UK.
[86] R. Stevenson, *Chem. Brit.* **26**, 731 (1990).

13 Excimer Laser Chemical Etching of Silicon and Copper

J. C. S. KOOLS, T. S. BALLER and J. DIELEMAN

Philips Research Laboratories, Eindhoven, The Netherlands

13.1 INTRODUCTION

Dry etching techniques such as plasma etching, reactive ion etching or magnetron ion etching play a crucial role in current VLSI technology. In particular, the decrease in characteristic dimensions, in both lateral and transverse directions, and the rapidly increasing complexity of integrated circuits have motivated studies on laser chemical etching as an alternative to plasma-based techniques. The possible advantages of the technique are evident at first sight: (i) the ion-induced damage associated with plasma techniques might be avoided, and (ii) the number of processing steps might be reduced by using direct patterning. If commercial UV excimer lasers are used, a resolution in the submicron range can be achieved; for example, 130 nm wide lines have been produced by ablation of "diamond-like" carbon resist using 193 nm excimer laser radiation [1] (see also Chapter 4).

In this chapter a review is given by excimer laser chemical etching of two materials that play a central part in integrated circuits, namely silicon and copper. A second reason for this choice is the availability of a rather comprehensive set of data, enabling the delineation of a reasonably complete picture of the phenomenology as well as the mechanisms of laser chemical etching. At 300 K, spontaneous etching of Cu and Si with chlorine gas Cl_2 is negligible. Since this makes Cl_2 a promising candidate for high-resolution laser chemical patterning, the review is restricted to experiments with this gas.

To avoid confusion, some definitions of terms to be used in this paper will first be given.

The term *laser ablation* is used when the energy absorbed from a short laser pulse by a small volume of material is considerably greater than a certain threshold, which is around the total binding energy in that small

Photochemical Processing of Electronic Materials
ISBN 0-12-121740-X

volume. The ablated atoms and molecules are strongly concentrated along the normal, and have kinetic energies of at least several electronvolts (see Chapters 14 and 15).

The term *laser etching* describes the phenomenon where the laser is simply applied to heat the surface of the material to a temperature high enough for thermal etching to take place.

Laser chemical etching is laser etching in the presence of a reactive gas that adsorbs on the surface of the material and thus reduces the surface temperature required for etching. (In the extreme case of a pure photolytical process, this temperature is reduced to 0 K, since no thermal excitation is needed.)

13.2 LASER CHEMICAL ETCHING OF SILICON

The threshold for vaporization of silicon with nanosecond UV excimer laser pulses lies typically around a laser pulse fluence $F = 1$ J cm^{-2} (e.g. 1.5 J cm^{-2} for 308 nm/20 ns pulses [2]).

Several authors have shown that addition of chlorine gas permits removal of material at considerably lower values of F than for pure Si [2–12]. In a series of elegant experiments Bauerle *et al.* [3] have shown that below the melting threshold laser chemical etching of high-resistivity Si requires both Cl atoms, produced by dissociation of the chlorine gas, and photogenerated carriers in the Si.

Since it was found earlier [6] that etching of n-type Si with Cl atoms increases with increasing conduction electron concentration, whereas p-type Si is reported not to etch without irradiation by UV, the laser-generated conduction electrons are the crucial carriers for laser chemical etching.

Etching by photogenerated Cl atoms is essentially isotropic. A large fraction of the Cl atoms produced by laser photodissociation in the Cl$_2$ gas travel over a fairly large distance before they reach the Si surface to be etched. Both effects degrade the resolution of laser chemical patterning. Furthermore, etch rates are quite low: $\leqslant 0.1$ Å per pulse.

If one wants to laser chemical etch high-resolution patterns with a reasonable etch yield per pulse [5], it is more appropriate to use the effect that adsorption of small quantities of chlorine (up to one monolayer) and subsequent laser irradiation at fluences well below the ablation threshold of pure Si causes anisotropic etching with a reasonable etch yield [4]. The exact mechanistic explanation of this second etching mechanism is nontrivial, and has been the subject of a number of studies [3, 4, 7–9, 11, 12]. It is our aim to give an overview of the laser chemical etching of silicon in the latter regime. Both technological (empirical characterization of the process) and scientific (investigation of the etching mechanism) aspects will be considered.

13.2.1 Process characterization

The first property to consider in judging the utility of an etching process is the etching rate R (in Å s^{-1}). Absolute etch rates have been published by several groups [2, 3, 7]. Their dependence on process parameters will be discussed below for a Si substrate temperature of 300 K.

13.2.1.1   Chlorine pressure

For a sufficiently high value of F (e.g. 0.5 J cm^{-2} of 25 ns excimer laser pulses at 308 nm) and for a very low Cl$_2$ dose between two pulses (10^{-2}–10 L, where 1 L $\equiv$ 1 langmuir $= 10^{-6}$ Torr s) the etch rate is approximately proportional to the dose [7, 8]. As a matter of course, for a constant dose between two laser pulses, R increases linearly with the pulse repetition rate r. Above a dose of 10 L, the etch rate saturates at about 0.6 Å per pulse [7] and remains constant over six orders of magnitude in the Cl$_2$ pressure. It is only for relatively high gas pressures ($p \geqslant 10$ Torr) that one sees an increase in the rate for low fluences [2, 3]

13.2.1.2   Laser fluence

The dependence of etch rate on F is shown in Fig. 13.1, which is a compilation of data [2, 3, 7] for nanosecond pulses from a XeCl excimer laser (308 nm). All data have been taken from the pressure range where, at sufficiently high values of F, etch rate saturation is observed, but where photolytic effects in the gas phase do not yet play a role. As can be seen from this figure, the etch rate increases sharply at the melting threshold F_m. There then follows a plateau until ablation sets in. Because accurate measurement of F on the substrate is difficult, it is not constant over the laser spot, and it varies from pulse to pulse, some care should be taken in interpreting these data.

   Roughly speaking, the etch rate is nearly zero below F_m, and has a value of approximately one monolayer per pulse for fluences between F_m and the ablation threshold.

   For pressures in the range where gas phase photodissociation does play a role, the saturation etch yield is not influenced. For $F < F_m$ some etching is seen, but the actual rates are orders of magnitude lower than the rates of $F > F_m$. The former rates scale approximately with $F^{1.0}$ [2].

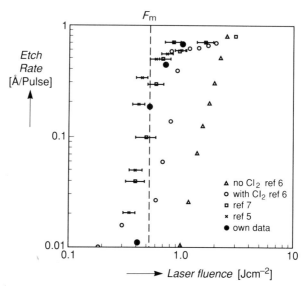

Fig. 13.1 Etch rate R as a function of laser fluence F at 308 nm. The melting threshold lies around 0.5 J cm^{-2}. All data have been taken with full monolayer coverage, but in the regime where photolytic effects are absent. (Data from [2, 3, 7].)

13.2.1.3 Wavelength

Most experiments have been performed with 308 nm radiation from a XeCl excimer laser. However, Kullmer and Bauerle [3] studied the effect of longer wavelengths. They used 423 and 583 nm light from an excimer pumped dye laser. It was found that the same stepwise increase occurred near F_m as for 308 nm.

The absorption cross-section for photolytic dissociation of gaseous Cl_2 is much larger for 308 than for 423 nm. It is zero above 480 nm. Therefore, etching due to photogenerated Cl atoms does not occur for wavelengths longer than 480 nm.

If 248 or 193 nm excimer laser radiation is used [2, 9], it is found that considerable etching occurs below F_m. It is very interesting to study this phenomenon further, since this opens perspectives for damage-free etching.

13.2.2 Microscopic mechanisms

To obtain a good understanding of the mechanisms of excimer laser chemical etching of Si with Cl_2, it is essential to describe briefly what is known about the separate interaction of a laser beam or a Cl_2 beam with the Si.

13.2.2.1 The interaction of excimer laser pulses with Si

Since all excimer lasers generate photons with an energy much higher than the Si bandgap, and strong absorption occurs at these wavelengths, electron–hole (e–h) pairs are created close to the surface (absorption depth about 70 Å). The energy contained in these e–h pairs is converted to heating of the lattice on a timescale short compared with the laser pulse (typically of the order of picoseconds). Therefore heating and melting of the silicon are described well by solution of the heat flow equation [13].

Although the absorption of radiation occurs within the near-surface area, melting may be found up to considerable depths (up to 1 μm).

For fluences over 1.5 J cm^{-2} ablation is found [2, 14] (see Fig. 13.1). Considerable etch rates are reported (up to 0.4 μm), but the etched surface has a poor morphological structure.

13.2.2.2 The reaction of chlorine gas with silicon

For temperatures near 300 K, Cl_2 has an initial sticking coefficient on clean Si surfaces of approximately 0.1. This sticking coefficient is independent of the coverage until a saturation coverage of about a monolayer is reached. Then it drops quickly over several orders of magnitude [15]. It was found from soft X-ray photo-emission [16, 17], and confirmed by scanning tunnelling microscopy [18], that absorption of Cl_2 on clean reconstructed silicon surfaces causes the formation of a chemisorbed overlayer. This does not imply that every Si surface atom is bonded to at most one Cl atom. Upon opening of the reconstruction on, for example, a perfect Si(111) surface, there will be sites where Si atoms have more than one dangling bond. However, diffusion of Cl atoms in the Si matrix is not observed. This is explained by a high activation energy for this diffusion step, as found from *ab initio* cluster calculations [19]. Therefore, at saturation coverage and relatively low substrate temperatures, newly arriving Cl_2 molecules will only find sites where they can be bound on chemisorbed Cl atoms by weak van der Waals interactions.

Since in-diffusion is obligatory for etching to occur, Si etching by Cl_2 at 300 K is reported to be slow [20] or negligible [21]. However, at higher substrate temperatures [$T \geqslant 600$ K), or with a supply of Cl atoms instead of molecules, one is able to etch Si with chlorine at a realistic rate.

Thermal desorption spectrometry of chlorinated Si surfaces [17, 20] shows a peak around 600 K, interpreted as Si atoms with several adatoms reacting with $SiCl_2$ or $SiCl_4$, and a contribution around 900 K, attributed to the reaction of the remaining Cl atoms to form $SiCl_2$. Irradiation with UV from a discharge lamp can change the occupation ratio of both states, indicating the occurrence of photochemical processes on the surface [20].

Steady state etching (under continuous gas exposure) gives $SiCl_4$ (for $T \leqslant 800$ K) and $SiCl_2$ (for $T \geqslant 800$ K) as the main etch products [21].

13.2.2.3 Experimental results on the mechanism of laser chemical etching

The technique most frequently applied to gather information on the mechanism of 308 nm excimer laser chemical etching of Si in a Cl_2 environment is mass spectroscopy combined with measurements of the time-of-flight (TOF) distributions of the desorption products [4, 7, 8, 11, 12]. Such measurements have proved that the dominant etch product is SiCl, with smaller contributions of Cl and $SiCl_2$. For low desorption yields, i.e. below about one percent of a monolayer, where collisions between desorbing molecules are negligible, angular-resolved TOF (AR TOF) distributions may yield definitive information about the desorption mechanism. If the desorption yield becomes higher, a so-called "Knudsen layer" will be formed [8, 22, 23]. In this layer, particles will have post-desorption collisions. Monte Carlo simulations have shown that as little as three collisions per particle are sufficient to transform every angular and velocity distribution into a so-called elliptical Maxwell–Boltzmann distribution superposed on a stream velocity:

$$f(v_x, v_y, v_z) \, \mathrm{d}v_x \, \mathrm{d}v_y \, \mathrm{d}v_z$$

$$= \frac{v^2}{c_1} \exp\left[-\frac{m(v_x^2 + v_y^2)}{2kT_{xy}} \right] \exp\left[-\frac{m(v_z - u)^2}{2kT_z} \right] \mathrm{d}v_x \, \mathrm{d}v_y \, \mathrm{d}v_z, \quad (13.1)$$

where z is the direction of the surface normal, and T_{xy}, T_z and u are fitted parameters. Transformation to polar coordinates and integration over the velocity gives the polar angle distribution:

$$f(\theta) \, \mathrm{d}\Omega = \frac{1}{c_2} \frac{\mu^3}{\cos^3 \theta} \, [\pi^{1/2} \, e^{\mu^2} (\mu^2 + \tfrac{1}{2}) \, \mathrm{erfc} \, (-\mu) + \mu] \mathrm{d}\Omega, \quad (13.2)$$

where μ is a generalized Mach number

$$\mu = \left(\frac{\tfrac{1}{2} m u^2 \cos^2 \theta}{T_z} \right)^{1/2} \left(\frac{T_{xy}}{T_z \sin^2 \theta + T_{xy} \cos^2 \theta} \right)^{1/2} \quad (13.3)$$

It is self-evident that AR TOF distributions are essential for determining the value of the above three fitting parameters. Knowledge of the fitting parameters, and thus the distribution, allows the calculation of the average kinetic energy of the particles. Such measurements have been performed [8], and show that, although the effect of the Knudsen layer is less drastic than in the case of copper etching, to be discussed later, it certainly cannot be ignored.

The effect of chlorine exposure has been studied at a fixed fluence $F > F_m$ [8]. At exposures $< 10^{-1}$ L, where the Knudsen layer is not yet developing, the AR TOF distributions are fitted well by Maxwell–Boltzmann (MB) distributions and $\cos \theta$ angular distributions. The temperature to be inserted in these MB distributions is around 1500 K. This is evidence that, at least under these conditions, the desorption process is thermal evaporation.

At high exposures, mean kinetic energies have been extracted from the experimental AR TOF data. These energies show a steady increase with the dose, given between two laser pulses, up to a saturation value around 10 L (see Fig. 13.2). This saturation occurs at energies associated with a surface temperature of 3800 ± 200 K, assuming the particles to be evaporated by a thermal process.

In another series of measurements F was varied in the range between F_m and the ablation threshold for a fixed Cl_2 exposure. Within experimental errors, no effect on the AR TOF distribution or the average kinetic energy has been observed [22].

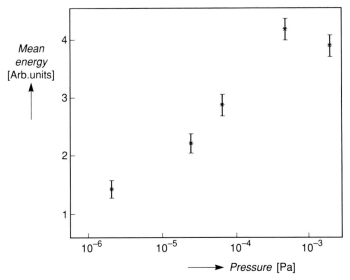

Fig. 13.2   Mean kinetic energy per particle as a function of p_{Cl_2}; $r = 4$ Hz.

In order to investigate the effect of melting of the Si on the etch rate, mass spectrometry has been combined with time-resolved reflectivity measurements. Since molten Si has a reflectivity of 0.7 for HeNe laser light, and solid Si has only a reflectivity of 0.3 for this wavelength, melting and solidification are easily monitored on a nanosecond timescale. Etch products have been detected only for laser pulses that induce melting.

It should be mentioned that, under very different conditions, namely low fluences (10 mJ cm^{-2}) and substrate temperatures (100 K), evidence for photolytic effects has been found from TOF-measurements [12].

In order to elucidate the role of surface temperature, time-resolved pyrometry has been applied [9]. The behaviour of the surface temperature as a function of F and gas dose is the same as that of the average kinetic energy of the desorbing particles derived from the AR TOF distributions.

For constant moderate values of F the surface temperature increases with dose between two laser pulses, and, as expected from the average kinetic energy derived from the AR TOF distributions, saturates for 10 L around 3500 K.

For constant dose and increasing F one sees a slight decrease in surface temperature. Surprisingly, an increase in intensity is observed. This cannot be explained by the assumption of a single hot layer on a cold substrate. It must be concluded that there are two radiation sources. First there is the actual surface layer, below which is a second and less hot layer that increases in depth with larger fluences.

13.2.2.4 A model of excimer laser chemical etching of silicon by chlorine gas at 308 nm

From the extensive set of experimental data compiled above, it is possible to deduce a model for the mechanism of the etching process. It is clear from the yield measurements, and is confirmed by the reflectivity measurements, that melting is necessary to get significant laser chemical etching. The main etch product is SiCl, which is quite different from the main etch products of pure laser etching. This is not surprising. Also, a clear difference from the main products observed for "slow" (i.e. non-laser-induced) thermal reaction is found. Therefore it is reasonable to seek a different thermal desorption mechanism.

A possible explanation could be that the extra entropy introduced by the melting enhances the frequency factor for the reaction of chemisorbed SiCl (state 1) to an energetically low lying state 2 from which the SiCl is readily desorbed. In this picture, the energy gained by the transition from state 1 to state 2 could be used to produce extra heating of the surface.

13.2.3   Implications for a practical etch process

Using a repetition rate of 50 Hz and a chlorine pressure in the 10^{-2} Torr range, it is possible to attain etch rates of 30 Å s^{-1}, which is very competitive with plasma-related etch techniques. The etch depth can be controlled very accurately by the number of laser pulses.

For average pulse fluences close to the threshold, it was found [2] that roughening of the etched surface occurred. This is related to pulse-to-pulse and spatial variations in the laser beam profile, translated as inhomogeneous etch behaviour. Increasing the pulse fluence gave smooth etch profiles. Therefore it is necessary to work with sufficiently high average fluences (typically 1 J cm^{-2}). Such a fluence causes melting (and thus dopant redistribution—see Chapter 11) or other structure degradation over depths of the order of 2000 Å. It must be concluded that, although there are no ion-bombardment-induced atomic displacements, or laser-stimulated chlorine diffusion, excimer laser chemical etching of Si is not a damage-free etch process.

13.3   LASER CHEMICAL ETCHING OF COPPER

The threshold for significant laser vaporization of Cu using nanosecond excimer laser pulses lies around $F = 5$ J cm^{-2} (e.g. 4.5 J cm^{-2} for 248 nm/15 ns pulses [24]), which is significantly higher than for Si. As in the case of Si, the addition of chlorine gas lowers the value of F that induces considerable vaporization to the range of a few 0.1 J cm^{-2} [25–27].

Formation of relatively thick (at least a few hundred ångströms) nonstoichiometric chlorinated Cu films is imperative to achieve etching at useful rates [26]. The required thicknesses can be obtained readily in a short time by using relatively high chlorine partial pressures at substrate temperatures between about 300 and 400 K [25, 27, 28] or by simultaneously applying a relatively low chlorine pressure and pulsed excimer radiation [26]. In the latter case, it takes some time (incubation period) before steady state etching is reached.

Photolytic effects, i.e. dissociation of the Cl$_2$ in the gas phase followed by reaction of the Cl atoms with the (chlorinated) Cu surface, have been shown to be negligible [27].

Diffusion of chlorine into the copper is isotropic, except for pulsed laser-enhanced diffusion [26]. Thus the use of high chlorine pressure results in great loss of lateral resolution. For example, 10 s at 300 K in a chlorine ambient of 1 Torr produces a resolution loss of about 1 μm.

If high-resolution patterns are desired, combined with useful etch yields per pulse, the simultaneous application of relatively low Cl_2 pressures and excimer laser pulses seems much more fitting [26]. In that case etch rates of a few monolayers per pulse, resulting in etch rates of 10^2–10^3 Å s^{-1}, are easily within reach.

The mechanism of excimer laser chemical etching of Cu in the presence of Cl_2 has recently been studied in some detail [26, 29]. Specifically, recent angular-resolved time-of-flight (AR TOF) studies of the laser desorbed products, and Monte Carlo simulations of this desorption, as a function of chlorine pressure, have shed much light on this mechanism.

The following sections will first describe technological aspects of the process and then the results of studies on the etching mechanism.

13.3.1   Process characterization

Depending on the choice of process conditions, the laser chemical etching rate R of Cu ranges from around one to a few hundred ångströms per pulse [26, 27]. The choice for a relatively low or a relatively high etching rate will depend very much on the specific application. Below, the dependence of R on some process parameters is discussed for the substrate temperature range between about 300 and 400 K.

13.3.1.1   Chlorine pressure

Using 308 nm nanosecond excimer laser pulses at F values of a few 0.1 J cm^{-2} and simultaneous exposure to very low chlorine pressures ranging from about 10^{-6} to 10^{-4} Torr at 300 K, the steady state etching rates vary roughly with the square root of the chlorine pressure. For a pulse repetition rate $r = 4$ Hz, these values range from about 0.3 to 3.0 Å per pulse. The time needed to reach steady state (length of the incubation period) is approximately inversely proportional to p_{Cl_2}. For the above chosen conditions, steady state etching at $p_{Cl_2} = 10^{-4}$ Torr is approached after 10^2 s. At this pressure, and with a repetition rate of 4 Hz, the time required to etch through 1 μm of Cu is about 800 s [26]. If only thermal diffusion is important, this will result in a loss of lateral resolution by sideways indiffusion of Cl of about 70 Å [28]. Since the laser will also stimulate diffusion on the crater edges, practical values of this resolution loss are expected to be higher. As will be shown below, the loss of resolution is, under these conditions, at maximum 500 Å.

Using 308 nm excimer laser pulses at F values of a few tenths of a joule per square centimetre (e.g. 0.3 J cm^{-2}) and simultaneous exposure to very high chlorine pressures, ranging from about 10^{-1} to 1 Torr, higher etch rates may be achieved. At 300 K, R is hardly dependent on p_{Cl_2}, and for $r = 10$ Hz etching rates of about 80 Å per pulse have been observed. R increases roughly as $r^{0.5}$. For $p_{Cl_2} = 0.1$ Torr and $r = 4$ Hz, the time required to etch through 1 μm of Cu is about 50 s [27]. The loss of lateral resolution due to sideways indiffusion of Cl during these 50 s at 0.1 Torr is about 1.6 μm [28].

13.3.1.2   Substrate temperature

The etch rate R increases monotonically by increasing the Cu substrate temperature from 300 K to about 400 K. The enhancement is small for $p_{Cl_2} = 0.1$ Torr and at most a factor of 2.5 for $p_{Cl_2} = 1$ Torr. Above about 400 K, R decreases again.

13.3.1.3   Wavelength

Extensive data on excimer laser chemical etching of Cu only exist for a wavelength of 308 nm [26, 27].

13.3.1.4   Fluence

For low as well as for high p_{Cl_2}, etch rates exhibit a practical threshold of about 0.1 J cm^{-2}. Between $F = 0.1$ J cm^{-2} and 0.3 J cm^{-2}, R increases rapidly with increasing F. Above the etch rate maximum at about 0.3 J cm^{-2}, R decreases gradually with increasing F.

13.3.2   Microscopic mechanisms

As in the case of Si, a description of the separate interaction of a laser beam or a Cl$_2$ atmosphere will precede the discussion of the mechanisms of laser chemical etching of Cu with Cl$_2$.

13.3.2.1   The interaction of excimer laser pulses with copper

Bloembergen [30] has suggested that energy absorbed from a short laser pulse by materials like Si or Cu in ordinary linear one-photon absorption processes is rapidly (within a picosecond) converted into heating of the material. Using the well known optical reflection and absorption properties as well as the specific heat, the mass density and the thermal conductivity, it was estimated that a laser pulse fluence of about 4 J cm^{-2} from a nano-second excimer laser pulse will heat Cu to its boiling point and thus cause evaporation. The experimentally determined threshold for observable vaporization of Cu in ultrahigh vacuum (UHV) using 15 ns, 248 nm excimer laser pulses is 4.5 J cm^{-2} [24], which is very close to the above estimate. The main species desorbed at, and slightly above, this threshold are Cu atoms. The TOF distributions of the desorbed Cu atoms can be fitted quite well in MB distributions with fit temperatures, measured along the surface normal, as a few tens of thousands kelvin. (Cu$^+$ ions start to be emitted at F around 7.5 J cm^{-2}). The spatial distribution of Cu vaporized under these conditions fits to a cos$^8 \theta$ distribution. Etch rates have not been mentioned.

13.3.2.2 The reaction of chlorine gas with copper

The initial sticking probability of Cl$_2$ on a clean Cu surface of 300 K is close to unity [31]. Reported doses needed to reach a first saturation coverage vary from 1 to 4 L for single-crystalline Cu(001) [32], 5 L for single-crystalline Cu(111) [33] and 2 L for polycrystalline Cu [32]. When exposure is prolonged after the first saturation coverage has been reached, the sticking probability drops below 0.002 [32]. Recently, the effects of prolonged exposure of Cu at 300 K to Cl$_2$ pressures ranging from 10^{-5} to 1 Torr have been investigated in great detail [25]. It is suggested that chlorine diffuses into the Cu. Below a Cl$_2$ pressure of about 10^{-3} Torr, the chlorinated surface film is relatively thin (at most a few hundred ångströms for long, i.e. 10^3 s, exposures), and the average Cl concentration in this film is small, i.e. considerably below that in stoichiometric CuCl. At higher Cl$_2$ pressures and for long exposure times, very thick chlorinated films are

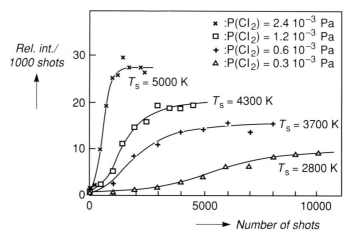

Fig. 13.3 Demonstration of the build-up effect. Etch rate is plotted versus the number of pulses for different chlorine pressures at constant F and r.

formed (up to several microns at 1 Torr for times longer than a few seconds), and especially the surface part is close to stoichiometric CuCl. The diffusion behaviour is quite complicated: two diffusion coefficients are necessary, a large one for short diffusion times, and a smaller one for longer times; furthermore, the values of these diffusion coefficients generally increase with Cl_2 pressure, with a particularly large step between 10^{-3} and 10^{-2} Torr.

Thermal desorption from chlorinated Cu surfaces [34] leads to gaseous Cu_3Cl_3 for temperatures below about 850 K, while CuCl is the dominant desorption product above about 920 K. Desorption of Cu_3Cl_3 follows zeroth-order desorption kinetics, with an activation energy equal to the heat of sublimation of Cu_3Cl_3 from solid CuCl. This suggests that the chlorine concentration in the surface remains approximately constant until the chlorine in the bulk is depleted.

13.3.2.3 Experimental results on the mechanisms of laser chemical etching

The first most striking observation made for excimer laser chemical etching of Cu by Cl_2 at low p_{Cl_2} (i.e. 10^{-6} to about 10^{-4} Torr) is that adsorption of one to a few monolayers of chlorine on the Cu does not lead to an instantaneous large enhancement of the etch rates as in the case of Si. In contrast, as illustrated by Fig. 13.3, there is a slow build-up to a steady state etch rate, which is only reached after the Cu surface has been exposed to a large dose of Cl_2. Even then, the steady state etch rate is low compared with this total dose. Also illustrated in Fig. 13.3 is the fact that, for a constant moderate value of F, the incubation period (i.e. the time needed to reach steady state etching) decreases in inverse proportion to the applied p_{Cl_2}. Figure 13.4 shows a Rutherford back-scattering analysis of the Cu surface at various stages during the build-up to this steady state, again for a rather low value of F. Clearly, a deep and relatively high chlorine profile is being built up during this incubation period. Extensive studies of this phenomenon have shown that each combination of F and p_{Cl_2} leads to a unique steady state chlorine profile, irrespective of the previously chosen steady state chlorination/etching situation. For constant p_{Cl_2}, the profile is deep and high for a low value of F and increasingly low and shallow for increasing values of F (see for illustration Fig. 13.5). Since the profiles obtained in steady state laser chemical etching are about one to two orders of magnitude deeper than observed for pure Cl_2 exposure at the same pressure for the same time, laser-induced diffusion into the copper is an important aspect of low-pressure excimer laser chemical etching of Cu by Cl_2.

Interestingly, for etching at higher pressures, it is also observed that the thickness of material removed per laser pulse is much less than the thickness of the chlorine profile built up during diffusion of the chlorine between two laser pulses. Evidently, in order to get a significant etch rate, a relatively deep and high chlorine profile is needed at first.

In this profile, the concentration of chlorine at the Cu surface is close to 50 at%, and decreases gradually from this CuCl surface to pure Cu in the bulk.

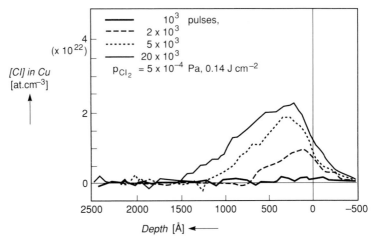

Fig. 13.4 Cl profiles in single-crystal Cu(001) as obtained from RBS measurements at various stages of the build-up.

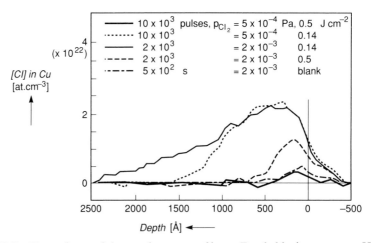

Fig. 13.5 Dependence of the steady state profile on F and chlorine pressure. High pressures and low fluences tend to give deeper diffusion profiles.

As mentioned above, for both low and high p_{Cl_2}, the etch rate is quite low for $F = 0.1$ J cm^{-2}, increases rapidly to a maximum at about 0.3 J cm^{-2}, and then decreases gradually with increasing F. The low etch rate at low F may be explained by insufficient heating of the surface during the laser pulse. The rapid increase of R with F is then related to a rapidly increasing surface temperature. The decrease after the maximum is, in all probability, due to the fact that only relatively pure CuCl is etched. This conclusion is consistent with the observation that a relatively Cl-rich surface is needed for significant etching to commence.

Mass spectrometry combined with TOF measurements has shown that, except for relatively high p_{Cl_2} or a very large dose of Cl$_2$ between two laser pulses, the main products desorbed are CuCl and Cu$_3$Cl$_3$. As mentioned above, these molecules are also the main products of thermal desorption [34]. When, for a constant F, at a value where R is at a maximum, the effect of using increasing values of p_{Cl_2} or increasing dose between two laser pulses is studied, observations largely similar to those also made for Si are seen. For a start, pulsed excimer laser radiation of Cu containing a small amount of chlorine in the surface film in UHV is used to desorb Cu$_3$Cl$_3$. AR TOF studies show that, under these conditions, this molecule is desorbed with a cos θ angular distribution and TOF distributions fit well to MB distributions with fit temperatures corresponding to the surface temperature calculated from the optical absorption and reflection and heat capacity and conductive properties of Cu and the value of F. This is illustrated by Fig. 13.6. This clearly demonstrates that Cu$_3$Cl$_3$ is thermally

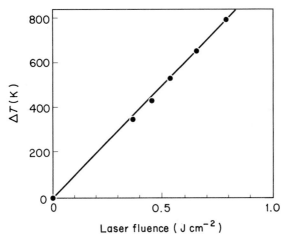

Fig. 13.6 Comparison of the temperature obtained from AR TOF measurements (•) with the temperature predicted by a heat flow simulation (———) for a low Cl$_2$ exposure.

desorbed and that collisions between desorbing particles are negligible under these conditions.

Next, the effects of increasing the p_{Cl_2} or Cl_2 dose between two pulses on the AR TOF distributions are investigated. As demonstrated by Figure 13.7 two things happen:

(1) the average kinetic energy of the desorbing molecules increases quite strongly with p_{Cl_2}; and

(2) collisions between desorbing particles cause a peaking along the normal to the surface (see Fig. 13.7).

The latter phenomena can be described well by the formulae given in the description of the Knudsen layer behaviour (see above). At very high p_{Cl_2} or equivalently large dose between two laser pulses, ablation-type AR TOF distributions with peak kinetic energies up to several electronvolts are observed. There is also a tendency to desorption of smaller molecules and, for sufficiently high p_{Cl_2}, ionized species are detected.

13.3.2.4   A model of excimer laser chemical etching of copper by chlorine gas

Since, apart from the ablation regime, the main products of laser chemical etching are the same as for the "slow" thermal reaction and, as the energies extracted from the TOF distributions correspond to plausible values for the surface temperature, a reasonable conclusion is that laser chemical etching of Cu by Cl_2 follows a thermal desorption mechanism. As shown above, the temperatures achieved at the Cu surface for F of a few tenths of a joule per square centimetre are quite low. This is for a large part due to the high heat conductivity of Cu. It also means that significant evaporation of Cu_3Cl_3 (CuCl) does not occur for, for example, monolayers of chlorine absorbed on pure Cu. To get significant etch rates, a relatively thick chlorinated film with its lower heat conductivity first has to be formed to isolate the surface film where most energy is deposited from the bulk Cu and thus achieve higher surface temperatures or ablation behaviour.

13.3.3   Implications for a practical etch process

Considering the fact that plasma etching of copper is relatively difficult, and the high etch rates achieved at moderate fluences in laser chemical etching of copper, it is clear that the latter process must be seen as an attractive alternative for high-resolution patterning of Cu in various applications. An important drawback is the loss of resolution in etching poly-

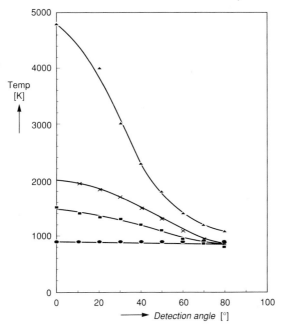

Fig. 13.7   Angular variation of the kinetic energy of Cu₃Cl₃ for different doses. The effect of the upcoming Knudsen layer is clearly demonstrated.

crystalline material. This is caused by enhancing chlorine diffusion along grain boundaries [1]. Since metal films in current IC-related technology are polycrystalline, these resolution losses are inevitable in this kind of application. Nevertheless, some micro-engineering problems can be solved on single-crystal substrates.

The problem of chlorine contamination in the etched layer can be reduced considerably. If the etching sequence ends with an irradiation without chlorine addition (some thousand pulses), most of the chlorinated layer is removed. However, some tail of the chlorine profile will remain in the copper film.

REFERENCES

[1] J. Brannon, *J. Vac. Sci. Technol.* **B7**, 1064 (1989).
[2] W. Sesselmann, E. Hudeczek and F. Bachmann, *J. Vac. Sci. Technol.* **B7**, 1284 (1989);
     W. Sesselmann and F. Bachmann, *Proc. SPIE* **998**, 90 (1988).

[3] R. Kullmer and D. Bauerle, *Appl. Phys.* **A43**, 227 (1987);
     P. Mogyorosi, K. Piglmayer, R. Kullmer and D. Bauerle, *Appl. Phys.* **A45**,
     293 (1988);
     R. Kullmer and D. Bauerle, *Appl. Phys.* **A47**, 377 (1988).
[4] T. Baller, D. J. Oostra, A. de Vries and G. N. A. van Veen, *J. Appl. Phys.*
     60, 2321 (1986).
[5] J. Brannon, *Mater. Res. Soc. Symp. Proc.* **101**, 27 (1988).
[6] Y. Horiike, M. Sekine, K. Horioka, T. Arikado and H. Okano, in *Abstracts
     of the 16th Conference on Solid State Devices and Materials, Kobe Japan*,
     p. 441 (Toshiba Corp., Kawasaki, 1984);
     H. Okano, Y. Horiike and M. Sekine, *Japan. J. Appl. Phys.* **24**, 68 (1985);
     Y. Horiike, N. Hayasaka, M. Sekine, T. Arikado, M. Nakase and H. Okano,
     Appl. Phys. **A44**, 313 (1987).
[7] J. Boulmer, B. Bourguignon, J. Budin and D. Debarre, *Appl. Surf. Sci.* **43**,
     424 (1989);
     J. Boulmer, B. Bourguignon, J. Budin and D. Debarre, *Chemtronics* **4**, 165
     (1989).
[8] T. S. Baller, J. van Zwol, S. T. de Zwart, G. N. A. van Veen, H. Feil and
     J. Dieleman, *Mater. Res. Soc. Symp. Proc.* **129**, 299 (1989).
[9] T. S. Baller, J. C. S. Kools and J. Dieleman, *Appl. Surf. Sci.* **46**, 292 (1990).
[10] S. van Nguyen, S. Fridmann and J. Rembetski, *Mater. Res. Soc. Symp. Proc.*
     101, 33 (1988).
[11] Y. Li, Z. Zhang, Q. Zheng, Z. Jin, Z. Wu and Q. Qin, *Appl. Phys. Lett.* **53**,
     1955 (1988).
[12] M. Kawasaki, H. Sato and N. Nishi, *J. Appl. Phys.* **65**, 792 (1989).
[13] S. de Unamuno and E. Fogarassy, *Appl. Surf. Sci.* **36**, 1 (1989).
[14] G. Shinn, F. Steigerwald, H. Stiegler, R. Sauerbrey, F. Tittel and W. Wilson,
     J. Vac. Sci. Technol. **B4**, 1273 (1986).
[15] J. V. Florio and W. D. Robertson, *Surf. Sci.* **18**, 398 (1969).
[16] R. D. Schnell, D. Rieger, A. Bogen, F. J. Himpsel, K. Wandelt and
     W. Steinmann, *Phys. Rev.* **B32**, 8057 (1985).
[17] L. J. Whitman, S. A. Joyce, J. A. Yarmoff, F. R. McFeely and
     L. J. Terminello, *Surf. Sci.* **32**, 297 (1990).
[18] J. S. Villarubia and J. J. Boland, *Phys. Rev. Lett.* **63**, 306 (1990).
[19] M. Seel and P. S. Bagus, *Phys. Rev.* **B18**, 2023 (1983).
[20] R. B. Jackman, H. Ebert and J. S. Foord, *Surf. Sci.* **176**, 183 (1986).
[21] F. Sanders, A. W. Kolfschoten, J. Dieleman, R. Haring, A. Haring and
     A. de Vries, *J. Vac. Sci. Technol.* **A2**, 481 (1984).
[22] T. S. Baller, PhD Thesis, Twente (1990).
[23] I. Noorbatcha, R. Luchese and Y. Zeri, *J. Chem. Phys.* **89**, 5251 (1988), and
     references therein.
[24] I. Hussla and R. Viswanathan, *Surf. Sci.* **145**, L488 (1984);
     R. Viswanathan and I. Hussla, *J. Opt. Soc. Am.* **B3**, 796 (1986).
[25] W. Sesselmann, E. E. Marinero and T. J. Chuang, *Appl. Phys.* **A41**, 209
     (1986).
[26] T. S. Baller, G. N. A. van Veen and J. Dieleman, *J. Vac. Sci. Technol.* **A6**,
     1409 (1988);
     G. N. A. van Veen, T. S. Baller and J. Dieleman, *Appl. Phys.* **A47**, 183
     (1988).
[27] J. J. Ritsko, F. Ho and J. Hurst, *Appl. Phys. Lett.* **53**, 78 (1988).
[28] W. Sesselmann and T. J. Chuang, *Surf. Sci.* **176**, 32 and 67 (1986).

[29] T. S. Baller, J. C. S. Kools, H. Feil and J. Dieleman, To be published.
[30] N. Bloembergen, in *Laser–Solid Interactions and Laser Processing* (ed. S. D. Ferris, H. J. Leamy and J. M. Poate), p. 1 (AIP Conf. Proc. Vol. 50, 1979).
[31] P. J. Goddard and R. M. Lambert, *Surf. Sci.* **67**, 180 (1977).
[32] D. Westphal and A. Goldmann, *Surf. Sci.* **131**, 113 (1983).
[33] S. Park, T. N. Rhodin and L. C. Rathbun, *J. Vac. Sci. Technol.* **A4**, 168 (1986).
[34] H. F. Winters, *J. Vac. Sci. Technol.* **A3**, 786 (1985).

14   Laser Ablation of Polymers

P. E. DYER

Department of Applied Physics, University of Hull, UK

14.1   INTRODUCTION

In the early 1980s Srinivasan and co-workers [1–4] reported that the deep-ultraviolet (193 nm) ArF excimer laser could be used to directly etch polymeric materials with unprecedented spatial resolution. The process was termed ablative photodecomposition [2, 3], it being conjectured that photo-induced bond breaking by the 193 nm laser led to the rapid removal of surface layers of the polymer in the form of low-molecular-weight fragments. A particularly notable feature of the interaction using the short-wavelength excimer laser was the apparent absence of significant thermal damage to the remaining polymer. Subsequently, many groups [5–14] reported studies of the effect, extending both the range of excimer wavelengths that could be used and polymers that could be successfully "ablated". Microscopic models describing the process were also developed [15–17].

UV photo-ablation of organic polymers has two important ingredients that have helped to maintain a lively interest in it over the past decade: First, there has been much uncertainty and debate over the fundamental mechanisms responsible for ablation—in particular, whether bond breaking proceeds via a direct photochemical route or by highly localized thermal reaction [2, 5, 18–20]. Secondly, the phenomenon has given birth to a powerful technique for processing organic polymers using lasers, which has potentially important practical applications in micro-electronics, opto-electronics and micromechanical machining. Related work on bio-polymers has also led to new developments in laser surgery [21, 22].

Reviews of UV laser photo-ablation have been published by Yeh [23], Srinivasan [24–26], Srinivasan and Braren [27] and Lazare and Granier [28], concentrating largely on the basic findings and mechanistic aspects. In view of this existing coverage, this chapter starts with a summary and

Photochemical Processing of Electronic Materials
ISBN 0-12-121740-X

discussion of the salient features of the topic and then concentrates more closely on its applications.

14.2 BASIC CHARACTERISTICS

The excimer laser photo-ablation of polymers is characterized by

(i)   a highly localized spatial interaction;

(ii) a threshold fluence for significant removal of material;

(iii) minimum heat-affected zone.

The first of these is easily understood. In a well designed experiment using contact or projection imaging techniques to define the irradiated area, the minimum feature size is set by diffraction and varies directly as the wavelength λ used [29]. For the common rare gas halide excimer lasers this lies in the range $\lambda = 0.193-0.351$ μm (0.157 μm for the F_2 laser), and it is possible with suitable optics to attain submicron definition. The third, fourth and fifth harmonics of the Q-switched Nd:YAG laser provide potentially useful alternative wavelengths, although at lower average power. In the third dimension the depth of material removed by a single pulse is set by

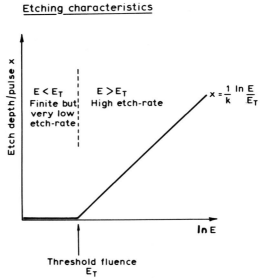

Fig. 14.1 Etch depth per pulse x as a function of ln (fluence) for ablation of a polymer when Beer's law is applicable. E_T marks the threshold for ablation, and the slope of the line defines the inverse of the effective absorption coefficient k.

the optical attenuation, and, since this is large for many polymers in the UV or deep-UV (of order 10^4–10^5 cm^{-1}), it is possible to etch the surface in a highly controllable way, typically removing $\leqslant 1$ μm per pulse. The poor thermal conductivity of polymers is also important. since it restricts the extent of heat transfer during the short laser pulse (of order 10 ns) to a very small value ($\leqslant 0.1$ μm). This spatial localization gives a unique advantage to UV lasers over other, longer-wavelength, lasers for many polymer micromachining applications.

It has been observed in essentially all UV laser–polymer ablation experiments [3, 5, 6, 8, 9, 30, 31] that a threshold fluence E_T (energy per unit area per pulse) for *significant* material removal exists (Fig. 14.1). This threshold typically lies in the range of a few tens to a few hundred millijoules per square centimetre, being dependent on the type of polymer involved and laser wavelength [30–32], but only weakly influenced by the pulse duration [33]. The term "significant" is stressed since in reality there is not a precise transition to the ablation regime and the threshold is then to some extent dependent on how accurately the experimentalist is able to determine the loss of material [34–39]. Although the details of this region are of great interest for modelling the interaction, defining a threshold for significant etching (for example by extrapolating to zero etch rate on the fluence axis [5, 30]) proves very useful for practical purposes. It is important to note that the ablation process does not obey reciprocity; thus exposure to a large number of low-fluence (subthreshold) pulses does not lead to the same removal rate as the equivalent single-pulse exposure [5].

As a broad observation, it has been found that the absorbed energy density needed to produce ablation, expressed in the product of the threshold fluence E_T and the (effective) UV absorption coefficient k, is approximately constant for a given polymer [8, 30, 31, 40]. This is exemplified by results for the ablation of polyimide and poly(ethylene terephthalate) (PET) with XeCl (308 nm), KrF (248 nm) and ArF (193 nm) lasers having pulse durations of 10–20 ns that are given in Table 14.1. Here kE_T lies in the range 3–4 kJ cm^{-3} [31].

Thermal coupling experiments using pyroelectric calorimetric techniques [40] or miniature thermocouples contacted to polymer films [31] have revealed that in the subthreshold region essentially all absorbed UV laser energy appears as heat. Above threshold, the thermal loading tends to saturate, however, at a value given approximately by AE_T, where A is the irradiated area, because the ablation products carry away the excess energy. This behaviour is illustrated schematically in Fig. 14.2, where the ideal case of a sharply defined threshold is assumed. This behaviour has been approximately confirmed by experiment [31, 40].

From the mechanistic aspect, the need to exceed a minimum energy density loading kE_T for ablation of a polymer would at first sight appear

Table 14.1 Threshold fluence for ablation of polyimide and polyethylene terephthalate (PET) derived from etch rate (E_e) and thermal coupling (E_T) measurements using the XeCl (308 nm), KrF (248 nm) and ArF (193 nm) lasers. The effective absorption coefficient k derived from etch rate measurements is also given, together with kE_e the threshold energy density for ablation [31].

PET

Polyimide

Laser wavelength (nm)	Polymer	E_e (mJ cm^{-2})	E_T (mJ cm^{-2})	k (μm^{-1})	kE_e (kJ cm^{-3})
308	PET	170	190	2	3.4
308	Polyimide	42	45	8	3.4
248	PET	30	37	10	3.0
248	Polyimide	31	36	14	4.3
193	PET	28	25	12	3.3
193	Polyimide	~30	45	13	3.2

to support a primarily photothermal degradation model. The surface temperature rise at threshold ΔT can be estimated as [31, 40]:

$$\Delta T = \frac{(1 - R)kE_T}{\rho c},$$ (14.1)

where R is the surface reflectivity, ρ the density and c the specific heat of the polymer. Equation (14.1) is obtained by assuming that on the timescale of the laser pulse (of order 10 ns), photo-excited states relax rapidly to produce heating, and conduction loss from the shallow heated surface layer can be neglected. For polymers such as PET and polyimide (Table 14.1), ΔT is found to be $\geqslant 1000$ K [31]. A substantial temperature rise has also been confirmed in the case of ArF laser ablated poly(methyl methacrylate) (PMMA) using spectroscopic [41] and mass spectroscopic [42] techniques. The question then remains as to whether the temperature attained is high enough to promote degradation by purely thermal means on the timescale involved in the ablation. It has been established by a variety of techniques such as photo-acoustic recording [38] of the stress wave generated by the ablation products (Fig. 14.3), spectroscopy [9, 41], time-resolved transmission and reflection [43–46] and laser imaging [39, 47–49] that ablation is a prompt process, occurring within a few nanoseconds of the start of the laser pulse. In some but not all [38] cases it appears that thermal degradation might be sufficiently rapid to account for these observations.

A single-pulse threshold also follows from a photochemical bond breaking description of ablation [50]. If n bonds per unit volume must be broken to achieve removal then the required fluence is

$$E_B = nh\nu/qk, \tag{14.2}$$

where q is the quantum efficiency for chain scission and $h\nu$ is the photon energy, which is assumed to exceed the bond energy of the polymer chain.

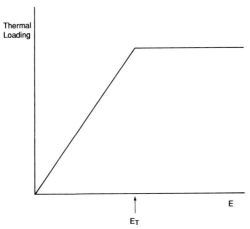

Fig. 14.2   Variation of thermal loading on polymer film with fluence. Below the threshold fluence E_T, all absorbed energy appears as heat, but above this, in the idealized case, the loading is "clamped" because the ablation products carry away excess energy.

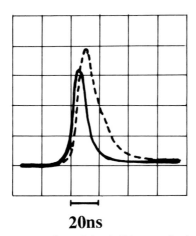

20ns

Fig. 14.3   Synchronized recording of the ArF laser pulse (——) and recoil stress wave (---) produced by ablation of a 2.5 µm thick poly(ethylene terephthalate) film (fluence 83 mJ cm^{-2}).

Assuming for simplicity that one bond per repeat unit must be broken, for a material such as PET, with $n \approx 4.5 \times 10^{21}$ cm^{-3} and $k \approx 10^5$ cm^{-1} at 248 nm, $E_B \approx 35/q$ mJ cm^{-2}. At this wavelength, the (low-signal) yield for chain scission has been estimated as $q = 10^{-3}$ [51], so a value $E_B \approx 35$ J cm^{-2} is predicted from (14.2), which is greatly in excess of the experimental threshold (Table 14.1). However, if it is postulated that, under the large signal conditions that obtain in ablation, a sufficiently high density of bonds is simultaneously broken so that the polymer mobility increases [52] and bond reformation is inhibited then the quantum yield might grow dramatically. If $q = 1$ is attained then $E_B \approx 35$ mJ cm^{-2} is predicted—in good agreement with experiment. If this is the case then ablation might picturesquely be described as a "switched cage effect", in that at threshold the quantum yield jumps to near unity as sufficient bonds are broken to increase the polymer mobility.

Above the ablation threshold, the experimental etch rate per pulse x can often be fitted over a limited range of fluence E by an equation of the form [3, 5, 6, 10, 30]

$$x = k^{-1} \ln\left(\frac{E}{E_T}\right), \tag{14.3}$$

where k is the (effective) polymer attenuation coefficient at the laser wavelength (Fig. 14.1). For relatively weakly absorbing polymers the appropriate value of k may be substantially larger than that measured under low-signal conditions [5], presumably due to nonlinear effects. In fact, with ultrashort excimer laser pulses and high irradiance levels, it has been found that even essentially transparent polymers can be efficiently ablated because of strong nonlinear absorption [53, 54]. It has also been found that doping with a suitable species to enhance the absorption coefficient can dramatically improve the ablation quality of some polymers, particularly at longer wavelengths, for example in 308 nm XeCl laser ablation of PMMA [55–60]. A phenomenon termed "incubation" has also been reported [61–64] in which ablation does not commence until the polymer has been exposed to a sufficient number of laser pulses to enhance its absorption by photochemical modification. This is, in effect, a form of self-doping [63].

Equation (14.3) follows from modelling a threshold process in which a Beer's law description of beam attenuation is applicable. This requires that the ablated material, which, on the timescale of the laser pulse, forms a thin "plume" ahead of the ablated surface, retains the same absorption coefficient as the parent polymer [65]. In many experiments [61, 66–68] it is found that (14.3) is not obeyed, and various models have been proposed to explain the form of the etch curves obtained [61, 66, 69–73] (a comparison of ablation models has recently been made in [74]). The simplest and most convincing explanation of this rests on the extent to which the ablation plume provides shielding of the underlying polymer [75]. If, for example, the plume has an absorption coefficient k_p that differs from that of the polymer (k), and during the laser pulse lateral motion of the plume is neglected, it is found that [75, 76]

$$x = k_p^{-1} \ln\left(\frac{k_p E}{k E_T} - \frac{k_p}{k} + 1\right). \tag{14.4}$$

This reduces to (14.3) when $k_p = k$, but in other cases the etch rate may (as shown in Fig. 14.4) increase more rapidly ($k_p < k$) or less rapidly ($k_p > k$) with $\ln (E/E_T)$. For a transparent plume ($k_p = 0$) the etch rate becomes linear in E:

$$x = \frac{E - E_T}{k E_T}.$$

Since k_p is likely to be fluence-dependent, it is possible for the etch curve to move within these extremes, leading to a relatively complex variation with fluence. Schildbach [75] has extended (14.4) to include a "memory" effect and has found excellent agreement with etch curves for polyimide over a range of fluence and wavelengths. Plume shielding in polyimide has been studied by delayed beam probing [77] and has been recognized as an important factor in short-pulse [78] and variable-pulse-length [79] etching experiments.

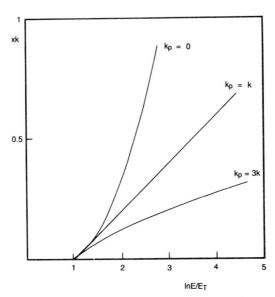

Fig. 14.4 Product of etch depth x and polymer attenuation coefficient k as a function of the logarithm of the normalized fluence, $\ln (E/E_T)$, where E_T is the ablation threshold, for various plume absorption coefficients k_p.

For most polymers the ablation plume consists of a complex mixture of species at elevated temperature [41, 80], which leave the surface at high speeds (of order 10^5–10^6 cm s^{-1}) [9, 41, 55, 81–83]. Laser-produced plasma formation is not important at the modest fluences usually employed, and the bulk of material is ejected as neutral species [84]. Analysis of the ablation products has provided considerable insight into the chemical processes involved; this has been well reviewed by Srinivasan and Braren [27]. Techniques employed include gas chromatography [2, 30, 55, 81, 83, 85], IR spectroscopy [5, 30], optical spectroscopy [8, 9, 41, 86], laser-induced fluorescence [55, 81, 83, 87] and mass spectroscopy [42, 88–97]. A substantial fraction of the products appears in the form of low-molecular-weight species; solid matter in the form of small particulates is also observed in most cases [81, 83, 85]. Even when ablation is carried out in vacuum, it is difficult to obtain first-hand information on the primary ablation products because of reactions occurring in the plume [92]. The situation becomes still more complex in air, where reactions take place with nitrogen and oxygen and the plume dynamics becomes governed by shock formation in the ambient gas [98, 99].

In many applications it is the state of the polymer surface remaining following ablation that is of interest rather than the precise mechanism whereby material is removed. As already mentioned, the energy per unit area remaining in the surface as heat is approximately E_T. If fast photo-excited state relaxation occurs and heat loss by conduction can be neglected on the timescale of the laser pulse then this is initially confined to a depth of order k^{-1} defined by the radiation absorption profile [31]. Following the pulse, cooling by conduction into the underlying polymer bulk occurs on a timescale given by $t = (4Dk^2)^{-1}$, where D is the thermal diffusivity (of order 10^{-3} cm^2 s^{-1} for typical polymers). Any radiation damage will clearly be limited to a depth of order k^{-1}.

There are thus distinct advantages in having a large absorption coefficient k, since

(i) this reduces the thermal loading on the polymer ($E_T \propto k^{-1}$);

(ii) the initial depth k^{-1} heated to a high temperature is small, as is any radiation damage zone;

(iii) conduction cooling to the bulk polymer is fast, since $t \propto k^{-2}$.

The apparent absence of thermal damage or restriction of this to micron or submicron depths is not necessarily inconsistent with a thermal model for ablation, but is rather a manifestation of the highly spatially localized interaction of UV lasers. Short wavelengths with concomitantly strong absorption such as provided by the ArF laser are clearly beneficial in this respect.

Whereas the intrinsic quality of surfaces produced by ablation in vacuum or in, some cases, rare gases is generally good, processing in air often leads to particulate contamination, which is highly undesirable in many applications. Particulates have been deduced to be carbon or carbon-rich compounds [30, 83, 85, 100] and are probably formed by condensation of small fragments in the plume [82, 97] or by secondary reactions with the ambient oxygen [99]. Redeposition apparently occurs because ablation leads to the formation of a density cavity in the gas, which, on cooling, causes a net flow of material, including particulates, back to the surface [98]. Removal of this deposited material without damage to the etched polymer has been demonstrated using CO_2 laser ablation [101]. Electrostatic collection of the ablated material [100] and control of redeposition using flowing gases [102] have also been reported.

14.3   APPLICATIONS

A variety of applications have emerged from the ability to efficiently ablate polymers using UV lasers. These mainly stem from the excellent material removal properties observed for nearly all polymers studied, which allow significantly better quality features to be defined than is possible with CO_2 or Nd : YAG lasers [103]. Excimer lasers are principally used in these applications, although the third, fourth and fifth harmonics of Nd : YAG offer possible alternatives, albeit currently with lower pulse energy and average power.

Most effort has been concentrated on polyimide and PMMA because of their importance in semiconductor packaging technologies and as a resist base respectively. PET has also been extensively studied probably because of its ready availability in the laboratory (e.g. as du Pont Mylar^{TM}, or ICI plc Melinex^{TM}). Many other polymers have been investigated, however, including polycarbonate [10, 12, 95, 104], cellulose acetate [12], polyetheretherketone (PEEK) [105–107], polytetrafluoroethylene (PTFE) [54, 108], plasma-polymerized tetrafluoroethylene (PPTFE) [109], poly(ethylene-2,6-naphthalate) [110], polyethersulphone [111], polystyrene [43, 67, 88], poly(α-methylstyrene) [95], polyacetylene [112], poly(N-vinylcarbazole) [113], nylon-6,6 [108], polyethylene [114, 115], polybenzimidazole [95], nitrocellulose [6, 116], polypropylene [117], ethylene–tetrafluoroethylene copolymer [117], poly(phenylene sulphide) [118], polyphenylquinoxaline [73], parylene [119], poly(dimethyl glutarimide) [57], chlorinated polymethylstyrene [57] and various commercial photoresists [5, 11, 12, 32].

14.4 LITHOGRAPHY

Early work on polymer ablation [1, 5, 6, 11, 12, 116, 120] recognized its potential for use in single-step lithography, i.e. a dry development process for patterning polymer resists that eliminates the need for wet chemical etchants. The direct etching characteristics of various photoresists have been studied [11, 12, 32] and high-spatial-resolution capabilities demonstrated [6, 11, 120] for contact and projection printing. As is also pointed out in Chapter 4, with ArF and F_2 lasers it has been possible to directly write features of sizes as small as 0.2 μm [114, 120, 121] with sharply defined edges because of the highly nonlinear response of ablation [121]. Even with these successes, however, the use of excimer lasers in a direct etching mode does not appear to have penetrated into commercial lithography systems—unlike their use as sources for conventional exposure, which is now becoming established [121]. As is highlighted in Chapter 4, this can, in part, be traced to problem areas such as mask and optics damage at the higher fluences involved in ablation and the formation of debris from the ablation products, which produces unacceptable particulate contamination at the substrate. Further research work on ablative resists may alter this situation [121].

In a related area, line-narrowed and spatially coherent excimer lasers have been used to produce high-quality short-period relief gratings by the holographic technique. In anthracene-doped PMMA it has been possible to produce 0.5 μm period structures by single-pulse ablation with a KrF laser [122]. Other recent trends in pattern projection applied to ablation and etching are reviewed in Chapter 3.

14.5 MICROMACHINING

A variety of structures can be produced in thin-deposited, free-standing or bulk polymer films by excimer laser ablation. As outlined in Chapter 3, a variety of approaches can be taken in this regard. The pattern can be defined by contact masks or, preferably, noncontact projection techniques in which a mask containing the desired pattern is imaged onto the surface (Fig. 14.5). The latter allows image reduction (with fluence gain) to be built in and avoids mask fouling by the ablation products. As an illustration of the delicacy of this technique, Fig. 14.6(a) shows a grid pattern projection-etched into a free-standing 1.5 μm thick PET film by the KrF laser [105]. The choice of laser wavelength to minimize thermal loading on the film is crucial in this application, as can be seen by comparing Fig. 14.6(a) with 14.6(b), where the XeCl laser was used. At the longer wavelength, the threshold fluence (about 170 mJ cm^{-2} compared with about 30 mJ cm^{-2}

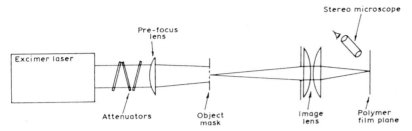

Fig. 14.5 Simple projection etching scheme for noncontact pattern definition by ablation.

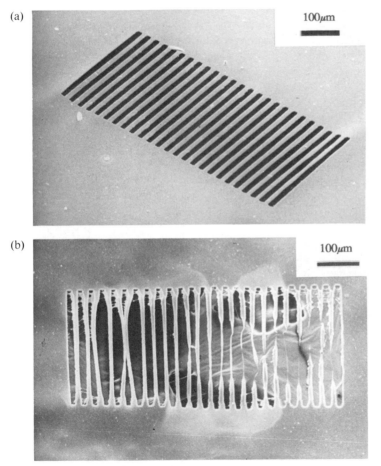

Fig. 14.6 (a) Projection-etched grid pattern in 1.5 μm thick free-standing polyethylene terephthalate film: KrF laser at fluence of 180 mJ cm^{-2} [105]. (b) Similar pattern produced by a XeCl laser in which melting destroys the structure.

for KrF) is high enough to raise the film to above its crystalline melting temperature, leading to collapse of the structure by melting [105].

As another example, a shallow etch pattern produced in polyetheretherketone (PEEK) using the XeCl laser is shown in Fig. 14.7. This demonstrates the highly controllable material removal that can be obtained when the effective absorption coefficient is suitably high ($k \geqslant 10^5$ cm^{-1} for PEEK at 308 nm [107]).

Film cutting with a high-repetition-rate XeCl laser has been investigated, with rates as high as 1.3 m s^{-1} being achieved at 1 kHz for 12 μm thick PET films [123]. At high repetition rates, thermal loading effects become important, which increase the cut-rate efficiency but also cause meltback of the cut edge [123], as shown in Fig. 14.8.

For drilling holes (not necessarily circular) in polymers the excimer laser has demonstrated advantages over CO_2 or Nd : YAG lasers in terms of edge quality [103]. As an example, Fig. 14.9(a) shows an ophthalmic-grade plastic, poly(2-hydroxyethyl methacrylate), used in contact lenses, where burr-free, smooth-edged, holes have been produced with the ArF laser [124]. In this case wall taper occurs (Fig. 14.9b), even over the relatively thin section (about 0.2 mm) involved, a feature that has been studied by Srinivasan and Braren [104]. They have observed a relatively complex behaviour for the profiles of high-aspect-ratio holes produced with the ArF and KrF lasers in several polymers.

One of the most important applications that has emerged for the excimer

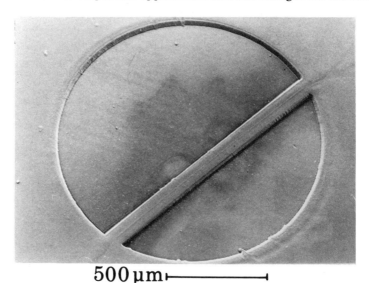

500 μm ⊢————————————┤

Fig. 14.7 Shallow etch crater produced in PEEK by XeCl laser ablation. The line is due to a fine wire placed on the circular mask to assist focusing. (From [107].)

laser is in drilling vias (holes) in polyimide layers used in semiconductor device packaging. These are needed to allow connections to be made to the underlying circuitry, and the laser approach allows avoidance of wet chemical etchants. The resolution requirements are less demanding than for lithography, although particulate contamination is equally undesirable. Much of the work on polyimide stems from this interest, although few details have emerged because of its proprietory nature. In a closely related

(a)

(b)

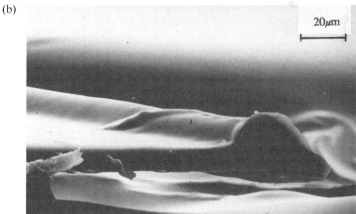

Fig. 14.8   (a) Edge of 12 μm thick poly(ethylene terephthalate) film cut using a 700 Hz XeCl laser. (b) Close up showing detail of edge. (From [123].)

area, however, what is essentially the first production line application of excimer lasers has been reported by Bachmann [125]: a KrF laser is used to drill layers of acrylic resin and polyimide in a multilayer printed circuit board, allowing interconnections between conductors to be made [125]. Holes opened in the copper layers using conventional lithography define a

(a)

(b)

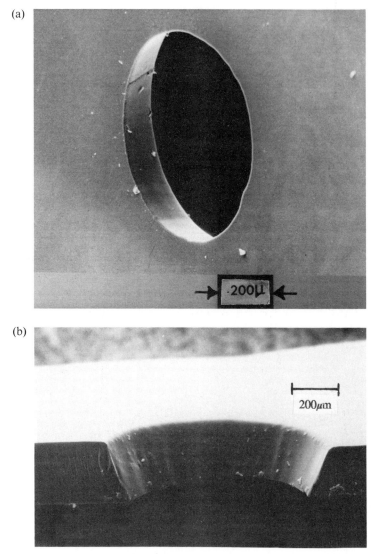

Fig. 14.9 (a) Hole drilled in boundary of plastic contact lens using an ArF laser. (b) Cross-section showing hole taper. (From [124].)

mask for the excimer ablation (Fig. 14.10). Up to 30 000, 80 μm diameter, 65 μm deep vias between adjacent layers are produced, the laser addressing 60 of these at one time.

Other applications for micromachining in the electronics industry include resist stripping for improved registration-mark clarification [126], stripping polymer coating from fine wires [127] and the removal of polymide passivation layers from chips to allow testing of the underlying circuitry [128].

Applications have also been found in the opto-electronics area, where increasing use is being made of polymer components. For example, machining to allow the mechanical interfacing of fibre and waveguide components [129] and to provide reflecting end facets on polymer waveguides [130] has been reported.

Excimer lasers have recently been demonstrated to provide an excellent method for drilling, cutting and etching polymer composites, materials that are often difficult to process by conventional techniques [131–133]. Inorganic (e.g. carbon) and polymer (e.g. Kevlar) fibres in a polymer matrix have been investigated [131], with good quality cuts being observed for both components of the composite. It is also possible, by correct choice of the fluence, to selectively remove the polymer matrix by ablation but leave the fibres undamaged [118, 133]. An example of this is shown in Fig. 14.11, where XeCl laser ablation has been used to remove the matrix from the surface of a PEEK–carbon-fibre composite to reveal the underlying fibre layout [133]. Studies of biological materials, which

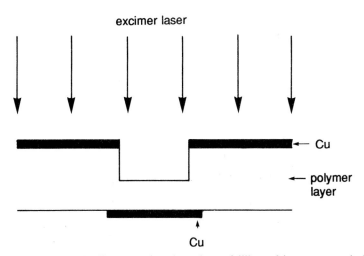

Fig. 14.10 Schematic diagram of excimer laser drilling of interconnect holes in multilayer printed circuit board [125].

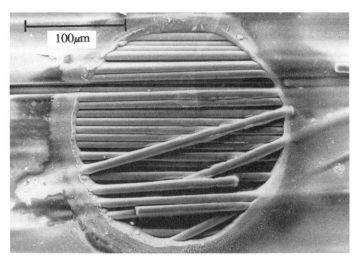

Fig. 14.11 Selective removal of PEEK matrix in PEEK–carbon-fibre composite using XeCl laser ablation [133].

provide an analogous form of composites, are also of great interest (see, e.g. [134]).

14.6 SURFACE MODIFICATION

Ablation can lead to physical [5, 8, 9, 135] and chemical [135–137] alteration of polymer surfaces. This is of interest both in terms of gaining improved understanding of the interaction and for its potential application in modifying polymer surfaces, which are often smooth and chemically relatively unreactive. Wetability, adhesion, biocompatibility and optical properties might be controlled in this way [136, 137].

In studies first reported by Lazare *et al.* [136] X-ray photo-electron spectroscopy (XPS) was used to investigate the surface composition of PET, polyimide and PMMA following excimer laser ablation. Unlike low-irradiance lamp exposure in the deep-UV, ablation with the ArF laser produced a decrease in the O/C ratio for PET and polyimide; the PMMA surface remained essentially unchanged. This behaviour has been confirmed in other studies of polyimide [138] and PMMA [139]. XPS has also been used to investigate chemical modification of poly(α-methylstyrene) [139], PEEK [106], poly(ethylene-2,6-naphthalate) [110], polyether-sulphone [111], polypropylene and ethylene–tetrafluoroethylene copolymer [117] ablated surfaces.

Recent work on PET shows [140] that its surface reflection is decreased by subablation threshold irradiation with KrF or ArF lasers. This reduction itself exhibits a threshold fluence, and is attributed to amorphization of a thin melted surface layer ($\leqslant 100$ nm) of the semicrystalline film.

A variety of processes that lead to the physical modification of ablated surfaces have been discovered. These fall broadly into three groups: relaxation of heated surfaces of oriented polymers; particulate-induced microstructures; and surface–scattered-wave interference effects.

In the first of these, strong surface roughness develops as a result of repeated above-threshold irradiation of an oriented polymer [1, 5, 137]. In

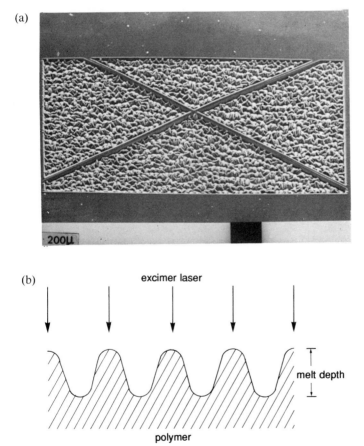

Fig. 14.12   (a) Rectangular pattern with cross-wires ablatively etched into biaxially oriented polyethylene terephthalate film, showing highly granular texturing. (b) The origin of texturing is thought to be related to stress relaxation in a shallow heated surface layer of the polymer.

biaxially oriented sheets (i.e. sheets that are stretched in two orthogonal directions in the film plane during manufacture) the structures are predominantly two-dimensional [141–143]. A typical example is shown in Fig. 14.12(a) for biaxially oriented PET irradiated using the XeCl laser. For uniaxial polymers (e.g. PET fibres or uniaxial sheets) the structure is approximately undirectional and runs perpendicular to the stretch direction [141–143]. A similar effect is observed for polyimide surfaces when ablated while simultaneously under high mechanical stress [144].

The characteristic scale size of the structures depends on laser wavelength, insofar as this influences the ablation threshold, and on the number of irradiating pulses [145, 146]. Mechanisms for the formation of these structures have been proposed based on selective etching of the amorphous and crystalline regions in the polymer [5, 141] and on a convection instability in the heated surface layer [142], although neither are fully satisfactory. It is likely that stress relaxation in a melted surface layer is the dominant effect (Fig. 14.12b). In unaxial fibres the texturing that results may provide a valuable way of modifying surface properties of these materials, which are used extensively in the textile industry.

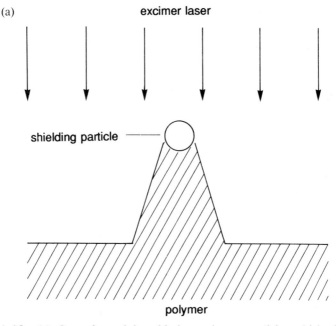

Fig. 14.13 (a) Cone formed by ablation-resistant particle, which shields the underlying polymer. (b) Cone in polyimide (unseeded) with normal-incidence irradiance. (c) Cones in seeded polymide irradiated at 50° to the normal using the ArF laser at 180 mJ cm^{-2}. (From [145].)

Particulate-induced surface microstructures arise from small particles that are redeposited on the surface from the ablation products or deliberately added to polymer films [147, 148]. If such particles are resistant to ablation they act to shield the underlying polymer, leading to the development of well defined and stable conical structures as etching proceeds (Figs 14.13a,b). A sloping wall develops as a result of local diffraction, and this wall cannot be ablated since the local fluence falls below the threshold value [36, 147]. Deliberate seeding with, for example, small alumina or rare earth compound particles allows a high areal density of cones to be formed, and their orientation can be controlled by the angle of incidence

(b)

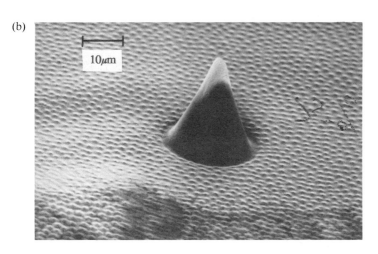

(c)

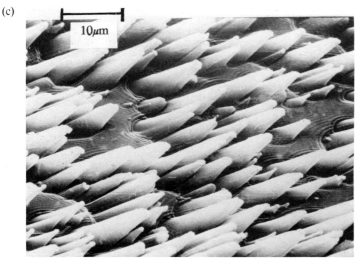

of the laser (Fig. 14.13c). In certain cases nonconical structures are observed; these are less well understood [149, 150].

A third effect, originating from interference between the incident laser beam and surface-scattered waves (Fig. 14.14), has recently been observed [151]. With plane polarized excimer beams, this leads to the growth by selective etching of well ordered grating-like structures with a period in the

(a)
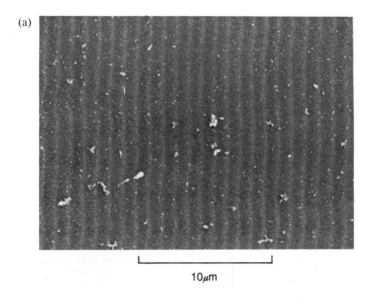

10μm

(b)
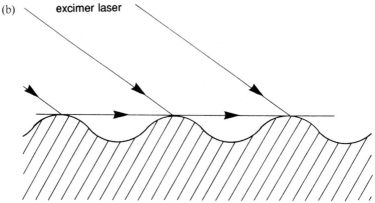

Fig. 14.14 (a) Grating-like structures etched into polyethersulphone using a KrF laser (angle of incidence 50° with s-polarized beam). (b) Model for formation of gratings based on interference between the incident and surface-scattered waves. If interference produces a local fluence minimum at the peaks then selective etching enhances the pattern. (From [151].)

micron range. The effect is most clearly observed on amorphous polymers such as polyethersulphone and polyimide, and does not require high spatial coherence from the excimer laser [151].

Regular surface microstructures that do not appear to fall into any of the above categories have been observed on polyethersulphone ablated using the XeCl laser [111]. These controllable microstructures have been shown to be effective for the alignment of nematic liquid crystals [152].

Laser surface modification of polymers can be used to form indelible marks on components. These may be produced in relief by direct etching [153] or through colour changes produced by photoreduction of additives in the polymer when irradiated in the subthreshold regime [154]. The latter technique has important applications for marking cables used in the aerospace industry [154].

14.7 MISCELLANEOUS APPLICATIONS

Thin-film deposition by the ablation of polymers using various UV lasers, including ArF, KrF and tripled Nd:YAG, has been reported [108]. This is of interest as a means of producing thin films from intractable polymers [108] and because it provides information on the nature of the ablation products and reactions that take place when they condense [155]. The ablative etching of polymer foams, which have potential application in several areas [156] but are difficult to work on a small scale, has been studied. Polystyrene [157, 158] and TPX [157] foams have been micromachined in this way.

UV laser ablation on the microscale has recently been reported for individual small-diameter polymer particles trapped using laser techniques [159]. Microfabrication and micromodification of small particles, possibly for use in chemical studies, might be achieved in this way.

14.8 DISCUSSION

Although a considerable wealth of data has now been gathered on the UV laser ablation of polymers (and biological materials [160]), a completely satisfactory general model that explains all the major features is still lacking. It seems likely, however, that both photothermal and photochemical components play a role in ablation, with an increasing photochemical contribution at shorter wavelengths, and that models founded solely on one mechanism will not adequately describe all the experimental findings. Photo-induced bond breaking in the deep-UV may, for example, act to efficiently initiate high-activation-energy steps, which are then propagated by thermal reactions. Further data on the degradation of polymers at the high temperatures (≥ 1000 K) that accompany ablation are required in order to help resolve these issues [19, 20]. For this, ablation studies using long pulse CO_2 [161] or short-pulse TEA CO_2 lasers [162–164], which should act purely thermally, or combined UV and IR laser experiments [165] may prove useful. Further experiments along the lines of those described by Dijkkamp et al. [18], which are designed to untangle the thermal and photochemical contributions in UV laser ablation, will also be valuable.

There is still a growing range of applications for polymer ablation, and this may be expected to continue, particularly as the often unique capabilities become more widely known among various manufacturing industries. With correct choice of UV laser wavelength, almost all common and many uncommon polymers can be successfully processed in this way. However, as Srinivasan [166] has pointed out, UV laser photons are expensive, and commercial success will depend on use in the manufacture of high-added-value components. The development of low-cost UV beam delivery optics, reliable UV lasers that are industry-acceptable, and improved understanding and control of particulate debris will help speed the uptake of this technology.

ACKNOWLEDGEMENTS

Grateful acknowledgements go to G. Bishop, R. J. Farley, S. D. Jenkins, S. Lau, G. A. Oldershaw, D. Schudel, J. Sidhu and B. L. Tait for providing published and unpublished material.

REFERENCES

[1] R. Srinivasan and V. Mayne-Banton, *Appl. Phys. Lett.* **41**, 576 (1982).
[2] R. Srinivasan and W. J. Leigh, *J. Am. Chem. Soc.* **104**, 6784 (1982).
[3] R. Srinivasan, *J. Vac. Sci. Technol.* **B1**, 923 (1983).
[4] R. Srinivasan, *J. Rad. Curing* **4**, 12 (1983).
[5] J. E. Andrew, P. E. Dyer, D. Forster and P. H. Key, *Appl. Phys. Lett.* **43**, 717 (1983).
[6] T. F. Deutsch and M. W. Geis, *J. Appl. Phys.* **54**, 7201 (1983).
[7] G. M. Geis, J. N. Randall, T. F. Deutsch, N. N. Efremow, J. P. Donelly and J. D. Woodhouse, *J. Vac. Sci. Technol.* **B1**, 1178 (1983).
[8] G. Koren and J. Yeh, *Appl. Phys. Lett.* **44**, 1112 (1984).
[9] G. Koren and J. Yeh, *J. Appl. Phys.* **56**, 2120 (1984).
[10] R. Srinivasan and B. Braren, *J. Polym. Sci.* **22**, 2601 (1984).
[11] M. Latta, R. Moore, S. Rice and K. Jain, *J. Appl. Phys.* **56**, 586 (1984).

[12] S. Rice and K. Jain, *Appl. Phys.* **33**, 195 (1984).

[13] G. Koren, *Appl. Phys. Lett.* **45**, 10 (1984).

[14] R. L. Melcher, in *Laser Processing and Diagnostics* (ed. D. Bauerle), p. 418 (Springer-Verlag, Berlin, 1984).

[15] B. J. Garrison and R. Srinivasan, *Appl. Phys. Lett.* **44**, 849 (1984).

[16] B. J. Garrison and R. Srinivasan, *J. Appl. Phys.* **57**, 2909 (1985).

[17] B. J. Garrison and R. Srinivasan, *J. Vac. Sci. Technol.* **A3**, 746 (1985).

[18] D. Dijkkamp, A. S. Gozdz, T. Venkatesan and X. D. Wu, *Phys. Rev. Lett.* **58**, 2142 (1987).

[19] R. Srinivasan, *Phys. Rev. Lett.* **60**, 381 (1988).

[20] T. Venkatesan, A. S. Gozdz, X. D. Wu and D. Dijkkamp, *Phys. Rev. Lett.* **60**, 382 (1988).

[21] S. T. Trokel, R. Srinivasan and B. Braren, *Am. J. Ophthalmol.* **96**, 710 (1983).

[22] R. Linsker, R. Srinivasan, J. J. Wynne and D. R. Alonso, *Laser Surg. Med.* **4**, 201 (1984).

[23] J. T. C. Yeh, *J. Vac. Sci. Technol.* **A4**, 653 (1986).

[24] R. Srinivasan, in *Photophysics and Photochemistry above 6 eV* (ed. F. Lahmani), p. 595 (Elsevier, Amsterdam, 1985).

[25] R. Srinivasan, *Science* **234**, 559 (1986).

[26] R. Srinivasan, in *Interfaces under Laser Irradiation* (ed. L. D. Laude, D. Bauerle and M. Wautelet), p. 359 (Martinus Nijhoff, Dordrecht, 1987).

[27] R. Srinivasan and B. Braren, *Chem. Rev.* **89**, 1303 (1989).

[28] S. Lazare and V. Granier, *Laser Chem.* **10**, 25 (1989).

[29] I. Brodie and J. J. Muray, *The Physics of Microfabrication*, pp. 267–346 (Plenum, New York, 1982).

[30] J. H. Brannon, J. R. Lankard, A. I. Blaise, F. Burns and J. Kaufman, *J. Appl. Phys.* **58**, 2036 (1985).

[31] P. E. Dyer and J. Sidhu, *J. Appl. Phys.* **57**, 1420 (1985).

[32] G. M. Davis and M. C. Gower, *Appl. Phys. Lett.* **50**, 1286 (1987).

[33] R. S. Taylor, D. L. Singleton and G. Paraskevopoulas, *Appl. Phys. Lett.* **50**, 1979 (1987).

[34] S. Lazare, J. C. Soulignac and P. Fragnaud, *Appl. Phys. Lett.* **50**, 624 (1987).

[35] J. A. Sell, D. M. Heffelfinger, P. Ventzek and R. M. Gilgenbach, *Appl. Phys. Lett.* **55**, 2435 (1989).

[36] P. E. Dyer, S. D. Jenkins and J. Sidhu, *Appl. Phys. Lett.* **42**, 1880 (1988).

[37] S. Lazare and V. Granier, *J. Appl. Phys.* **63**, 2110 (1988).

[38] P. E. Dyer and R. Srinivasan, *Appl. Phys. Lett.* **48**, 445 (1986).

[39] R. Srinivasan, K. G. Casey, B. Braren and M. Yeh, *J. Appl. Phys.* **67**, 1604 (1990).

[40] G. Gorodetsky, T. G. Kazyaka, R. L. Melcher and R. Srinivasan, *Appl. Phys. Lett.* **46**, 828 (1985).

[41] G. M. Davis, M. C. Gower, C. Fotakis, T. Efthimiopoulos and P. Argyrakis, *Appl. Phys.* **A36**, 27 (1985).

[42] B. Danielzik, N. Fabricus, M. Rowenkamp and D. von der Linde, *Appl. Phys. Lett.* **48**, 212 (1986).

[43] J. Meyer, J. Kutzner, D. Feldmann and K. H. Welge, *Appl. Phys.* **B45**, 7 (1988).

[44] G. M. Davis and M. C. Gower, *J. Appl. Phys.* **61**, 2090 (1987).

[45] D. L. Singleton, G. Paraskevopoulos and R. J. Taylor, *Appl. Phys.* **B50**, 227 (1990).

[46] P. Klopotek, B. Burghardt and W. Muckenheim, *J. Phys.* **E20**, 1269 (1987).
[47] P. Simon, *Appl. Phys.* **B48**, 253 (1989).
[48] R. Srinivasan, B. Braren, K. G. Casey and M. Yeh, *Appl. Phys. Lett.* **55**, 2790 (1989).
[49] R. Srinivasan, K. G. Casey and B. Braren, *Chemtronics* **4**, 153 (1989).
[50] H. H. G. Jellinek and R. Srinivasan, *J. Chem. Phys.* **88**, 3048 (1984).
[51] F. B. Marcotte, D. Campbell, J. A. Cleaveland and D. T. Turner, *J. Polymer Sci.* **A1**, 481 (1967).
[52] J. Guillet, *Polymer Photophysics and Photochemistry* (Cambridge University Press, 1987).
[53] R. Srinivasan, E. Sutcliffe and B. Braren, *Appl. Phys. Lett.* **51**, 1285 (1987).
[54] S. Kuper and M. Stuke, *Appl. Phys. Lett.* **54**, 4 (1989).
[55] R. Srinivasan, B. Braren, R. W. Dreyfus, L. Hadel and D. E. Seeger, *J. Opt. Soc. Am.* **B3**, 785 (1986).
[56] H. Masuhara, H. Hiraoka and K. Domen, *Macromolecules* **20**, 450 (1987).
[57] H. Hiraoka, T. J. Chuang and H. Masuhara, *J. Vac. Sci. Technol.* **B6**, 463 (1988).
[58] T. J. Chuang, H. Hiraoka and A. Modl, *Appl. Phys.* **A45**, 277 (1988).
[59] R. Srinivasan and B. Braren, *Appl. Phys.* **A45**, 289 (1988).
[60] A. Itaya, A. Kurahashi, H. Masuhara, Y. Taniguchi and M. Kiguchi, *J. Appl. Phys.* **67**, 2240 (1990).
[61] E. Sutcliffe and R. Srinivasan, *J. Appl. Phys.* **B60**, 3315 (1986).
[62] S. Kuper and M. Stuke, *Appl. Phys.* **B44**, 199 (1987).
[63] S. Kuper and M. Stuke, *Appl. Phys.* **A49**, 211 (1989).
[64] P. Heszler, Zs. Bor and G. Hajos, *Appl. Phys.* **A49**, 739 (1989).
[65] P. E. Dyer, in *Photoacoustic and Photothermal Phenomena* (ed. P. Hess and J. Pelzl), p. 164 (Springer-Verlag, Berlin, 1988).
[66] V. Srinivasan, M. A. Smrtic and S. V. Babu, *J. Appl. Phys.* **59**, 3861 (1986).
[67] H. S. Cole, V. S. Lin and H. R. Phillip, *Appl. Phys. Lett.* **48**, 76 (1986).
[68] B. Braren and D. Seeger, *J. Polym. Sci. Polym. Lett.* **24**, 371 (1986).
[69] G. D. Mahan, H. S. Cole, Y. S. Liu and H. R. Phillip, *Appl. Phys. Lett.* **53**, 2377 (1988).
[70] R. Sauerbrey and G. H. Pettit, *Appl. Phys. Lett.* **55**, 421 (1989).
[71] N. P. Furzikov, *Appl. Phys. Lett.* **56**, 1638 (1990).
[72] L. B. Kiss and P. Simon, *Solid State Commun.* **65**, 1253 (1988).
[73] S. Lazare and V. Granier, *Appl. Phys. Lett.* **54**, 862 (1989).
[74] D. S. Singleton, G. Paraskevopoulos and R. S. Taylor, *Chem. Phys.* **144**, 415 (1990).
[75] K. Schildbach, in *Proceedings of SPIE International Congress on Optical Science and Engineering, The Hague, 1990*, Paper 1279-07.
[76] P. E. Dyer, Unpublished work.
[77] G. Koren, *Appl. Phys. Lett.* **50**, 1030 (1987).
[78] M. Chuang and A. C. Tam, *J. Appl. Phys.* **65**, 2591 (1989).
[79] R. Srinivasan and B. Braren, *Appl. Phys. Lett.* **53**, 1233 (1988).
[80] G. Koren, *Appl. Phys.* **B46**, 147 (1988).
[81] R. Srinivasan, B. Braren, D. E. Seeger and R. W. Dreyfus, *Macromolecules* **19**, 916 (1986).
[82] P. E. Dyer and R. Srinivasan, *J. Appl. Phys.* **66**, 2608 (1989).
[83] R. Srinivasan, B. Braren and R. W. Dreyfus, *J. Appl. Phys.* **61**, 372 (1987).
[84] R. E. Walkup, J. M. Jasinski and R. W. Dreyfus, *Appl. Phys. Lett.* **48**, 1690 (1986).

[85] D. L. Singleton, G. Paraskevopoulos and R. S. Irwin, *J. Appl. Phys.* **66**, 3324 (1989).

[86] S. Deshmukh, E. W. Rothe and G. P. Peck, *J. Appl. Phys.* **66**, 1370 (1989).

[87] R. W. Dreyfus, R. Kelly, R. E. Walkup and R. Srinivasan, *Proc. SPIE* **710**, 46 (1987).

[88] D. Feldmann, J. Kutzner, J. Laukemper, S. MacRobert and K. H. Welge, *Appl. Phys.* **B44**, 81 (1987).

[89] R. C. Estler and N. S. Nogar, *Appl. Phys. Lett.* **49**, 1175 (1986).

[90] R. C. Estler and N. S. Nogar, *J. Vac. Sci. Technol.* **B5**, 1465 (1987).

[91] R. Larciprete and M. Stuke, *Appl. Phys.* **B42**, 181 (1987).

[92] C. E. Otis, *Appl. Phys.* **B49**, 455 (1989).

[93] E. E. B. Campbell, G. Ulmer, K. Bues and I. V. Hertel, *Appl. Phys.* **A48**, 543 (1989).

[94] E. E. B. Campbell, G. Ulmer, B. Hasselberger and I. V. Hertel, *Appl. Surf. Sci.* **43**, 346 (1989).

[95] S. G. Hansen, *J. Appl. Phys.* **66**, 1411 and 3329 (1989).

[96] P. Goodwin and C. E. Otis, *Appl. Phys. Lett.* **55**, 2286 (1989).

[97] W. R. Creasy and J. T. Brenna, *Chem. Phys.* **126**, 453 (1988).

[98] G. Koren and U. P. Oppenheim, *Appl. Phys.* **B42**, 41 (1987).

[99] P. E. Dyer and J. Sidhu, *J. Appl. Phys.* **64**, 4657 (1988).

[100] R. J. von Gutfeld and R. Srinivasan, *Appl. Phys. Lett.* **51**, 15 (1987).

[101] G. Koren and J. J. Donelon, *Appl. Phys.* **B45**, 45 (1988).

[102] W. Spiess and H. Strack, *Semicond. Sci. Technol.* **4**, 486 (1989).

[103] T. A. Znotins, D. Poulin and J. Reid, *Laser Focus* **23**, 54 (1987).

[104] B. Braren and R. Srinivasan, *J. Vac. Sci. Technol.* **B3**, 913 (1985).

[105] P. E. Dyer and J. Sidhu, *Opt. Laser Engng* **6**, 67 (1985).

[106] O. Occhiello, F. Garbassi and V. Malatesta, *Angew. Makromol. Chem.* **169**, 143 (1989).

[107] P. E. Dyer, G. A. Oldershaw and D. Schudel, *Appl. Phys.* **B51**, 314 (1990).

[108] S. G. Hansen and T. E. Robitaille, *Appl. Phys. Lett.* **52**, 81 (1988).

[109] J. H. Brannon, *J. Vac. Sci. Technol.* **B7**, 1064 (1989).

[110] H. Niino, A. Yabe, S. Nagano and T. Miki, *Appl. Phys. Lett.* **54**, 2159 (1989).

[111] H. Niino, M. Nakano, S. Nagano, A. Yabe and T. Miki, *Appl. Phys. Lett.* **55**, 510 (1989).

[112] M. Golombok, M. C. Gower, S. J. Kirkby and P. T. Rumsby, *J. Appl. Phys.* **61**, 1222 (1987).

[113] A. Masuhara, W. Hiraoka and E. E. Martinero, *Chem. Phys. Lett.* **135**, 103 (1987).

[114] P. E. Dyer and J. Sidhu, *J. Opt. Soc. Am.* **B3**, 792 (1986).

[115] R. Srinivasan, E. Sutcliffe and B. Braren, *Laser Chem.* **9**, 147 (1988).

[116] M. W. Geis, J. N. Randall, T. F. Deutsch, P. D. De-Graff, K. E. Krohn and L. A. Stern, *Appl. Phys. Lett.* **43**, 74 (1983).

[117] S. Kawanishi, Y. Shimizu, S. Sugimoto and N. Suzuki, *Polymer* **32**, 979 (1991).

[118] H. Niino, M. Nakano, Nagano, H. Nitta, K. Yano and A. Yabe *J. Photopolym. Sci. Technol.* **3**, 53 (1990).

[119] Y. J. Yang, S. Lee and S. D. Allen, in *Proceedings of Conference on Lasers and Electro-optics*, p. 264 (Technical Digest Series, Vol. 11, Optical Society of America, Washington, D.C., 1989).

[120] D. Henderson, J. C. White, H. G. Craighead and I. Adesida, *Appl. Phys. Lett.* **46**, 900 (1985).
[121] M. Rothschild and D. J. Ehrilich, *J. Vac. Sci. Technol.* **B6**, 1 (1988).
[122] T. Ahlhorn, H. Pohlmann and J. P. Kotthaus, *Proc. SPIE* **1023**, 231 (1989).
[123] P. E. Dyer and G. J. Bishop, *Appl. Phys. Lett.* **47**, 1229 (1985).
[124] P. E. Dyer and B. L. Tait, Report on hole drilling in a plastic contact lens using excimer lasers, *Appl. Phys. Dep. Rep. Univ. Hull* (1986).
[125] F. Bachmann, *Mater. Res. Soc. Bull.* **14**, 49 (1989).
[126] Technical Report/Lasers, *Laser Focus* **22**, 22 (1986).
[127] J. H. Brannon, R. Kurth and A. C. Tam, in *Proceedings of Conference on Lasers and Electro-Optics*, p. 546 (Technical Digest Series, Vol. 76, Optical Society of America, Washington, D.C., 1990).
[128] C. J. Jones, S. Rolt and K. G. Snowden, in *Proceedings of IEE Colloqium on Laser Processing of Materials, IEE Digest* **129**, 13/1 (1986).
[129] J. L. Hohman, K. B. Keating, B. L. Booth and J. E. Marchegiano, in *Proceedings of Conference on Lasers and Electro-optics*, p. 210 (Technical Digest Series, Vol. 11, Optical Society of America, Washington, D.C., 1989).
[130] J. M. Trewhella and M. M. Oprysko, in *Proceedings of Conference on Lasers and Electro-Optics*, p. 308 (Technical Digest Series, Vol. 7, Optical Society of America, Washington, D.C., 1990).
[131] G. M. Proudley and P. H. Key, *Proc. SPIE* **1132**, 111 (1989).
[132] M. Wehner, R. Poprawe and F. J. Trasser, *Proc. SPIE* **1023**, 179 (1988).
[133] P. E. Dyer, S. T. Lau, G. A. Oldershaw and D. Schudel, in *Proceedings of Conference on Lasers and Electro-Optics*, p. 154 (Technical Digest Series, Vol. 7, Optical Society of America, Washington, D.C., 1990).
[134] M. J. Berry and G. M. Harpole (eds), *Proc. SPIE* **1064** (1989).
[135] R. Srinivasan and S. Lazare, *Polymer* **26**, 1297 (1985).
[136] S. Lazare, P. D. Hoh, J. M. Baker and R. Srinivasan, *J. Am. Chem. Soc.* **106**, 4288 (1984).
[137] S. Lazare and R. Srinivasan, *J. Phys. Chem.* **90**, 2124 (1986).
[138] F. Kotai, H. Saito and T. Fujioka, *J. Appl. Phys.* **66**, 3252 (1989).
[139] M. C. Burrell, Y. S. Liu and H. S. Cole, *J. Vac. Sci. Technol.* **A4**, 2459 (1986).
[140] D. S. Dunn and A. J. Ouderkirk, *Macromolecules* **23**, 770 (1990).
[141] Y. Novis, J. J. Pireaux, A. Brezini, E. Petit, R. Caudano, P. Lutgen, G. Feyder and S. Lazare, *J. Appl. Phys.* **64**, 365 (1988).
[142] T. Bahners and E. Schollmeyer, *J. Appl. Phys.* **66**, 1884 (1989).
[143] T. Bahners and E. Schollmeyer, *Angew. Makromol. Chem.* **151**, 39 (1987).
[144] K. Tonyali, L. C. Jensen and J. T. Dickinson, *Mater. Res. Soc. Symp. Proc.* **100**, 665 (1988).
[145] S. D. Jenkins, PhD Thesis, University of Hull (1989).
[146] H. Niino, M. Nakano, S. Nagano, A. Yabe, H. Moriya and T. Miki, *J. Polym. Sci. Tech.* **2**, 133 (1989).
[147] P. E. Dyer, S. D. Jenkins and J. Sidhu, *Appl. Phys. Lett.* **49**, 453 (1986).
[148] R. S. Taylor, K. E. Leopold, D. Singleton, G. Paraskevopoulas and R. S. Irwin, *J. Appl. Phys.* **64**, 2815 (1988).
[149] J. Marshall, S. Trokel, S. Rothery and R. R. Krueger, *Lasers in Ophthalmol.* **1**, 21 (1986).
[150] T. Znotins, D. Poulin and J. Reid, *Proc. SPIE* **737**, 132 (1987).
[151] P. E. Dyer and R. J. Farley, *Appl. Phys. Lett.* **57**, 765 (1990).

[152] H. Niino, Y. Kabawata and A. Yabe, *Japan. J. Appl. Phys.* **28**, 2225 (1989).
[153] G. D. Poulin, P. A. Eisele and T. A. Znotins, *Proc. SPIE* **1023**, 202 (1988).
[154] S. W. Williams and P. C. Morgan, in *Proceedings of 7th International Congress on Applications of Lasers and Electro-Optics (Santa Clara, Ca)*, p. 148 (1988).
[155] S. G. Hansen and T. E. Robitaille, *J. Appl. Phys.* **64**, 2128 (1988).
[156] J. H. Aubert and R. L. Clough, *Polymer* **26**, 2047 (1985).
[157] P. J. Hargis, in *Proceedings of Symposium on Interaction of Laser Radiation with Biological Tissue, BAPS* **32**, 307 (1987).
[158] W. W. Duley and G. Allan, *Appl. Phys. Lett.* **55**, 1701 (1989).
[159] H. Misawa, M. Koshioka, K. Sasaki, N. Kitamura and H. Masuhara, *Chem. Lett.* **8**, 1247 (1990).
[160] R. Srinivasan, *Ber. Bunsenges. Phys. Chem.* **93**, 265 (1989).
[161] R. F. Cozzens and R. B. Fox, *Polym. Eng. Sci.* **18**, 900 (1978).
[162] J. H. Brannon and J. R. Lankard, *Appl. Phys. Lett.* **48**, 1226 (1986).
[163] P. E. Dyer, G. A. Oldershaw and J. Sidhu, *Appl. Phys.* **B48**, 489 (1989).
[164] R. Braun, R. Nowak, P. Hess, H. Oetzmann and C. Schmidt, *Appl. Surf. Sci.* **43**, 352 (1989).
[165] R. K. Al-Dhahir, P. E. Dyer, J. Sidhu, C. Foulkes-Williams and G. A. Oldershaw, *Appl. Phys.* **B49**, 435 (1989).
[166] R. Srinivasan, *Polym. Degrad. Stabil.* **17**, 193 (1987).

15 Laser Ablation of Electronic Materials

FRANK BEECH and IAN W. BOYD

Department of Electronic and Electrical Engineering, University College London, UK

15.1 INTRODUCTION

As can be seen elsewhere in this book, unique processing capabilities are achievable using the specific wavelength emissions or the intense photon fluxes available from present-day laser and lamp systems. Radiation from the vacuum UV through the visible to the IR, either monochromatic or broadband, in pulses from below the picosecond range or in continuous mode, gives rise to a host of physical and chemical possibilities that can be applied towards a myriad of processing operations. In particular, photon-induced absorption, heating, melting and evaporation, as well as thermal and photolytic bond dissociation, can all be seen to give rise to thin-film modification, removal and growth.

Film growth can be broadly classified into two regimes: chemical deposition and physical deposition. In the better known techniques of the former category, the constituents are generally provided in the gas phase, as in metal–organic chemical vapour deposition (MOCVD), or metal–organic vapour phase epitaxy (MOVPE), and chemical reactions such as pyrolysis, reduction and oxidation are initiated to liberate the required reactants to form the film. Elsewhere in this book, the use of photons to stimulate such reactions is discussed.

In the latter category, the film-forming atoms are provided by a physical rather than a chemical mechanism. Depending upon the actual kinetics involved, these techniques may be either equilibrium or non-equilibrium processes. Well known equilibrium film-forming techniques include molecular beam epitaxy (MBE), and resistive, eddy current or e-beam thermal evaporation in vacuum. Nonequilibrium methods include RF, magnetron, or ion beam sputtering.

Laser radiation can also provide the energy to initiate these physical

Photochemical Processing of Electronic Materials
ISBN 0-12-121740-X

reactions. When continuous work (CW) beams are used, the equilibrium regime is accessed. When short pulsed laser beams are employed, the technique enters the predominantly nonequilibrium domain. However, as will be discussed below, the precise kinetics governing pulsed laser removal of material are at present rather poorly understood. This is in part reflected in the use of a wide variety of descriptive terms that have been historically associated with the technique, which has been referred to as laser evaporation [1–4], laser-assisted deposition and annealing (LADA) [5], laser flash evaporation [6], laser-assisted sputtering [7], laser MBE [8], hydrodynamic sputtering [9], laser ablation [10–12], laser ablation deposition [13] and laser evaporation deposition (LEDE) [14].

The necessary conditions for laser ablation of inorganic films are similar to those required to achieve similar material removal from organics such as plastics, photoresists and general polymeric materials. In fact, there are many similarities associated with both cases (see previous chapter). For example, although generally intense radiation initiates material removal, short-wavelength and certainly short-duration pulses are most favourable to the process. Removal is often accompanied by a plasma or plume of emitted particles just above the sample surface, as well as a snapping sound as the velocity of some of the species exceeds the speed of sound in the immediate environment. However, the mechanisms by which the absorbed energy enters and is redistributed within the target in each case appears to be different. Thus, despite the similarities, the structural arrangement and correspondingly the optical, electronic and thermodynamic properties of the irradiated material are crucial in determining the identity of the ejected species as well as the manner in which they are removed.

Long-pulse and even CW radiation can be used to remove material from an irradiated target. In this case, conventional thermal evaporation is expected to occur, as can be judged by the nonstoichiometric manner in which the species leave the surface. However, for many of the multicomponent materials, such as high-temperature superconductors, these conditions of thermal equilibrium are unsuccessful in producing good quality films, since congruent removal is often not achieved. Rather surprisingly, it is only under short-pulse ablation conditions that the material collected at a nearby substrate is identical in composition with that of the target. Hence one must reasonably invoke a nonequilibrium mechanism in determining the process by which material is removed.

Here we shall follow the most common terminology and consistently use the term *laser ablation (LA)* to describe the removal phenomenon in the situations where short-pulsed lasers (Q-switched, 50 ns or less pulse duration) are employed. Under other conditions, we prefer to use the term *laser evaporation*. Conventional sputtering and evaporation, although widely used, tend to suffer from the individual source atoms exhibiting different vapour pressures or geometrical trajectories, or widely varying ion-sputter yields. Hence their application to multicomponent film preparation is fairly limited. Laser ablation does not suffer from these problems. Indeed, it is quite insensitive to small changes in the processing parameters, such as beam flux, and the nature of the ejected material does not depend upon deposition rate in the 1–100 Hz regimes used predominantly to date. LA also offers several other advantages in addition to the congruency described above. It is conceptually very simple, relatively cheap and adaptable.

In this chapter, we shall present the background theory, as far as it is presently understood, and some key results that support the current models. We shall briefly summarize the wide variety of films that have been grown using laser ablation. In particular, however, the application of laser ablation to the growth of the new classes of "high-temperature" superconducting oxides in thin-film form will be reviewed.

15.2 BACKGROUND AND THEORY

The intense photon flux available with lasers has traditionally enabled a great variety of machining processes to be achieved [15]. By focusing the radiation down to very small spot sizes and moving it across the workpiece using modern control technology, various drilling, cutting and heating operations can be performed on metals, cloth, semiconductors and ceramics [16]. Depending on the nature and properties of the material being irradiated (either in gas, solid or liquid form) and its immediate environment, the photons can induce either chemical or physical reactions. The photochemical mode has not yet been widely explored in superconductor processing, and here we shall concentrate on the physical processing regime, mainly involving removal of material from the irradiated sample surface, to be used for subsequent thin-film growth.

In order to remove atoms from a solid, sufficient energy must be given to the material to overcome the potential barrier to the removal process. Optical absorption must therefore be very strong so that the energy of the incident photons can be transferred to the atoms. In semiconductors and some insulators this usually constrains the incident radiation to be at a wavelength that is greater than the bandgap energy so that electronic transitions can accommodate the energy. However, other methods are also available, and, by utilizing the superior intensity of longer wavelengths available from the pulsed CO_2 laser, multiphoton absorption by the lattice or the free carriers within the material can also consume the incident photons. Thus a vast number of materials in principle can be encouraged to absorb the intense radiation required to induce ablation.

Since the new generations of superconductors have a shiny black appearance, light in the visible and near-visible (IR and UV) regions of the electromagnetic spectrum can satisfy the initial absorption requirements for ablation. In this spectral regime, where α (the optical absorption coefficient) is very high, typically as much as 10^5 cm^{-1}, the absorption depth α^{-1} can be much smaller than 100 nm. With most lasers operable in the Q-switched mode, therefore, the Nd:YAG (1.064 µm, 532 nm, 355 nm, 266 nm), ruby (694 nm), and excimers (XeCl at 308 nm, KrF at 248 nm and ArF at 193 nm), large amounts of energy can be deposited in a thin region close to the target surface. With the recent improvements in performance (e.g. reliability and stability), the excimer and Nd:YAG lasers have been most widely used.

15.2.1   The ablation threshold

As mentioned already, the precise mechanism associated with laser ablation has yet to be fully established. Nevertheless, it is clear from the many reported data to date that the energies involved are close to those required to heat the target materials well beyond their melting temperatures and initiate significant evaporation. As a first approximation, and in the absence of strong indicators to the contrary, it is instructive to model the energy input to the system as a conventional laser heating problem. It is beyond this review to produce a detailed model of the situation, but nevertheless it is valuable to address several simplified limiting situations.

To couple energy into a solid as efficiently as possible, it is desirable to increase the surface temperature extremely rapidly using radiation that is not only intense, but also in short-pulse form. In this way, the energy absorbed within the irradiated target zone does not diffuse from the surface region, and is mostly used to energetically excite atoms and particles. Thus one may establish that the relationship between pulsewidth τ_p and thermal diffusion time in the solid τ_{th} is important.

The parameter τ_{th} can be transformed to an equivalent distance L_{th}, known as the thermal diffusion length, defined as $(2K\tau_{th}/C)^{1/2}$, and also $(2D\tau_{th})^{1/2}$, where C is the heat capacitance, D the thermal diffusivity and K the thermal conductivity. When $L_{th} > \alpha^{-1}$, the temperature induced at the surface of the sample ($z = 0$) is given by

$$T(0, t) = \frac{2P(1 - R)}{\pi K r^2} \left(\frac{D\tau_p}{\pi}\right)^{1/2}, \qquad (15.1)$$

where P is the incident laser power, r the beam radius and R the sample reflectivity. Hence the laser power required to achieve a particular temperature in this situation is proportional to the square root of the pulsewidth. The most energy-efficient method for removing material is clearly the regime accessed most readily by short-pulsed lasers.

When $L_{th} < \alpha^{-1}$, the temperature will be governed more by where the photons are initially situated in the solid than by the subsequent diffusion of heat. In this case, we can derive

$$T(z, t) = P(1 - R)\alpha \exp(-\alpha z) \frac{\tau_p}{\pi \rho c_p r^2}, \qquad (15.2)$$

where ρ is the density of the sample. In this case, the heating *rate* is independent of the pulse duration, although clearly the maximum temperature is directly proportional to it. These equations are useful for situations where the beam is homogeneous and the optical and thermodynamic constants are independent of temperature. However, by taking temperature-averaged values, one can obtain order-of-magnitude estimates of the parameters involved.

The incident fluence required to melt a target, F_m, can be estimated by assuming that all the energy from the beam is transferred to the atoms in one temperature-averaged absorption depth of the solid, α_s^{-1}, where α_s is the optical absorption coefficient of the solid, and disregarding all losses due to heat diffusion or re-radiation (except the temperature-averaged reflection from the solid, R_s). In this case, F_m can be calculated from:

$$F_m = \rho_s \frac{C_s(T_m - T_r) + L_m}{(1 - R_s)\alpha_s}, \qquad (15.3)$$

where C_s is the temperature-averaged specific heat of the solid, $T_m - T_r$ is the temperature increase required to melt the material, and L_m is the latent heat of melting. Similarly, one can estimate the *additional* fluence necessary to thermally evaporate the atoms once they are in the liquid phase to be

$$F_e = \rho_m \frac{C_m(T_e - T_m) + L_e}{(1 - R_m)\alpha_m}, \qquad (15.4)$$

where the symbols have the same meanings, and the subscripts refer to the melt-to-evaporate phase transition. Thus the total fluence to thermally evaporate the irradiated target, F_{total}, is given by

$$F_{total} = F_m + F_e. \qquad (15.5)$$

The optical properties of the high-T_c superconductor YBaCuO in the solid phase are known [17]. As the irradiating wavelength is decreased, the reflectivity is reduced, and, together with the fact that the optical absorption coefficient is increased, more energy is coupled to the lattice. Thus, in general terms, the threshold for material removal should go down with increasing wavelength.

In the case of the high-T_c superconductor YBaCuO, values for the necessary thermodynamical constants can also be obtained from the literature. For D we use a temperature-averaged value of 2.6×10^{-3} cm^2 s^{-1} from the mean of 1.2 [18] and 4×10^{-3} (an extrapolation from the data of [19]), and for C_s we estimate a value of 1.5 J g^{-1} K^{-1}. We can calculate $\rho_s \approx 6.7$ g cm^{-3}, and use $T_m - T_r = 1000$ K and $L_m \approx 360$ J g^{-1} [19]. For the reflectivity we estimate R_s (1064, 532, 248 nm) $\approx 0.27, 0.10, 0.08$ respectively from the data of Aspnes and Kelly [17], and for optical absorption we obtain α_s (248 nm) $\approx 2 \times 10^5$ cm^{-1} [20]. By substituting these into the equations, we assess the fluence required to melt this material, F_m, to be at least 70 mJ cm^{-2} for 248 nm radiation and 260 mJ cm^{-2} at 1064 nm.

We expect the optical constants to be much less wavelength-dependent for the molten phase, and thus the fluence thresholds for ablation to be predominantly governed by the solid phase characteristics. Although we do not yet have sufficient knowledge of their equivalent values to determine F_e, we do not expect its value to be significantly different from F_m, Hence it would be a fair approximation at this stage to suggest that $1.5 F_m < F_{total} < 2 F_m$. Thus values up to 140 and 560 mJ cm^{-2} for 248 and 1064 nm ablation are the *minimum* fluences necessary to energize the thermal removal process. In reality, good ablation will not necessarily occur near the threshold, but at higher incident energy levels. These estimates are nevertheless in good agreement with the reported observations [21–24].

A more complete model clearly must take into account the proper nonlinear behaviour of thermal properties with temperature, as well as the precise extent of the laser beam in space and time. Furthermore, it is known that material already begins to leave the irradiated surface during the pulse, and so interactions between the photons and the ablated plasma must also be considered.

It is well known that, at high temperatures, thermionic emission of ionic species (and electrons) occurs at surfaces. Traditionally, the fluxes of these species have been estimated using the Richardson–Dushman (Langmuir–Saha) equations respectively. In the presence of large photon intensities, however, such standard relationships have to be modified to accommodate additional phenomena such as photo-excitation and photo-ionization and their consequential effects. When electrons are present near the surface, they strongly couple the electromagnetic radiation, and are accelerated and collide with any plasma ions or any nearby solid or gas phase atoms (inverse bremsstrahlung). The electron population can then be increased and excited further, causing extremely rapid cascade ionization. The collective properties of the plasma thus strongly determine what happens to the remainder of the incident light. The energy absorbed in the plasma is rapidly shared among the individual particles, raising their kinetic energy, i.e. temperature, and shielding the surface from further exposure to the laser beam. If the optical absorption of the plasma falls, its increased transparency will enable further surface exposure to repopulate the plasma and again increase absorption and surface shielding. Thus it could be theoretically deduced [25] that an instantaneous equilibrium plasma density can be obtained that, once formed, would attempt to self-regulate against any changes in density, temperature or irradiance.

The temperature generated within such as plasma have been estimated to be as much as 20 000 K (see later), while it must be noted that the surface temperature of the irradiated solid beneath is only a modest amount above the evaporation point (about 2500 K).

We conclude this section by noting that one can achieve a fair indication of the levels of energy required for ablation to occur by treating the problem initially as a thermally dominated process. This, of course, in no way suggests that a purely equilibrium process controls the kinetics of material removal. The mechanisms by which the material actually leaves the surface seem to be much more complex than can be predicted using thermal evaporation, as will be discussed in the following sections.

15.2.2 Material removal

In attempting to model the ejection of particles from a laser-ablated target, it seems to be unjustified to use sputtering principles, since the incident particles (photons) do not contain any significant momentum with which to sputter atoms from the irradiated surface. Rather, one is tempted to use the simple Knudsen model normally applicable to thermal evaporation. This relates the arrival rate of atoms R at a collecting substrate to the separation distance r between substrate and source. The source must be strictly an infinitesimally small area. If the angle between the normal to the substrate and the radial vector joining the source to the point being considered is θ then

$$R = \frac{\cos \theta}{r^2}. \qquad (15.6)$$

A real evaporating source is finitely extended. In addition, the dense collection of atoms that builds up just above the centre of the evaporated region interferes with the theoretical cosine trajectory. However, the predicted profile is not affected too severely for conventional thermal evaporation.

For laser ablation using visible or UV wavelengths, it has been shown
[26] that, within a cone of acceptance, the angle of incidence of the incident laser beam does not affect the trajectory of the ejected material, which
is virtually always positioned at right-angles to the irradiated surface.
When the deposits obtained by short-wavelength laser ablation are analysed, it is found that there are two components to the contour of the film
grown [25, 26]. As shown in Fig. 15.1, the expected broad $\cos \theta$ profile is
found, superimposed upon which is a more centrally located deposit,
whose profile can be fitted by a separate $\cos^n \theta$ curve, where $9 < n < 11$.
This outline is known to be dependent upon the pressure of the gas surrounding the irradiated surface, as is shown in Fig. 15.2, which has the
effect of de-emphasizing the narrow peak and producing values of n closer
to 4–6 at high pressures [27]. Such profiles have also been reported for
laser ablation by 1064 nm radiation for both millisecond [28] and nanosecond pulses [29], where it has been suggested there is more of a $\cos \theta$
contribution to the process. Interestingly, for shorter wavelengths but at
fluences near the threshold for material removal, only the pure $\cos \theta$ form
is observed, while the $\cos^{10} \theta$ deposition is found only at higher fluences
[26]. These peaked films may be detrimental to any applications desired of
them, but one possibility of homogenizing the profile is to tilt the rotating
target during ablation in order to "spray" the deposit across the substrate.
At an angle of wobble of about $30°$, this has the effect of producing a
uniform film over an area of several square centimetres, as is shown in
Fig. 15.3 for Nd : YAG (532 nm) laser-ablated BiSrCaCuO layers.

Rutherford back-scattering (RBS) studies have shown that the central

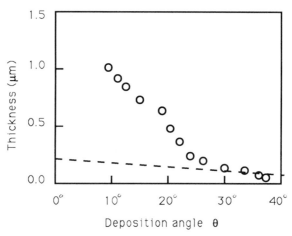

Fig. 15.1 Angular distribution of the laser-ablated film using 248 nm radiation at
$F = 1.5$ J cm^{-2}. The dashed line represents a $\cos \theta$ fit. (After [26].)

portions of the $\cos^{10}\theta$ films grown by ablation are essentially stoichiometric, while the outer edges (containing the $\cos\theta$ component only) are not [26]. This is highlighted in Fig. 15.4 for the case of excimer laser-ablated YBaCuO films. In attempting to explain the overall shape of the deposits obtained not only for superconductors but for many other types of

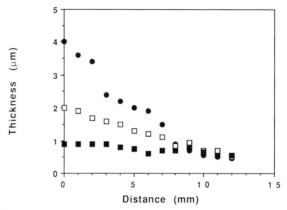

Fig. 15.2   Thickness profile of thin films, laser-ablated under different pressures of nitrogen: •, 0.3 mbar; □, 0.5 mbar; ■, 0.6 mbar.

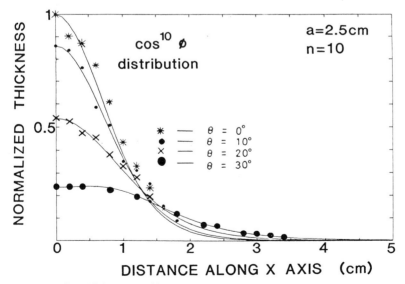

Fig. 15.3   Film thickness profile obtained as a function of angle of wobble during deposition (see text) [163].

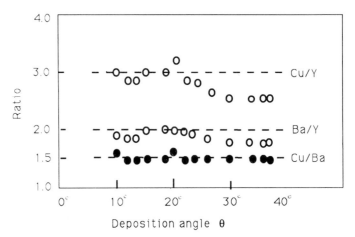

Fig. 15.4 Compositional ratio of the atomic species laser ablated from a $YBa_2Cu_3O_{7-x}$ target by a fluence of 1.5 J cm^{-2} at 248 nm. (After [26].)

material, Kelly and Rothernberg [9] have proposed hydrodynamic sputtering, which assists in the formation of craters on the irradiated surface and confines the ablated material to a $\cos^4 \theta$ profile. The power value could, however, be increased by enhancing the number of collisions among the particles. Others have discussed the effects of shock waves, surface superheating and subsurface explosions [30–33]. Singh *et al.* [34] have modelled a plasma expanding isotropically into vacuum, and have been able to obtain similar atomic profiles to those mentioned above. In their case, however, they prefer to describe the growth contour as a Gaussian profile.

The consequences of these observations appear to suggest that there is an unavoidable thermal component to the removal process containing nonstoichiometric matter. More intriguing is the mechanism behind the highly peaked stoichiometric material whose existence is associated with plasma formation above the sample surface, and subsequent interactions of the species within with each other, with any surrounding species and with the nearby substrate. In order to begin to understand the process, it is informative to study the various physical phenomena that accompany the material removal as well as the plasma properties themselves.

15.2.3 Particle ejection

The most spectacular accompaniment to the ablation growth of films is the colourful plume that is often formed between the target and the substrate.

This plume envelops a volume containing high densities of neutrals, electrons and excited and ionized atoms and molecular clusters moving at very high velocities away from the irradiated surface. Figure 15.5 shows a typical time-integrated image of a plume formed in front of a superconductor target irradiated in oxygen, while Fig. 15.6 shows a streak photograph of a similar event.

The mechanism by which this plume is formed and what determines its

TARGET MATERIAL : $Bi_2 Sr_2 Ca_2 Cu_3 O_y$

IN VACUUM

IN OXYGEN

Fig. 15.5   Time integrated photograph during deposition of BiSrCaCuO in O_2 using a 532 nm Nd : YAG laser pulse.

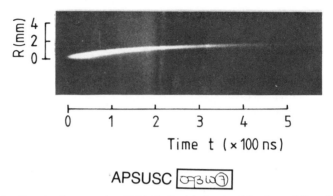

Time t (× 100 ns)

APSUSC

Fig. 15.6   Streak photograph of YBaCuO ablation in 6.5 mbar of O_2 using a KrF laser at a fluence of 4 J cm^{-2} [40].

content is not yet well understood. The phenomena are reminiscent of dielectric breakdown, caused by the highly localized electric field strengths at the sample surface. Early theories suggested that clusters large enough to be essentially stoichiometric could be ejected by the laser beam. Either through their inherent thermodynamic instability, or by collision with themselves or any surrounding molecules, or interaction with the remnants of the laser beam, these then decay into smaller components prior to reaching the substrate. It is clear from post-deposition examination of the films that even droplets (occasionally several tens of microns) can be formed at some time during the process (see Fig. 15.7). Interestingly, the vast majority of early experiments that have examined the plume can mostly only find diatomic and suboxide species present [35]. These include laser ion mass spectroscopy (LIMS) [35, 36] and optical emission spectroscopy [37–40]. In such experiments, however, one must ascertain whether the technique is inherently capable of detecting larger species.

Cheung and Sankur have compared the mass spectra of the species removed from the bulk by evaporative heating and by pulsed laser ablation of the semiconductor CdTe [41]. Figure 15.8 shows that, while Cd^+ ions are present in both cases, the Te_2^+ species are not present in the thermal evaporant from the 650°C bulk. At 1300°C, thermal evaporation and dissociation of Te_2^+ molecules became apparent, but only when the laser was used to induce a surface temperature of a few thousand degrees was complete dissociation observed.

Neutral species form a significant fraction of the ablatant. It is also known that more neutral species are formed when the longer-wavelength lasers are used (i.e. 1064 nm rather than 193 nm) [23]. This is most likely due to the restricted variety of interactions available to the less energetic photons, thus eliminating, for example, the possibility of photon-

(a)

(b)

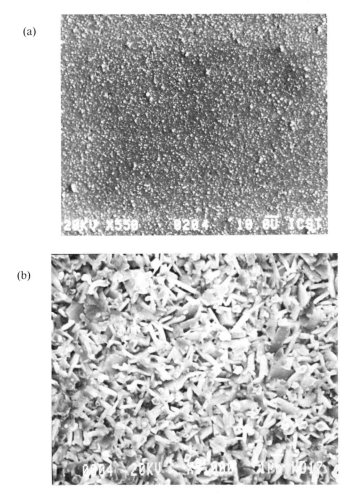

Fig. 15.7   SEM micrographs of a laser-ablated BiSrCaCuO film prepared under conditions where a large density of surface lumps are achieved: (a) as-deposited; (b) after annealing. Although the needles of the 2212 phase and the plates of the 2223 phase are apparent, the lumps are still clearly observable.

ionization of the species in the plasma. Electron impact ionization or cascade ionization will also be much less, and so generally, the plasmas associated with the shorter-wavelength induced ablation will be the most energetic and reactive. For the case of the CO_2 laser, extremely efficient inverse bremsstrahlung photon coupling can be achieved in the plasma. Particle velocities can thus be much higher than for even 248 nm ablation, as shown in Fig. 15.9 [42], though it is not yet clear why stoichiometric

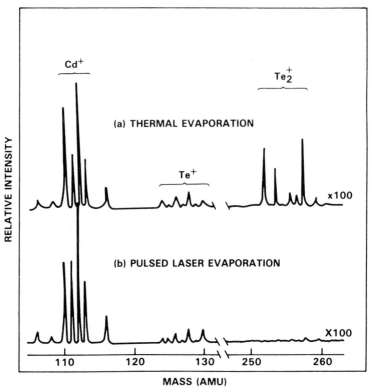

Fig. 15.8 Mass spectra of CdTe evaporated by (a) resistive heating and (b) pulsed laser heating, where the temperature was higher than 3000°C [41].

ablation can only be achieved at this wavelength for only low and not high fluences [43].

It is therefore clear that the plasma can consist of electrons, ions, atoms, molecules, clusters and droplets. Because of the usual irradiation geometries used, it will initially be thin and flat, extending over the area of the surface exposed to the laser beam. Soon after its formation and until the end of the laser pulse, it can be considered to be isothermal, with temperatures exceeding 10 000 K (7.5 eV). Immediately after irradiation, the amount of material augmenting the plasma will drop considerably. It will then expand preferentially normal to the sample surface, this direction being where the greatest density gradients are, into the surrounding vacuum accelerating the species to top speeds of several kilometres per second (up to 1000 times the usual velocities encountered during thermal evaporation). This expansion can push any existing gases away from the target setting up a pressure wave. As the particles slow down and lose energy, they give rise

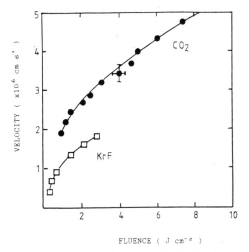

Fig. 15.9   Ion expansion velocity as a function of ablation fluence using TEA CO_2 (•) and KrF (□) lasers [42].

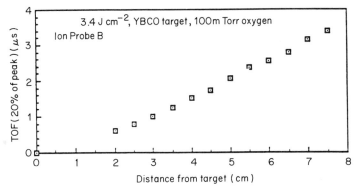

Fig. 15.10   Plasma front time-of-flight data showing a constant slope, giving a constant velocity of 2×10^6 cm s^{-1} [45].

to characteristic UV and visible emission patterns that can be studied *in situ*.

The explosive velocity distributions associated with the particles leaving the ablated surface have been measured directly using time-of-flight (TOF) measurements [44]. There are extremely fast particles, atoms and ions, which are ejected at speeds of around 2 km s^{-1} in pressures of 100 mTorr [44–46]. Figure 15.10, for example, shows plasma front data taken by an ion probe at positions up to 7 cm away from the ablated target. The slope indicates a constant velocity of the particles of 2×10^6 cm s^{-1} [45]. The velocity distributions compare well with the formation of supersonic molecular beams in chemical studies, where the velocities can be described by [14].

$$f(v) = Av^3 \exp\left[\frac{m(v - v_0)^2}{2kT_s}\right].$$ (15.7)

Singh *et al.* [34] have modelled many of the characteristics of the plume by treating the partially ionized plasma as an adiabatically expanding high-temperature high-pressure gas. The group have also shown that each ablated species will accelerate and achieve a characteristic velocity that is dependent upon its mass as well as the incident fluence applied (i.e. the plasma temperature induced). While their model predicted that the asymptotic velocity of the ionic species should be proportional to the square root of their molecular weight, this did not agree with observations, as they suggested, because of possible interactions of different species in the plume. Figure 15.11 shows the predicted, TOF spectra. Since the emission depends on the presence of energetic electrons, the measured emission TOF is a convolution of the true TOF and the electron velocity spectrum. Absorption spectroscopy can correct for this, and, as shown in Fig. 15.12 for ablated BaI at 13 515 cm^{-1}, a slow tail develops that is not present on previously observed emission TOF data.

By using a rotating substrate to gather a circular distribution of particles landing some 43 mm from the ablation surface, Dupendent *et al.* [12] have also been able to examine the post-deposition morphologies associated with particles of different speeds. A wide range of velocity components is associated with the ablated particles. In addition to the rapid (2 km s^{-1}) compo-

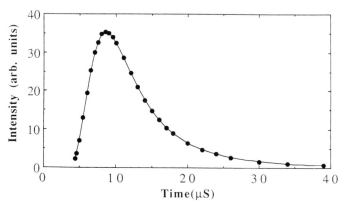

Fig. 15.11 Calculated time-of-flight spectra from a laser-ablated target of YBCO for an initial plasma temperature of 10 000 K and a target–substrate distance of 3 cm [34].

nents mentioned above and seen by others, the group found much slower components, which exhibited typical peaked velocities around $(1-2) \times 10^4$ cm s^{-1}, for samples of YBaCuO superconductor, as well as for metals such as Al, Au, Y, and Fe (see Fig. 15.13). These slower moving particles resulted in lumps forming on the film surface, measured to be

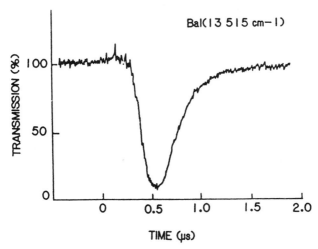

Fig. 15.12 Transmission time of flight of Ba atoms at 5 mm above the irradiated target, showing a deep absorption minimum [14].

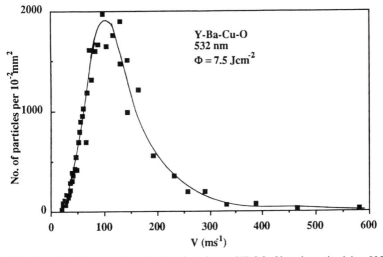

Fig. 15.13 Particle velocity distribution for a YBCO film deposited by 532 nm pulses at a fluence of 7.5 J cm^{-2} [13].

from a few hundred nanometers in diameter up to a few tens of microns. These were found either when ablating metals or superconductors, and generally the slowest (about 50 m s^{-1}) species resulted in the largest lumps.

Several techniques have been suggested to eliminate these undesirable quantities. In general, more lumps tend to be formed at the highest fluences used and also for picosecond rather than nanosecond ablation. By far the most successful method applied to date involves the simple rotation of the target. This ensures that the remnants of any previous ablation event, which are known to be resolidified off-stoichiometric mixtures, are not re-irradiated and ablated toward the substrate. Barr [47] and many other groups have used a 10 000 rev min^{-1} rotating deflector that selectively deflects slow particulates and allows faster moving atoms through. Some of the smaller and faster clusters could still be transmitted in this case, though.

Gaponov et al. [48] have used two colliding beams to generate a higher-pressure zone that expanded towards the substrate. The transmitted beam containing the heavier lumps actually missed the substrate. A negative substrate bias of several hundred volts can also considerably reduce heavy clusters that are similarly charged, and this improved the growth of Ge by laser ablation [49]. Koren et al. [170] and Chiba et al. [171] have also used two laser beams for ablation of YBaCuO, where the second beam excites and dissociates the larger particles within the plasma. Sankur [50] has also found that using melts as target sources can virtually eliminate particulate.

Spectroscopic studies of the plasma radiation have apparently revealed the predominant existence of excited elemental and monoxide species, in agreement with the mass spectrometry results mentioned above. Figure 15.14 shows the spectra obtained when YBaCuO was ablated with a KrF laser (248 nm) and a CO_2 laser (10.6 μm) [42]. Although quite similar, fewer lines are present in the KrF case. Time-resolved studies of the Y and YO emissions have revealed that, while the Y emission occurs only 10 ns after the start of the laser pulse, the YO emission is only apparent after 20 ns [35]. Also using emission spectroscopy, Dyer et al. [40] similarly suggest that diatomic oxides may form during the expansion. The general conclusion therefore is that the oxide species are most likely formed only with collisions near the surface. Venkatesan et al. [35] have used this information to optimize the background oxygen pressure during deposition. Interestingly, the best quality films were grown when the substrate was located at the position of the plume boundary, defined as the point in space where the local pressure of the ablation products falls to that of the background [40]. The mechanisms behind preferred formation of certain small cluster ions are known to be very complex, and clearly require considerable study [36].

In summary, the picture of the ablation process may be considered as follows. Under the extreme and sudden absorption of optical energy in a

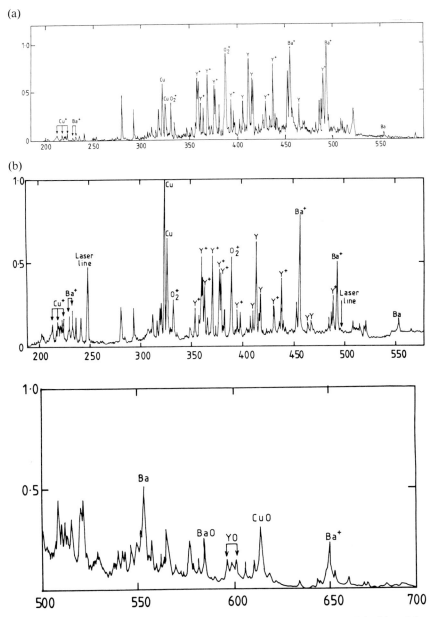

Fig. 15.14 Comparison of the emission spectra of YBCO material ablated by CO_2 (a) and KrF (b) lasers. (After [42, 163].)

Table 15.1 Summary of various films grown by laser ablation or laser evaporation using different lasers. (Updated from [41]).

Material	Laser	Year	Ref.
Ag	Nd : glass, pulsed	1973	[100]
	TEA–CO₂, pulsed	1978	[108]
Al	Ruby, pulsed	1967	[99]
	TEA–CO₂, pulsed	1978	[108]
	Nd : glass, pulsed	1979	[101]
	Nd : YAG, CW	1990	[91]
Al₂O₃	Ruby, pulsed	1970	[3]
	CO₂, CW	1970	[131]
	TEA–CO₂, pulsed	1978	[108]
	Nd : glass, pulsed	1982	[118]
	Nd : YAG, CW	1990	[91]
As₂S₃	Ruby, pulsed	1965	[1]
Au	Nd : YAG, pulsed	1989	[13]
	Nd : YAG, CW	1990	[91]
B₂O₃	CO₂, pulsed	1988	[149]
BaCO₃	Nd : YAG, pulsed	1990	[129]
	CO₂, pulsed	1990	[129]
BaF₂	CO₂, CW	1984	[96, 124]
BaTiO₃	Nd : glass, pulsed	1969	[120]
BeO	Nd : glass, pulsed	1982	[134]
Bi	TEA–CO₂, pulsed	1978	[108]
BiF₃	Ruby, pulsed	1984	[111]
Bi₂Te₃	Q-switched	1974	[135]
C	CO₂, CW	1982	[132]
	TEA–CO₂	1986	[125]
	Nd : YAG, pulsed	1985	[159]
		1989	[160]
		1990	[112]
		1990	[167]
	XeCl excimer, pulsed	1989	[161]
CaF₂	CO₂, CW	1984	[124]
CaTiO	Ruby, pulsed	1970	[3]
	CO₂, CW	1970	[3]
CeF₃	CO₂, CW	1984	[124]
Cd	TEA–CO₂, pulsed	1978	[108]
Cd₃As₂	Nd : YAG, pulsed	1984	[95]
CdCr₂S₄	Ruby, pulsed	1970	[3]
CdCr₂Se₄	Ruby, pulsed	1970	[3]
Cd₁₋ₓMnₓTe	XeCl excimer, pulsed	1989	[150]
CdO–SnO₂ mixture	Nd : glass, pulsed	1982	[118]
CdS	Ruby, pulsed	1970	[3]
	TEA–CO₂, pulsed	1978	[108]
	Nd : glass, pulsed	1985	[136]
CdSe	Ruby, pulsed	1970	[3]
CdSnAs₂	Nd : glass, pulsed	1981	[137]
CdSₓSe₁₋ₓ	Excimer, pulsed	1988	[39]

Table 15.1 (continued).

Material	Laser	Year	Ref.
CdTe	Ruby, pulsed	1965	[1]
	CO_2, CW	1974	[135]
	Nd : glass, pulsed	1981	[137]
	Nd : YAG, pulsed	1985	[90]
CeO_2	CO_2, pulsed	1989	[152]
Cr	Nd : glass, pulsed	1969	[120]
	Nd : glass, pulsed	1973	[100]
	TEA–CO_2, pulsed	1987	[108]
CuO	Excimer, pulsed	1989	[153]
	CO_2, pulsed	1990	[129]
	Nd : YAG, pulsed	1990	[129]
Dy	TEA–CO_2, pulsed	1978	[108]
Er	Nd : glass, pulsed	1978	[103]
Fe	TEA–CO_2, pulsed	1978	[108]
Fe/Cr	Ruby, pulsed	1963	[99]
Fe_2O_3	TEA–CO_2, pulsed	1978	[108]
GaAs	Nd : glass (?), pulsed	1968	[139]
	CO_2, CW	1970	[3]
	Ruby, pulsed	1970	[3]
	Nd : glass, pulsed	1981	[137]
	Nd : glass, pulsed	1981	[117]
	Nd : glass, pulsed	1982	[118]
GaP	Ruby, pulsed	1970	[3]
	CO_2, CW	1970	[3]
	TEA–CO_2, pulsed	1978	[108]
	Nd : glass, pulsed	1981	[137]
GaSb	Ruby, pulsed	1970	[3]
	CO_2, CW	1970	[3]
Gd	TEA–CO_2, pulsed	1963	[108]
Ge	Ruby, pulsed	1965	[1]
	CO_2, CW	1969	[133]
	Nd : glass, pulsed	1977	[140]
	KrF excimer, pulsed	1985	[49]
	KrF excimer, pulsed	1990	[93]
	XeCl excimer, pulsed	1990	[92]
Hf	TEA–CO_2, pulsed	1978	[108]
HfC	Nd : glass, pulsed	1975	[121]
HfO_2	Nd : glass, pulsed	1979	[106]
	CO_2, CW	1984	[124]
$Hg_{1-x}Cd_xTe$	Nd : glass, pulsed	1983	[5]
HgTe	Nd : glass, pulsed	1968	[137]
	Nd : YAG, pulsed	1986	[138]
HgTe/CdTe mixtures	Nd : YAG, pulsed	1982	[87]
In	TEA–CO_2, pulsed	1970	[3]
	Nd : glass, pulsed	1981	[137]
InAs	Ruby, pulsed	1968	[3]
	Nd : glass, pulsed	1981	[137]

(*Continued*)

F. Beech and I. W. Boyd

Table 15.1 (continued).

Material	Laser	Year	Ref.
InSb	TEA–CO_2, pulsed	1978	[108]
	Nd : glass, pulsed	1981	[137]
	Nd : YAG, pulsed	1985	[98]
	Nd : glass, pulsed	1987	[117]
Ir	Ruby, pulsed	1967	[99]
$LaAlO_3$	CO_2, CW	1968	[131]
LaF_3	CO_2, CW	1984	[124]
La_2O_3	CO_2, CW	1985	[119]
$MgAl_2O_4$	CO_2, CW	1969	[133]
	CO_2, CW	1970	[3]
	Ruby, pulsed	1970	[3]
MgF_2	CO_2, CW	1984	[124]
MgO	CO_2, CW	1985	[119]
	ArF Excimer, pulsed	1990	[116]
	Nd : YAG (532 nm), pulsed	1991	[143]
Mo	Nd : glass, pulsed	1977	[102]
	TEA–CO_2, pulsed	1978	[108]
MoO_3	Ruby, pulsed	1965	[1]
Na_3AlF_6	TEA–CO_2, pulsed	1978	[108]
	CO_2, CW	1984	[124]
$Na_5Al_3Fl_4$	CO_2, CW	1984	[124]
NaF	CO_2, CW	1984	[124]
Nb_2O_5	Nd : glass, pulsed	1981	[118]
NdF_3	CO_2, CW	1984	[124]
$Nd_2Fe_{13}B$	Nd : YAG, pulsed	1990	[155]
Ni	Nd : YAG, pulsed	1990	[169]
Ni_3Mn	Nd : glass, pulsed	1975	[121]
Ni–dimethylglyoxime	Ruby, pulsed	1965	[1]
Os	Nd : YAG pulsed and CW	1979	[106]
Oxide and carbide mixtures	Nd : glass, pulsed	1982	[118]
Pb	Nd : YAG, CW	1990	[91]
PbC_{12}	Ruby, pulsed	1965	[1]
$Pb_{1-x}Cd_xSe$	Nd : glass, pulsed	1986	[7]
PbF_2	CO_2, CW	1986	[131]
	TEA–CO_2, pulsed	1986	[131]
PbO	Nd : YAG (532 nm) pulsed	1990	[151]
PbO_2	TEA–CO_2, pulsed	1978	[108]
PbS	Ruby, pulsed	1988	[157]
	TEA–CO_2, pulsed	1988	[157]
	Nd : glass, pulsed	1988	[157]
		1988	[122]
PbSe	Nd : glass, pulsed	1977	[102]
		1988	[122]
PbTe	Ruby, pulsed	1965	[1]
	Q-switched	1977	[109]
	Nd : glass, pulsed	1981	[137]

Table 15.1 (continued).

Material	Laser	Year	Ref.
PbTiO$_3$	XeCl, ArF Excimer, pulsed	1990	[116]
PbZrO$_3$	XeCl, ArF Excimer, pulsed	1990	[116]
Pb(Zn$_{0.5}$Ti$_{0.48}$)O$_3$	ArF excimer, pulsed	1989	[148]
PEI	Argon ion, CW	1990	[115]
PI	Argon ion, CW	1990	[115]
Pt	Nd : YAG, CW	1972	[107]
	Nd : YAG, CW	1990	[91]
PTFE	Argon ion, CW	1990	[115]
PZT	Nd : YAG, pulsed	1990	[110]
Rh	Nd : YAG, pulsed	1990	[169]
ReBe$_{22}$	Ruby and Nd: glass, pulsed	1975	[121]
RM$_5$	Nd : YAG, pulsed	1990	[114]
Ru	Nd : glass, pulsed	1973	[100]
Sb	Nd : glass, pulsed	1979	[101]
Sb$_2$S$_3$	Ruby, pulsed	1965	[1]
	Nd : glass, pulsed	1969	[120]
Sc$_2$O$_3$	CO$_2$, CW	1984	[124]
Se	Ruby, pulsed	1965	[1, 123]
	Nd : YAG (266 → 1064 nm), pulsed	1987	[141]
Si	CO$_2$, CW	1969	[133]
	TEA−CO$_2$, pulsed	1978	[108]
	Nd : YAG, pulsed	1983	[128]
	KrF excimer, pulsed	1985	[49]
a-Si	Nd : YAG, pulsed	1990	[126]
SiC	ArF excimer, pulsed	1989	[130]
	CO$_2$, pulsed	1990	[129]
	Nd : YAG, pulsed	1990	[129]
SiO	CO$_2$, CW	1986	[131]
	ArF, KrF excimers, pulsed	1990	[88]
SiO$_2$	Ruby, pulsed	1970	[3]
	TEA−CO$_2$, pulsed	1978	[108]
	ArF, KrF excimers, pulsed	1990	[89]
	CO$_2$, CW	1984	[124]
Si$_3$N$_4$	CO$_2$, CW	1968	[131]
Sn	TEA−CO$_2$, pulsed	1978	[108]
	Nd : glass, pulsed	1979	[101]
SnO$_2$	Nd : glass, pulsed	1981	[118]
	Nd : YAG, pulsed	1982	[142]
	CO$_2$, CW	1984	[124]
SrF$_2$	CO$_2$, CW	1984	[124]
SrTiO$_3$	Nd : glass, pulsed	1969	[120]
Ta	Ruby, pulsed	1963	[99]
	TEA−CO$_2$, pulsed	1978	[108]
	Nd : glass, pulsed	1981	[104]
	Nd : YAG, CW	1990	[91]
Ta$_2$O$_5$	CO$_2$, CW	1969	[119]

(Continued)

Table 15.1   (continued).

Material	Laser	Year	Ref.
Te	Ruby, pulsed	1965	[1]
	CO_2, CW	1969	[133]
	TEA–CO_2, pulsed	1978	[108]
Ti	TEA–CO_2, pulsed	1978	[108]
	Nd : glass, pulsed	1979	[101]
	Nd : YAG, CW	1990	[91]
	Nd : YAG, pulsed	1990	[167]
TiN	XeCl excimer, pulsed	1989	[162]
TiO_2	Nd : glass, pulsed	1981	[118]
	CO_2, CW	1985	[119]
Ti_2O_3	CO_2, CW and TEA	1984	[124]
Triglycine sulphate	CO_2, CW	1974	[144]
V_2O_5	CO_2, CW	1985	[119]
W	Ruby, pulsed	1967	[97]
	Nd : glass, pulsed	1969	[120]
	Nd : YAG, CW	1990	[91]
Y	Nd : YAG pulsed	1990	[13]
YCo_5	Nd : YAG, pulsed	1990	[154]
Y_2Fe_{15}	Nd : YAG, pulsed	1990	[155]
YNi_3	Nd : YAG, pulsed	1990	[155]
YNi_5	Nd : YAG, pulsed	1990	[155]
	Excimer, pulsed	1990	[111]
	CO_2, pulsed	1990	[111]
Y_2O_3	CO_2, CW	1984	[124]
	Nd : YAG, pulsed	1990	[129]
Y_2O_3–$BaCO_3$ mixture	TEA CO_2, pulsed	1990	[94]
Y_2O_3–CuO mixture	TEA CO_2, pulsed	1990	[94]
ZnO	Ruby, pulsed	1970	[3]
	CO_2, CW	1983	[4]
ZnS	Nd : glass, pulsed	1969	[120]
	Ruby, pulsed	1970	[3]
	CO_2, CW	1971	[108]
	TEA–CO_2, pulsed	1978	[108]
ZnSe	CO_2, CW	1969	[133]
	Ruby, pulsed	1970	[3]
	TEA–CO_2, pulsed	1978	[108]
ZnTe	Ruby, pulsed	1965	[1]
Zr (in O_2)	TEA–CO_2, pulsed	1978	[108]
	Nd : glass, pulsed	1979	[105]
ZrC	Nd : glass, pulsed	1975	[121]
ZrO_2	CO_2, CW	1977	[102]
	TEA–CO_2, pulsed	1978	[108]
	Nd : glass, pulsed	1982	[118]
	TEA–CO_2, pulsed	1984	[124]
	Nd : YAG (532 nm), pulsed	1987	[113]
	Nd : YAG, CW	1990	[91]

Table 15.1 (continued).

Material	Laser	Year	Ref.
Superlattices/multilayers			
Cr/C	Nd : glass	1981	[158]
HfO$_2$/SiO$_2$	CO$_2$, CW	1986	[156]
HgTe–CdTe	Nd : YAG, pulsed	1986	[138]
HgCdTe$_{x_1}$– HgCdTe$_{x_2}$	Nd : YAG, pulsed	1988	[145]
InSb–PbTe	Nd : glass, pulsed	1979	[146]
PbTe–CdTe	Nd : glass, pulsed	1979	[146]
InSb–CdTe	Nd : glass, pulsed	1980	[147]
Bi–CdTe	Nd : glass, pulsed	1980	[147]
(InSb, PbTe, CdTe) with (C, Ge, GaAs, Ta)	Nd : glass, pulsed	1981	[104]
Ni/C	Nd : YAG, pulsed	1990	[168]
TiN/TiC	KrF excimer, pulsed	1990	[92]
W/C	XeCl excimer, pulsed	1990	[92]
	Nd : YAG, pulsed	1991	[168]
ZrO$_2$/SiO$_2$	CO$_2$, CW	1986	[156]

very thin layer of the sample, all the component atoms are ejected in an energetic and violent rupture of the surface. This ensures that stoichiometry is conserved. Once formed, this plasma begins to absorb the remaining light in the incident laser pulse and attains a very high temperature. Many clusters that are ejected most likely rapidly fragment as a result of inherent instability, collisions or photonic interactions. Nevertheless, some small fraction may reach the substrate surface. After undergoing collisions with themselves and with any background gas species, the plasma expands preferentially in the direction away from the surface and emits its own characteristic and detectable light. After some distance the particles rapidly run out of kinetic energy, and at this point may form particulates or a smooth film layer on a nearby substrate. Thus it is critical that the substrate is placed at an optimum position with respect to the plume. It may also be essential to eliminate the most slowly moving particles (i.e. droplets) that are the remnants of the ablation process, which can proceed to form so called 'blobs' on the film surface.

In the remainder of this chapter, we shall focus our attention on laser ablation growth of the thin-film superconductors, rather than any of the myriad of other materials collected in Table 15.1. The technique, although clearly useful in the formation of other layers, has been shown to be best suited for producing multicomponent structures, and the new high-critical-temperature (T_c) superconductors contain as many as six different atoms in unit cells approaching 1000 amu in mass. Figure 15.15 shows the

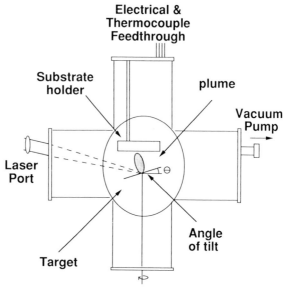

Fig. 15.15 Schematic of a typical laser ablation set-up for deposition of thin films [162].

proportion of thin-film superconductor papers published in the scientific literature that contain studies of laser-ablation-produced films. This therefore gives a broad indication of the high proportion of high-T_c films worldwide that have been manufactured by the laser ablation method.

15.3 EXPERIMENTAL APPROACH

A typical experimental configuration applied to laser ablation deposition is shown in Fig. 15.16. It basically consists of an evaporation chamber with laser beam handling optics usually positioned outside the cell. The optical entry window is often geometrically shielded from the ejected material, although this is not always necessary because of the directionality of the particle emission. The target pellet need not be oriented at exactly $90°$ to the beam, as a wide cone of acceptance angles is possible [25], and is commonly irradiated at an angle of $45–60°$. The substrate upon which the film grows is usually positioned parallel to the target, and it has the capability of being heated to sufficiently high temperatures to help promote good film growth. To achieve good stoichiometry, it is essential to rotate or move the target during successive exposures to the laser beam, or to scan the impinging beam relative to the target. This minimizes the possibility of formation

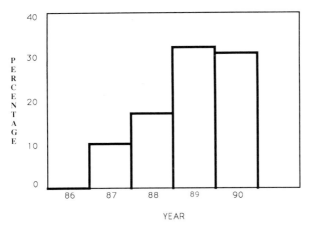

Fig. 15.16 Percentage of papers published annually concerning the growth of thin-film superconductors.

of a damage crater in the target, from which undesirable melting and resolidified non-stoichiometric material may be ablated.

A vacuum pump controls the atmosphere within the cell, and, although processing pressures around 10^{-6} Torr are commonly maintained, precise amounts of O_2, N_2O or other gases may be pumped through the chamber during deposition. Where this is necessary, a plasma ring can be introduced close to the growth substrate in order to assist in activating the reacting gas.

The incident fluences usually applied to initiate superconductor ablation are around $1-4\,\text{cm}^{-2}$, although occasionally higher fluences are used, and to date most of the lasers used operate at 10 Hz. The repetition rate is simply a reflection of what is commercially available, and in principle rates into the kilohertz regime would be more desirable. Typical deposition rates are around $1-10\,\text{Å}$ per pulse, and, in 10−20 min irradiation time, films around 1 μm in thickness can be grown.

Depending upon the nature of the target irradiated, and more importantly the properties of the host substrate and its temperature, the deposited films may be amorphous, or can be grown epitaxially. Often they do not contain any memory of the original crystallinity of the target, although, as will be discussed later, this may not always be the case. In the vast majority of optimized cases reported, nevertheless, the ratio of the metal atoms in the evaporated film is extremely close to the initial stoichiometry. This only seems to fail when the target material is heated for an extended period, such as when very long pulses or CW lasers are used [10].

The thermal and chemical characteristics of the substrate are crucial to the eventual quality of the deposit, or the subsequent properties achieved after annealing. The degree of lattice and thermal expansion mismatch is as important as in any heterogeneous film growth. It is also important to ensure that chemical reactivity at the interface is not detrimental to the superconducting film. Substrates of strontium titanate ($SrTiO_3$), magnesium oxide (MgO) and yttrium-stabilized zirconia (YSZ) have been host to amongst the best laser growth thin-film superconductors reported to date. Although slightly better properties have occasionally been achieved with Al_2O_3 and $LiNbO_3$ at $500°C$, thermally induced stress cracks generally hamper reproducibility.

To further improve the quality of superconductor films, cheaper and better substrates than $SrTiO_3$ are needed, particularly if microelectronic integration is desirable. Attempts to limit the interdiffusion of atoms in Si (which can be as high as 1000 Å) or Si-based materials have not been successful because of the tendency of Si, SiO_2 and Si_3N_4 to react strongly with the superconductors, affecting the electrical properties of films even up to around 1 μm thick. There is presently a great deal of investigation into the possible use of buffer layers that might in future enable successful combination of the superconducting films with Si devices.

15.4   HIGH-T_c FILMS

Since the discovery of the original high-T_c materials based on the alkaline earth doping of La_2CuO_4, several oxide compounds have been isolated that also display superconducting properties. The currently known inorganic high-T_c superconducting systems are reviewed in Table 15.2.

A detailed description of the chemistry of these phases is beyond the scope of this chapter. Interested readers are referred to the excellent reviews of Cava [51] and Sleight [52]. Instead, we intend to highlight a few of the properties that influence the processing of these materials. In particular, we shall concentrate on the information obtained from the most studied material, $Ba_2YCu_3O_7$.

The requirements for device-quality thin films have been formulated by Venkatesan [53] and are reviewed below.

(a)   A high T_c

This requirement can be thought of as the necessity to grow the films with the appropriate composition in the correct crystal phase. The potential complexity of this problem is apparent when it is considered that the accurate control of three to five separate cation species is required to grow films of the materials in Table 15.2. The problem of obtaining the correct crystal phase will be highlighted later in Section 15.4.3, in which the growth of the 2212 and 2223 phases in the bismuth-based films is discussed.

Table 15.2   The various high-temperature superconductor phases and their associated transition temperatures T_c.

Material	T_c (K)
$(Ba_{1-x}K_x)BiO_3$	30
$(La_{2-x}Sr_x)CuO_4$	40
$(Nd, Ce, Sr)_2CuO_4$	28
$(Nd_{2-x}Ce_x)CuO_4$	24
$Ba_2YCu_3O_7$	92
$Ba_2YCu_4O_8$	80
$Ba_4Y_2Cu_7O_{14}$	40
$Bi_2Sr_2CuO_8$	20
$Bi_2Sr_2CaCu_2O_x$	80
$Bi_2Sr_2Ca_2Cu_3O_y$	110
$Pb_2Sr_2Nd_{0.5}Ca_{0.5}Cu_3O_8$	70
$TlBa_2CaCu_2O_7$	103
$TlBa_2Ca_2Cu_3O_9$	120
$Tl_2Ba_2CuO_6$	90
$Tl_2Ba_2CaCu_2O_8$	112
$Tl_2Ba_2Ca_2Cu_3O_{10}$	125

(b) A small transition width ΔT between the normal and superconducting states

This implies that we need all of the film to enter the superconducting state over a very narrow temperature range. This problem can also be highlighted by reference to $Ba_2YCu_3O_{6+x}$, which displays a range of oxygen content between $0 < x < 1$. There are remarkable changes in electronic properties associated with this nonstoichiometry. At $x = 0$, the material is semiconducting, while with increasing oxygen content the material becomes metallic and superconducting. At $x = 1$, it has a superconducting transition at 92 K.

(c) A high critical current J_c

This requirement states that we need good current-carrying capabilities in the superconducting state if device applications are to be realized, and can be influenced by the nature of the flux-pinning sites in the film. Thus the introduction of defects into the crystal structure or features such as grain boundaries in the microstructure introduced by the processing technique is very important.

(d) A smooth film surface

This constraint is a function of the need to be able to fabricate working device structures within any surface inhomogeneities, and clearly requires as uniform and featureless a surface as possible.

(e) Stable films on semiconductor substrates

This ability would enable the integration of semiconducting and super-conducting technologies and is discussed in detail in Section 15.4.2.

(f) Sharp interfaces between films and film and substrate

This property requires that little interdiffusion or chemical reaction take place, since such effects tend to act so as to lower the superconducting transition temperature in the cuprate-based materials. The existence of sharp interfaces between films is a key requirement for the successful fabrication of Josephson junctions.

15.4.1 $Ba_2YCu_3O_7$ film growth

This system is by far the most intensively studied of the superconducting cuprates, and the growth of films of this material is now an essentially trivial operation. The growth conditions used by the vast majority of groups are based on the original work of Inam and co-workers [54]. Their basic conditions consist of irradiating a stoichiometric target with an approximate fluence of $1.5 \, J \, cm^{-2}$ delivered from a pulsed excimer laser operating with 30 ns pulses of 248 nm radiation. The deposition is performed in 100 mTorr of oxygen on a substrate heated to a surface temperature of between 700 and $750°C$ (corresponding to a heater stage temperature of approximately $800°C$). The substrates are chosen to match the superconductor structure and are either rocksalt-structured, such as (100) MgO, or perovskites, such as (100) $SrTiO_3$. The characterization of these films by X-ray, Rutherford back-scattering (RBS) and helium ion channeling studies shows that the films grow epitaxially with an orientation consistent with the c axis being normal to the substrate surface. The electronic characteristic of films grown under these conditions are generally excellent. Films with reproducible zero-resistance temperatures of 89 K and critical current densities can now be made with J_c values up to $5 \times 10^6 \, A \, cm^{-2}$ at 77 K.

It is instructive at this point to discuss some of the parameters that influence the growth and microstructure of the films. Among these are the substrate orientation and temperature, which determine the epitaxy, and also the oxygen partial pressure, which determines the composition and crystal structure (and hence T_c) of the film. Eibl and Ross [55] have investigated the role of the oxygen partial pressure and temperature on the uptake of oxygen into the film, and they suggest two key regimes for the optimal processing sequence. The first, during deposition, requires a reasonable oxygen pressure (about 0.3 mbar). This should then be increased to 1 bar during an additional post-deposition annealing stage. The sharpest transitions (0.6 K) into the superconducting state have been observed during this stage when the films were slowly cooled by $10-100°C$ min^{-1}.

The evolution of the microstructure during the early stages of growth has been investigated by Norton and Carter [56] using both bright- and dark-field imaging TEM techniques and films grown on electron transparent thin-foil specimens of (001)-oriented MgO. The results of these investigations suggest that the ratio of c-axis oriented grains to a- (or b-) axis oriented grains appears to be a function of the growth rate and also the surface mobility of the ablated species. The critical role of the growth rate was underpinned by the finding that at high growth rates (about 1.5 nm s^{-1}) a mixture of c-axis and a- (or b-) axis oriented grains were formed, whereas at lower growth rates (<0.3 nm s^{-1}) the film was oriented exclusively with the c axis perpendicular to the substrate. It was also observed in these studies that the c-axis orientation could be produced preferentially by either the use of high laser pulse energies or shorter wavelengths.

The orientation of the as-grown film is of great concern, since, as a consequence of the anisotropic nature of the crystal structure of the superconducting cuprates, the electronic properties of these materials are also anisotropic. As discussed by Worthington and co-workers [57], this effect is particularly manifested in the superconducting coherence length ξ, which can be more than one order of magnitude larger in the (a, b)-axis directions lying *in* the Cu–O planes than along the c axis *normal* to the Cu–O planes. In constructing devices based on the Josephson effect, a long coherence length is considered a definite advantage, since this length sets the scale over which the order parameter can vary. A value of ξ_a large compared with the lattice parameters will be relatively insensitive to the modest amount of disorder expected at the interfaces in a Josephson tunnel junction or weak link. Thus the small value of ξ_c compared with the large c-axis length in $Ba_2YCu_3O_7$ suggests that it would be difficult to maximize the value of the order parameter in this direction. Hence it would be advantageous to isolate a set of growth conditions that produce films with the CuO_2 planes presenting edges at the surface of the film, i.e. displaying a (or b) orientation.

As we have already discussed, the general growth conditions produce predominantly c-axis oriented grains. There have been a number of reports of efforts to systematically grow out-of-plane a- (or b-) axis growth. Generally, two approaches have been investigated [58, 59]. An enhancement of the required growth over c-axis orientation has been observed by simply lowering the growth temperature by roughly $100°C$. The second technique involves changing the orientation of the substrate. For example, (110)-oriented $SrTiO_3$ has been used, and the resultant films have been shown to have mixed (103) and (110) orientations. The relative amount of each orientation depends critically on the deposition temperature.

An interesting potential solution to this problem has been observed by Iman et al. [60] during the growth of heterostructures of the type $Ba_2YCu_3O_7-Ba_2PrCu_3O_7-Ba_2YCu_3O_7$. It is generally possible to replace the Y in the $Ba_2YCu_3O_7$ phase with another rare earth element and still retain the superconducting properties. However, the substitution with praseodymium (Pr) results in a nonsuperconducting material that retains the host crystal structure and very similar lattice parameters. Films of this material can be grown under the same ablation conditions as for $Ba_2YCu_3O_7$ [60]. Thus these materials are ideal candidates for the hetero-epitaxial growth of multilayer junction structures. Indeed, the group at Bellcore have grown a wide variety of structures varying in complexity from trilayer weak links to full superlattices [61–63]. During these experiments, detailed analysis of TEM micrographs revealed an increased tendency for the Pr phase to form a- (or b-) oriented grains, and that once this orientation was obtained it propagated into a layer grown on to it. This phenomenon was reproducible even under growth conditions that would normally result in c-axis growth. Prompted by the observation of this template effect, Inam et al. [60] studied the growth of the Pr-substituted material to isolate conditions favourable to the growth of a (or b) orientation. They found that the simple expedient of lowering the growth temperature from $800°C$, where high-quality c-axis oriented films grow, to $700°C$ resulted in the growth of a- (or b-) oriented films at a quality comparable to the c-axis films. Thus the growth of a- (or b-) oriented multilayer structures can be envisaged as the deposition of a layer of the Pr phase at $700°C$ followed by the Y layer deposited at $800°C$. The interpretation of X-ray diffraction data taken on films grown in this manner suggest that they show less than 5% c-axis orientation. These layers show acceptable superconducting properties, with a reproducible transition temperature around 80 K.

15.4.2   Growth of $Ba_2YCu_3O_7$ on Si

One potential area for the application of the high-T_c materials is in hybrid superconductor–semiconductor devices. Since the electrical and optical properties of these materials span the range from an insulating dielectric to a superconductor, there is considerable opportunity for fabricating novel optoelectronic and microelectronic devices based on these materials. However, such devices will require the growth of high-quality films on Si or GaAs, and there are several difficulties associated with this process. As pointed out earlier, among the major problems are

(a) intermixing the chemical reaction at the superconductor—semiconductor interface;

(b) thermal mismatch between the two components;

(c) strain introduced by mismatch in the lattice constants of the two materials.

Initial attempts to grow films on Si, such as those performed by Wu *et al.* [64] required a high-temperature ($850°C$) anneal in order to crystallize the film. This heat treatment resulted in films that were cracked (detrimental for high J_c), and more seriously, promoted interdiffusion and chemical reaction at the film—substrate interface. These effects led to a poor transition temperature into the superconducting state, with the best results of 45 K being obtained on relatively thick films, of about 5–8 μm in extent.

One solution to this problem is to grow a buffer layer onto the Si before the deposition of the superconducting film. The most intensively studied candidate material for the buffer layer is yttria-stabilized zirconia (YSZ). Impressive progress has been made on this problem. For example, Tiwari *et al.* [65] and Fork *et al.* [66] have reported the *in situ* growth of the YSZ and $Ba_2YCu_3O_7$ structure. The latter report highlights the importance of the strain introduced into the film due to the large difference in thermal expansion coefficients between the two materials. In fact, they note that thin (< 500 Å) films are under enough tensile strain that the unit cell is distorted. Nevertheless, it is still possible to grow ensembles that have good crystalline quality and electronic characteristics ($T_c(r = 0)$ of 88 K associated with a J_c of 2.2×10^6 A cm^{-2} at 77 K).

The problem of minimizing the lattice—film thermal mismatch has been addressed in a study by Miura *et al.* [67]. They have considered the growth of a two-layer buffer structure in which the values of the mismatched materials are bridged by the intermediate values of the buffer materials. In particular, they investigated the growth of a structure of the type Si–$MgAl_2O_4$–$BaTiO_3$–$Ba_2YCu_3O_7$. The aluminate was grown onto the Si by chemical vapour deposition and the barium titanate and superconducting layers by RF magnetron sputtering, rather than laser ablation. A combination of RHEED, SEM and X-ray techniques showed that the superconducting layer had grown epitaxially, with the *a* axis being preferentially oriented normal to the substrate surface. This film displayed a T_c onset at 80 K and a zero resistance at 70 K, associated with a critical current density of 4×10^4 A cm^{-2} at 40 K. This work has been extended by Wu *et*

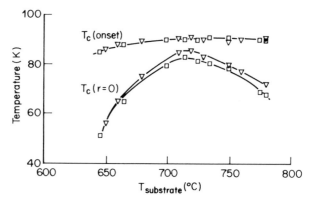

Fig. 15.17 Deposition-temperature dependence of the values of T_c(onset) and $T_c(r = 0)$ for YBCO films grown using a XeCl laser at 2 J cm^{-2} on two different substrates: □, ZrO$_2$; ▽, ZrO$_2$ on NiSi$_2$ [165].

al. [68], who have ablated a superconducting film onto a buffer layer structure provided by Miura. The interpretation of X-ray diffraction studies combined with TEM and ion channelling measurements revealed that the epitaxial film had grown in this case with *c*-axis preferential orientation. The as-deposited films had reasonably good metallic characteristics, but at room temperature their resistivities were around 2–3 times those of the best films grown on SrTiO$_3$ substrates. They had a zero-resistance temperature of 86 K and a significantly improved critical current density of 5.8×10^4 A cm^{-2} at 77 K.

Blank *et al.* [165] have similarly used buffer layers of ZrO$_2$ and of NiSi$_2$ with ZrO$_2$. Although the former single buffer layer minimized interdiffusion, which was only found along the grain boundaries, the latter structure, which contained microcrystallites up to 10 times larger, provided the best interface for the superconductor film with the Si. $T_c(r = 0)$ values of 86 K and J_c levels of 4×10^4 A cm^{-2} have been obtained in the best samples. Figure 15.17 shows how also the deposition temperature is less critical to the value of ΔT for the Si–NiSi$_2$–ZrO$_2$–YBCO geometry. A more detailed review of these and other types of buffer layer can be found elsewhere [166].

The results of these studies strongly suggest that the superconducting properties of the layers are limited by the quality and degree of crystallinity of the buffer layers and that there are still possible improvements to be made to this approach.

15.4.3 Growth of bismuth-containing films

The homologous series Bi$_2$Sr$_2$Ca$_n$Cu$_{n+1}$O$_x$ contains three members at $n = 0$,

1 and 2, with superconducting transitions of 20, 80 and 110 K. These materials are, of course, structurally interrelated and differ only by the insertion of the $CaCuO_2$, which is accommodated by extra layers. Thus the materials are often described in terms of the number of Cu–O planes, these being 1 ($n = 0$), 2 ($n = 1$) or 3 ($n = 2$). The chemistry of these phases is unfortunately more complex than the previous statement indicates, and the isolation of pure single-phase ceramic samples has proven to be a major challenge to synthetic chemists. One apparent problem is the existence of a large amount of cross-substitution on the cation lattice sites which leads to the creation of large composition variations that apparently correspond to the superconducting phase. This, of course, is reflected in a wide range of "optimum" target stoichiometries. A far more complex problem is that the 2223 ($n = 2$, $T_c = 110$ K) phase is only stable over a range of approximately 10 K near its melting point. In general, the thermodynamically stable 2212 ($n = 1$, $T_c = 80$ K) phase is preferentially synthesized, and the highest-T_c phase is then generated by the decomposition of this phase.

Attempts to grow layers of this material have nonetheless been performed by several groups, although the reported level of activity on this material is substantially lower than that for $Ba_2YCu_3O_7$. Initial reports concerned the deposition of films onto room temperature substrates (generally $SrTiO_3$, ZrO_2 or MgO) and the isolation of a post-annealing schedule to produce the 2212 phase. This presupposes that the correct stoichiometry has been transferred from the target to substrate, which strictly may not always be the case, as mentioned in Section 15.2.3. For example, layers displayed a $T_c(r = 0)$ of 78 K after post-annealing, when using relatively long 1 ms pulses from an 800 W CO_2 laser, although RBS showed that large composition gradients existed within the film [69]. A similar effect has been observed when using 6 J cm^{-2} CO_2 pulses with a 100 ns initial spike, and a 2 µs tail containing around 4 J cm^{-2} [70]. In this case EDX revealed a wide variation in the film composition with angle from the target normal, even though (neglecting Ca) the average composition across the layer was found to be the same as in the target. Thus, although it is well known that the amount of superconducting material needed to obtain a convincing T_c value is only a few percent, great care must be taken in using solely one method, or an averaging technique to study the proportion of different phases in multiphase systems such as with Bi films.

Stoichiometric transfer of the Bi system has been possible with both Nd:YAG [71] and excimer lasers [72–74]. Fork et al. [72] have determined that the fluence required for stoichiometric transfer of all the components is in the region of 3 J cm^{-2}, which is considerably higher than used for $Ba_2YCu_3O_7$. In the studies in which stoichiometric transfer is obtained, the 2212 phase is formed by a post-annealing process. Each study uses its own temperature–time processing cycle and it is difficult to accurately compare these since the furnaces used can typically vary by tens of degrees.

Nevertheless, there seems to be a ballpark consensus for temperatures in the region of 850°C for short annealing periods. These schedules produce films that show c-axis orientation with full superconducting onsets between 70 and 80 K. No critical-current measurements were reported in any of these studies.

There have been attempts to grow the 2212 phase by *in situ* techniques. Schmitt and co-workers [75] used a XeCl laser operating at 2 J per pulse with a 60 ns pulse duration and 5 Hz repetition rate. The beam was focused onto a 2212 target to a fluence of 3 J cm^{-2}. The substrates, (100) SrTiO$_3$, were heated to within 10 K of the decomposition temperature of the 2212 phase, and the deposition was performed in an oxygen atmosphere of 0.4–0.8 mbar. After deposition, the films were cooled in this oxygen partial pressure at a rate of 100 K min^{-1} to a final temperature of 100°C. Films grown under these conditions show an onset temperature for superconductivity of 76 K and an allied transition width into the fully superconducting state of 8 K. The best of these films have a highly textured microstructure with the c axis oriented perpendicular to the substrate surface. The critical currents, measured at 4.2 K in zero magnetic field, showed a wide scatter of values reaching a maximum value of 10^6 A cm^{-2}.

Further experiments were performed in this study to observe the effect of heating the as-deposited films. The conditions in the first experiments were recreated, and in this case the films were heated *in situ* for a period of 50 min. These films displayed an increase for the onset temperature for superconductivity to 83 K, but also displayed much broader transitions to the completely superconducting state. This effect is suggested to result from film–substrate interactions. X-ray analysis of these films shows the presence of the 2201 and 2223 phases, as well as other unidentified impurity materials.

There have been many reports that the incorporation of lead (Pb) onto the bismuth site promotes the formation of the 2223 phase over the 2212 phase. This effect has been clearly demonstrated in bulk ceramic samples in several studies [76, 77]. Perhaps the most systematic study of the incorporation of Pb in the growth of thin films is that of Tabata *et al.* [78]. In the first part of this work, undoped films were grown using conditions similar to those discussed above, with an ArF laser delivering 1–10 J cm^{-2} on to the target. The as-grown films, deposited at room temperature in an oxygen partial pressure, were then post-annealed to optimize the production of the 2212 phase. A target with a composition of Bi$_{1.2}$Pb$_{0.35}$Sr$_{0.8}$Ca$_{1.0}$Cu$_{1.6}$O$_y$ was also used in this study. Post-annealing of these room-temperature-deposited layers for 15 min between 840 and 850°C led to a small resistivity changes at 110 K. The amount of 2223 material required to produce this effect, however, was sufficiently small that it could not be detected by X-ray techniques. Prolonged annealing for up to 15 h produced films with higher concentrations of the desired phase, although, on the basis of X-ray studies, it was still a minority phase. The increase in the 2223 content of these layers is reflected in the superior electrical properties, with a T_c onset value of 100 K and a $T_c(r = 0)$ of 90 K.

The relationship between the annealing time and the proportions of 2223 compared with other phases in films has been investigated by a number of groups, including Tseng et al. [79] and Levoska et al. [80]. The consensus of opinion is that nearly single-phase material can be produced by very long annealing times (on the order of 100 h) provided that during annealing the film is contained within an upturned crucible along with some high-quality target material.

An apparent precursor state to the superconducting phase has been observed when the films were deposited onto substrates heated to 400°C [78]. Structural characterization of these films has shown that they are oriented and that the precursor state appears to be an incomplete version of the double Cu—O layer structure of the 2212 phase. This suggests that clusters of the double-layer structure are emitted from the target by the ablation process. This appears to be supported by LIMA, which has recently detected very large clusters, consistent with unit cells, existing in the ablation plume [81].

Variations to the standard laser ablation growth process have been proposed by several groups. Kanai et al. [82] attempted to grow the desired material by successive deposition of each individual layer of the structure. Starting from targets of $Bi_7Pb_3O_y$, $SrCuO_y$ and $CaCuO_y$, they grew films that appear to contain, in addition to the expected materials containing two or three Cu—O planes, phases that apparently contain four and five Cu—O

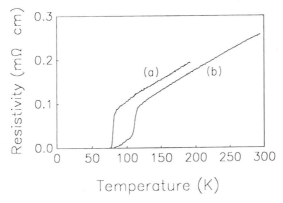

Fig. 15.18 Resistivity curve of Bi(Pb)SrCaCuO films grown by interleaving layers of PbO during ablation by 532 nm Nd:YAG laser pulses, and annealed for 1 h ($T_c \approx 80$ K) (a) and 24 h ($T_c \approx 110$ K) (b) [164].

planes. These structures are unstable and do not survive the high-temperature anneal required to generate the superconductivity. This technique of layer-by-layer successive deposition has been extended to study the effect that chemical substitutions onto the Ca and Sr sites have on the transition temperature [83]. The most interesting substitution was that of Ba onto the Ca site, which resulted in an increase in the onset temperature from 80 to 93 K. Alternate laser ablation deposition of PbO layers with the Bi superconductor in a sandwich structure has been shown to assist in the formation of the 2223 phase [151]. Figure 15.18, for example, shows the resistivity of such a multilayered film, annealed for 1 h ($T_c \approx 80$ K), and for 24 h ($T_c \approx 110$ K). This method has certain advantages over the prolonged annealing technique described above, since it enables independent control over the Pb content in the film and the temperature cycling.

15.4.4 Growth of thallium-containing films

At first sight, the existence of phases in the thallium-containing materials that display some of the highest transition temperatures is an attractive inducement to study the system. The distinct lack of any major activity in this area can probably be explained by reference to its well publicized toxicity problems, in addition to the high volatility of thallium, which leads to synthetic difficulties.

In the first attempt to grow layers of these phases, Johs et al. [84] adopted a two-part strategy. In the first stage, thallium-free films were grown and the thallium was diffused in separately. The films were grown at room temperature on YSZ from $Ba_2Ca_2Cu_3O_7$ and $Ba_2Ca_3Cu_3O_8$ targets irradiated by 532 nm radiation from a Nd:YAG laser. The Tl was introduced via the vapour pressure created by thermal decomposition of Tl_2O_3. The as-grown films were placed on top of a Pt boat containing the Tl_2O_3, which was in turn contained in a quartz boat. The entire ensemble was then placed into a preheated furnace at $900°C$ for 3 min under flowing O_2 and then allowed to cool. The best films produced under this scheme showed zero resistances at 115 K, but degraded rapidly with only small changes in the annealing temperature or duration. X-ray diffraction studies of the best films showed them to be multiphase, $Tl_2Ba_2Ca_1Cu_2O_7$ being the dominant phase, with $Tl_1Ba_2Ca_1Cu_2Cu_3O_9$ among the minority constituents. SEM studies of the microstructure showed plate-like grains of size 5–10 μm, with a predominantly c-axis orientation. Critical-current measurements at 77 K showed that the films can only carry approximately 100 A cm^{-2} in zero magnetic fields and that this value dropped dramatically in the presence of magnetic fields. These characteristics have been improved by Liou et al. [85] by including the Tl in a target of $Tl_2Ba_2Ca_2Cu_3O_y$ but with otherwise identical deposition conditions. The films that were post-annealed at $870^{\circ}C$ in an atmosphere of O_2 for 5 min contained mainly the 2223 and 2212 phases. A further 5 min anneal produced a nearly pure 2223 phase, although X-ray fluorescence microprobe spectroscopy suggested that the films were somewhat deficient in Tl. While the electronic characteristics of these layers were superior to those produced by the vapour processing route they are still not of device quality. The best films showed an onset temperature of 125 K and a $T_c(r = 0)$ of 110 K, with an associated J_c of 10^4 A cm^{-2} measured at 77 K with a zero magnetic field.

One reason for this somewhat disappointing value is that the microstructure of these films consists of poorly connected grains and that in addition broad composition inhomogeneities exist within the film. Attempts to improve their quality have centred on optimizing the post-deposition annealing conditions. To date, the best layers exhibit a T_c of 115 K and J_c of 3.2×10^6 A cm^{-2} measured at 77 K without an applied magnetic field [86]. However, in common with previous films, a major degradation in the value of J_c is observed with the application of a magnetic field.

15.5 SUMMARY

The process of laser ablation of inorganic materials, particularly supercondutors, has been described. Although not well understood, a simple outline of the perceived mechanisms has been presented in terms of absorption, particle ejection and plume dynamics, along with some relevant experimental results.

We have seen how the method is conceptually very simple and relatively insensitive to minor fluctuations in the processing conditions. Growth rates of the order of several ångströms per pulse lead to good quality films of the order of microns in thickness in only minutes. The transition temperatures reported for high-T_c superconductors are usually very close to those achieved in the bulk materials.

Given the choice of electron beam evaporation or reactive magnetron sputtering, from one or more sources, or deposition by molecular or laser beam, we have seen the latter method holding several distinct advantages. It is relatively inexpensive and requires little or no specialist knowledge. The environment does not necessarily have to be a vacuum, as in electron beam processing, and the species arriving at the substrate are totally incorporated into the growing film. There appear to be no problems associated with the control of the film stoichiometry, besides the occasional oxygen deficiency found in many other deposition techniques. A variety of lasers can be used at different wavelengths and pulse durations, although it is clear that the shorter wavelengths and pulse durations are superior.

There are several disadvantages to the laser ablation approach. First, the incident fluence collectively controls the ablation rate, the variety and the energy of ablated species. Particulates remain a mischief rather than an insurmountable problem, but the film areas that can be currently processed by lasers is severely limited by the beam sizes available with present-day technology.

Probably one of the strongest advantages of laser ablation, however, is that it is compatible with the *"in situ"* processing concept mentioned elsewhere in this book. In any subsequent processing steps such as oxidation patterning, annealing and metallization, which would by necessity be low-temperature and "dry", the laser would play an important role.

REFERENCES

[1] H. M. Smith and A. F. Turner, *Appl. Optics* **4**, 147 (1965).
[2] P. D. Zavitsanos, L. E. Brewer and W. E. Sauer, *Proc. Natl Electron. Conf.* **24**, 864 (1968).
[3] V. S. Ban and D. A. Kramer, *J. Mater. Sci.* **5**, 978 (1970).
[4] H. Sankur and J. T. Cheung, *J. Vac. Sci. Technol.* **A1**, 1806 (1983).
[5] J. T. Cheung and T. Magee, *J. Vac. Sci. Technol.* **A1**, 1604 (1983).
[6] C. Cali, V. Daneu, A. Orioli and S. Riva-Sanseverino, *Appl. Optics* **15**, 1327 (1976).
[7] M. I. Baleva, M. H. Maksimov, S. M. Metev and M. S. Sendova, *J. Mater. Sci. Lett.* **5**, 533 (1986).
[8] J. T. Cheung and J. Madden, *J. Vac. Sci. Technol.* **B5**, 705 (1987).
[9] R. Kelly and J. E. Rothenberg, *Nucl. Instrum. Meth. in Phys. Res.* **B7/8**, 755 (1985), and references therein.
[10] R. Srinivason and V. Mayne-Banton, *Appl. Phys. Lett.* **41**, 576 (1982).
[11] D. Dijkkamp *et al.*, *Appl. Phys. Lett.* **51**, 619 (1987); X. D. Wu *et al.*, *Appl. Phys. Lett.* **51**, 861 (1987).
[12] J. E. Andrew, P. E. Dyer and D. Forster, *Appl. Phys. Lett.* **43**, 717 (1983).
[13] H. Dupendant, J. P. Gavigan, D. Givord, A. Lienard, J. P. Rebouillat and Y. Souche, *Appl. Surf. Sci.* **43**, 369 (1989).
[14] H. S. Kwok, D. T. Shaw, Q. Y. Ying, J. P. Zheng, S. Witanachchi, E. Petrou and H. S. Kim, *Proc SPIE* **1187**, 161 (1989).
[15] I. W. Boyd, *Laser Processing of Thin Films and Microstructures* (Springer-Verlag, Berlin, 1987).
[16] J. F. Ready, *Industrial Applications of Lasers* (Academic Press, New York, 1978).
[17] D. E. Aspnes and M. K. Kelly, *IEEE J. Quantum Electron.* **25**, 2378 (1989).
[18] J. V. Armstrong, PhD Thesis, Trinity College Dublin (1990).
[19] L. Gomes, M. M. F. Vieira, S. L. Baldochi, N. B. Lima, M. A. Novak, N. D. Vieira, S. P. Morato, A. J. P. Braga, C. L. Ceaser, A. F. S. Penna and J. Mendes Filho, *J. Appl. Phys.* **63**, 5044 (1988).
[20] A. Inam, X. D. Wu, T. Venkatesan, S. B. Ogale, C. C. Chang and D. Dijkkamp, *Appl. Phys. Lett.* **51**, 1112 (1987).

[21] D. Dijkkamp, T. Venkatesan, X. D. Wu, S. A. Shaheen, N. Jisrawa, Y. H. Min-Lee, W. L. McLean and M. Croft, *Appl. Phys. Lett.* **51**, 619 (1987).

[22] J. Narayan, N. Biunno, R. Singh, O. W. Holland and O. Auchiello, *Appl. Phys. Lett.* **51**, 1845 (1987).

[23] H. S. Kwok, P. Mattocks, L. Shi, X. W. Wang, S. Witanachchi, Q. Y. Ying, J. P. Zheng and D. T. Shaw, *Appl. Phys. Lett.* **52**, 1825 (1988).

[24] N. Savva, K. F. Williams, G. M. Davis and M. C. Gower, *IEEE J. Quantum Electron.* **25**, 2399 (1989).

[25] A. Caruso and R. Gratton, *Plasma Phys.* **10**, 867 (1968).

[26] T. Venkatesan, X. D. Wu, A. Inam and J. B. Wachtman, *Appl. Phys. Lett.* **52**, 1193 (1988).

[27] M. Brown, M. Shiloh, R. B. Jackman and I. W. Boyd, *Appl. Surf. Sci.* **43**, 382 (1989).

[28] L. Lynds, B. R. Weinberger, G. G. Peterson and H. A. Krasinski, *Appl. Phys. Lett.* **52** (1988).

[29] W. Marine, M. Peray, Y. Mathey and D. Pailharey, *Appl. Surf. Sci.* **43**, 377 (1989).

[30] A. M. Bonch-Bruevich and Y. A. Imas, *Sov. Phys. Tech. Phys.* **12**, 1407 (1968).

[31] R. K. Singh and J. Narayan, *Phys. Rev.* **B41**, 8843 (1990).

[32] F. P. Gagliano and U. C. Paek, *Appl. Optics* **13**, 274 (1974).

[33] P. E. Dyer, A. Issa and P. H. Key, *Appl. Phys. Lett.* (1990).

[34] R. K. Singh, P. Tiwari and J. Narayan, *Proc. SPIE* **1187**, 182 (1989).

[35] T. Venkatesan, W. Wu, A. Inam, C. C. Chang, M. S. Hegde and B. Dutta, *IEEE J. Quantum Electron.* **25**, 2388 (1989).

[36] A. Mele, D. Consalvo, D. Stranges, A. Giardini-Guidoni and R. Teghil, *Appl. Surf. Sci.* **43**, 398 (1989).

[37] W. A. Weimer, *Appl. Phys. Lett.* **52**, 2171 (1982).

[38] O. Auciello, S. Athavale, O. E. Hankins, M. Sito, A. F. Schriener and N. Biunno, *Appl. Phys. Lett.* **53**, 72 (1988).

[39] H. S. Kwok, P. Mattocks, L. Shi, X. W. Wang, S. Witanachchi, Q. Y. Ying, J. P. Zang and J. T. Shaw, *Appl. Phys. Lett.* **52**, 1825 (1988).

[40] P. E. Dyer, A. Issa and P. H. Key, *Appl. Surf. Sci.* **46** (1990).

[41] J. T. Cheung and H. Sankur, *CRC Crit. Rev. Solid State Mater. Sci.* **15**, 68 (1988).

[42] P. E. Dyer, R. D. Greenough, A. Issa and P. H. Key, *Appl. Surf. Sci.* **43**, 387 (1989).

[43] S. Miura, T. Yoshitake, T. Satoh, Y. Miyasaka and N. Shohata, *Appl. Phys. Lett.* **52**, 1008 (1988).

[44] J. P. Zheng, Z. Q. Huang, D. T. Shaw and H. S. Kwok, *Appl. Phys. Lett.* **54**, 280 (1989).

[45] D. N. Mashburn and D. B. Geohegan, *Proc. SPIE* **1187**, 172 (1989).

[46] Yu. A. Bykovskii, S. M. Sil'nov, E. A. Sotnichenko and B. A. Shestakov, *Sov. Phys. JETP* **66**, 285 (1987).

[47] W. P. Barr, *J. Phys.* **E2**, 1024 (1969).

[48] S. V. Gaponov, A. A. Gudkov and A. A. Freeman, *Sov. Phys. Tech. Phys.* **27**, 1130 (1982).

[49] D. Lubben, S. A. Barnett, K. Suzuki, S. Gorbatin and J. E. Greene, *J. Vac. Sci. Technol.* **B3**, 968 (1985).

[50] H. Sankur, Personal communication.

[51] R. J. Cava, *Science* **247**, 656 (1990).

[52] A. W. Sleight, *Science* **242**, 1519 (1988).

[53] T. Venkatesan, X. D. Wu, A. Inman, M. S. Hedge, E. W. Chase, C. C. Chang, P. England, D. M. Hwang, R. Krchnavek, J. B. Wachtman, W. L. McLean, R. Levi-Setti, J. Chabala and Y. L. Wang, in *Chemistry of High Temperature Superconductors 11* (ed. D. Nelson and T. F. George) (ACS Symp. Ser. No. 377, 1988).

[54] A. Inman, M. S. Hedge, X. D. Wu, T. Venkatesan, P. England, P. F. Micelli, E. W. Chase, C. C. Chang, J. M. Tarascon and J. B. Wachtman, *Appl. Phys. Lett.* **53**, 908 (1988).

[55] O. Eibl and B. Roas, *J. Mater. Res.* **5**, 2620 (1990).

[56] M. G. Norton and C. B. Carter, *Physica* **C172**, 47 (1990).

[57] T. K. Worthington, W. J. Gallagher and T. R. Dinger, *Phys. Rev. Lett.* **59**, 1160 (1987).

[58] M. Matsumoto, H. Akoh and S. Takada, *J. Appl. Phys.* **66**, 3907 (1989).

[59] A. Gupta, G. Koren, R. J. Baseman, A. Segmuller and W. Holber, *Physica* **C162–164**, 127 (1989).

[60] A. Inam, C. T. Rogers, R. Ramesh, K. Remschnig, L. Farrow, D. Hart, T. Venkatesan and B. Wilkins, *Appl. Phys. Lett.* **57**, 2484 (1990).

[61] C. T. Rogers, A. Iman, M. S. Hedge, B. Dutta, X. D. Wu and T. Venkatesan, *Appl. Phys. Lett.* **55**, 2032 (1989).

[62] T. Venkatesan, A. Inman, B. Dutta, R. Ramesh, M. S. Hedge, X. D. Wu, L. Nazer, C. C. Chang, J. B. Barner, D. M. Hwang and C. T. Rogers, *Appl. Phys. Lett.* **56**, 391 (1990).

[63] Q. Li, X. X. Xi, X. D. Wu, A. Iman, S. Vadlamannati, W. L. McLean, T. Venkatesan, R. Ramesh, J. A. Martinez and L. Nazar, *Phys. Rev. Lett.* **64**, 804 (1990).

[64] X. D. Wu, A. Inman, T. Venkatesan, C. C. Chang, W. E. Chase, P. Barboux, J. M. Tarascon and B. Wilkens, *Appl. Phys. Lett.* **52**, 754 (1988).

[65] P. Tiwari, S. M. Kanetkar, S. Sharan and J. Narayan, *Appl. Phys. Lett.* **57**, 1578 (1990).

[66] D. K. Fork, D. B. Fenner, R. W. Barton, J. M. Phillips, G. A. N. Connell, J. B. Boyce and T. H. Geballe, *Appl. Phys. Lett.* **57**, 1161 (1990).

[67] S. Miura, T. Yoshitake, S. Matsubara, Y. Miyasaka, N. Shohata and T. Satoh, *Appl. Phys. Lett.* **53**, 1967 (1988).

[68] X. D. Wu, A. Inman, M. S. Hedge, B. Wilkins, C. C. Chang, D. M. Hwang, L. Nazar, T. Venkatesan, S. Muira, S. Matsubara, Y. Miyasaka and N. Shohata, *Appl. Phys. Lett.* **54**, 754 (1989).

[69] M. Meskoob, T. Honda, A. Safari, J. B. Wachtman, S. Danforth and B. J. Wilkins, *J. Appl. Phys.* **67**, 3069 (1990).

[70] N. K. Jaggi, M. Meskoob, S. F. Wahid and C. J. Rollins, *Appl. Phys. Lett.* **53**, 1551 (1988).

[71] H. J. Dietze, S. Becker and D. Hirsch, *Z. Phys.* **B78**, 361 (1990).

[72] D. K. Fork, C. B. Eom, G. B. Anderson, G. A. N. Connell, J. B. Boyce, T. H. Gabelle, R. I. Johnson and F. A. Ponce, *Appl. Phys. Lett.* **53**, 337 (1988).

[73] B. F. Kim, J. S. Wallace, L. J. Swartzendruber, J. Bohandy, L. H. Bennett, F. J. Adrian, E. Agostinelli, W. J. Green, K. Moorjani, R. D. Schull and T. E. Phillips, *Appl. Phys. Lett.* **53**, 321 (1988).

[74] C. R. Guarnieri, R. A. Roy, K. L. Saenger, S. A. Shivashanker, D. S. Yee and J. J. Cuomo, *Appl. Phys. Lett.* **53**, 532 (1988).

[75] P. Schmitt, L. Schulz and G. Saemann-Ischenko, *Physica* **C168**, 475 (1990).

[76] M. Takano, J. Takada, K. Oda, H. Kitaguchi, Y. Miura, Y. Ikeda, Y. Tomii, and H. Mazaki, *Japan. J. Appl. Phys.* **27**, L1041 (1988).

[77] M. Shiloh, I. Wood, M. Brown, F. Beech and I. W. Boyd, *J. Appl. Phys.* **68**, 2304 (1990).

[78] H. Tabata, T. Kawai, M. Kanai, O. Murata and S. Kawai, *Japan. J. Appl. Phys.* **28**, L430 (1989).

[79] M. R. Tseng, J. J. Chu, Y. T. Huang, P. T. Wu and W. N. Wang, *J. Appl. Phys.* **29**, 2657 (1990).

[80] J. Levoska, T. Murtoniemi and S. Leppavouri, *J. Less Common Metals* **164/165**, 710 (1990).

[81] F. Saba, A. Sajjadi, F. Beech and I. W. Boyd, *J. Mater. Sci. Engng* B. (1991).

[82] M. Kanai, T. Kawai, S. Kawai and H. Tabata, *Appl. Phys. Lett.* **54**, 1802 (1989).

[83] H. Tabata, O. Murata, T. Kawai and S. Kawai, *Appl. Phys. Lett.* **56**, 1576 (1990).

[84] B. Johs, D. Thompson, N. J. Ianno, J. A. Woollam, S. H. Liou, A. M. Hermann, Z. Z. Scheng, W. Kiehl, Q. Shams, X. Fei, L. Sheng and Y. H. Liu, *Appl. Phys. Lett.* **54**, 1810 (1989).

[85] S. H. Liou, K. D. Aylesworth, N. J. Ianno, B. Jons, D. Thompson, D. Mayer, J. A. Wollam and C. Barry, *Appl. Phys. Lett.* **54**, 760 (1989).

[86] H. Itozaki, S. Tanaka, K. Higaki, K. Harada, S. Yazu and K. Tada, in *Proceedings of 1st ISTEC Workshop on Superconductivity, Oiso, Japan*, p. 63 (1989).

[87] J. T. Cheung and D. T. Cheung, *J. Vac. Sci. Technol.* **21**, 182 (1982).

[88] E. Fogarassy, A. Salaoui, C. Fuchs and J. P. Stoquert, *Appl. Surf. Sci.*, **46**, 195 (1990).

[89] E. Fogarassy, E. Fuchs, A. Salaoui and J. P. Stoquert, *Appl. Phys. Lett.* **57**, 664 (1990).

[90] J. J. Dubowski, D. F. Williams, P. B. Sewell and P. Norman, *Appl. Phys. Lett.* **46**, 1081 (1985).

[91] R. Pielmeier, D. Bollmann and K. Haberger, *Appl. Surf. Sci.* **46**, 163 (1990).

[92] N. Biunno, J. Krishnaswamy, S. Sharan, L. Ganapathi and J. Narayan, *Proc. Mater. Res. Soc.* **158**, 477 (1990).

[93] C. N. Afonso, R. Serna, F. Catalina and D. Bermejo, *Appl. Surf. Sci.* **46**, 249 (1990).

[94] A. Giardini-Guidoni, A. Morone, M. Snels, E. Desimoni, A. M. Salvi, R. Fantoni, W. C. M. Berden and M. Giorgi, *Appl. Surf. Sci.* **46**, 321 (1990).

[95] J. J. Dubowski and D. F. Williams, *Appl. Phys. Lett.* **44**, 339 (1984).

[96] H. Sankur, F. Woodberry, R. Hall and W. J. Gunning, *Proc. Mater. Res. Soc.* **77**, 727 (1987).

[97] J. A. R. Samson, J. P. Padur and A. Sharma, *J. Opt. Soc. Am.* **57**, 966 (1967).

[98] J. J. Dubowski, *Proc. SPIE* **668**, 97 (1986).

[99] Honig, R. E., *Appl. Phys. Lett.* **3**, 8 (1963).

[100] J. Kliwer, *J. Appl. Phys.* **44**, 490 (1973).

[101] S. V. Gaponov, A. A. Gudkov, B. M. Luskin, V. I. Luchin, and N. N. Salaschenko, *Sov. Tech. Phys. Lett.* **5**, 195 (1979).

[102] S. V. Gaponov, B. M. Luskin, B. A. Nesterov and N. N. Salashchenko, *Sov. Phys. Solid State* **19**, 1736 (1977).

[103] H. Osterreicher, H. Bittner and B. Kothari, *J. Solid State Chem.* **26**, 97 (1978).
[104] S. V. Gaponov, B. M. Luskin and N. N. Salashchenko, *Solid State Commun.* **39**, 301, 1981; *JETP Lett.* **33**, 517 (1981).
[105] S. V. Gaponov, E. B. Klyuenkov, B. A. Nesterov, N. N. Salashchenko and M. I. Kheifets, *Sov. Tech. Phys. Lett.* **5**, 193 (1979).
[106] P. Maier-Komor, *Nucl. Instrum. Meth.* **167**, 73 (1979).
[107] M. S. Hess and J. F. Milkosky, *J. Appl. Phys.* **43**, 4680 (1972).
[108] Yu. A. Nokolaev and A. N. Orevskii, *Sov. Phys. Tech. Phys.* **23**, 578 (1978).
[109] P. E. Dyer, S. A. Ramsden, J. A. Sayers and M. A. Skipper, *J. Phys.* **D9**, 373 (1976).
[110] P. Schenck, L. Cook, J. Zhao, J. Hastie, E. Farabaugh, C. Chiang and M. Vaudin, *Proc. Mater. Res. Soc.*, 157 (1989).
[111] J. Gavigan, in *Proceedings of NATO ASI on the Science and Technology of Nanostructured Magnetic Materials, June–July 1990.*
[112] J. Martin, L. Vasquez and F. Bernard, *Appl. Phys. Lett.* **57**, 1742 (1990).
[113] P. Murray, J. Wolf, J. Mescher and J. Grant, *Mater. Lett.* **5**, 250 (1987).
[114] J. P. Gavigan, D. Givord, A. Lienard, J. Rebouillat and Y. Souche, *J. Appl. Phys.* (1991).
[115] A. Baver, J. Ganz, K. Hesse and E. Koher, *Appl. Surf. Sci.* **46**, 113 (1990).
[116] K. Schmatjko, B. Roas, G. Enders and L. Schults, *Proc. SPIE*, **1279**, 135 (1990).
[117] R. Sheftal and I. Cherbakov, *Cryst. Res. Tech.* **16**, 887 (1981).
[118] D. Dimitrov, S. Metev, I. Gugov and V. Kozhkharov, *J. Mater. Sci. Lett.* **1**, 334 (1982).
[119] H. Sankur and H. Tourtellotte, *Appl. Optics* **24**, 3343 (1985).
[120] H. Schwartz and H. Tourtellotte, *J. Vac. Sci. Technol.* **6**, 3763 (1969).
[121] J. Desserre and J. Eloy, *Thin Solid Films* **29**, 29 (1975).
[122] S. M. Metev and M. S. Sendova, *Proc. SPIE* **1033**, 260 (1988).
[123] B. Knox, *Mater. Res. Bull.* **3**, 329 (1986).
[124] H. Sankur, *Mater. Res. Soc. Symp. Proc.* **29**, 373 (1984).
[125] H. Sankur, *Ann. Rep.* No. 2, *Air Force Office of Scientific Research Contract* No. F49620-84-C-0091 (1986).
[126] W. Marine, J. M. Scotto.D'Aniello and J. Marfaing, *Appl. Surf. Sci.* **46**, 239 (1990).
[127] P. Wei, D. Nelson and R. Jall, *J. Chem. Phys.* **62**, 3050 (1975).
[128] M. Hanabusa, S. Moriyama and H. Kikuchi, *Thin Solid Films* **107**, 227 (1983).
[129] A. Giardini-Guidoni, A. Morone and M. Snels, *J. Mater. Sci.*
[130] S. Otsubo, T. Maeda, T. Minamikawa, Y. Yonezawa, A. Morimoto and T. Shimizu, *Japan. J. Appl. Phys. Lett.* **29**, L133 (1990).
[131] G. Groh, *J. Appl. Phys.* **39**, 5804 (1968).
[132] S. Fujimori, T. Kasai and T. Inamura, *Thin Solid Films* **92**, 71 (1982).
[133] G. Hass and J. Ramsey, *Appl. Optics* **8**, 1115 (1969).
[134] M. Mitrikov, D. Dimitrov and S. Metev, *J. Phys.* **D17**, 1563 (1984).
[135] Yu. Bykovskii, A. Dudolandov, V. Kozlenkov and P. Leont'ev, *JETP Lett.* **20**, 135 (1974).
[136] B. Pashmakov, D. Nesheva, E. Vateva, D. Dimitrrov and S. Metev, *J. Mater. Sci. Lett.* **4**, 442 (1985).
[137] S. Gaponov, A. Gudkov, B. Luskin and V. Luchin, *Sov. Phys. Tech. Phys.* **26**, 598 (1981).

[138] J. Cheung, G. Niizawa, J. Moyle, N. Ong, B. Paine and T. Vreeland, *J. Vac. Sci. Technol.* **A4**, 2086 (1986).
[139] P. Zavitsanos and W. Sauer, *J. Electrochem. Soc.* **115**, 109 (1968).
[140] Yu. Bykovskii, A. Dudoladov, V. Kozlenkov and P. Leont'ev, *Sov. Phys. Tech. Phys.* **22**, 1041 (1977).
[141] S. Hansen and T. E. Robitaille, *Appl. Phys. Lett.* **50**, 359 (1987).
[142] H. Yang and J. Cheung, *J. Cryst. Growth* **56**, 429 (1982).
[143] A. Archer, B. Tanguy, S. Amirhaghi and I. W. Boyd, To be published.
[144] A. Stephen, T. Zrebiec and V. Ban, *Mater. Res. Bull.* **9**, 1427 (1974).
[145] J. Cheung, E. Cirlin and N. Otsuka, *Appl. Phys. Lett.*
[146] S. Gaponov, B. Luskin and N. Salaschchenko, *Sov. Phys. Tech. Phys.* **5**, 210 (1979).
[147] S. Gaponov, B. Luskin and N. Salaschchenko, *Sov. Phys. Semicond.* **14**, 873 (1980).
[148] P. E. Pehrsson and R. Kaplin, *J. Mater. Res.* **4**, 1480 (1989).
[149] H. Sankur and J. T. Cheung, *Appl. Phys.* **A47**, 271 (1988).
[150] J. M. Wrobel and J. T. Duboski, *Appl. Phys. Lett.* **55** (1989).
[151] A. Sajjadi, S. Kilgallon, F. Beech, F. Saba and I. W. Boyd, *Electron. Lett.* **27**, 345 (1991).
[152] H. O. Sankur and W. Gunning, *Appl. Optics* **28**, 2806 (1989).
[153] K. L. Saenger, *J. Appl. Phys.* **66**, 4435 (1989).
[154] J. P. Gavigan, D. Givord, A. Lienard, J. P. Rebouillant and Y. Souche, To be published.
[155] J. P. Gavigan, D. Givord, A. Lienard, O. F. K. McGrath, J. P. Rebouillant and Y. Souche, To be published.
[156] B. Brauns, D. Schafer, R. Wolf and G. Zscherpe, *Optica Acta* **33**, 545 (1986).
[157] M. Popescu, I. Apostol, I. N. Mihaelescu, T. Botila, E. Pentia, M. L. Ciurea, M. Dinescu, J. Jaklovski, G. Aldica, L. Miu, C. Rusu, A. Hening, S. Mihai, C. Constantin, T. Stoica, P. Pausescu, E. Cruceanu, W. Pompe, R. Wunsch, A. Richter and H. J. Scheibe, *Proc. SPIE* **1033**, 287 (1988).
[158] S. V. Gaponov and S. A. Gusev, *Optics Commun.* **38**, 7 (1981).
[159] C. L. Marquardt, R. T. Williams and D. J. Nagel, *Proc. Mater. Res. Soc.* **38**, 325 (1985).
[160] C. B. Collins, F. Davanloo, E. M. Juengerman, D. R. Jander and T. J. Lee, *Proc. SPIE* **1190**, 78 (1989).
[161] J. Krishnaswamy, A. Rengan, A. R. Srivatsa, G. Matera and J. Narayan, *Proc. SPIE* **1190**, 109 (1989).
[162] N. Biunno, J. Narayan and A. R. Srivatsa, *Proc. SPIE* **1190**, 118 (1989).
[163] A. Sajjadi, K. Kuen-Lau, F. Saba, F. Beech and I. W. Boyd, *Appl. Surf. Sci.* **46**, 85 (1990).
[164] P. E. Dyer, R. D. Greenough, A. Issa and P. H. Ke, *Appl. Phys. Lett.* **53**, 534 (1988).
[165] D. H. A. Blank, W. A. M. Aarnink, J. Flokstra, H. Rogalla and A. van Silfhout, *J. Less Common Metals* **164/165**, 1178 (1990).
[166] R. de Reus, F. W. Saris, G. J. van der Kolk, C. Witmer, B. Dam, D. H. A. Blank, D. J. Adelerhof and J. Flokstra, *Mat. Sci & Eng.* **B7**, 135 (1990).
[167] P. Macquart, J. Corno and C. Mahe, *J. Optics (Paris)*, **21**, (1990).
[168] P. Macquart, F. Bridou and B. Pardo, *Thin Solid Films*, (1991).
[169] P. Macquart, Fabrication par ablation laser et caractérisation de couches

minces de matérieux utilisables par le fabrication de multicouches pour les optiques X-UV. PhD thesis, University of Paris XI, Orsau (1990).

[170] G. Koren, R. J. Baseman, A. Gupta, M. I. Lutwyche and R. B. Laibowitz, *Appl. Phys. Lett.*, **56**, 2144 (1990).

[171] H. Chiba, K. Murakami, O. Eryu, K. Shihoyama, T. Mochizuki and K. Masuda, *Japan. J. Appl. Phys.* (1991)

16 Fast *In Situ* Metallization: A Comparison of Several Methods with Possible Applications in High-Density Multichip Interconnects

MARCEL WIDMER, PATRIK HOFFMANN,
BAUDOUIN LECOHIER, HERBERT SOLKA,
JEAN-MICHEL PHILIPPOZ
and HUBERT VAN DEN BERGH

Laboratoire de Chimie Technique, Ecole Polytechnique Fédérale de Lausanne, Switzerland

16.1 INTRODUCTION

Interconnecting integrated circuits can be done either by costly wafer-scale integration, or by loading a printed circuit board with multiple individually packaged chips. Concerning the latter case, the decrease in feature size at the printed circuit board level has lagged behind size reductions in VLSI technology in recent years. This may result in the chip-to-chip interconnect lengths becoming too large, and interconnect delay times may limit the maximum chip operating frequency. Furthermore, large distances between chips cause power losses and increase the possibility of cross-talk. An alternative solution to multichip interconnection is the so-called high-density multichip module (MCM) [1]. Placing chips close together in clusters with compact interconnections in three dimensions, using multiple alternating layers of insulator and metal contacts (above or below the chips), has many advantages. As MCMs with high-density interconnects (HDI) have as high as possible line, via hole and chip packing density, the interconnect lengths are minimized. Cross-talk is further reduced by choosing low-dielectric-constant insulating material. Power distribution is also simplified in such HDIs, although care must be taken in the thermal design, since several watts may be dissipated per chip. In some cases, using

Photochemical Processing of Electronic Materials
ISBN 0-12-121740-X

laser writing, the interconnects may be programmed for quick turnaround of small series and prototyping. Repair of interconnects and replacing chips is then also facilitated. Finally, the overall size reduction and the possibility of hermetic sealing are also advantages.

In mutiple-layer structures as shown in Fig. 16.1, via holes can be laser-drilled and subsequently filled with metal. The in-plane interconnects between via holes, at least in part, could be made by flexible computer-controlled laser writing. As of the order of one metre or more of such metal lines are needed per module, economical writing speeds are in excess of 1 cm s^{-1}. The material of these connecting lines would be mainly copper or aluminium, with dimensions of width, height and pitch respectively of the order of 10, 5 and 25 μm. If the insulating layers are polyimide (PI), the direct writing process is limited to substrate temperatures below 400°C.

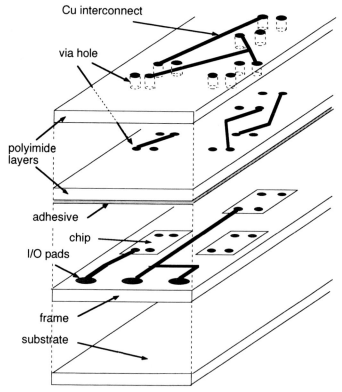

Fig. 16.1 Schematic of a multilayer multichip module. Bottom: substrate and frame serve as a support for the chips; I/O lines and power are supplied by the frame. Middle and top: interconnects between the different chips are made respectively on and through polyimide overlay layers.

Low-dielectric-constant ceramics or dielectrics, which can sustain higher temperature processes like fast pyrolytic laser chemical vapour deposition (LCVD), may also be considered as insulating layers.

Three different approaches to the fast laser writing of the "in-plane" interconnects have been tried in our laboratory [2] and are reported below. The first is classical pyrolytic LCVD of copper from its bis(hexafluoro-acetylacetonate) [Cu(hfa)$_2$]. Compared with previous LCVD experiments with this compound [3–7], in this paper we shall show how this process is pushed to the limits of high writing speeds, while trying to maintain acceptable electrical resistivity, morphology and adhesion to the substrate. Clearly, this is a high-temperature process that cannot be carried out on a PI substrate. The second approach is laser writing of metal lines from a metal-containing precursor that is in the form of a solid film placed on the surface of the substrate. With respect to previous similar studies on different metals including Au [8–12], Cu [13–16], Pd [17, 18] and Ag [19], the novelty of this work is the use of metal cluster coordination complexes as precursors, mainly [Ir$_6$(CO)$_{15}$] [N(CH$_3$)$_3$(C$_8$H$_{17}$)]$_2$ and Au$_{55}$(PΦ_3)$_{12}$Cl$_6$ (where Φ stands for the phenyl group $-$C$_6$H$_5$). Such compounds have been chosen for their high metal content, which implies less "shrinkage" upon transformation to metal, and favourable thermochemistry for this decomposition. The latter implies that there may be a "window" of conditions at which fast *in situ* metallization may occur at temperatures low enough to be feasible even on PI. The third approach consists of dividing the metallization into two steps. In the first step, a thin "prenucleation" film of metal is deposited on selected areas of the substrate. This has already been done by laser [15, 20–22], but other laser beam [23], electron [24–28] or ion beam [18, 29–32] processes seem to be feasible and can be quite fast since only minute quantities of precursor are changed to metal. In a second step, a thick metal film is built up selectively on top of those parts of the surface that have the prenucleation layer. The second step can be done by relatively slow parallel processing of several substrates simultaneously. Different methods for prenucleating the surface and for "developing" these thin films to micron-thick conducting metal lines are discussed.

16.2 LASER CHEMICAL VAPOUR DEPOSITION (LCVD)

Laser CVD of metal from metal bis(hexafluoroacetylacetonates) was pioneered in our laboratory [33, 34]. Similar and other coordination complex compounds were already well established for low-pressure CVD of metals [35–38]. Platinum bis(hexafluoroacetylacetonate) [Pt(hfa)$_2$] was chosen to study the mechanism of photolytic and pyrolytic LCVD [39–44].

The $Cu(hfa)_2$ may still be today the best precursor for pyrolytic LCVD of copper, although another copper complex appears to be also of interest [45]. For the present purpose, we want to produce copper lines of 10×5 μm^2 cross-section at writing speeds of the order of 1 cm s^{-1}. The resistivity ratio should be 2 or less, and the lines should adhere well to the substrate surface. This is by no means a trivial task, and the pyrolytic LCVD of $Cu(hfa)_2$ must be pushed close to its practical limits to reach these goals. The conditions of line cross-section and writing speed indicated above necessitate very fast pyrolysis indeed which implies quite high temperatures. Thus a PI substrate is excluded and oxide or ceramic substrates must be sought. Here, as throughout in this paper, we have used Pyrex glass, which was heated in concentrated nitric acid for two hours and finally extensively rinsed with doubly distilled water. This procedure takes off all organic impurities and soluble parts of the glass surface in order to render it chemically comparable to SiO_2 surfaces.

For this work the substrate is placed in a heatable thermostated cell [6], which contains the solid $Cu(hfa)_2$ loaded on a U-shaped tube. The cell is then evacuated to below 10^{-2} mbar, closed and heated to a temperature T_c, while the U-shaped tube is heated to T_u. The substrate temperature is equal to the cell temperature T_c. T_u is measured with a thermocouple and determines the vapour pressure of the $Cu(hfa)_2$. It is always inferior to T_c in order to avoid condensation of the copper complex on the cell window, the cell walls and the substrate in the cell. The $Cu(hfa)_2$ is either synthesized [46–48] or purchased commercially (Strem Chemicals). The latter is grey-blue and does not contain water according to the producer. Before use, the compound is exposed to air and its colour changes to green due to formation of an adduct with water. This green compound is used for the LCVD experiments reported here. As the exact composition of the $Cu(hfa)_2$ is not known (unhydrated, monohydrated or dihydrated) as a function of the experimental conditions (temperature, quantity present in the cell, etc.), it is hard to determine the vapour pressure as a function of the temperature. Indications of the vapour pressure of $Cu(hfa)_2$ were given in [49, 50], but no comments on the degree of hydration were made. At the highest temperatures attained, condensation of the $Cu(hfa)_2$ (green) on the cell window becomes somewhat of a problem, and is avoided by shielding the window with a hot metal cap with a small slit through which the laser beam passes. Additional heating by means of IR lamps is also applied. A CW argon ion laser is used in the visible. The beam is expanded and subsequently focused by a single lens. For the nickel-seeded experiments (see below), the beam waist was determined by moving a thin slit perpendicular to the beam with a stepper motor and measuring the intensity of the transmitted light. This technique allows us to record directly the intensity distribution of the beam [51–53], and we obtained a $1/e^2$ diameter of

13 μm. Due to the nonideal properties of the optical system, the shape and diameter of the focused beam differ substantially from a Gaussian intensity distribution, and from the beam waist calculated by the paraxial theory using the equation

$$2\omega_0 = \frac{4}{\pi} \frac{\lambda f}{d}, \qquad (16.1)$$

where $2\omega_0$ is the $1/e^2$ beam diameter, λ the wavelength, f the focal length of the lens and d the diameter of the laser beam in the focal plane of the lens. If not mentioned explicitly, the beam diameter given in this chapter are calculated. For direct writing, the cell is moved perpendicular to the laser beam either by a stepper motor or by a synchronous motor. The lines are written between two gold-coated nickel dots. The resistivity ρ of the deposited copper lines is obtained from their resistance measured between these two points contacted with Cu/Be needles, and their average profile measured with a stylus profilometer (Tencor Instruments). For low values, a correction for the resistance of the cables was applied. The reported resistivity ratios ρ/ρ_0 give the line resistivity divided by that of pure bulk copper. All resistivities are given at room temperature. In some LCVD experiments, a thin layer of an optically absorbing substance (in our case nickel) is vacuum-evaporated on the surface of the Pyrex substrate. The thickness of this thin layer is measured with a quartz microbalance.

An essential variable for the fast pyrolytic Cu deposition from $Cu(hfa)_2$ is the vapour pressure of the precursor controlled by T_u [6]. In Fig. 16.2 we show the extension of the measurements of the resistivity ratio of the copper lines to higher vapour pressures and consequently higher writing speeds. Similar "U-shaped" curves of resistivity ratio against writing speed have been observed in the pyrolytic LCVD of Pt from $Pt(hfa)_2$ and also in the writing of gold lines from a film of cluster coordination compounds on the substrate surface (see below). The high resistivity ratios obtained for a set of experimental conditions (vapour pressure, laser beam intensity at the surface) at the lower and higher writing speeds can be due to poor morphology and/or incomplete removal of ligand material from the copper deposit. It should be noted in Fig. 16.2 that the writing speed at which the best (lowest) resistivity ratio is found (i.e. the lowest point in the U-shaped curve) increases essentially proportionally with the vapour pressure of $Cu(hfa)_2$ over the range of conditions applied. For the time being, at the highest pressure indicated of about 100 mbar, writing speeds of several millimetres per second are possible while retaining reasonable ρ/ρ_0 values. LCVD of copper from $Cu(hfa)_2$ at the wavelengths between 457 and 514.5 nm used here is essentially a pure "pyrolytic" process, in the sense that gaseous $Cu(hfa)_2$ does not absorb light at these wavelengths in a single-photon process to a significant extent. On an essentially transparent

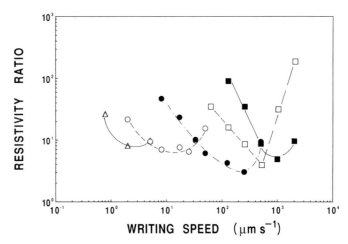

Fig. 16.2 Resistivity ratio as a function of the writing speed at five vapour pressures of Cu(hfa)$_2$; the laser power is 0.3 W: $\triangle$, $P(\text{Cu(hfa)}_2) = 0.1$ mbar, λ = all lines visible, $2\omega_0' \approx 4$ μm; $\bigcirc$, $P(\text{Cu(hfa)}_2) = 1.5$ mbar, λ = all lines visible, $2\omega_0 \approx 4$ μm; $\bullet$, $P(\text{Cu(hfa)}_2) = 12$ mbar, λ = all lines visible, $2\omega_0 \approx 4$ μm; $\square$, $P(\text{Cu(hfa)}_2) = 30$ mbar, $\lambda = 514.5$ nm, $2\omega_0 \approx 2$ μm; $\blacksquare$, $P(\text{Cu(hfa)}_2) = 100$ mbar, $\lambda = 514.5$ nm, $2\omega_0 \approx 2$ μm. The beam waists are calculated.

substrate like Pyrex, it is thus not clear how the process starts. In our case, laser-induced pyrolysis is generally initiated on a gold-coated Ni dot. Thus relatively long lines (several millimetres) are written so that the line can be taken to be characteristic of having been written on a transparent glass, in the absence of a strongly absorbent metal. It is not well understood how the writing of metal lines on a transparent substrate progresses [54]. To some extent—and this may depend on the metal and the applied conditions—part of the deposited metal line is so strongly heated by the laser beam that it is driven off. Thus small particles of metal might "seed" the surface ahead of the laser beam and provide some degree of optical absorption necessary for surface heating and triggering the pyrolytic process. Otherwise, of course there is always a growth of the *hot* copper deposit, which contributes to the line growth in all directions. From our previous experiments with so-called hybrid LCVD with Pt(hfa)$_2$ [39, 40, 42, 44], one could surmise that some extent of additional surface seeding, i.e. a covering of the surface with a minute thin film of a highly optically absorbing substance, might assist a pyrolytic process [12, 54]. Such surface coverings need not significantly alter the electrical properties of the substrate, as they may be done with metals, semiconductors or even nonconducting ceramic films, but influences on electrical properties of the written lines cannot be excluded. As mentioned above, these thin layers have the task of absorbing the light strongly and thus heating the surface.

They may also be used in principle to improve the adhesion between metal line and substrate.

Hence we have seeded Pyrex surfaces with different substances and the first results showing the influence of the seeding on the line height, width and resistivity ratio are shown in Figs 16.3, 16.4 and 16.5. Figure 16.3

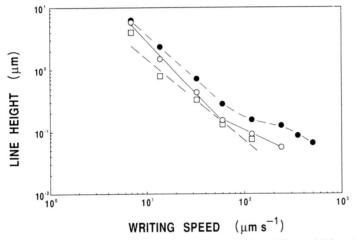

Fig. 16.3   Line height as a function of the writing speed at different Ni layer thicknesses on the Pyrex substrate; the laser power is 0.3 W, λ = all lines visible and $2\omega_0 = 13$ μm: □, 0 Å Ni; ○, 15 Å Ni; ●, 45 Å Ni.

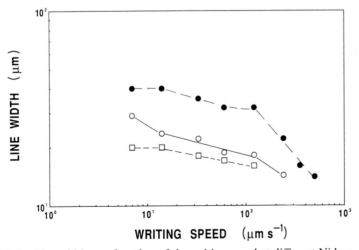

Fig. 16.4   Line width as a function of the writing speed at different Ni layer thicknesses on the substrate. Other conditions as in Fig. 16.3.

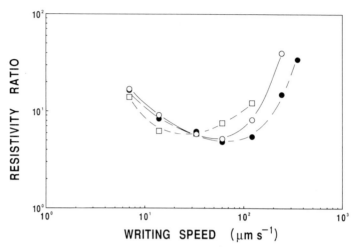

Fig. 16.5 Resistivity ratio as a function of the writing speed at different Ni layer thicknesses on the substrate. Other conditions as in Fig. 16.3.

shows the line height of a Cu line written with a laser power of 0.3 W (all lines visible) focused to a $1/e^2$ diameter of 13 μm at the substrate surface. The three sets of data are taken on a "clean" Pyrex surface, and on a Pyrex surface onto which respectively 15 and 45 Å of Ni have been evaporated under vacuum. It can clearly be seen that the line height increases significantly with increasing surface seeding. Typically, an average line height of 0.08 μm was found at a writing speed of about 100 μm s^{-1} on the unseeded surface, whereas on the surface seeded with 45 Å Ni the same line height was found at 500 μm s^{-1}. The same holds for the linewidth, as can be seen from Fig. 16.4. Finally, Fig. 16.5 shows the shift in the resistivity ratio minimum to higher writing speeds with surface seeding. The optimum resistivity ratio under the conditions applied occurs at about 20 μm s^{-1} on an unseeded surface, whereas on both seeded surfaces the minimum is marginally lower (i.e. improved resistivity ratio) and at significantly higher writing speed (about 60 μm s^{-1}). We are presently testing to see if these surface seeding effects can also be found under different experimental conditions, in particular at higher precursor vapour pressures and higher writing speeds.

An extreme effect of how surface seeding can assist pyrolytic LCVD is shown in Fig. 16.6. Here again the substrate is Pyrex glass. The "clean" surface shows up as the dark part on the photograph, whereas the greyish parts are those that have been seeded with a thin layer of silver clusters that were "soft-landed" on the surface [55]. The picture indicates how pyrolytic LCVD (in this case Pt from Pt(hfa)$_2$) can be turned on and off by the presence of this seeding layer. The thin silver layer has been scratched by hand-

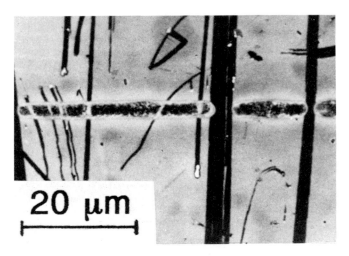

Fig. 16.6 Photograph of a Pt line deposited on a substrate partially covered with silver clusters. The vapour pressure of Pt(hfa)$_2$ is 1 mbar, the laser power is 0.105 W at 514.5 nm, $2\omega_0 = 4$ µm and the writing speed is 250 µm s^{-1}. The deposition is turned on and off by the scratches in the silver coating layer.

ling; the weak laser power (105 mW) and high writing speed (250 µm s^{-1}) used provide conditions under which the minute scratches on the surface, where the Ag clusters have been removed, turn the pyrolytic process on and off.

Another parameter that was varied in these pyrolytic LCVD experiments is the gas added in the cell. Several partial pressures of different gases (He, Ar, H$_2$, O$_2$) were added in the hope of finding effects of scavenging of ligand material or thermodiffusion. Added gases do influence growth rates and morphology, as can be seen in Fig. 16.7. Here the copper line formation in the presence of two different H$_2$ pressures (10 and 100 mbar) is compared with the line formation without H$_2$. All lines were written under identical experimental conditions. The addition of 10 mbar of H$_2$ already reduces the line width from 45 to 22 µm; at 100 mbar, the width is further reduced to 18 µm. For the line height, a similar decrease is observed. This reduction in size is accompanied by a small improvement in the resistivity ratio (11 at 0 mbar H$_2$, 7.8 at 10 mbar and 7.1 at 100 mbar).

As stated above, the vapour pressures cited here are uncertain to some extent. Since the increased vapour pressure is attained by increasing both T_u and T_c, at some point a temperature is reached at which the Cu(hfa)$_2$ decomposes on the timescale of the experiment. To avoid this occurring at too low temperatures, we are now coating the cell with gold so that the Cu(hfa)$_2$ will be in contact only with gold, quartz and Kalrez O-rings.

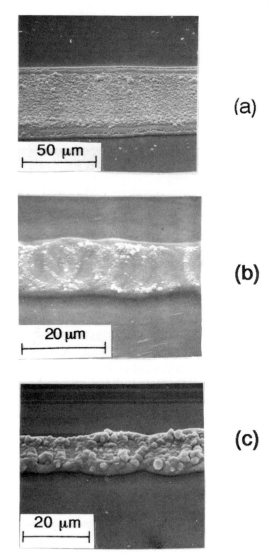

Fig. 16.7 Scanning electron micrographs (tilt angle $50°$) of copper stripes taken at three different H_2 pressures; $P(Cu(hfa)_2) = 1$ mbar, λ = all lines visible, the laser power is 0.3 W, $2\omega_0 = 4.5$ μm (calculated) and the writing speed is 12 μm s^{-1}: (a) 0 mbar H_2, height 2 μm, width 45 μm, resistivity ratio 11; (b) 10 mbar H_2, height 1.5 μm, width 22 μm, resistivity ratio 7.8; (c) 100 mbar H_2, height 0.8 μm, width 18 μm, resistivity ratio 7.1.

All in all, the desired copper lines with a height of a few microns, a width of about ten microns and a resistivity of less than a few times the bulk value may be attainable by pyrolytic LCVD from a precursor like Cu(hfa)$_2$ at writing speeds of the order of 1 cm s^{-1}, and tests are under way to try to demonstrate this.

16.3 LASER-INDUCED DECOMPOSITION OF A SOLID LAYER ON THE SUBSTRATE

In pyrolytic LCVD the deposition rate is limited in principle by two main processes [56]: first, the kinetics of the surface decomposition reaction, which includes effects like the temperature dependence of the sticking coefficient and of the surface decomposition itself; and, secondly, the transport of the precursor molecules and decomposition products to and from the surface. The fact that the LCVD rate may be limited by insufficient transport of the precursor to the surface as shown for deposition of Pt [33] and Cu [6] indicated that it might be worthwhile to replace the vapour deposition process by one in which the precursor is put directly *on* the surface in the form of a solid layer. One way to approach this is to take metal-containing screen inks that upon laser heating decomposes to volatile organic material and leave a metal deposit on the surface [8–12]. Another approach is to take a coordination complex or metal salt precursor in solution and to paint, spray or spin it onto the surface [13–19]. The precursor is transformed into metal and volatile products by locally heating the layer with a focused laser beam. Moving the laser across the surface thus results in a metal line. The remaining nondecomposed film is then dissolved. The requirements for such a process are that (a) the film is fairly homogeneous, (b) the precursor film does not evaporate prior to decomposition, (c) decomposition leads to volatile products, (d) it decomposes rapidly at relatively low temperatures, and (e) the precursor film is transparent enough for the laser beam to start the metallization at the surface–layer interface. Point (a) implies that we want films if possible without microcrystallites since these can lead to inhomogeneities in the metal film upon metallization. Copper formate was one of the substances that appeared to fulfil these requirements and that gave interesting and promising results [14, 15], even at high writing speeds (in the order of several centimetres per second) [16]. Several other compounds in this class were tried without success, including copper oxalate and copper acetate. Some of these failures were due to problems in the layer preparation. It is interesting to note that this process, in contrast with LCVD, can be performed in air, so that the substrate covered with the precursor solid film does not need to be enclosed in a controlled-atmosphere cell. Possibly, heating and subsequent cooling by the rapidly moving laser beam is sufficiently fast to avoid oxidation of the copper deposited. Due to differences of the copper density in the $Cu(HCOO)_2$ film and in bulk metal copper, a volume "shrinkage" by a factor of 11 is observed during metallization. Less shrinkage may be obtained by using cluster compounds, which have a higher metal content. The first experiments were done with Ir_4 and Ir_6 cluster carbonyl compounds ($[Ir_4(CO)_{11}Br][N(C_2H_5)_4]$ and $[Ir_6(CO)_{15}]$ $[N(CH_3)_3(C_8H_{17})]_2$ respectively) [16]. Volume shrinkage upon laser-induced pyrolytic metallization of the Ir_6 cluster carbonyl compound is about a factor of 6.7. Iridium was chosen for several reasons. First it has high melting and boiling points of respectively 2410 and 4130°C, so that removal of the metal at the line centre, where the light intensity and consequently the temperature are highest, is relatively unlikely. Furthermore, its electrical resistivity is only 5.3 $\mu\Omega$ cm at room temperature. Also, IrO_2 is not formed at low temperature and is unstable at high temperature; even if formed in small quantities, this oxide is still a fair conductor and hence it is not very harmful to the overall electrical properties of the Ir line. Finally, Ir has no known carbide, so that, even if ligand material is decomposed to carbon, the latter may be not too strongly bound to the metal. Results of fast laser writing in these Ir complex films have been reported previously [16]. In summary, resistivity ratios of about 20 can be attained at speeds close to 1 cm^{-1} s. The laser power in these experiments is low (15–250 mW at 488 nm). Line heights between 400 and 1200 Å are obtained for films 0.8 μm high, and line widths are about 5–30 μm. In these preliminary experiments, neither the wavelength and power density nor the choice of ligands and counterions were varied over a wide range to optimize conditions. Such Ir compounds can also be used for surface seeding, as will be discussed below. It will be worthwhile to see if with such a compound laser surface seeding can be achieved under a wider window of conditions than is the case for palladium acetate films [20, 57].

The main emphasis in this chapter is on new results obtained with an even larger coordination cluster compound: $Au_{55}(P\Phi_3)_{12}Cl_6$ [58]. Experimental results are reported for $\lambda = 1.32$ μm where the precursor films are quite transparent and for $\lambda = 514.5$ nm, where they already absorb significantly, as can be seen on the absorption spectrum of such a film shown in Fig. 16.8.

The cluster compound $Au_{55}(P\Phi_3)_{12}Cl_6$ is dissolved in CH_2Cl_2 and the solution is centrifuged and filtered (P4) before use. Thick layers (0.5–7 μm) are produced by dropping the saturated solution of $Au_{55}(P\Phi_3)_{12}Cl_6$ in CH_2Cl_2 onto the Pyrex or silicon substrate and evaporating the solvent at room temperature with the sample covered by a glass funnel. Thin layers (0.02–0.3 μm) are spun on from different diluted solutions: 20 μl of sol-

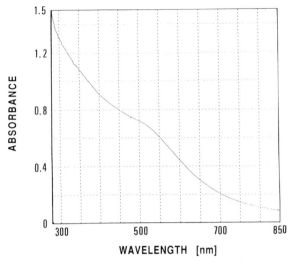

Fig. 16.8 UV–visible absorption spectrum of a layer of the Au_{55} cluster compound on a Pyrex substrate. The layer thickness is 0.23 ± 0.04 μm. On the long-wavelength side, the absorbance remains low up to 3 μm.

ution were dropped on the sample rotating at 1000–4000 rev min⁻¹ from a distance of 3 mm above the centre of the sample. This procedure resulted in homogeneous layers, both at high dilution (thin layers) and low dilution (thicker layers). Increasing the rotational speed decreased the layer thickness. An influence on the layer quality of the polyethylene syringe used to dose the solution onto the sample cannot be excluded.

The thickness of the layer was measured with a stylus profilometer after locally removing it by scratching with a metal needle. Selective washing with CH_2Cl_2 was also tried for thickness measurement, but abandoned because it tends to crystallize the gold cluster compound. The stylus pressure on the sample must be chosen as small as possible (1 mg in our case) to avoid errors in the thickness measurement due to scratching of the layer by the stylus itself.

Samples were irradiated with IR and visible laser beams. A CW Nd:YAG laser operating at 1.32 μm was used for the IR.

For the irradiations in the visible, the 514.5 nm line of a CW argon ion laser was used. The beam was focused by means of a laser microscope onto the sample mounted on a motorized three-axis translation system. The beam intensity profile in the focal plane was measured by scanning a "knife edge" perpendicular to the propagation axis and recording the transmitted light with a photodiode; this technique provides the integral of the intensity distribution, and the $1/e^2$ diameter ω can be calculated from the distance between the 10% and 90% of the measured increase or decrease in intensity [59]. One advantage of our knife edge compared with our slit technique is that two mutually perpendicular beam diameters can be easily measured by moving a square "knife edge" along one of its diagonals; this allows monitoring of the astigmatism of the optical system. In our case, some astigmatism was present, the image of the beam was an ellipse that changed its orientation by $90°$ when the sample was moved along the beam propagation axis from the sagittal to the tangential focus. The values of the $1/e^2$ diameter measured for the two foci (1.9 μm) agree with the result calculated by the paraxial theory, whereas at the circle of least confusion we measured a diameter of 3 μm. We have located the source of this optical error, and it will be corrected in the near future.

The electrical resistance of the resulting lines was determined as above using the two-point method. The line height was measured with the stylus profilometer and its width by optical and/or scanning electron microscopy.

Figure 16.9 shows some pictures of the Au_{55} cluster compound layers. (a) is an enlargement of the surface where the slight pressure of the tip of the stylus profilometer initiated premature crystallization upon measuring the layer height (2 μm in this case). (b)–(d) show Au_{55} cluster compound layers prepared with different spinning speeds from the same CH_2Cl_2 solution; the best layer (d) with the smallest number of visible microcrystallites was spun on at the highest speed. (e) shows gold lines written at different speeds in the layer of (d). The first set of experiments described was done with a CW YAG laser. The height of the spun-on precursor layer was 0.23 ± 0.04 μm. Figures 16.10, 16.11 and 16.12 show respectively the height, width and resistivity ratio of the gold lines as functions of the writing speed at two laser beam powers, while Fig. 16.13 shows a SEM picture of such a gold line written at a speed of 10 mm s^{-1}. At writing speeds of a few centimetres per second, line heights close to 0.04 μm were obtained, which corresponds to about 17% of the precursor layer height. It has not yet been proved if this "shrinkage", which is close to the theoretically estimated value, will be the same for much thicker precursor layers, and this will be tested in the near future. Since at wavelengths between 1 and 1.3 μm the precursor layer is quite transparent, much higher metal lines with improved aspect ratio (i.e. the ratio between the height and width of the structure) may well be possible. The larger line heights and increasing error bars observed at the highest writing speeds may be due to incomplete decomposition of the precursor; this hypothesis is supported by the high resistivities found under these conditions. In the lower range of speeds tested here, the resistivity ratio of the gold lines is also relatively high. In this case, this could well be due to poor morphology. In the range between 1 and 2 cm s^{-1}, the resistivity ratio is well below 10, and may still be significantly optimized—for instance by improving the metallization conditions or by another choice of ligands on the cluster compound.

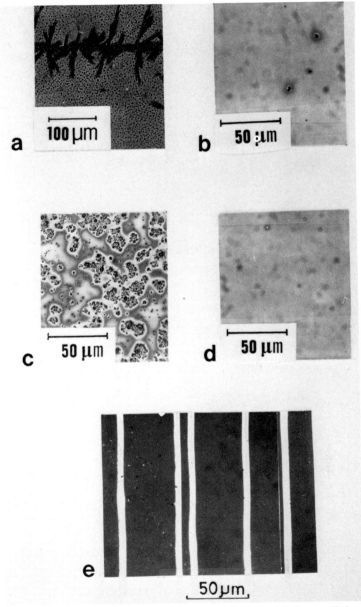

Fig. 16.9 Layers of $Au_{55}(P\Phi_3)_{12}Cl_6$ on a Pyrex substrate: (a) the tip of the stylus profilometer initiated crystallization during the measurement of its 2 μm height; (b) Au_{55} layer spun on at 800 rev min^{-1} (thickness 0.22 ± 0.04 μm); (c) Au_{55} layer spun on at 1100 rev min^{-1} (thickness 0.23 ± 0.08 μm); (d) Au_{55} layer spun on at 3550 rev min^{-1} (thickness 800 Å); (e) lines written in the layer shown in (d) at speeds of 50, 50, 50, 10 and 5 mm s^{-1} (left to right), with λ = 514.5 nm, laser power 15 mW and measured focal diameter $2\omega_0$ = 3 μm.

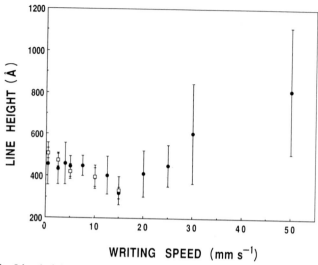

Fig. 16.10 Line height as a function of the writing speed. The height of the Au_{55} cluster compound layer on a Pyrex substrate is 0.23 ± 0.04 μm. The lines are written by a CW Nd:YAG laser at $\lambda = 1.32$ μm with laser power 0.87 W(●) and 0.67 W(□). The error bars represent the 67% confidence intervals on five measurement for each line.

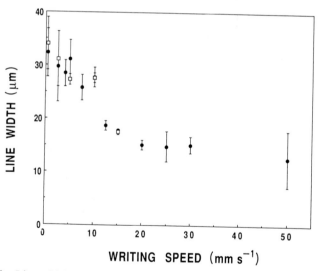

Fig. 16.11 Line width as a function of the writing speed. Other conditions as in Fig. 16.10.

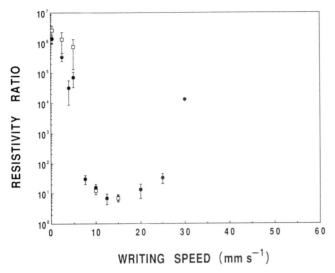

Fig. 16.12 Resistivity ratio as a function of the writing speed. Other conditions as in Fig. 16.10.

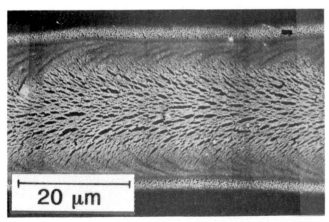

Fig. 16.13 Scanning electron micrograph of a gold line written in a Au_{55} cluster compound layer on a Pyrex substrate. The writing speed is 10 mm s^{-1}. Other conditions as in Fig. 16.10.

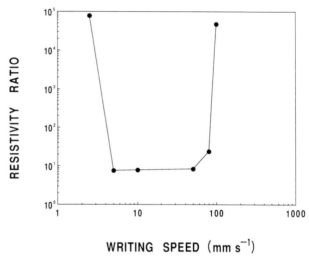

WRITING SPEED $(mm\ s^{-1})$

Fig. 16.14 Resistivity ratio as a function of the writing speed. The height of the Au_{55} cluster compound layer on a Pyrex substrate is 800 ± 20 Å. The lines are written by a CW Ar^+ laser at 514.5 nm, laser power 15 mW and $2\omega_0 = 3$ μm.

The second series of experiments shows results obtained at 514.5 nm with the CW Ar^+ laser. These still quite preliminary results indicate that, at the writing speeds of interest for HDI, line widths between 2 and 10 μm can be obtained for lines that show a "shrinkage" close to the expected value. Resistivity ratios between 7 and 25 can be obtained over a wide range of writing speeds (see Fig. 16.14). The increase at low and high writing speeds is extremely pronounced. An investigation of this behaviour is in progress. Some typical lines written under these conditions are shown in Fig. 16.9(e). It should be noted that, in all these experiments with precursor surface films, the laser powers necessary for fast writing are quite low compared with the LCVD case.

For both the gold and the iridium compounds, there is a window of conditions where fast laser writing on polyimide substrates is possible. The line widths attained and the real possibility of writing much higher lines than those reported here simply by starting from thicker precursor layers implies that this approach may be feasible, in principle, for HDI.

16.4 FAST *IN SITU* PRENUCLEATION FOLLOWED BY SELECTIVE METAL BUILD UP

The methods described in Sections 16.2 and 16.3 are both hampered by the fact that relatively large amounts of the metal-containing precursor must be transformed into pure metal in the *in situ* processing step. This is difficult to do at the high writing speeds of interest, especially if good morphology, low chemical contamination due to remaining ligand material, and good adhesion to the surface are required. A further restriction is that the process should be carried out at the lowest temperature. One solution to this problem is to make the metal lines in two stages. In the first stage, the substrate surface is prepatterned with a thin metal film, typically of order of 1 nm thick, for instance by decomposing a few surface monolayers of a cluster compound like the iridium complexes mentioned above. Hence, in the direct writing step, only a minute quantity of precursor has to be transformed to metal, so that high speeds may be possible even at relatively low processing temperatures. The remaining nonprocessed compound is then removed with a solvent. In the second part of the metallization process, the essentially nonconducting thin prenucleation lines must now be built up to high-quality metal lines of a few microns in height. The important requirement of the build-up process is that it occurs selectively, i.e. exclusively on top of the prenucleated parts of the surface. This selectivity implies that many devices that have been sequentially *in situ* patterned with prenucleation lines can be built up simultaneously in a parallel processing unit. Consequently, this parallel processing step may be relatively slow compared with the *in situ* processing step. The prenucleation step need not be carried out using a laser to decompose a surface precursor layer: other possibilities might be, for instance, laser deposition from the gas phase (LCVD) or writing with electron [24–28] or ion [18, 29–32] beams. In the latter case, it may be possible to write prenucleation lines at quite high resolution. The aspect ratio that can be obtained will depend, as in all selective deposition processes, on the preferential build up perpendicular to the surface, a phenomenon that is not yet understood. Treating larger areas using the prenucleation technique and projection patterning, either with UV [21] or possibly even X-rays, also appears to be possible. The build up of the thin prenucleation lines to thick metallic conductors can be done by several techniques. One possibility is electroless deposition from a solution [15, 20, 57]. In this process, the metal build up may be extendable by using the metal lines as electrodes in an electroplating step. This is somewhat related to the method of laser-induced electrodeposition [60–65]. Another possibility is decomposition from a solid precursor film [66] that has been deposited on top of the partially prenucleated surface of the device as will be described below. A third possibility is the selective low-pressure CVD using a gaseous precursor compound, and was done, for example, for Al [21], Mo [67], W [68], and as will be described briefly below for copper.

In our laboratory, we have made, among others, thin iridium films by laser-decomposing a surface layer of the Ir_4 cluster compound $[Ir_4(CO)_{11}Br]$

[N(C$_2$H$_5$)$_4$]. After dissolving the nonprocessed metal complex with acetone, such prenucleated Ir lines could be selectively developed by low-pressure CVD with Cu(hfa)$_2$ or by electroless deposition from a copper bath (Shipley).

In order to study the selective build-up process in detail, we made thin prenucleation film by evaporation of pure metals under vacuum. These films show highly reproducible and homogeneous properties over large area. The first selective metal build-up process discussed here is the decomposition of a solid copper formate layer on a platinum prenucleation film [66]. It is shown schematically in Fig. 16.15. For these experiments, 10 Å of Pt was vacuum-evaporated through a contact mask (2 mm thick with 300 μm wide slits) onto a Pyrex substrate. To improve adhesion [69], the locally prenucleated Pyrex surface was then covered with γ-aminopropyltriethoxysilane and dried. A room temperature saturated solution of Cu(HCOO)$_2$ in water was then sprayed onto the surface held at 50°C during the repetitive

a) Metal deposition

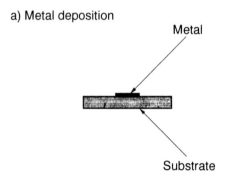

b) Precursor layer deposition

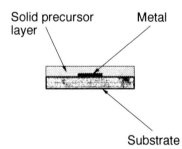

Fig. 16.15 Schematic of two-step metallization via delineated Pt prenucleation and selective build up by decomposition of copper formate. (a) First step: metal deposition by, for example, laser- or particle-beam-assisted metal deposition (in our

spraying/solvent evaporation cycle. The resulting homogeneous layer (probably $Cu(HCOO)_2 \cdot 2H_2O$) was about 10 μm thick. The substrate treated in this way was placed in a cell, which was evacuated to 10^{-2} mbar and then filled with slightly more than 1 bar of H_2 (>99.99%). The substrate was then heated to the desired "developing" temperature somewhere in the range of 70–130°C. In this temperature range, the build up of a copper metal line of about 0.3 μm in height was always found to be selective, i.e. the decomposition of $Cu(HCOO)_2 \cdot 2H_2O$ to copper took place exclusively on top of the prenucleated Pt films. The remaining nonmetallized film was then removed by dissolving in water or by blowing it away with compressed air. Resistivity ratios ρ/ρ_0 of lines obtained in this way were as low as 1.9 [66]. The height of these copper lines was measured with a stylus profilometer. If the copper formate film is not completely converted to metal above the prenucleated surface, the height of the copper line attained after a certain time may be representative of the growth rate

c) Development

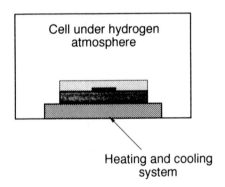

Cell under hydrogen atmosphere

Heating and cooling system

d) Cleaning

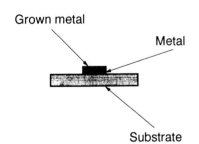

Grown metal

Metal

Substrate

experiment, Pt evaporation was used). (b) Deposition of the solid precursor layer. (c) Second step: selective development in a reducing H_2 atmosphere. (d) Cleaning. For further details, see text.

for the decomposition of copper formate on essentially a copper surface. In that case, a logarithmic plot of the line height divided by the development time as function of the inverse temperature shows Arrhenius-like behaviour. Such a graph is shown in Fig. 16.16 for two different development times. At the highest temperatures and the longest development times used, we are probably approaching the limiting line height attainable under these conditions. It was not possible with our apparatus to reach the temperature limit for loss of selectivity. The apparent activation energies E_{act} observed for both development times are close to 50 kJ mol^{-1}, which is a factor of two to three lower than the value measured for the decomposition of anhydrous crystalline copper formate [70]. Similar catalytic effects are also found in LCVD processing of metals [71]. Among the interesting questions raised are the influences of varying seeding material and morphology on the catalytic decomposition of different precursors. Cluster compounds, similar to those described in Section 16.3, may also in principle be used for this development step.

Development of micron-high metal lines on prenucleated Pt with a commercial electroless copper bath was also successful, but will not be discussed here in detail.

As mentioned above, another way of depositing metal lines several microns in height selectively on top of a locally prenucleated surface is by using low-pressure CVD. Here we take as an example the selective

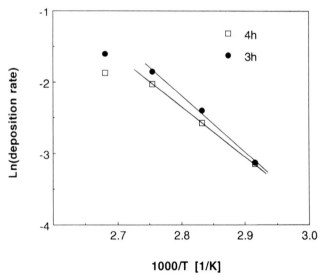

Fig. 16.16 Arrhenius plot. The deposition rate is defined as the copper deposit thickness divided by the development time.

decomposition of copper from gaseous Cu(hfa)$_2$ on Pyrex surfaces. Two electron microscope grids were placed on these surfaces prior to the evaporation of 2.5–20 Å of Pt. One served as an evaporation mask, while the other was a grid covered by 400 Å of amorphous carbon to determine the morphology of the Pt prenucleation films by transmission electron microscopy (TEM). Evaporated Pt on amorphous carbon may not behave similarly to Pt on a Pyrex surface. In later experiments, we plan to deposit a thin layer of SiO$_2$ on the electron microscope grid to make a surface more similar to that of Pyrex or quartz but that can still be used for TEM observations. The evaporations are done using the pendant drop method [72], where a high-purity metal rod (Métaux Précieux) hanging vertically above the substrate is heated by an electron beam up to the melting point. This method provides well controlled and predictable evaporation rates [73], typically of about 90 Åh^{-1} for Pt under our conditions, and the evaporation rate was checked by measuring some thick layers with the stylus profilometer. Thus we can easily deposit small quantities of metal, corresponding to thicknesses of a few angstroms. On the carbon surface, island building is not observed for most Pt layers tested. Only for layers as thin as 2.5 Å islands become apparent to some degree.

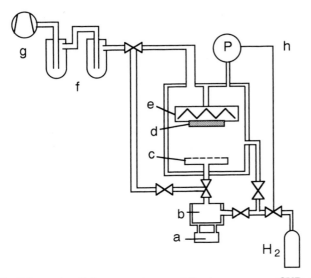

Fig. 16.17 Schematic of the apparatus used for low-pressure CVD experiments with Cu(hfa)$_2$: (a) thermostat for the cell containing the MO precursor; (b) thermostated cell containing the precursor; (c) "shower"-type gas diffuser; (d) substrate, which has been locally prenucleated; (e) heating element for substrate; (f) liquid nitrogen traps; (g) rotary pump; (h) pressure measuring device, which controls the valve regulating the carrier flow rate.

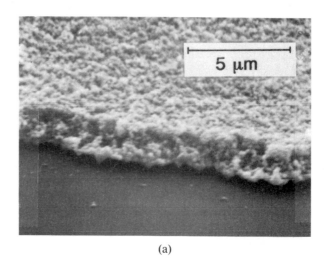

(a)

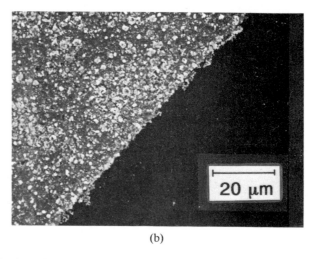

(b)

Fig. 16.18 Scanning electron micrographs of two copper deposits obtained by selective low-pressure CVD from $Cu(hfa)_2$. The prenucleation layer was 5 Å Pt vacuum-evaporated. Copper build up was done in 2.5 h with a flow of 25 sccm of a mixture of H_2 and $Cu(hfa)_2$ at a total pressure of 1 mbar. The substrate temperature was $250°C$ (a) and $290°C$ (b); the sample (a) was cleaned in an ultrasonic bath filled with acetone. The deposit thicknesses are 1.2 μm (a) and 0.67 μm (b).

The development of the Pyrex substrate with the Pt prenucleation pads is as follows. The substrate is placed in the reactor cell of the low-pressure CVD apparatus shown in Fig. 16.17. While the substrate is heated to a temperature between 250 and 300°C, hydrogen gas (>99.99%) at a pressure of 1 mbar is flown through the LPCVD cell, by-passing the cell containing the $Cu(hfa)_2$. When the system has reached thermal equilibrium, the bypass is closed and the hydrogen gas is passed through the cell containing the $Cu(hfa)_2$, which is at 60°C. The vapour pressure of the $Cu(hfa)_2$ at this temperature is estimated at 0.2 mbar. The total flow rate through the LPCVD chamber is 25 sccm. The copper deposition times are about 2.5 h to build-up a micron-high copper film on the Pt-seeded pads. The apparatus is then cooled to room temperature, again using the H_2 by-pass. In

I) Cover surface with catalytic layer

II) Pattern by laser ablation : note ! you only have to ablate
   ~ a monolayer

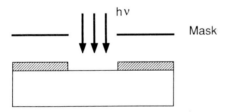

III) Selective build-up

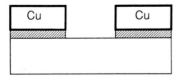

Fig. 16.19   Schematic of the two-step metallization by large-area patterning with an excimer laser through a mask followed by selective build up.

some cases, the substrate is then cleaned with diluted HCl or cleaned in an
ultrasonic bath. The deposited copper adheres well to the surface (Scotch
tape test). The resistivity is measured using a four-point method [74]. The
resisitivity ratios obtained are about 1.5. Selectivity in the deposition was
found for all Pt prenucleation layers between 2.5 and 25 Å.

Scanning electron micrographs of the approximately 1 μm high copper
film formed by selective LPCVD on Pt are shown in Fig. 16.18. Note that
the edge of the Cu on top of the Pt appears to be quite vertical. This is presently under investigation. If vertical development is really possible, this
implies that copper or other metallic features may be written at quite high
resolution, so that the creation of submicron nucleation lines with particle
or X-ray beams may be of interest.

Large-area patterning using the prenucleation technique and, for
instance, lamps or excimer lasers may also be feasible. Here the whole
surface would be covered with a thin layer of metal or metal-organic. In
the latter case, exposure through a mask would lead to partial metallization, and the nonphotodecomposed material would be washed away. The
whole surface is then developed selectively. In the former case, the surface
is coated with, for instance, a Pt metal monolayer, and parts of this
monolayer are ablated through a mask, leaving the unexposed parts to be
built up in a second step. This is shown schematically in Fig. 16.19.

16.5 FAST *IN SITU* PROCESSING BY A PARTIAL CHANGE OF THE PRECURSOR LAYER

Another possible approach to fast *in situ* processing is briefly mentioned
here. Again, this is a multistep approach. However, rather than changing
a small quantity of precursor to metal by complete decomposition of a very
thin layer of precursor, as has been described above, we now suggest only
a partial change of a thick layer of some metal-containing precursor in the
in situ step. This partial conversion should not require high intensity of a
particle or photon beam. In the *in situ* step, some property of the precursor
layer exposed to the beam must change, such as its vapour pressure or its
solubility in some solvent. Following exposure, the nonirradiated part of
the layer can then be selectively removed by taking advantage of the
modified property. The remaining lines of partially developed precursor
can then be further metallized in a slow parallel processing step. A hypothetical example of such a process is shown in Fig. 16.20. A preliminary
test using iridium-containing layers as described above and a 50 kV electron
beam appeared to indicate that such a process may be feasible.

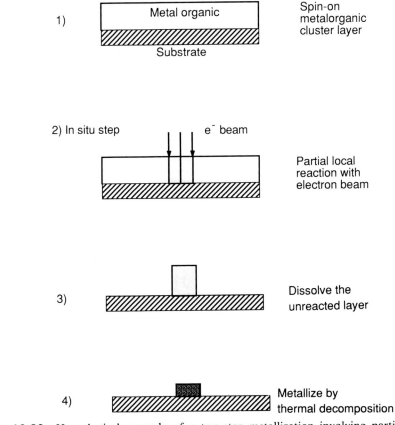

Fig. 16.20   Hypothetical example of a two-step metallization involving partial conversion of the precursor layer followed by removal of the nonirradiated parts and complete metallization of the remaining patterns.

16.6   CONCLUSIONS

Several approaches have been discussed for fast processing, with possible applications to high-density interconnects. Pyrolytic copper deposition from gaseous copper bis(hexafluoroacetylacetonate) leads to writing speeds of a few millimetres per second, with acceptable quality of copper. Further improvements that have been indicated above could lead to 1 cm s^{-1} or faster speeds. This is clearly a high-temperature process, which could not be carried out on polyimide. The use of metal cluster compounds in the form of solid layers with thermal laser beam *in situ* processing also appears quite promising for the interconnect purpose. Both the chemicals used and the applied experimental conditions remain to be optimized. Such a process might even be possible on polyimide. The cluster compounds appear to present several advantages, such as less shrinkage on development and favourable thermochemistry. The multistep processes also look promising. Several metals have been used as prenucleation agents for later selective build up in a second step. It is not clear at present which other materials can be selected for prenucleation. Adhesion may play a role in the selection of the optimal prenucleating agent; this remains to be investigated. All three build-up steps discussed here (deposition from a solid layer, electroless deposition from solution and low-pressure CVD from the gas phase) seem to lead to reasonably high-quality copper. Further experiments are required to assess which process would be the most convenient, the most selective, and would give the best aspect ratios. Finally, we have suggested the possibility of making metal lines by using a partial conversion of a metal-containing precursor in the *in situ* step to attain high writing speeds. Some of the processes mentioned may also be applicable to processing at much higher resolution.

ACKNOWLEDGEMENTS

The authors are grateful to G. Schmid for providing the Au_{55} cluster compound, to R. Roulet for helpful discussions on metal-containing precursors, to W. Harbich for the deposition of silver clusters on Pyrex substrates, to Y. S. Liu and H. S. Cole for stimulating discussions on high-density interconnects, to E. sz Kováts for recommendations concerning the cleaning of Pyrex substrate. Several figures in this chapter have previously appeared in the same or in a slightly modified form in [2] and are reproduced here with the kind permission of SPIE. We thank the Swiss National Fonds for supporting this work in both Divisions II and IV (PN 13 and PN 24). We also thank the CERS and Asea Brown Boveri for financial and material support.

REFERENCES

[1] (a) Y. S. Liu, in *Laser Microfabrication* (ed. D. J. Ehrlich and J. Y. Tsao), p. 3 (Academic Press, Boston, 1989).
(b) R. R. Johnson, *IEEE Spectrum* No. 34 (1990).
[2] P. Hoffmann, M. Widmer, B. Lecohier, H. Solka, H. van den Bergh, *Proc. SPIE* **1352**, 29 (1990).
[3] F. A. Houle, C. R. Jones, T. Baum, C. Pico and C. A. Kovac, *Appl. Phys. Lett.* **46**, 204 (1985).

[4] D. Braichotte, K. Ernst, R. Monot, J.-M. Philippoz, M. Qiu and H. van den Bergh, *Helv. Phys. Acta* **58**, 879 (1985).

[5] F. A. Houle, R. J. Wilson and T. H. Baum, *J. Vac. Sci. Technol.* **A4**, 2452 (1986).

[6] B. Markwalder, M. Widmer, D. Braichotte and H. van den Bergh, *J. Appl. Phys.* **65**, 2470 (1989).

[7] F. A. Houle, T. A. Baum and C. R. Moylan, in *Laser Chemical Processing for Microelectronics* (ed. K. G. Ibbs and R. M. Osgood), p. 25 (Cambridge University Press, 1989).

[8] G. J. Fisanik, M. E. Gross, J. B. Hopkins, M. D. Fennell, K. J. Schnoes and A. Katzir, *J. Appl. Phys.* **57**, 1139 (1985).

[9] G. J. Fisanik, J. B. Hopkins, M. E. Gross, M. D. Fennell and K. J. Schnoes, *Appl. Phys. Lett.* **46**, 1184 (1985).

[10] M. E. Gross, G. J. Fisanick, P. K. Gallagher, K. J. Schnoes and M. D. Fennell, *Appl. Phys. Lett.* **47**, 923 (1985).

[11] M. E. Gross, A. Appelbaum and K. J. Schnoes, *J. Appl. Phys.* **60**, 529 (1986).

[12] K. W. Beeson and N. S. Clements, *Appl. Phys. Lett.* **53**, 547 (1988).

[13] R. B. Gerassimov, S. M. Metev, S. K. Savtchenko, G. A. Kotov and V. P. Veiko, *Appl. Phys.* **B28**, 266 (1982).

[14] A. Gupta and R. Jagannathan, *Appl. Phys. Lett.* **51**, 2254 (1987).

[15] H. G. Müller, *Appl. Phys. Lett.* **56**, 904 (1990).

[16] P. Hoffmann, B. Lecohier, S. Goldoni and H. van den Bergh, *Appl. Surf. Sci.* **43**, 54 (1989).

[17] M. E. Gross, A. Appelbaum and P. K. Gallagher, *J. Appl. Phys.* **61**, 1628 (1987).

[18] M. E. Gross, W. L. Brown, J. Linnros and H. Funsten, in *High Energy Processes in Organometallic Chemistry* (ed. K. S. Suslick), p. 290 (American Chemical Society, Washington D.C., 1987).

[19] A. Auerbach, *J. Electrochem. Soc.* **132**, 1437 (1985).

[20] H. S. Cole, Y. S. Liu, J. W. Rose and R. Guida, *Appl. Phys. Lett.* **53**, 2111 (1988).

[21] G. E. Blonder, G. S. Higashi and C. G. Fleming, *Appl. Phys. Lett.* **50**, 766 (1987).

[22] G. S. Higashi, *Appl. Surf. Sci.* **43**, 6 (1989).

[23] D. J. Ehrlich and J. Y. Tsao (eds), *Laser Microfabrication* (Academic Press, Boston, 1989).

[24] S. Matsui and K. Mori, *J. Vac. Sci. Technol.* **B4**, 299 (1986).

[25] R. B. Jackman and J. S. Foord, *Appl. Phys. Lett.* **49**, 196 (1986).

[26] R. R. Kunz and T. M. Mayer, *Appl. Phys. Lett.* **50**, 962 (1987).

[27] H. W. P. Koops, R. Weiel, D. P. Kern and T. H. Baum, *J. Vac. Sci. Technol.* **B6**, 477 (1988).

[28] H. G. Craighead and L. M. Schiavone, *Appl. Phys. Lett.* **48**, 1748 (1986).

[29] G. M. Shedd, H. Lezec, A. D. Dubner and J. Melngailis, *Appl. Phys. Lett.* **49**, 1584 (1986).

[30] L. R. Harriott, K. D. Cummings, M. E. Gross and W. L. Brown, *Appl. Phys. Lett.* **49**, 1661 (1986).

[31] J. S. Ro, A. D. Dubner, C. V. Thompson and J. Melngailis, *J. Vac. Sci. Technol.* **B6**, 1043 (1988).

[32] P. G. Blauner, J. S. Ro, Y. Butt and J. Melngailis, *J. Vac. Sci. Technol.* **B7**, 609 (1989).

[33] M. Qiu, R. Monot and H. van den Bergh, *Scientia Sinica* **A27**, 531 (1984).
[34] D. Braichotte and H. van den Bergh, in *Laser Processing and Diagnostics* (ed. D. Bäuerle), p. 183 (Springer-Verlag, Berlin, 1984).
[35] M. J. Rand, *J. Electrochem. Soc.* **120**, 686 (1973).
[36] J. M. Morabito and M. J. Rand, *Thin Solid Films* **22**, 293 (1974).
[37] M. J. Rand, *J. Electrochem. Soc.* **122**, 811 (1975).
[38] L. A. Ryabova, in *Current Topics in Materials Science*, Vol. 7 (ed. E. Kaldis), p. 586 (North-Holland, Amsterdam, 1981).
[39] D. Braichotte and H. van den Bergh, in *Proceedings of International Conference on Lasers '85* (ed. C. P. Wang), p. 688 (Society for Optical and Quantum Electronics) (STS Press, McLean, 1986).
[40] D. Braichotte and H. van den Bergh, in *Proceedings of NATO Workshop on Emerging Technologies for* in-situ *Processing, Cargèse, 1987.*
[41] D. Braichotte and H. van den Bergh, *Appl. Phys.* **A44**, 353 (1987).
[42] D. Braichotte and H. van den Bergh, *Appl. Phys.* **A45**, 337 (1988).
[43] C. Garrido-Suarez, D. Braichotte and H. van den Bergh, *Appl. Phys.* **A46**, 285 (1988).
[44] D. Braichotte and H. van den Bergh, *Appl. Phys.* **A49**, 189 (1989).
[45] C. G. Dupuy, D. B. Beach, J. E. Hurst and J. M. Jasinski, *Chem. Mater.* **1**, 16 (1989).
[46] J. T. Adams and C. R. Hauser, *J. Am. Chem. Soc.* **66**, 1220 (1944).
[47] R. L. Belford, A. E. Martell and M. Calvin, *J. Inorg. Nucl. Chem.* **2**, 11 (1956).
[48] J. A. Bertrand and R. I. Kaplan, *Inorg. Chem.* **5**, 489 (1966).
[49] W. R. Wolf, R. E. Sievers and G. H. Brown, *Inorg. Chem.* **11**, 1995 (1972).
[50] D. Braichotte and H. van den Bergh, in *Integrated Optics* (ed. H. P. Nolting and R. Ulrich), p. 38 (Springer-Verlag, Berlin, 1985).
[51] L. D. Dickson, *Opt. Engng.* **18**, 70 (1979).
[52] E. C. Brookman, L. D. Dickson and R. S. Fortenberry, *Opt. Engng.* **22**, 643 (1983).
[53] D. K. Cohen, B. Little and F. S. Luecke, *Appl. Optics* **23**, 637 (1984).
[54] M. M. Oprysko and M. W. Beranek, *J. Vac. Sci. Technol.* **B5**, 496 (1987).
[55] W. Harbich, S. Fedrigo, F. Meyer, D. M. Lindsay, G. Lignières, J.-C. Rivoal and D. Kreisle, *J. Chem. Phys.* **93**, 8535 (1990).
[56] H. J. Zeiger, D. J. Ehrlich and J. Y. Tsao, in *Laser Microfabrication* (ed. D. J. Ehrlich and J. Y. Tsao), p. 285 (Academic Press, Boston, 1989).
[57] Y. S. Liu and H. S. Cole, *Proc. Mater. Res. Soc. Symp.* **154**, 11 (1989).
[58] G. Schmid, R. Pfeil, R. Boese, F. Bandermann, S. Meyer, G. H. M. Calis and J. W. A. van der Velden, *Chem. Ber.* **114**, 3634 (1981).
[59] Y. Suzaki and A. Tachibana, *Appl. Optics* **14**, 2809 (1975).
[60] R. H. Micheels, A. D. I. Darrow and R. D. Rauh, *Appl. Phys. Lett.* **39**, 418 (1981).
[61] R. J. von Gutfeld, M. H. Gelchinski and L. T. Romankiw, *J. Electrochem. Soc.* **130**, 1840 (1983).
[62] A. K. Al-Sufi, H. J. Eichler, J. Salk and H. J. Riedel, *J. Appl. Phys.* **54**, 3629 (1983).
[63] F. Friedrich and C. J. Raub, *Galvanotechnik* **77**, 2658 (1986).
[64] H. R. Khan, M. U. Kittel and C. J. Raub, *Plating and Surface Finishing*, Vol. 75, p. 58 (1988).
[65] L. Nanai, I. Hevesi, F. V. Bunkin, B. S. Luk'yanchuk, M. R. Brook, G. A.

Shafeev, D. A. Jelski, Z. C. Wu and T. F. George, *Appl. Phys. Lett.* **54**, 736 (1989).

[66] B. Lecohier and H. van den Bergh, *Appl. Surf. Sci.* **43**, 61 (1989).

[67] N. Lifshitz, D. S. Williams, C. D. Capio and J. M. Brown, *J. Electrochem. Soc.* **134**, 2061 (1987).

[68] R. H. Wilson and A. G. Williams, *Appl. Phys. Lett.* **50**, 965 (1987).

[69] M. Wick, G. Kreis and F. H. Kreuzer, in *Ullmans Enzyklopädie der modernen Chemie*, 4. Auflage, Bd 21, p. 485 (Verlag Chemie, Weinheim).

[70] A. K. Galwey and D. M. Jamieson, *J. Phys. Chem.* **78**, 2664 (1974).

[71] D. Bäuerle, *Chemical Processing with Lasers* (Springer-Verlag, New York, 1986).

[72] L. I. Maissel and R. Glang, *Handbook of Thin Film Technology*, pp. 1–50 (McGraw-Hill, New York, 1983).

[73] O. S. Heavens, *J. Sci. Instrum.* **36**, 95 (1959).

[74] F. M. Smits, *Bell Syst. Tech. J.* **37**, 711 (1958).

17 Laser-Assisted Fabrication of Integrated Circuits in Gallium Arsenide

FRANÇOIS FOULON and MINO GREEN

Department of Electrical Engineering, Imperial College of Science, Technology and Medicine, London, UK

17.1 INTRODUCTION

Integrated circuits based on gallium arsenide have been under intensive investigation over the past decade. GaAs has several properties that make it a very attractive semiconducting material. It has a high electron mobility and high saturation velocity, which give it the potential for high-speed devices. The easy fabrication of high-resistivity substrate decreases parasitic capacitance and considerably simplifies device isolation. Furthermore, the opto-electronic properties of GaAs offers the possibility of monolithic opto-electronic integration, including both digital, microwave and optical circuits.

In spite of its great potential, GaAs does have some disadvantages that restrict the use of several integrated circuit processing techniques. Since it is a binary semiconductor, growing high-quality crystals is difficult. Care must be taken because GaAs is subject to chemical dissociation at the surface of the crystal at temperatures in excess of $600°C$. It is more susceptible to stress-induced dislocations and also more easily attacked by chemicals used in semiconductor processing than is silicon. Finally, since GaAs does not have a stable and easily grown native oxide like silicon, sophisticated methods for its surface passivation are necessary. So although, in principle, the processing technologies for GaAs and silicon are relatively similar, they do differ in many important aspects. It is for these reasons that the development of low-temperature processing of GaAs has been necessary. The basic device in GaAs integrated circuits is the metal semiconductor field effect transistor (MESFET), instead of the metal oxide

Photochemical Processing of Electronic Materials
ISBN 0-12-121740-X

semiconductor FET (MOSFET) or bipolar transistor employed in silicon integrated circuits.

The MESFET is basically a voltage-controlled resistor that uses a Schottky barrier gate metal contact to modulate the flow of carriers through a doped layer, called the channel, between the source and drain Ohmic contacts [1]. To exploit the high electron mobility, the GaAs MESFET is based on a n-type doped channel. There are two basic modes of operation, the depletion mode (D-MESFET) and the enhancement mode (E-MESFET). A depletion mode, or normally-on, device has a conductive channel with zero bias applied to the gate contact (Fig. 17.1a). The device is turned off by applying a sufficiently negative bias to the gate contact in order to cause the depletion layer to extend across the whole n layer. Threshold voltages of normally-on MESFETs are typically of the order of 1 V. In contrast, an enhancement mode, or normally-off, device does not have a conducting channel with zero gate bias (Fig. 17.1b). This is achieved by reducing the active-layer depth sufficiently so that it is fully depleted by the built-in potential of the Schottky barrier. When a positive gate bias larger than the threshold voltage (typically 0.2 V) is applied, a channel current begins to flow.

D-MESFET devices have been successfully applied to the fabrication of GaAs integrated circuits with MSI and LSI level complexity. However, E-MESFETs are considered to be more suitable for VLSI devices because of

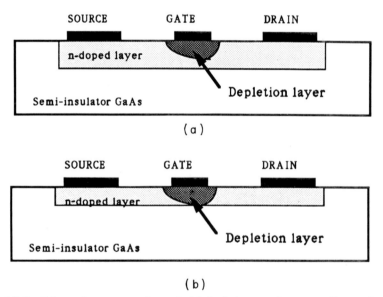

Fig. 17.1 Schematic cross-section of (a) depletion mode, normally-on device (D-MESFET); (b) enhancement mode, normally-off device (E-MESFET).

their lower power dissipation than D-MESFETs. From a technological point of view, enhancement mode devices are more difficult to produce because the n-doped layer must be very thin (50–100 nm). Moreover, the small voltage swing (typically 0.5 V) of E-MESFETs places severe demands on the uniformity of the pinch-off voltage across the wafer and therefore the uniformity of the fabrication process. Thus E-MESFETs need more complex and expensive technology than D-MESFETs. In addition, the use of a thin active channel layer that is virtually totally depleted by the gate potential results in high open channel resistance values, which degrade the circuit performance.

High-speed circuit performance, both for D-MESFETs and E-MESFETs, are limited by their open channel resistance and also by the Ohmic contact resistances. The contact resistance is usually reduced by the introduction of a heavily n^+-doped layer (10^{18}–10^{19} at cm^{-3}) under the Ohmic metal contacts. Two approaches may be taken in order to reduce the open channel resistance. One approach is to recess the gate into the GaAs surface by etching into a thicker channel region (Fig. 17.2c). This allows a relatively thicker material to be used between the gate and the Ohmic contacts and reduces the resistance between these two contacts. Another way to reduce the channel resistance is to decrease the gate to Ohmic contact separation. This is achieved by using a self-aligned gate process that allows the formation of an n^+-doped contact very close to the gate contact (Figs 17.2d, e). The optimum n^+ implant–gate separation, from the point of view of the open channel resistance and the gate capacitance, has been determined by Yamasaki [2] to be about 0.2–0.3 μm.

Since the first GaAs digital integrated circuits were reported in 1974 by Hewlett-Packard [3], using mesa epitaxial MESFETs (Fig. 17.2a), many GaAs FET structures have been developed in order to improve the operating speed and increase the circuit complexity and process yield. Figure 17.2 also shows some of these GaAs device structures, including (b) the planar MESFET with selective ion implantation [4, 5], (c) the recessed gate MESFET [6, 7], (d) the refractory gate self-aligned (SAG) MESFET [8–11] and (e) the self-aligned implantation for n^+ layer technology (SAINT) MESFET [12, 13]. The two latter structures have been mainly developed in order to achieve the fabrication of E-MESFET devices with good performances and reproducibility.

At the same time, various GaAs logic families have been proposed. They can be divided into two groups: the D-MESFET-based families and the E/D-MESFET-based family. The former include buffered FET logic (BFL) [14] and Schottky diode FET logic (SDFL) [15]. These logic circuits, based on a normally-on mode, require a voltage level shifting between stages and need an additional negative power supply. At the opposite extreme, the E/D-MESFET-based family, usually designed as

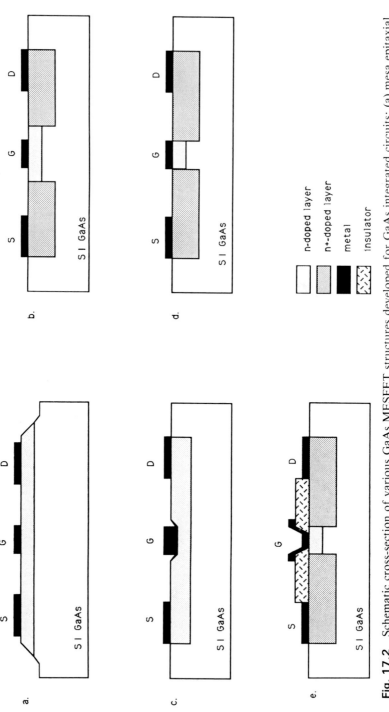

Fig. 17.2 Schematic cross-section of various GaAs MESFET structures developed for GaAs integrated circuits: (a) mesa epitaxial MESFET, (b) planar MESFET with selective ion implantation, (c) recessed gate MESFET, (d) refractory gate self-aligned (SAG) MESFET, (e) self aligned implantation for n^+ layer technology (SAINT) MESFET.

directly coupled FET logic (DCFL) [16], do not need an additional power supply because they use a normally-off FET.

A comparison of the performances of these different families is given in Table 17.1. One can note that BFL logic has the highest operating speed and the highest noise margin, which mean that it is relatively immune to process variations. However, its high power dissipation limits its use to circuits with MSI level complexities. In fact, LSI and VLSI level complexities can be reached only by using SDFL and DCFL logics, which in turn need better control of the fabrication process. This is particularly true for DCFL logic, which involves E-MESFETs requiring lower logic swing and having lower noise margins. Consequently, new technologies, such as SAG or SAINT, have been developed. Unfortunately, these technologies need more complex fabrication processes and a larger number of steps, increasing at the same time the risk of semiconductor degradation by chemical or thermal treatment, and corresponding loss of production yield.

In this context, research on new fabrication processes that can reduce the high-temperature treatments and the number of processing steps as well as improve device characteristics are of great interest. Laser chemical processing of GaAs is a good candidate for this objective. It offers some advantages directly linked to the properties of the laser light. Directionality and coherence of the laser radiation permit strong spatial localization of the heat and/or chemical treatment of semiconductors, with a resolution theoretically limited by the diffraction to the laser wavelength. These properties open the way to projection-patterned processes that thereby avoid resist patterning and reduce the number of processing steps. The directionality of the laser light makes possible *in situ* processing of the material placed in a controlled ambient, limiting contamination problems. The monochromaticity and the high energy densities of laser light as well as a large choice in laser wavelength from infrared to far-ultraviolet (CO_2 laser 10 600 nm, F_2 excimer laser 157 nm) allow access to a large number of

Table 17.1 Characteristics of various GaAs logic families. (After [17].)

Logic family	BFL	SDFL	DCFL
Gate speed	Fast	Medium	Medium
Fan-out capability	Large	Small	Medium
Power dissipation	High	Low	Lowest
Circuit complexity	MSI	LSI	LSI/VLSI
Packing density	Poor	Good	Excellent
Power supplies	$+ V_{DD}, - V_{SS}$	$+ V_{DD}, - V_{SS}$	$+ V_{DD}$
Noise margins	High	Medium	Low
Process-control requirement	Loose	Medium	Stringent

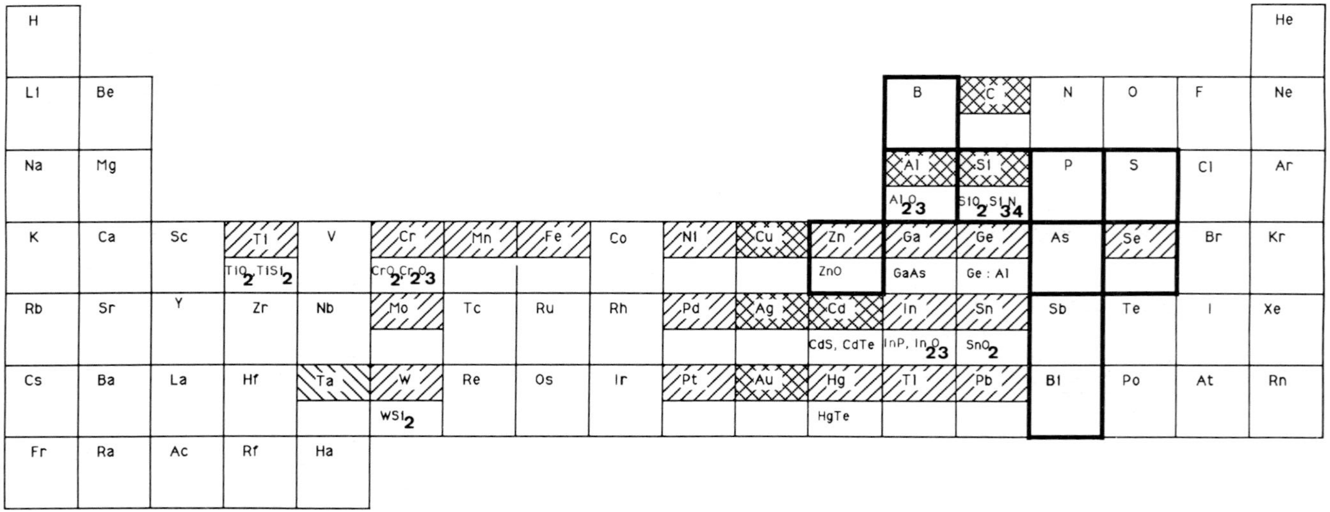

Fig. 17.3 Elements and compounds that have already been: deposited (▨), removed by ablation or etching (▧) or used as dopants (☐).

laser-induced physical and chemical processes on various materials with a high degree of process control. Deposition, etching, surface modification and ablation of various materials such as GaAs, metals, insulators or organic materials have been reported, as summarized in Fig. 17.3 [18–24]. Consequently, one can hope to gather together all of these laser processes in order to fabricate a complete integrated circuit with a corresponding unification of the processing techniques. This has been done already in silicon by McWilliams [25], who has reported the fabrication of N MOS transistors and simple logic functions (NOR, NAND and NOT gates) with a resolution of about 1 μm using only pyrolytic laser direct writing. Furthermore, laser processes can reduce the damage to materials that is inherent to conventional low-temperature techniques such as plasma and ion or electron beam processing. Laser-prepared films exhibit electrical and structural properties that compare favourably with those of conventional techniques [20].

In this chapter, we discuss the prospects for the utilization of laser chemical processing in GaAs technology. For this purpose, we describe in the second section the fabrication of an invertor in SDFL logic by conventional techniques. In the third section, we review the problems associated with these conventional techniques and propose alternative laser-induced processes that could improve the fabrication process. This concerns particularly laser projection-patterned chemical processing of materials. In the last section, the characteristics of this technique, including laser sources, optical systems and processing mechanisms, are discussed.

17.2 CONVENTIONAL PROCESSES

We describe in this section the process flow chart of an inverter made in SDFL logic. Figure 17.4 shows a schematic representation of such an inverter, which involves two stages. The level shifter consists of a pull down FET (T1) and a Schottky diode (S1). To ensure that the buffer stage can be turned off, an additional negative supply line is used. The second stage involves two D-MESFETs, one active pull-up (T2) and one switching FET (T3). The symbolic layout of such an inverter is shown in Fig. 17.5: it includes four active areas to form the three MESFETs and the Schottky diode, and three bias lines. A second-level metallization is required for the negative supply line. As discussed in Section 17.1, self-aligned technologies have some advantages compared with the other basic technologies. There are various processes for achieving self-aligned gates (SAG) [8–13].

We have chosen for our fabrication process a SAG technology with a T-gate structure shown in Fig. 17.6. In this technique, after channel formation, the gate contact is created and a second metal covering the gate

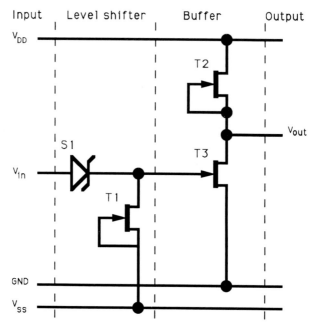

Fig. 17.4 Schematic representation of an inverter made in SDFL (Schottky diode FET logic).

is used as a mask for the n^+ layer contact ion implantation. The overhang of the T structure is used to provide the optimum separation between the n^+ implant and the edge of the gate contact. This technique has been successfully applied to the fabrication of high-speed submicron GaAs MESFETs [10].

The process flow chart of such an SDFL inverter made in a self-aligned T-gate structure technology is shown in Fig. 17.7. This figure also shows a cross-section of the D-MESFET and a plan view of the complete inverter at the end of the principal stages of the fabrication process.

The process steps are carried out as follows.

17.2.1 Substrate specification

Undoped semi-insulating material, having a resistivity above $10^7 \, \Omega$ cm is commonly used. Low dislocation densities ($< 5 \times 10^4 \, cm^{-2}$) are required in order to avoid scatter in threshold voltage of closely spaced FETs. Thus good uniformity of properties across the wafer is important. A high electron mobility is required, with values about 5000 $cm^2 \, V^{-1} \, s^{-1}$. One-face-

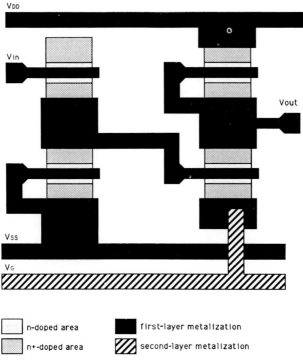

Fig. 17.5 Symbolic layout of an SDFL inverter.

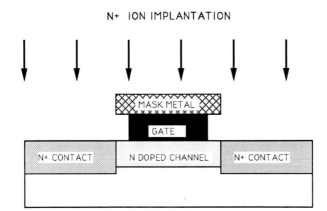

Fig. 17.6 Principle of the self-aligned structure made using a T gate.

Fig. 17.7 Process flow chart of a SDFL inverter made using a self-aligned T-gate structure technology by conventional processing and by laser processing. The cross-section of a D-MESFET and plan view of the main stages of the fabrication process are shown.

Conventional technique	Laser processing
	Number of processing steps

1. Substrate specification   **1. Substrate specification**

2. Surface preparation   **2. Surface preparation**

3. Active-layer formation   **3. Active-layer formation**

Conventional	Laser
(a) resist patterning: resist deposition and backing resist exposition by resist development (b) channel implantation (c) resist removing	(a) projection patterned deposition of the resist (b) channel implantation (c) resist removing (d) rapid thermal annealing

                      7   4

4. T-gate structure formation   **4. T-gate structure formation**

Conventional	Laser
(a) refractory metal deposition (b) resist patterning (c) etch-mask metal deposition (d) etch-mask metal lift-off (e) refractory metal etching	(a) projection patterned deposition of the refractory metal (b) projection patterned deposition of the etch-mask metal (c) refractory metal etching

                      16   7

5. n^+ contact formation   **5. n^+ contact formation**

Conventional	Laser
(a) resist patterning (b) n^+ contact implantation (c) resist removing (d) etch-mask metal removing (e) wafer encapsulation (f) furnace thermal annealing (g) encapsulant removing	(a) projection-patterning doping (b) etch-mask metal removing

                      27   9

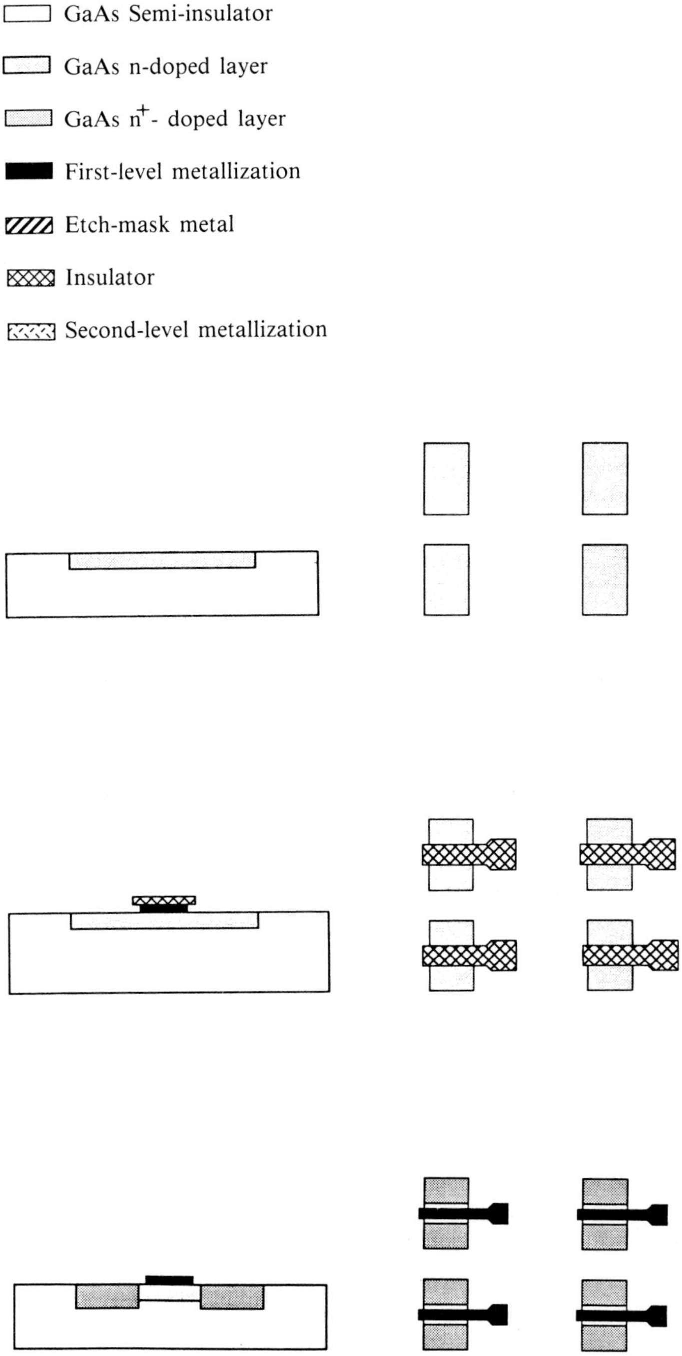

GaAs Semi-insulator

GaAs n-doped layer

GaAs n^+- doped layer

First-level metallization

Etch-mask metal

Insulator

Second-level metallization

6. Ohmic contact formation

(a) resist patterning
(b) metal deposition
(c) metal lift-off
(d) metal alloying

6. Ohmic contact formation

(a) projection-patterned deposition
of the Ohmic contact metal
(b) metal alloying

35   11

7. First insulating layer

(a) dielectric deposition
(b) resist patterning
(c) selective dielectric etching

7. First insulating layer

(a) projection pattern deposition
of the insulator (polymer)

42   12

8. Second level metallization

(a) metal deposition
(b) resist patterning
(c) selective metal etching

8. Second level metallization

(a) projection-patterned deposition
of the metal

49   13

**9. Second insulating layer
deposition**

(a) dielectric deposition
(b) resist patterning
(c) selective dielectric etching

**9. Second insulating layer
deposition**

(a) projection-patterned deposition
of the insulating layer

52   14

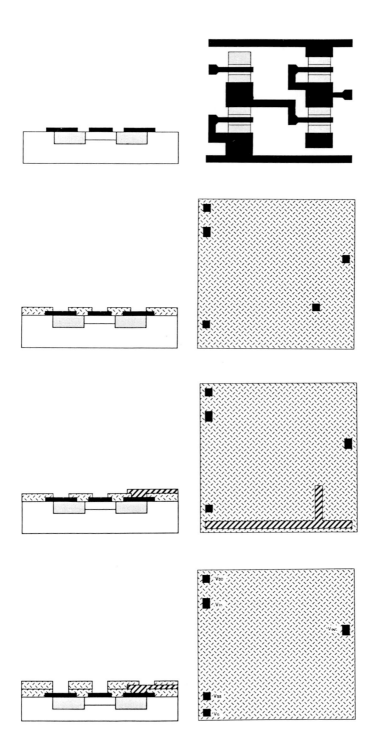

polished (100)-oriented wafers are used. Flatness of about 1.5 μm cm^{-1} is required. Thermal consideration in the design of the integrated circuit generally requires substrate thickness of less than 200 μm, due in part to the rather poor thermal conductivity of GaAs. However, the fragile nature of GaAs makes handling extremely difficult during fabrication processes. Consequently, wafers having thicknesses of about 500 μm are generally required during processing, and they are reduced in thickness after the end of the fabrication process.

17.2.2 Surface preparation

The GaAs surface has to be cleaned in order to remove organic substances before any fabrication process. This cleaning can be achieved in different ways. A simple treatment by vigorous rinsing in acetone, methanol and deionized water seems to be as good as any [26]. The GaAs surface residual layer, which is essentially an oxide layer, degrades the device characteristics: Schottky barriers are found to be very sensitive to the surface dielectric thickness [27–29]. In order to remove the surface residual layers on GaAs, NH$_4$OH [28, 30, 31] or HCl [27, 31] etches are commonly used with good results.

17.2.3 Active-layer formation

Formation of the n-doped channel of the MESFET is commonly achieved by an ion implantation doping technique. Before the implantation stage, the active area has to be defined. This is achieved by opening windows (i.e. resist patterning) in a resist, which acts as an implantation mask. The successive operations resulting in resist patterning are described in Section 17.3. The resist patterning is repeated every time a new layer of material has to be patterned in the fabrication process.

Channel formation is mostly achieved by implantation with ^{29}Si$^+$. A typical implant suitable for the n channel of a D-MESFET has doses of $(1-5) \times 10^{12}$ Si cm^{-2} and an energy of 100–500 keV, resulting in Gaussian profiles with surface concentrations of a few 10^{17} cm^{-3} and energies projected ranges between 100 and 200 nm [32, 33]. In order to achieve more uniform concentration profiles, multi-implants can be performed [34]. Removal of the implantation damage and electrical activation requires the wafer to be annealed at high temperatures. This is generally achieved after the n$^+$ contact layer formation. The resist is removed before the next processing step.

17.2.4   T-gate structure formation

As pointed out before, gate formation is a critical stage in MESFET fabrication. The GaAs surface residual layer must be removed by an appropriate chemical etch. Almost all the metal deposited on low- and medium-doped GaAs results in a Schottky contact [35]. However, in the self-aligned gate technology, the Schottky contact must be stable during the high-temperature annealing that follows the n^+ contact implantation. For this purpose, refractory metals such as W, TiW or WSi [8–11, 36, 37] are used. To achieve a T-gate structure, the following procedure is carried out. A uniform film of refractory metal about 200 nm thick is deposited on the wafer. Metal deposition onto GaAs is generally carried out by evaporation or by sputtering methods. In the former technique, metal vaporization is achieved by hot filament or electron beam evaporation. This method of metal deposition is not recommended for metallizing over step edges, but is good for the lift-off technique. Deposition of thick metal films (>1 μm) or alloys is best carried out by sputtering, which ensures a good step coverage. In this technique, high-energy ions are directed onto a metal source, knocking metal atoms from the latter that redeposit on to the GaAs wafer. The gate lines are defined by resist patterning. A second layer of metal, Al or Ni, which will act as a mask for the etching of the refractory metal and for the n^+ implantation, is evaporated onto the surface. The metal etch mask is obtained by a lift-off technique. Reactive ion etching of the refractory metal is then performed in a CF_4 plasma at high pressure and power density to achieve the desired undercut in a reasonable time. Al and Ni are not etched in a fluorine-based plasma. The T-gate structure shown in Fig. 17.6 is finally obtained.

17.2.5   Formation of the n^+ contact layer

The area to be n^+ implanted is defined as for n-channel formation by resist patterning. The ion implantation is then performed through the resist at higher doses and energy densities, typically 10^{13} Si cm^{-2} and 125–200 keV respectively, in order to obtain peak dopant concentrations of $(2-5) \times 10^{18}$ cm^{-3}. The metal etch mask that has served as the top overhanging layer of the T gate is then stripped off. The wafer is annealed to remove the implantation damage and electrically activate the doping impurities. To prevent GaAs surface dissociation during the high-temperature treatment, three techniques have been developed: (i) surface passivation using a dielectric encapsulating film (SiO_2, Si_3N_4, AlN) [9, 38, 39]; (ii) annealing in an arsenic overpressure [40–43]; and (iii) transient annealing using incoherent light [44, 45], graphite strip heater [46], laser [47–49] or scanned electron beam [50]. In the transient techniques, the heat treatment is carried out for a very short time, so that surface loss is expected to be negligible. In industry, the first technique is generally used. When the sample is coated, furnace annealing is performed at 800–950°C for 10–30 min. The encapsulant is then removed by plasma or wet etching.

17.2.6 Ohmic contact formation

The ohmic contacts and the first-level metallization are formed by resist patterning, metal deposition and lift-off. In gallium arsenide technology, alloys of gold and germanium with additives such as nickel or indium are commonly used for the Ohmic contacts. A 200–500 nm thick film of AuGe in the proportion of the eutectic, which is the mixture of the two metals having the lowest melting point, is deposited by sputtering. The additive metal is used to prevent the alloy balling up during the subsequent heat treatment ($T \approx 450°C$) performed to obtain the desired low-resistance contact.

17.2.7 First isolation layer

Isolation between the first- and second-level metallizations is achieved by deposition of a dielectric film (SiO_2, Si_3N_4). The deposition is performed at low temperatures by plasma-enhanced chemical vapour deposition.

17.2.8 Contact cut

The contact cuts between the first- and second-layer metallizations are prepared by opening windows in the dielectric layer. This is achieved by resist patterning, followed by selective etching of the uncovered area by plasma etching, ion beam milling or wet etching.

17.2.9 Second-level metallization

Gold with a low percentage of alloyed Ti and Pt is commonly used for the second-layer metallization. It can be delineated either by evaporation onto patterned resist and lift-off or by sputter deposition followed by ion beam milling through a patterned resist.

17.2.10 Encapsulation

A final dielectric passivating layer, such as SiO_2, Si_3N_4 or polyimide, is deposited onto the wafer surface and etched using resist patterning to open the windows for wire bonding.

17.3 LASER PROCESSING

Laser processing of GaAs may have two main advantages for GaAs technology. It should lower the damage caused by conventional low-temperature processes, which are generally high-energy, and it should also significantly reduce the number of fabrication steps.

The most important and frequent processing step in integrated circuit fabrication is the transfer of specific patterns to the wafer. Patterned deposition, etching and doping of materials are generally achieved using photoresist layers. Developed pattern definition in photoresist material is a complex process, and pattern transfer involves not less than six operations in addition to the relevant basic process step, which might be deposition, etching or doping. First, one must coat the wafer surface with a uniform photosensitive polymer film, namely the photoresist. This is usually achieved by a spin-on technique. The photoresist is then baked at temperatures between 80 and $120°C$ before exposing it to UV light through a photomask by a lithographic technique. After exposure, the photoresist is developed in a solution that removes either the exposed (positive) or unexposed (negative resist) areas, depending on the polymer used. The resist is then baked again at $100–150°C$. After all these steps, the underlying material to be exposed to the relevant basic process is defined. Finally, after this active process, the resist is removed in order to repeat the same sequence for the next active process in the production operation. In the conventional fabrication process described in Section 17.2, this sequence, called resist patterning and removal, is repeated seven times, so the number of processing and handling steps is large. The risk of substrate defect creation is correspondingly high. In this context, laser projection-patterned processing of materials is a very attractive alternative. Indeed, it theoretically allows deposition, etching or doping of various materials by a one-step process, meaning only the active process, with a resolution down to the laser wavelength. This is achieved by projecting the laser light via a photomask onto the surface of the wafer placed under the appropriate reacting gas, as shown in Fig. 17.8. The photon–gas and photon–surface interactions in the illuminated areas lead to the patterned processing of the wafer. This strongly decreases the number of fabrication steps, as one can see in Fig. 17.7. Furthermore, the different fabrication steps such as gate

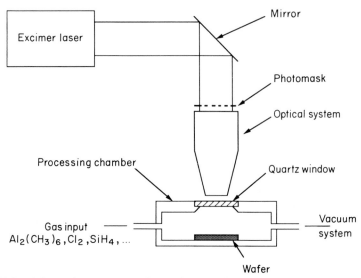

Fig. 17.8 Schematic representation of a typical apparatus used for laser projection-patterned processing of semiconductor wafers.

metal deposition, n^+ contact formation and metal and insulator deposition, can be achieved by *in situ* processing, merely by changing the reactive gas and the mask at each step, without taking the wafer out of the processing chamber.

A major problem in GaAs technology is the damage caused by the various dry processing techniques, which are generally high-energy processes. Indeed, plasma processing, sputtering or electron beam evaporation involve ion, electron or VUV photon bombardment of the wafer, which can easily cause damage to the surface layer or subsurface layers of GaAs or induce impurity sputtering, resulting in contamination of materials. Ion damage can cause serious degradation in the electrical properties of the material, and hence lead to changes in circuit performance. Thus laser processing, which allows low-temperature processing with lower-energy irradiation of the wafer, appears to be a good candidate for GaAs technology. Furthermore, this technology should reduce the degree of complexity and the production line cost compared with processing techniques based on the use of ion or electron beams.

So, taking into account the results reported in the literature concerning laser chemical processing of different materials, including metal and insulator deposition or etching and laser-induced doping of GaAs, we propose a process flow chart for the SDFL inverter made by laser processing. This process flow chart shown in Fig. 17.7 can be compared with the process flow chart using conventional techniques.

17.3.1 Substrate specification

On the assumption of complete *in situ* processing of GaAs devices using laser processing that reduces handling steps, GaAs wafers thinner than those needed with conventional processing might be used. This should reduce the cost of the final device by reducing the cost of the initial wafer and the cost of wafer thinning.

17.3.2 Surface preparation

Since laser processing is an *in situ* process, it reduces the risk of substrate contamination and at the same time the number of cleaning operations needed at each step in a conventional fabrication process. Furthermore, *in situ* laser cleaning of surfaces might be used to produce atomically clean substrate surfaces at the start of processing, or it could be used to remove the native oxide on the GaAs surface. In the case of silicon, for example, it has been shown that pulsed lasers can be used to prepare an atomically clean surface in an ultrahigh-vacuum ambient [51–53]. Wafer surface *in situ* cleaning could be achieved by irradiating the sample surface under an appropriate reactive gas atmosphere or by using the photochemical or photothermal desorption of impurities from the surface during laser irradiation.

17.3.3 Active-layer formation

Gas immersion laser doping of semiconductors, known as the GILD technique, has been under intensive investigation [54–59]. The principle of this technique is described in Chapter 11. This technique allows, in a one-step process, the formation of junctions with good control of the profile shape, surface concentration and junction depth. In addition, since the doping process takes place only in the irradiated areas, projection-patterned doping can be achieved so that the use of resist to define the doped areas is not necessary. Projection-patterned doping in silicon has already been performed by Ibbs [60]. In the case of silicon, it has been shown that the GILD technique is suitable for fabrication of submicron MOSFETs having good electrical characteristics [61, 62].

Doping of GaAs material, n and p type, has been reported using H_2S, H_2Se, SiH_4 and $Zn(C_2H_5)_2$ dopant gases [63–71]. Unfortunately, the volatility of As makes GaAs more susceptible than silicon to damage resulting from the surface melting process. Near-surface loss of stoichiometry and surface degradation has been observed after GaAs laser-induced surface melting [69, 72–74]. Therefore the GILD technique does not seem to be suitable as a means of forming the GaAs active layer, whose surface quality is critical in the formation of Schottky contacts with good electrical characteristics. Consequently, we have not included this technique in our considerations and we still use ion implantation for channel formation. However, as will be discussed later, the GILD technique does seem to be suitable for n^+ Ohmic contact layer formation.

Before the ion implantation step, the areas to be doped must be defined by resist patterning. In laser processing, this could be achieved by projection-patterned deposition of the resist. In this connection, two techniques have already been investigated. The first is photopolymerization of volatile surface-adsorbed MMA with an Ar^+ laser, which results in PMMA growth [75]. Resolution limited by the spot size, down to 0.8 μm, has been found for the PMMA-grown lines. The second technique is the spontaneous polymerization of organic monomer on patterned catalyst film predeposited from adsorbate mixtures [76, 77]. An alternative technique to resist patterning can be achieved by an *in situ* two-step process: uniform photodeposition of the resist and subsequent projection-patterned ablation. This last operation has already been carried out with a resolution down to 0.13 μm using an ArF excimer laser [78].

After ion implantation, the resist can be removed either by pyrolytic or photochemical ablation [79–83]. To date, almost all experiments have been performed in air, so investigation of etching under various reactive gases such as oxygen would be interesting.

In the conventional method, ion-implanted wafer annealing is achieved in a furnace, and the wafer must be encapsulated with a passivating layer. In order to avoid this encapsulation step, and the subsequent cap removal operation, a transient annealing process can be used. As laser annealing takes place by surface melting, it involves with the same problems as that of the GILD technique and therefore is not suitable for GaAs annealing. However, rapid thermal annealing, using incoherent light or graphite strip heaters, appears to be a promising technique. Indeed, it has various advantages over furnace annealing. It can limit the diffusion of the implanted ions as the heating time is reduced. Encapsulation was found to be required only for annealing temperatures greater than $950°C$. Moreover, the electrical properties of implanted samples processed using rapid thermal annealing are generally better than those processed in a furnace. Consequently, rapid thermal annealing has been included in the laser processing schedule rather than furnace annealing.

17.3.4 T-gate structure formation

As shown in Fig. 17.6, T-gate structure formation needs the deposition of two layers of metal onto the GaAs surface where the gate is to be formed, followed by the definition of an undercut by selective isotropic etching of the first metal layer. As can be seen from Fig. 17.3, various metals have already been deposited by laser processing, such as tungsten [84–87] and aluminium [88–92], which are used as first- and second-layer metals, respectively. Thus T-gate structure formation might be achieved by laser projection-patterned deposition of tungsten and subsequently aluminium with the same mask, just by changing the reactive gas in the processing chamber. Isotropic etching of the tungsten could then be achieved using laser irradiation to create etching species from the gaseous ambient. Fluorine-induced etching of tungsten using ArF excimer laser irradiation in a COF_2 atmosphere has already been reported [93].

To date, only a few studies have been carried out on laser projection-patterned deposition of metals (e.g. silver [94], chromium [95] and gold [96]). Excimer laser projection has also been used to photolyse adlayers of $Al(C_4H_9)_3$ to produce reactive Al sites that serve as nucleation centres for spatially selective growth of Al film patterns made by standard CVD [92]. The vast amount of reported work is concerned with the results of laser deposition of metal in the form of spots or the direct writing of stripes—both being achieved with perpendicular irradiation of the wafer surface through the reactive gas. Deposition of extended thin films of metals with laser irradiation using both perpendicular and parallel incidence has also been reported. It is clear that more extensive investigations should be carried out on projection-patterned deposition of materials.

17.3.5 Formation of n^+ contact layers

The formation of n^+ contact layers in GaAs by the GILD technique using a frequency-doubled Nd : YAG laser and H_2Se dopant gas has already been reported [69]. GaAs surface sheet resistances down to $30\,\Omega/\square$ were achieved and contact resistance down to $3 \times 10^{-6}\,\Omega\,cm^{-2}$ were obtained with an Ni/AuGe evaporated contact alloyed at $450°C$ for 5 min. Projection-patterned doping of GaAs under an SiH_4 atmosphere with a KrF excimer laser has been performed by Sugioka [71]. Nonalloyed copper contacts formed on these doped layers by electroplating in a $CuSO_4$ aqueous solution resulted in a specific contact resistance of about $2 \times 10^{-5}\,\Omega\,cm^{-2}$.

Thus projection-patterned doping of GaAs could be used for n^+ contact layer formation. In this process, the aluminium layer would act as a reflective mask for the GILD process. Aluminium films have already been used in this way as reflective masks to protect the gate during n^+ contact layer formation by the GILD technique in the fabrication of submicron silicon MOSFET [60]. It has been found that the lateral dopant diffusion under the gate during the laser doping process is small (about 0.05 μm).

After doping, the aluminium mask has to be removed by selective etching. Laser-induced etching of aluminium has been performed using Cl_2 gas [97, 98] or $HNO_3/H_3PO_4/K_2Cr_2O_7/H_2O$ solution [99]. The first method could be used, taking care that the GaAs surface is not etched.

17.3.6 Ohmic contact formation and second-layer metallization

As can be seen in Fig. 17.3, various metals have been deposited by laser chemical processing, so that a large range of metals is available for laser-processed devices. However, as pointed out already only a few studies have dealt with projection-patterned deposition of metals and with the deposition and formation of multilayers of metals and metal alloys. Thus formation of Ohmic contacts using alloys such as AuGe needs further investigation.

As far as second-level metallization is concerned, the projection-patterned deposition of gold has already been performed by the UV excimer photolysis of dimethylgold acetylacetonate in the vapour phase [96].

17.3.7 Insulating layers

Laser chemical vapour deposition of SiO_2 [84, 100–103], Si_3N_4 [84, 104–106], Al_2O_3 [84, 88, 105, 107] and TiO_2 [108, 109] films has been reported in the literature using parallel, perpendicular or combined parallel and perpendicular irradiation of the wafer surface in a gas atmosphere. Since insulating film formation comes about mainly by the deposition onto the surface of photochemical products generated in the surrounding gas phase, most of the studies concern only the deposition of extended films, and projection-patterned deposition of such dielectrics has not been reported. However, deposition of spots or strips of SiO_2 and Si_3N_4 has been performed [102, 104], so that projection-patterned deposition of insulating layers may be possible.

A very attractive approach in laser processing is the use of organic materials for the insulating layers. Projection-patterned deposition of polymers could be achieved as reported above for PMMA.

17.4 REVIEW OF LASER PROJECTION-PATTERNED PROCESSING

The main aspects of laser projection-patterned processing are discussed in this section. A brief summary of the most commonly used lasers along with their suitability for the above application is given. The design of optical systems developed for this technique is discussed. The chemical mechanisms are reviewed and classified according to the manner in which the spatial resolution is controlled by the chemical reactions induced by the laser irradiation.

17.4.1 Lasers

A great diversity of lasers and laser wavelengths are available and regularly applied to the processing of materials. While the optical properties of materials are strongly linked to the light wavelength, and photochemical reactions are also strongly dependent on the photon energy, microstructure formation requires a spatial resolution that is in the UV wavelength range. So, in order to obtain the required resolution and to make possible the direct photodissociation of many reactive molecules, lasers operating in the ultraviolet domain such as excimer lasers (XeF, $\lambda = 351$ nm; XeCl, $\lambda = 308$ nm; KrF, $\lambda = 248$ nm; ArF, $\lambda = 193$ nm) or frequency-doubled Ar^+ lasers ($\lambda = 257$ nm) are commonly used.

These lasers have different modes of operation. Ar^+ lasers are continuous devices with relatively low output power (of order 0.1 W at 257 nm), while excimer lasers are pulsed and deliver a large amount of energy over a short time (typically 0.5 J in 20 ns). Because of their stability and focusability, continuous lasers are mainly used for direct writing of microstructures by scanning the focused beam across the sample surface. On the other hand, because of their high power and repetition rate (up to 500 Hz) and their large beam cross-section (up to 3 cm × 3 cm), pulsed excimer lasers seem destined to become the most extensively adopted for projection-patterned processing of large areas. Moreover, excimer lasers have relatively poor coherence, since their output is highly multimode, thereby avoiding the interference effects that cause severe problems for the formation of microstructure with light of high spatial coherence.

17.4.2 Optical systems

An optical projection system is generally characterized by, among other parameters, the operating wavelength and its numerical aperture (NA). For an optical system free of aberrations, working with a highly incoherent source of light, the size of the minimum resolvable feature R than can be generated is ultimately limited by diffraction to $R = \lambda / (4\,NA)$. So improvement of the resolution might be achieved by reducing the wavelength or increasing the numerical aperture. However in practice, resolution is affected by other parameters—the tolerance on defocusing as well as the spatial coherence and the spectral width of the light source—which impose additional constraints on resolution as discussed by Goodall in Chapter 3.

In order to obtain good images, the wafer must be in the image plane. Defocusing, which results from several factors, such as wafer surface flatness or defocusing error, is a major source of image degradation. The range of poor focus allowed without serious degradation of the image is characterized by the depth of focus (DOF) which is generally given by $\lambda/(2\,NA)^2$, according to Rayleigh theory. In manufacturing practice, the depth of focus is typically of order 1 μm. Thus the resolution of an optical system is not only diffraction-limited, but is also limited by the tolerance on defocusing. Figure 17.9 represents the relation between R and DOF for different wavelengths and numerical apertures. It shows that the resolution of an optical system working at 351 nm (XeF excimer laser) with a numerical aperture of 0.5 is theoretically limited by diffraction to about 0.17 μm, but such a system would not be practical because its depth of focus DOF = 0.7 μm is too low. Thus it appears that the best way to reduce R, conserving an acceptable value of DOF, is to reduce the wavelength rather than increase the numerical aperture, Resolution down to 0.15 μm can be achieved with $\lambda = 193$ nm and NA = 0.3, conserving a depth of focus of 1 μm. This explains why excimer laser projection lithography has received such a great deal of attention recently (see [110–119] and Chapter 3).

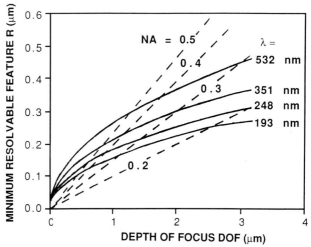

Fig. 17.9 Relation between the size of the minimum resolvable feature R, and the depth of focus DOF, for different wavelengths λ and numerical apertures NA. The diagram is used by noting the coordinates (DOF, R) for a particular intersection of the NA and λ curves.

The quality of the image at the wafer surface is also determined by the degree of spatial coherence of the optical system, S, which is defined as the ratio of the numerical aperture of the image of the source light to the numerical aperture of the imaging optics. The diffraction-limited minimal resolvable feature under a highly coherent source ($S < 0.3$), $R = \lambda/(2\,NA)$, is two times greater than that obtained using highly incoherent light ($S > 1$). Incoherent and coherent sources both have their advantages. Coherent imaging, for example, is less sensitive toward defocusing. In practice, optical systems working with lamps have S values between 0.4 and 0.8 in order to take advantages of the positive features of both types of sources.

The degree of spatial coherence of an optical system working with a laser beam without a condenser between the laser and the mask has been evaluated by Rothschild [112] to be $S \approx 3m\theta_{div}/NA$, where m and θ_{div} are respectively the lens magnification and the angular divergence of the laser beam (typically $\theta_{div} \approx 3$ mrad for an excimer laser). Thus, for an optical system with $m = 5$ and $NA = 0.2$, the degree of coherence $S \approx 0.2$. Such an optical system is too coherent. It is therefore necessary to reduce the coherence of the laser beam at the input. Several techniques can be used for this purpose [113, 116]; for example, a diffuser can be placed in the beam. However, this increases the complexity of the optical system and often reduces its output energy density.

Another major problem in the fabrication of optical systems working with excimer lasers is their significant spectral width (100 cm^{-1} for the ArF excimer laser) [120], which results in chromatic spread under the best focus conditions of the optical system. This effect is amplified by the significant variation in the refractive index, δn, of optical materials in the UV domain. In fused silica, for example, $\delta n \approx 6.2 \times 10^{-4}$ across the 100 cm^{-1} width at 193 nm for an ArF laser. This leads to a spread in focal length given by $\delta f = f\,\delta n/(n - 1)$, where f is the focal length and n the refractive index at 193 nm. Thus, with a short focal length $f = 2$ cm, $\delta f \approx 20$ μm, a value much larger than the current tolerance on defocusing. Several methods are employed to decrease the spread in focal length. The spectral width can be reduced with the use of intracavity etalons or with injection locking in an oscillator–amplifier configuration [113, 114]. Another method is achromatization of the projection lenses, which is a difficult task in the UV range, especially at 193 nm [115, 116, 118]. Finally, in order to eliminate chromatic aberrations, projection optics made of reflective elements (mirrors), can be used [117].

It appears that the fabrication of laser projection optical systems working in the UV range is not an easy task. Improvement of these optical systems is a major key to the development of laser projection-patterned processing of materials, but one can obtain minimum feature sizes down to approximately 0.15 μm with ArF excimer lasers working at 193 nm.

17.4.3 Laser chemical processing

When the surface of a material placed in a reactive gas atmosphere is irradiated through the gas with a laser beam, chemical reactions can occur as the result of three main types of interaction.

First, there are gas–surface interactions, which occur even without laser irradiation, before or between the laser pulses. When material is exposed to a gas, adsorption of that gas on the material surface can occur spontaneously. Two types of adsorption are generally distinguished—chemisorption and physisorption—corresponding to molecules respectively strongly (0.5–5 eV) or weakly (0.05–0.5 eV) bonded to the surface. This difference in surface bond strength results from the nature of the bonds concerned, respectively either van der Waals or chemical bonds. Chemisorption leads to a perturbation of the electronic energy level, which can, in the extreme case, lead to dissociation of the molecule. Chemisorbed molecules are expected to form only a single monolayer, whereas physisorbed molecules can form several layers, depending on the gas pressure.

Secondly, there are laser–gas interactions, which can take place either in the gas phase or in the adsorbed phase. Two kinds of mechanisms are distinguished—photolytic and pyrolytic. In pyrolytic processes, the laser serves as a heat source, and the decomposition of the gas molecules takes place by thermochemical reaction. In photolytic processes, laser light directly absorbed by the molecule breaks its chemical bonds when the photon energy is sufficiently high. The photodissociation yield of a molecule at a given wavelength is characterized by its dissociation cross-section. Excited molecules generated by laser beams may react with other gaseous species or diffuse to the material surface and react with it. Laser–gas interaction also has some influence on the adsorbed phase constitution. Laser irradiation may be able to resonantly excite the gas species and enhance their sticking probability [121–123]. It can also result in the photodesorption of the adsorbed species [124].

Thirdly, there are laser–surface interactions. Absorption of an incident photon in the material creates electron–hole pairs in the surface region, and, depending on the laser energy density, can also result in a substantial increase in the surface temperature and even surface melting. This gives rise to several processes. Photogenerated carriers can enhance surface chemical reactions. Surface heating can result in the excitation of gas phase molecules striking the surface or the adsorbed species by a pyrolytic mechanism. It can also induce the thermal desorption of adsorbed species.

In many cases, these different types of excitation and reaction contribute simultaneously to the laser processing mechanism. For example, a process may be initiated photolytically and proceed pyrolytically. However, it is frequently the case that one of the mechanisms is dominant and controls the processing rate and the spatial resolution for a given set of processing parameters. Depending on the manner in which the spatial resolution is controlled by the processing mechanism, one can distinguish three dominant sources of reactive species, as shown in Fig. 17.10.

17.4.3.1 Laser-excited gas phase species

Gas phase molecules excited or dissociated by laser irradiation can be the main source of reactive species. Those species resulting from laser–gas interactions taking place at the position r' in the gas volume traversed by the laser beam diffuse to the material surface, where they may react spontaneously on contact. Laser processing is controlled by the flux N_1 of reactive species reaching the surface. Assuming that the reactive species generated in the gas phase are redistributed in an isotropic manner, the value of N_1 at a point r on the surface can be expressed as

$$N_0(r, E, P, T) \propto \int \frac{\rho \cos \Phi}{4\pi |r' - r|^2} \exp\left[- \frac{|r' - r|}{L(P, T)} \right] dV,$$

where E is the laser energy density, P is the gas pressure, T is the processing temperature, ρ is the density of the reactive species generated in the gas phase, Φ is the angle between the vector $|r' - r|$ and the normal to the substrate, and L is the mean free path of the reactive species before their de-excitation or recombination with other gas phase species. At low laser energy densities, the processing rate is generally limited by the kinetics of the chemical reaction, while at high densities, it is limited by mass transport. In both cases, the spatial resolution is limited by the distance covered by the reactive species before their de-excitation or recombination. According to gas kinetics, the average speed v of a species with a molar mass M, at a temperature T, is given by

$$v = (3R_0 T / M)^{1/2},$$

where R_0 is the gas constant. A reactive species with a lifetime of 10^{-8} s can diffuse across a few microns before it recombines. In this case, the chemical reaction will not be limited to the irradiated area, resulting in a degradation of the pattern resolution. Such a phenomenon was observed by Brewer [125] for excimer laser projection etching of GaAs with HBr as the precursor gas. The etching process is controlled by HBr photolysis in the gas phase. SEM examination of the patterns revealed a spatial resolution of about 21 μm, although the resolution of the optical system was about 4 μm. The spatial resolution can be improved by reducing the mean free path of the reactive species by the addition of inert buffer gas. However, this improvement generally results in an increase in processing time.

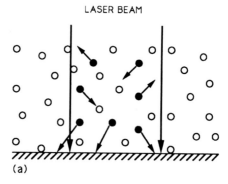

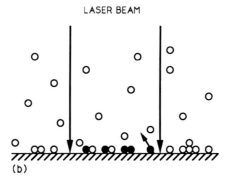

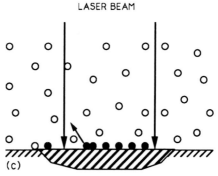

Fig. 17.10 Diagramatic representation of the dominant mechanisms that control processing rate and spatial resolution (○, parent species; ●, reactive species; ▨, excited material (carrier generation or heating)): (a) laser excitation of the gas phase species; (b) laser excitation of the adsorbed phase species; (c) laser excitation of the material surface.

17.4.3.2 Laser-excited adsorbed phase species

Laser−gas interaction in the adsorbed phase can result in an excitation or dissociation of the adsorbed species by a photolytic mechanism. The density of generated reactive species per unit time per unit surface area on the material surface can be expressed as

$$N_1(r, E, P, T) \propto N_{ads}\sigma_a I,$$

where N_{ads} and σ_a are respectively the density of the adsorbed parent species and their absorption cross-section and I is the laser beam intensity. As in the former case, the processing rate can be limited either by the kinetics of the chemical reaction or by the replenishment of the adsorbed layer. In both cases, the reactive species are located on the material surface, where they can react directly with the surface. For this reason, such a mechanism seems more appropriate than the excitation of gas phase species in order to achieve good spatial resolution. Tsao [74] showed, for example, that polymerization of MMA molecules adsorbed on Si and SiO_2, by absorption of UV laser light, results in a local deposition of PMMA only in the irradiated area. PMMA lines having widths down to 0.8 μm, corresponding to the minimal beam size, were deposited by this technique. In practice, irradiation of the sample surface through the gas atmosphere would give rise to the excitation of both adsorbed and gas phase species. There are two ways to favour the excitation of adsorbed species and thereby achieve good resolution. The first method is the use of parent species that are excited only in the adsorbed phase at the laser wavelength. This is sometimes possible because the absorption spectrum of adsorbed molecules is often shifted to longer wavelengths compared with those of gas phase molecules [126−129]. Secondly, one can use a parent gas having a low vapour pressure, so that a high coverage of the surface is achieved even at low gas pressure, limiting the supply of gas phase excited species.

17.4.3.3 Laser excitation of material surface

Direct laser excitation of the material gives rise to two types of re-action—thermal or photochemical. In thermal reactions, the laser beam is used to control localized material heating, which can result in a large range of thermally driven processes, such as adsorbed or gas phase molecular pyrolysis, and the activation of chemical reactions involving reactive species previously generated by gas phase photolysis. The processing rate R is expected to depend exponentially on the substrate temperature, according to the usual Arrhenius relation

$$R \propto \exp\left(-\frac{\Delta E}{k_B T_s}\right),$$

where ΔE is the activation energy of the thermal reaction, k_B is Boltzmann's constant and T_s is the temperature of the material surface. It should be pointed out that pyrolytic processing of materials gives generally higher processing rates than do the photolytic mechanisms. Spatial resolution is controlled primarily by the extent of the heated surface area. Thermal modelling in the case of silicon [20] predicts that, for a laser beam with a diameter of 5 μm and a pulse duration of 10^{-7} s, the surface temperature at a distance of about two times the beam radius would drop to $1/e$ of the peak temperature at the centre of the laser beam. Depending on the value of the activation energy ΔE of the thermal reaction, patterns with a size larger or smaller than the image size would be produced. High values of ΔE are needed to achieve good resolution. Thus thermal reaction takes advantage of the nonlinear dependence of the processing rate on the material temperature to increase the resolution. This can lead to minimal resolvable features down to approximately 0.6 times the image size in the case of laser projection-patterned processing [130].

Photochemical reactions at the gas–material interface can be controlled by carrier generation at the irradiated material surface. In this case, the spatial resolution is determined by carrier diffusion out of the irradiated area. At high laser intensities, this phenomenon can seriously degrade the spatial resolution.

From the above discussion, one can conclude that, whatever the dominant source of reactive species, by choosing appropriate laser–gas–material systems and processing conditions, spatial resolution could be limited by the image dimensions. In addition, for highly nonlinear processes, the resolution can be improved and results in resolvable features smaller than the image size. In the case of projection patterns performed with an ArF laser and an optimized optical system, one can expect an overall resolution approaching 0.1 μm.

17.5 CONCLUSIONS

We have reviewed the conventional processes involved in GaAs MESFET fabrication, and have compared a process flow chart for a SDFL inverter made in self-aligned T-gate structure technology by conventional processes with one for laser projection-patterned chemical processing. We have shown that laser chemical processing has many advantages over conventional processing. It considerably reduces the number of processing steps. It reduces the variety of processing techniques, since it can induce deposition, doping or removal of materials in the same processing chamber just by changing the reactive gas. In addition, one can expect a reduction in

GaAs surface damage as laser processing involves lower irradiation energies compared with conventional low-temperature processing techniques.

Thus laser processing of GaAs appears to be a promising technique. The principles and limitations of laser projection-patterned processing techniques have been discussed. UV excimer lasers seem to be particularly well adapted to laser projection-patterned chemical processing, as reviewed elsewhere in Chapter 3. The development of optical systems for excimer laser projection involves serious problems, as discussed by Goodall in Chapter 3. However, it is under intensive investigation, and one can expect to achieve optical spatial resolution as low as 0.15 μm.

Three basic mechanisms for reactive species formation have been distinguished, resulting in different limitations on the spatial resolution of the material processing. These mechanisms can give rise to spatial resolution limited by the image size and even better resolution if an appropriate laser–gas–material system and experimental conditions are chosen. Research on such systems is the main task in the development of laser projection-patterned processing of GaAs devices. The development of this technique is a challenge that may bring large benefits to GaAs device fabrication.

REFERENCES

[1] S. M. Sze (ed.), *Physics of Semiconductor Devices*, p. 312 (Wiley, New York, 1981).

[2] K. Yamasaki, Y. Yamane and K. Kurumada, *Electron. Lett.* **18**, 592 (1982).

[3] R. V. Tuyl and C. A. Liechti, *IEEE J. Solid State Circuits* **9**, 269 (1974).

[4] B. M. Welch and R. C. Eden, *IEDM Tech. Dig.* **18**, 205 (1987).

[5] R. C. Eden, *Proc. IEEE* **70**, 5 (1982).

[6] H. M. Macksey, F. H. Doerbeck and R. C. Vail, *IEEE Trans. Electron Devices* **27**, 467 (1980).

[7] K. Ohata, H. Itoh, F. Hasegawa and Y. Fujiki, *IEEE Trans. Electron Devices* **27**, 1029 (1980).

[8] N. Yokoyama, T. Mimura, M. Fukuta and M. Ishikawa, *ISSCC Dig. Tech. Papers*, p. 218 (1981).

[9] H. M. Levy and R. E. Lee, *IEEE Electron Devices Lett.* **4**, 102 (1983).

[10] R. A. Sadler and L. F. Eastman, *IEEE Electron Devices Lett.* **4**, 215 (1983).

[11] K. Imamura, T. Ohnishi, M. Shigaki, N. Yokoyama and H. Niski, *Electron. Lett.* **21**, 805 (1985).

[12] K. Yamasaki, K. Asai, T. Mizutani and K. Kurumada, *Electron. Lett.* **18**, 119 (1982).

[13] M. F. Chang, F. J. Ryan, R. P. Vahrenkamp and C. G. Kirkpatrick, *Electron. Lett.* **21**, 354 (1985).

[14] R. L. Vantuyl, C. Liechti, R. E. Lee and E. Gowen, *IEEE J. Solid State Circuits* **12**, 485 (1977).

[15] R. C. Eden, F. S. Lee, S. I. Long, B. M. Welch and R. Zucca, *ISSCC Dig. Tech. Papers*, p. 122 (1980).

[16] H. Ishikawa, *ISSCC Dig. Tech. Papers*, p. 200 (1977).

[17] A. G. Rode and J. G. Roper, *Solid State Technol.* **28**, 209 (1985).

[18] D. J. Ehrlich and J. Y. Tsao, *J. Vac. Sci. Technol.* **B1**, 969 (1983).

[19] Y. Rytz-Froidevaux, R. P. Salathe and H. H. Gilgen, *Appl. Phys.* **A37**, 121 (1985).

[20] D. Bauerle (ed.), *Chemical Processing with Lasers* (Springer-Verlag, Berlin, 1986).

[21] I. W. Boyd (ed.), *Laser Processing of Thin Films and Microstructures* (Springer-Verlag, Berlin, 1987).

[22] R. M. Osgood, *Mater. Res. Soc. Symp. Proc.* **75**, 3 (1987).

[23] M. Rothschild and D. J. Ehrlich, *J. Vac. Sci. Technol.* **B6**, 1 (1988).

[24] D. Bauerle, *Appl. Phys.* **B46**, 261 (1988).

[25] B. M. McWilliams, I. P. Herman, F. Mitlitsky, R. A. Hyde and L. L. Wood, *Appl. Phys. Lett.* **43**, 946 (1983).

[26] S. D. Mukherjee and D. W. Woodard, in *Gallium Arsenide* (ed. M. J. Howes and D. V. Morgan), p. 119 (Wiley, New York, 1985).

[27] A. C. Adams and B. R. Pruniaux, *J. Electrochem. Soc.* **120**, 408 (1973).

[28] C. M. Garner, C. Y. Su, W. A. Saperstein, K. G. Jew, C. S. Lee, G. L. Pearson and W. E. Spicer, *J. Appl. Phys.* **50**, 3376 (1979).

[29] B. R. Pruniaux and A. C. Adams, *J. Appl. Phys.* **43**, 1980 (1972).

[30] Y. Tarui, Y. Komina and T. Yamaguchi, *J. Japan. Soc. Appl. Phys. Suppl.* **42**, 78 (1973).

[31] H. Adachi and H. L. Hartnagel, *J. Vac. Sci. Technol.* **19**, 427 (1981).

[32] B. J. Sealy, in *Gallium Arsenide for Devices and Integrated Circuits* (ed. H. Thomas, D. V. Morgan, B. Thomas, J. E. Aubrey and G. B. Morgan), p. 128 (IEEE Electrical and Electronic Materials and Devices Ser., Vol. 3, 1986) (Peter Peregrinus Ltd, London, 1986).

[33] D. V. Morgan and F. H. Eisen, *Gallium Arsenide* (ed. M. J. Howes and D. V. Morgan), p. 161 (Wiley, New York, 1985).

[34] R. N. Thomas, H. M. Hopgood, G. W. Eldridge, D. L. Barrett, T. T. Barggins, B. Ta and S. K. Wang, *Semiconductors and Semimetals* **20**, 1 (1984).

[35] C. J. Palmstrom and D. V. Morgan, in *Gallium Arsenide* (ed M. J. Howes and D. V. Morgan), p. 195 (Wiley, New York, 1985).

[36] T. Ohnishi, N. Yokoyama, T. Onodera, S. Suzuki and A. Shibatomi, *Appl. Phys. Lett.* **43**, 600 (1983).

[37] H. Nakamura, Y. Sato, T. Nonaka, T. Ishida and K. Kaminishi, *GaAs IC Symposium. IEEE Galium Arsenide Integrated Circuit Symposium Technical Digest 1983*, Phoenix, AZ, 25–27 October 1983, p. 134 (IEEE, New York, NY, 1983).

[38] J. Gyulai, J. W. Mayer, I. V. Mitchell and V. Rodrigez, *Appl. Phys. Lett.* **17**, 332 (1970).

[39] F. H. Eisen, B. M. Welch, H. Muller, K. Gamo, T. Inada and J. W. Mayer, *Solid State Electron.* **20**, 219 (1977).

[40] A. A. Immorlica and F. H. Eisen, *Appl. Phys. Lett.* **29**, 94 (1976).

[41] R. M. Malbon, D. H. Lee and J. M. Whelan, *J. Electrochem. Soc.* **123**, 1413 (1976).

[42] J. Kasahara, M. Arai and N. Watanabe, *J. Appl. Phys.* **50**, 541 (1979).

[43] B. J. Sealy, *Microelectron. J.* **13**, 21 (1982).
[44] D. E. Davis, P. J. McNally, J. P. Lorenzo and M. Julian, *IEEE Electron Devices Lett.* **3**, 25 (1982).
[45] M. Arai, K. Nishiyama and N. Watanabe, *Japan J. Appl. Phys.* **20**, L124 (1981).
[46] J. C. C. Fan, B. Y. Tsaur and M. W. Geis, in *Laser and Electron Beam Interactions with Solids* (ed. B. R. Appleton and F. K. Celler), p. 741 (Elsevier, New York, 1982).
[47] J. L. Tandon and F. H. Eisen, *AIP Conf. Proc.* **50**, 616 (1979).
[48] D. H. Lowdes, *Semiconductors and Semimetals* **23**, 471 (1984).
[49] A. Ryss, Y. Shieh, A. Compaan, H. Yao and A. Bath, *Opt. Eng.* **29**, 329 (1990).
[50] B. J. Sealy, S. S. Kular, K. G. Stephens, R. Croft and A. Palmer, *Electron Lett.* **14**, 720 (1978).
[51] R. F. Pinizzotto, in *Ion Implantation and Ion Beam Processing of Materials* (ed. G. K. Hubler, O. W. Holland, C. R. Clayton and C. W. White), p. 265 (North-Holland, New York, 1984).
[52] D. M. Zehner, C. W. White and G. W. Ownby, *Appl. Phys. Lett.* **36**, 56 (1980).
[53] P. L. Cowan and J. A. Golovchenko, *J. Vac. Sci. Technol.* **17**, 1197 (1980).
[54] T. F. Deutsch, *Mater. Res. Soc. Symp. Proc.* **17**, 225 (1983).
[55] K. G. Ibbs and M. L. Lloyd, *Appl. Phys. Lett.* **36**, 243 (1980).
[56] T. W. Sigmon, *Mater. Res. Soc. Symp. Proc.* **75**, 619 (1987).
[57] S. Kato, H. Saeki, J. Wada and S. Matsumoto, *J. Electrochem. Soc.: Solid State Sci. Technol.* **135**, 1030 (1988).
[58] G. G. Bentini, M. Bianconi and S. Summonte, *Appl. Phys.* **A45**, 317 (1988).
[59] F. Foulon, A. Slaoui and P. Siffert, *Appl. Surf. Sci.* **43**, 333 (1989).
[60] M. L. Lloyd and K. G. Ibbs, *Mater. Res. Soc. Symp. Proc.* **29**, 35 (1984).
[61] P. G. Carey, K. Bezjian, T. W. Sigmon, P. Gildea and T. J. Magee, *IEEE Electron Device Lett.* **7**, 440 (1986).
[62] P. G. Carey, K. H. Weiner and T. W. Sigmon, *IEEE Electron Device Lett.* **9**, 542 (1988).
[63] T. F. Deutsch, D. J. Ehrlich, D. D. Rathman, D. J. Silversmith and R. M. Osgood, *Appl. Phys. Lett.* **39**, 825 (1981).
[64] T. F. Deutsch, J. C. C. Fan, D. J. Ehrlich, G. W. Turner, R. L. Chapman and R. P. Gale, *Appl. Phys. Lett.* **40**, 722 (1982).
[65] T. F. Deutsch, D. J. Ehrlich, D. D. Rathman, D. J. Silversmith and R. M. Osgood, in *Laser Diagnostics and Photochemical Processing for Semiconductor Devices* (ed. R. M. Osgood, S. R. J. Bruek and H. R. Schlossberg), p. 225 (North-Holland, New York, 1983).
[66] C. Cohen, J. Siejka, M. Berti, A. V. Drigo, G. G. Bentini, D. Pribat and E. Jannitti, *J. Appl. Phys.* **55**, 4081 (1984).
[67] H. Beneking, in *Laser Processing and Diagnostics* (ed. D. Bauerle), p. 274 (Springer-Verlag, Berlin, 1984).
[68] H. Krautle, W. Roth, A. Krings and H. Beneking, in *Laser Controlled Chemical Processing of Surfaces* (ed. A. W. Johnson, D. J. Ehrlich and H. R. Schlossberg), p. 353 (North-Holland, New York, 1984).
[69] H. Krautle, P. Roentgen, M. Maier and H. Beneking, *Appl. Phys.* **A38**, 49 (1985).
[70] H. Krautle and D. Wachenschwanz, *Solid State Electron.* **28**, 601 (1985).

[71] K. Sugioka and K. Toyoda, *Appl. Phys. Lett.* **55**, 619 (1989).
[72] P. A. Barnes, H. J. Leamy, J. M. Poate, S. D. Ferris, J. S. Williams and G. K. Celler, *Appl. Phys. Lett.* **33**, 965 (1978).
[73] D. H. Lowndes, J. W. Cleland, W. H. Christie and R. E. Eby, *Mater. Res. Soc. Symp. Proc.* **1**, 223 (1981).
[74] J. Fletcher, J. Narayan and D. H. Lowndes, in *Proceedings of 2nd Oxford Conference on Microscopy of Semiconductor Materials* (eds A. G. Cullis and D. C. Joy), p. 121 (Oxford University Press, 1981).
[75] J. Y. Taso and D. J. Erhlich, *Appl. Phys. Lett.* **42**, 997 (1983).
[76] D. J. Erhlich and J. Y. Tsao, *J. Vac. Sci. Technol.* **B1**, 969 (1983).
[77] D. J. Erhlich and J. Y. Tsao, *Appl. Phys. Lett.* **46**, 198 (1985).
[78] M. Rothschild and D. J. Erhlich, *J. Vac. Sci. Technol.* **B5**, 389 (1987).
[79] H. H. G. Jellinek and R. Srinivasan, *J. Phys. Chem.* **88**, 3048 (1984).
[80] R. Srinivasan, in *Laser Processing and Diagnostics* (ed. D. Bauerle) (Springer-Verlag, Berlin, 1984).
[81] B. J. Garrisson and R. Srinivasan, *J. Appl. Phys.* **57**, 2909 (1985).
[82] T. Keyes, R. H. Clarke and J. M. Isner, *J. Phys. Chem.* **89**, 4194 (1986).
[83] R. Srinivasan, in *Interfaces under Laser Irradiation* (eds D. Bauerle, L. D. Laude and M. Wautelet) (Martinus Nijhoff, Dordrecht, 1987).
[84] P. K. Boyer, C. A. Moore, R. Solanski, W. K. Ritchie, G. A. Roche and G. J. Collins, in *Laser Diagnostics and Photochemical Processing for Semiconductor Devices* (eds R. M. Osgood *et al.*) (North-Holland, New York, 1983).
[85] T. F. Deutsch and D. D. Rathman, *Appl. Phys. Lett.* **45**, 623 (1984).
[86] D. K. Flynn, J. I. Steinfeld and D. S. Sethi, *J. Appl. Phys.* **59**, 3914 (1986).
[87] D. Bauerle, *Appl. Phys.* **B46**, 261 (1988).
[88] R. Solanki, W. H. Ritchie and G. J. Collins, *Appl. Phys. Lett.* **43**, 454 (1983).
[89] D. J. Ehrlich and R. M. Osgood, *Chem. Phys. Lett.* **79**, 381 (1981).
[90] G. S. Higashi, G. E. Blonder and C. G. Fleming, in *Photon, Beam and Plasma Stimulated Chemical Processes at Surfaces* (ed. V. M. Donnelly, I. P. Herman and M. Hirose), p. 117 (Materials Research Society, Boston, 1987).
[91] J. E. Bouree and J. Flicstein, *Mater. Res. Soc. Symp. Proc.* **101**, 55 (1988).
[92] J. Flicstein and J. E. Bouree, *Appl. Surf. Sci.* **36**, 443 (1989).
[93] G. L. Loper and M. D. Tabat, *J. Appl. Phys.* **58**, 3649 (1985).
[94] R. K. Montgomery and T. D. Mantei, *Appl. Phys. Lett.* **48**, 493 (1986).
[95] H. Yokoyama, F. Uesugi, S. Kishida and K. Washio, *Appl. Phys.* **A37**, 25 (1985).
[96] T. H. Baum and E. E. Marinero, *Appl. Phys. Lett.* **49**, 1213 (1986).
[97] G. Koren, F. Ho and J. J. Ritsko, *Appl. Phys. Lett.* **46**, 1006 (1985).
[98] G. Koren, F. Ho and J. J. Ritsko, *Appl. Phys.* **A40**, 13 (1986).
[99] J. Y. Tsao and D. J. Ehrlich, *Appl. Phys. Lett.* **43**, 146 (1983).
[100] P. K. Boyer, K. A. Emery, H. Zarnani and G. J. Collins, *Appl. Phys. Lett.* **45**, 979 (1984).
[101] A. Tabe, K. Jinguji, T. Yamada and N. Takato, *Appl. Phys.* **A38**, 221 (1985).
[102] S. Szikora, W. Krauter and D. Bauerle, *Mater. Lett.* **2**, 263 (1984).
[103] T. Szorenyi, P. Gonzalez, M. D. Fernandez, J. Pou, B. Leon and M. Perez-Amor, in *Proceedings of 8th International Conference on Thin Films, San Diego, April 1990.*

[104] T. F. Deutsch, D. J. Silversmith and R. W. Mountain, in *Laser Diagnostics and Photochemical Processing for Semiconductor Devices* (eds R. M. Osgood *et al.*) (North-Holland, New York, 1983).

[105] Z. Yu, G. J. Collins and R. Solanki, in *Beam Induced Chemical Processes* (ed. R. J. von Gutfeld, J. E. Greene and H. Schlossberg), p. 93 (Materials Research Society, Boston, 1985).

[106] A. Sugimura and M. Hanabusa, *Japan. J. Appl. Phys.* **26**, L56 (1987).

[107] T. F. Deutsch, D. J. Silversmith and R. W. Mountain, in *Laser Processing and Diagnostics* (ed. D. Bauerle), p. 67 (Springer-Verlag, Berlin, 1984).

[108] M. Bass and J. R. Franchi, *J. Chem. Phys.* **64**, 4417 (1976).

[109] S. D. Allen, *J. Appl. Phys.* **52**, 6501 (1981).

[110] M. Rothschild and D. J. Ehrlich, *J. Vac. Sci. Technol.* **B5**, 389 (1987).

[111] M. Rothschild and D. J. Ehrlich, *J. Vac. Sci. Technol.* **B6**, 1 (1988).

[112] V. Pol, J. H. Bennewitz, T. E. Jewell and D. W. Peters, *Opt. Engng* **26**, 311 (1987).

[113] V. Pol. J. H. Bennewitz, G. C. Escher, M. Feldman, V. A. Firtion, T. E. Jewell, B. E. Wilcomb and J. T. Clemens, *Proc. SPIE* **633**, 6 (1986).

[114] M. Endo, M. Sasago, Y. Hirai, K. Ogawa and T. Ishihara, *Proc. SPIE* **774**, 138 (1987).

[115] N. Nakase, T. Sato, M. Nonaka, I. Higashikawa and Y. Horiike, *Proc. SPIE* **773**, 226 (1987).

[116] M. Kameyama and K. Ushida, *Proc. SPIE* **774**, 147 (1987).

[117] R. T. Kerth, K. Jain and M. R. Latta, *IEEE Electron Device Lett.* **7**, 299 (1986).

[118] F. Goodall and R. A. Lawes, *Microelectron. Engng* **6**, 61 (1987).

[119] F. Goodall and R. A. Lawes, *Microelectron. Engng* **11**, 187 (1990).

[120] T. R. Loree, K. B. Butterfield and D. L. Barker, *Appl. Phys. Lett.* **32**, 171 (1978).

[121] J. Heidberg, H. Stein, A. Nestmann, E. Hoefs and I. Hussla, *Surf. Sci.* **126**, 183 (1983).

[122] T. J. Chuang, *Surf. Sci.* **178**, 763 (1986).

[123] E. B. D. Bourdon, J. P. Cowin, I. Harrison, J. C. Polanyi, J. Segner and C. D. Stanners, *J. Phys. Chem.* **88**, 6100 (1984).

[124] T. J. Chuang, *J. Vac. Sci. Technol.* **21**, 798 (1982).

[125] P. D. Brewer, D. McClure and R. M. Osgood, *Appl. Phys. Lett.* **49**, 803 (1986).

[126] D. J. Ehrlich and R. M. Osgood, *Chem. Phys. Lett.* **79**, 381 (1981).

[127] C. J. Chen and R. M. Osgood, *Appl. Phys.* **A31**, 171 (1983).

[128] Y. Rytz-Froidevaux, R. P. Salathe, H. H. Gilgen and H. P. Weber, *Appl. Phys.* **A27**, 654 (1982).

[129] G. Radhakrishnan, W. Stenzel, H. Conrad and A. M. Bradshaw, in *Proceedings of European Materials Research Society, Symposium E* (eds I. W. Boyd, E. Fogarassy and M. Stuke) (Elsevier, Amsterdam, 1990).

[130] D. J. Ehrlich and J. Y. Tsao, *Appl. Phys. Lett.* **44**, 267 (1984).

18 Photon Probes for In-Process Control During Semiconductor Fabrication

GABRIEL M. CREAN

National Microelectronics Research Centre, Lee Malting, Prospect Row, Cork, Ireland

18.1 INTRODUCTION

The development of ultra large-scale integrated (ULSI) circuits requires both the successful application of novel material systems and fabrication processing techniques and also the application of what is now called "manufacturing science", which is essentially good manufacturing practice. A key factor in this area is nondestructive in-process control during fabrication [1, 2].

As semiconductor device geometries are reduced, the sensitivity of circuit performance to variations in the semiconductor manufacturing process increases. This sensitivity can be translated into poor manufacturing yields, resulting in higher production costs. As a result, stringent specifications are required for process tolerances such as layer composition and geometrical dimensions. To ensure that these and other process criteria (e.g. uniformity and reproducibility) are satisfied requires noncontact and nondestructive process control. This can be performed either *in situ* (during the process) or on-line (immediately after the process), at each step of the fabrication process.

A number of different physical mechanisms have been proposed to form the basis of these in-process measurements [3, 4]. Of these, photon-based probes are particularly suitable due to their noninvasive nature and their ability to operate in a broad range of environments from atmospheric pressure to ultrahigh-vacuum (UHV). The aim of this chapter is to review the applicability of photon-based probes to in-process control applications in modern semiconductor fabrication processes.

Photochemical Processing of Electronic Materials
ISBN 0-12-121740-X

18.2 PHOTON–SOLID INTERACTION

As new material systems and structures are developed, the material to be analysed has evolved to include bulk, thin-film and surface structures. In order to choose an appropriate photon probe for a given analysis problem, it is necessary to have an understanding of the interaction of the probe with the material system.

The physics of photon–solid interactions has been discussed in numerous reviews [5–8] and in Chapter 1. In the case of photon–semiconductor interaction, several phenomena can occur, ranging from absorption to scattering (Fig. 18.1).

From an optical science viewpoint, a classical point dipole model involving the interaction of a photon with a polarizable medium permits a phenomenological description of these interactions [7]. Photon interactions can thus be classified according to linear dipole origins (e.g. ellipsometry), first-order induced dipole effects (e.g. Raman spectroscopy) or second-order and high nonlinear polarizabilities (e.g. second-harmonic generation).

However, a more appealing classification for material scientists and semiconductor process control engineers is whether the technique is bulk-

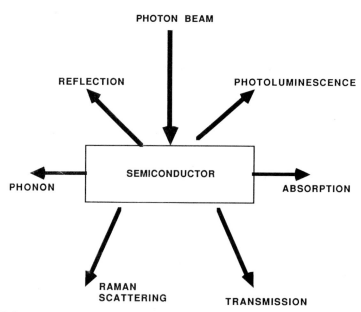

Fig. 18.1 Macroscopic interaction between a photon beam and a semiconductor material.

or surface-specific. One can define surface-specific techniques as those that take advantage of crystal symmetry to suppress the contribution to the photon response arising from the bulk. The principal surface-specific techniques demonstrated to have a submonolayer sensitivity include second-harmonic generation, reflectance difference spectroscopy and laser light scattering. These are discussed in Section 18.5.

18.3 PHOTON RESPONSE: MODELLING AND INTERPRETATION

Accurate extraction of material parameters such as film thickness, composition, crystal quality or uniformity from a measured photon response requires the use of physical models (Fig. 18.2). Significant modelling effort has been directed at linking a macroscopic photon probe response (reflectivity, absorption, dielectric function, etc.) to the compositional and microstructural parameters of the material system under investigation [8]. However, for the purposes of quantitative in-process control it is equally important that confidence limits can be assigned to these material properties; for several photon probe applications this requires further work.

An interesting example concerns the application of spectroscopic ellipsometry to the determination of the material properties of silicon-on-insulator (SOI) material. SOI technology has a number of advantages compared with bulk silicon (Si) technology for fabricating submicron integrated circuits [9]. However, acceptance of the emerging SOI technology is being impeded by the lack of availability of characterization technology for rapid and nondestructive qualification of the starting SOI material [10, 11]. A typical SIMOX (separation of implantation of oxygen) SOI material structure is presented in Fig. 18.3. The material parameters of interest are the thicknesses of the capping oxide, top silicon, buried oxide and interface layers and the quality (percentage crystallinity) of the top Si and interface layers.

Ellipsometry measures the change in polarization state that occurs upon interaction of an incident light beam on a sample [12]. The measured ellipsometric parameters Δ and ψ can be related to the complex reflectance ratio R of the material through the equation

$$R = e^{i\Delta} \tan \psi. \tag{18.1}$$

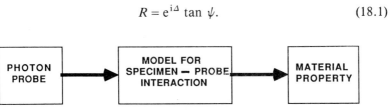

Fig. 18.2 Connection between photon probe and material property.

A measured spectroscopic ellipsometer (Δ, ψ) spectrum response for a typical SIMOX structure over the wavelength range 250–700 nm is presented in Fig. 18.4 [13]. In order to extract the material properties discussed above, we must relate the ellipsometric parameters to the SOI film structure. The level of sophistication of the model to be employed is determined by the number of model parameters utilized. The objective is to find an optimized set of parameters that minimize the deviation of a calculated ellipsometric spectrum from the experimental spectrum of Fig. 18.4. The calculated spectra from two such exercises are presented in Figs 18.4 and

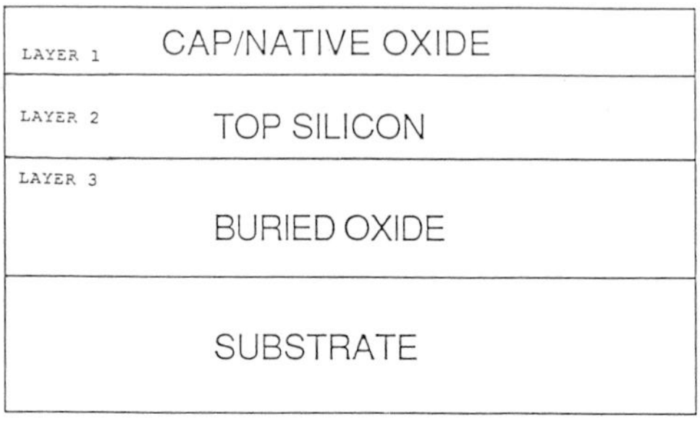

LAYER 1	CAP/NATIVE OXIDE
LAYER 2	TOP SILICON
LAYER 3	BURIED OXIDE
	SUBSTRATE

Fig. 18.3 Cross-section of a silicon-on-insulator SIMOX structure.

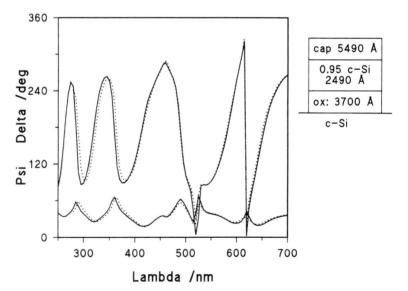

Fig. 18.4 Experimental ($\cdots\cdots$) and fitted (———) ellipsometric spectra with a three-layer SOI model. (From [13].)

18.5, together with the model parameters utilized. In both cases the top Si layer microstructure is assumed to comprise primarily crystalline silicon, with a fractional percent of amorphous silicon material. This heterogeneous layer can be approximated by a homogeneous medium with an effective dielectric response. The three such effective medium theories that exist are the Maxwell Garnet (MG) model [14, 15], the Lorentz–Lorentz (LL) model [16, 17] and the Bruggeman effective medium approximation (EMA) model [18]. In the above calculations the EMA model was employed.

The important point to note in Figs 18.4 and 18.5 is that both calculated spectra provide very good fits to the experimental data despite the significant difference in the level of sophistication of the SOI models utilized. Such a situation would clearly pose a dilemma for an incoming SOI material inspection/qualification engineer. This example demonstrates that to use spectroscopic ellipsometry effectively, one needs to address not only the physical model involved but also the subsequent decision tree for extracting a set of specific material parameters with predetermined confidence factors.

In general, before a given photon probe can be utilized in a microelectronics process control environment, several data interpretation factors require evaluation:

(i)   minimum required level of sophistication of the model;

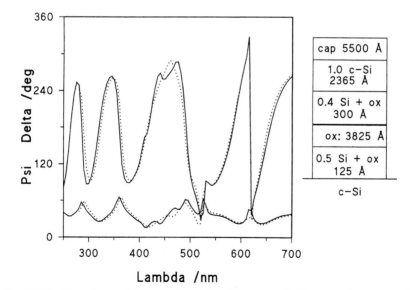

Fig. 18.5   Experimental (·······) and fitted (———) ellipsometric spectra with a five-layer SOI model. (From [13].)

(ii)   accuracy of the input optical data over the scanned wavelength range;

(iii) correlation of model parameters;

(iv)   correlation of the sensitivities of the measured parameters with the model parameters;

(v)   choice of minimization routine;

(vi)   decision tree.

18.4   *IN SITU* PROBE DESIGN CONSIDERATIONS AND TRENDS

Having selected on theoretical grounds the most suitable photon prove for a given material analysis or process control problem and an appropriate data interpretation–decision strategy, one must then evaluate the feasability of an *in situ* probe design and its subsequent integration with the process reactor. Several key parameters must be decided:

(i)   spectral range required for data acquisition (this will dictate whether a laser or lamp source must be employed);

(ii)   reactor geometry:
       (a) optical view port or fibre position(s);
       (b) working distance between sample and view port or fibre
       (both of these parameters will have an impact on the optics configuration);

(iii) reactor environment (will the interior face of the view port become contaminated or the presence of a fibre perturb the process?);

(iv)   availability of suitable view port material/window (the choice of view port material will be dictated by the spectral transmittance and retardance parameters required);

(v)   response time (this will have an impact on the signal-to-noise ratio attainable).

Examples of the modification of reactor geometries to facilitate the attachment of *in situ* photon probes are presented in several figures in Section 18.5. A design for an optimized view port for *in situ* optical characterization in ultrahigh vacuum is also available in the literature [19].

As photon-based probes become more widespread as *in situ* process control tools, it will no longer be acceptable for the semiconductor equipment vendor to add off-the-shelf optical systems to process equipment. The trend is increasingly towards semiconductor manufacturing companies working with processing equipment manufacturers to provide *in situ* process control capabilities rather than modifying them *a posteriori*. The optimal solution requires collaboration between semiconductor technologists and materials characterization scientists working in multidisciplinary teams to respond to the above issues.

One example of this type of initiative is the Commission of the European Community (CEC) sponsored research project ESPRIT 5004 SEDESES (selective deposition of epitaxial silicides and silicon). The objective of this project is the development of a multichamber multiprocessing system. One of the six work packages concerns the design and development of photon-based in-process control instrumentation. The project partners working on this include ASM (semiconductor equipment manufacturer), Siemens (semiconductor manufacturer) and NMRC (design of *in situ* photon probes).

18.5   APPLICATIONS

The aim of this section is to demonstrate the potential utility of photon-based probes to in-process control applications in modern microelectronics fabrication processes. The examples and discussion are organized according to the specific fabrication step monitored.

18.5.1   Plasma and reactive ion etching

The drive towards reduced device dimensions in both silicon and gallium arsenide technologies will depend upon the application of dry etching techniques for the purpose of high-fidelity pattern transfer [20]. As is discussed in Chapters 12 and 13, current research in the field of dry etching is focused on the attainment of etch selectivity, rate control and anisotropy of etch process and the minimization of etch-induced lattice and interface damage. This latter parameter is becoming increasingly important, since dry-etch induced surface modifications can directly affect device manufacturing yields via degradation mechanisms such as increased junction leakage and threshold voltage shifts. These degradation mechanisms have severely limited the applicability of dry etching techniques to nanometric fabrication [21, 22]. At present, several techniques are under examination for etch endpoint detection and surface damage quantification. Photon probe techniques include photoluminescence (PL) [23–27], Raman spectroscopy [28–30], ellipsometry [31–35], laser-induced fluorescence (LIF) [36] and photothermal radiometry [37–39].

18.5.1.1 Photoluminescence

Photoluminescence is an emission technique in which near-IR luminescence caused by the annihilation of an exciton bound to a neutral impurity is observed. The technique may be employed for plasma monitoring due to its following properties

(i) the PL spectrum is material-specific;

(ii) quantum yields are dependent on surface state density.

The former property enables endpoint detection, while the latter permits etch damage detection. While use of PL has been documented for characterization of etched samples, its use as an *in situ* real-time probe is recent [27]. Figures 18.6 and 18.7 present results obtained in this study. PL from a $Al_{0.3}Ga_{0.7}As$ (1 μm) epitaxial layer on semi-insulating GaAs is plotted as a function of etch depth during BCl_3 plasma etching in Fig. 18.6. It is observed that the photoluminescence from the GaAs layer decreases dramatically when the epilayer is removed. This is consistent with the GaAs surface being directly exposed to the BCl_3 plasma at this instance, and provides a distinct endpoint detection signal. Figure 18.7 demonstrates the recovery of the PL GaAs signal after passivation of the GaAs surface using a H_2 plasma, and indicates the surface-state-sensitive nature of the technique.

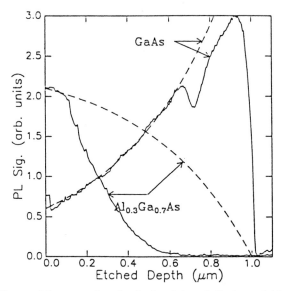

Fig. 18.6 Observed (———) and calculated (------) GaAs and $Al_{0.3}Ga_{0.7}As$ PL intensity induced by CW laser excitation during BCl_3 plasma etching. (From [27].)

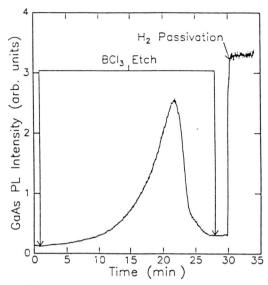

Fig. 18.7 Etch passivation sequence. GaAs PL excited using the pulsed laser during etching of $Al_{0.3}Ga_{0.7}As$ using a BCl_3 plasma and subsequent H_2 plasma passivation of the GaAs substrate. (From [27].)

18.5.1.2 Ellipsometry

Ellipsometry has been widely used for monitoring of plasma and reactive ion etching. A schematic of a single-wafer plasma-reactive ion etch reactor and associated *in situ* ellipsometer for monitoring etching of SiO_2 films on Si is presented in Fig. 18.8 [34]. (Δ, ψ) curves recorded during RIE etching are presented in Figs 18.9 and 18.10. The closed-loop curves are typical of the ellipsometric response from a homogeneous nonabsorbing thin film on an absorbing substrate [40]. For this configuration, as the thin-film thickness decreases, plots of ψ versus Δ produce cyclical closed-loop curves. It can be observed from the above figures that at pressures below 8 Pa the etch terminates with a bare Si substrate surface, while above this value a fluorocarbon layer is deposited. From a given position on the (Δ, ψ) curve, the thickness of the SiO_2 layer can be determined via a comparison with a calculated ellipsometric response for the thin-film on substrate structure.

Apart from endpoint detection and etch damage quantification, another parameter of interest is surface roughness. For example, roughening of silicon surfaces has been observed during chlorinated (Cl_2 or HCl) plasma etching [41]. Figure 18.11 shows a measured (Δ, ψ) curve recorded during *in situ* monitoring of a two-stage chlorinated plasma etch of a silicon surface that resulted in a roughening–smoothing transition [35]. Roughening of the Si surface (100 mTorr, 100 W) is characterized by the

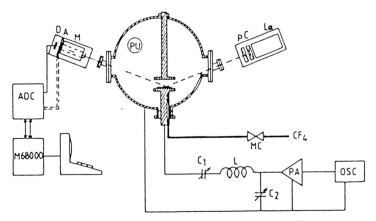

Fig. 18.8 Schematic of an *in situ* ellipsometry arrangement of the measurement of etch rate: A, rotating analyser; C, quarter-wave plate; C1, C2, matching capacitors; D, detector; L, matching inductor; La, light source; P, polarizer. (From [34].)

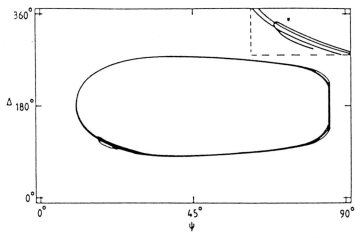

Fig. 18.9 Measured (Δ, ψ) contour for a pressure of 20 Pa. Each full revolution of the egg-shaped curve corresponds to an etched thickness of 280 nm. At the end of the process a deviation of the ideal curve is observed caused by the deposition of a fluorocarbon layer. [From [34].)

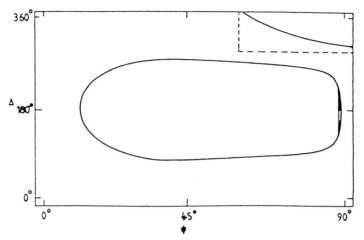

Fig. 18.10 Measured (Δ, ψ) contour for a pressure of 5 Pa. In this case the curve does not reverse direction after the endpoint. This indicates that no polymer film starts growing at lower pressures. (From [34].)

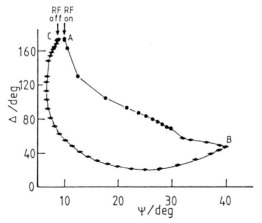

Fig. 18.11 Measured (Δ, ψ) contour during a two-stage etch process for silicon: AB recorded during roughening at 100 mTorr and 100 W; BC recorded during subsequent etching at 125 mTorr and 50 W. (From [35].)

curve segment AB on the (Δ, ψ) signature. After this etch sequence, the Si surface appeared black upon removal from the etch chamber. Subsequent etching at 125 mTorr and 50 W resulted in a smoothing effect and a highly reflective Si wafer surface. This latter etch is characterized by the curve segment BC. *In situ* ellipsometric curves such as Fig. 18.11 can provide real-time data that, when compared with a calculated ellipsometric response, allow quantitative interpretation of the extent and nature of the roughening process.

The above ellipsometric examples demonstrate that the technique is potentially very useful as an *in situ* process control tool for etch monitoring. It can provide information on thin-film etch rate, thin-film thickness, surface roughness and polymer film formation.

18.5.1.3 Photothermal radiometry

The use of another laser-based technique, photothermal radiometry [42–44], for monitoring defects induced in both Si and gallium arsenide (GaAs) wafers as a function of etch parameters in reactive ion etching (RIE) and reactive ion beam etching (RIBE) systems has been demonstrated [37–39].

The PTR system is presented in Fig. 18.12 and comprises a modulated semiconductor laser beam focused onto the surface of the semiconductor substrate with a spot size better than 2 μm. The laser wavelength is 0.83 μm, with a maximum power output of 30 mW. Infrared radiation emitted from the injected electron–hole plasma together with reflected laser light is focused onto a selective wavelength beamsplitter. The infrared radiation is subsequently focused onto the sensitive element of a cooled HgCdTe detector, where it is converted to an electrical signal.

A theoretical framework for modelling the photothermal radiometric response from a two-layer semiconductor system has been developed [44]. The introduction of a dense electron–hole plasma into a semiconductor sample will disrupt the equilibrium absorption and hence radiant emittance of the semiconductor. Using the above model, it can be demonstrated that, for a surface-modified semiconductor substrate, the change in emissivity due to the modified surface electronic properties is proportional to the integrated photo-induced carrier density, and gives rise to the dominant signal mechanism.

Figures 18.13 and 18.14 present the PTR measurements for the GaAs RIBE and Si RIE etched substrates [37]. For the Si etched substrates the PTR amplitude decreases monotonically as a function of increasing dry etch voltage bias. These experimental results are consistent with the assumption that increasing voltage bias tends to extend the range and severity of lattice damage in the near-surface region of the Si substrates.

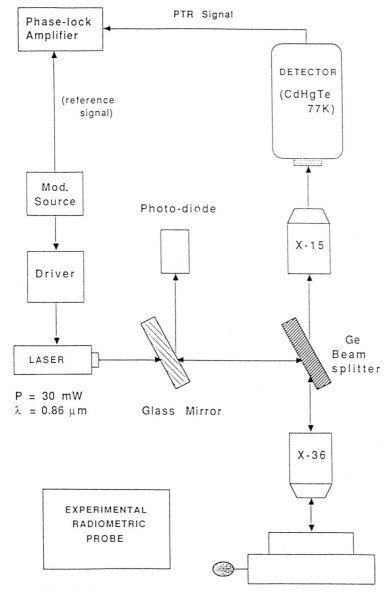

Fig. 18.12 Schematic of photothermal radiometric probe.

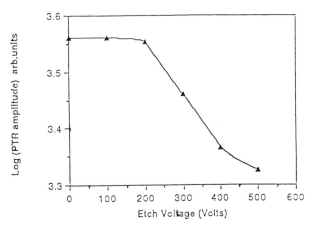

Fig. 18.13   Measured PTR amplitude variation as a function of applied dry etch voltage bias for 2×10^{18} cm^{-3} n-type GaAs substrates. (From [37].)

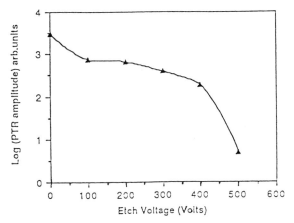

Fig. 18.14   Measured PTR amplitude variation as a function of applied dry etch voltage bias for 2–4 Ω n-type Si substrates. (From [37].)

However, for the GaAs substrates the experimental PTR results indicate that a bias voltage level between -200 and -300 V appears to represent a threshold for dry-etch-induced process damage with the stated RIBE system parameters. This result correlates strongly with recent data (Fig. 18.15) concerning dry-etch-induced damage in GaAs studied using Raman spectroscopy, where it was observed that Ar RIE induced damage in GaAs increased dramatically at a self-bias of approximately -300 V [30]. It is interesting to note that the maximum depletion depth as deter-

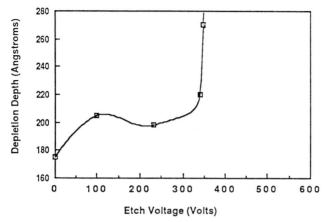

Fig. 18.15 Depletion depth versus self-bias voltage for RIE in argon determined using Raman scattering spectroscopy. (From [30].)

mined by Raman spectroscopy is less than 300 Å, thus indicating the high sensitivity of the PTR probe to near-surface lattice modifications.

The experimental PTR results presented for both material systems indicate the feasibility of rapid and noncontact monitoring of etch-induced process variations in semiconductor materials. Work is currently in progress on extending this technique to indium phosphide (InP) etch characterization [39].

18.5.2 Crystal growth

Epitaxial films fabricated using a variety of different techniques such as molecular beam epitaxy (MBE) [45, 46] or metal-organic chemical vapour deposition (MOCVD) [47] are increasingly used in both Si and compound semiconductor fabrication processes. New electronic and opto-electronic devices employing these materials impose severe tolerance requirements on layer thickness and composition. Consequently, a method of *in situ* control of such parameters is urgently required. In addition, basic research on epitaxial growth processes is currently hampered by the lack of real-time diagnostics during the growth sequence. Such considerations have led to the investigation of several photon-based techniques for monitoring crystal growth processes. These include laser-induced fluorescence [48], laser light scattering [49, 50], reflectance difference spectroscopy [51–56] and second-harmonic generation (SHG) [57–61].

18.5.2.1 Second-harmonic generation

As stated in Section 18.2, SHG is a surface-specific technique with a sub-monolayer sensitivity. This results from the fact that SHG is forbidden in bulk centrosymmetric crystals but is allowed where any inversion symmetry is broken—for example at the surface of a crystal or the interface between two dissimilar epitaxial layers [62, 63]. One example is presented below.

A schematic of an MBE chamber with *in situ* SHG and reflection high-energy electron diffraction (RHEED) probes is shown in Fig. 18.16 [61].

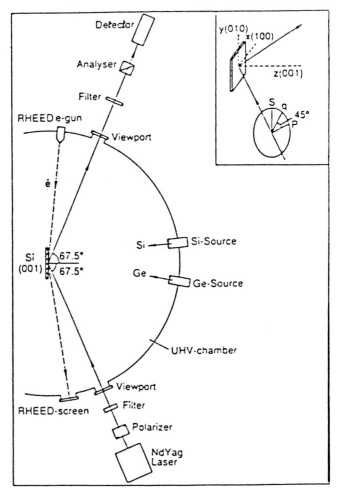

Fig. 18.16 Schematic of *in situ* second-harmonic generation probe attached to a UHV chamber. (From [61].)

This experimental arrangement was used to observe in real time the deposition of Si and germanium (Ge) on Si(100) as a function of deposition temperature. Figure 18.17 presents SHG intensity curves recorded using the above arrangement during deposition of Si on a Si(001) 2 × 1 reconstructed surface at deposition temperatures of 300 and 750 K. It is observed that growth at room temperature results in a decrease in SHG intensity *vis à vis* that for deposition at 750 K, indicating disordered growth at the lower deposition temperature.

The results indicate the surface sensitivity of the SHG probe. However, one of the disadvantages of single-wavelength SHG is that it is not species-specific. This has led to the proposal to use spectroscopic SHG and tune the frequency to sweep through an electronic transition of interest [64]. A complementary *in situ* photon probe technique that is species-specific is laser-induced fluorescence.

18.5.2.2 Laser-induced fluorescence

LIF is based upon monitoring the fluorescence from excited molecules that have been pumped from a ground state to a selected higher electronic state. This technique has recently been used to study the gas phase populations of vibrational and rotational states of As_2 emitted from a commercial As_4 MBE oven cracker source in order to monitor surface stoichiometry during

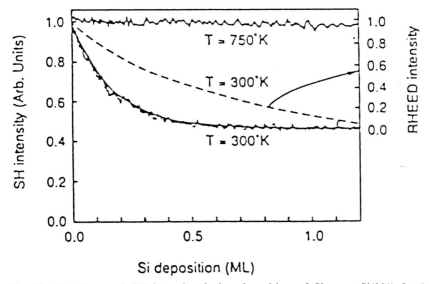

Fig. 18.17 Measured SH intensity during deposition of Si on a Si(001) 2 × 1 reconstructed surface at different deposition temperatures. (From [61].)

epitaxial growth. The MBE reactor and attached LIF instrumentation are presented in Fig. 18.18 [48]. Extraction of data is carried out by the fitting of calculated rotational line spectra and vibrational band intensities with measured AS_2 spectra. A typical result is presented in Fig. 18.19. Analysis of such spectra has led to the conclusion that both the rotational and vibrational states are thermalized at the cracker temperature.

The application of this technique in conjunction with other surface-specific techniques may lead to enhanced understanding and optimization of epitaxial growth processes.

18.5.2.3 Reflectance difference spectroscopy

In RDS the difference between normal-incidence reflectances of light polarized parallel and perpendicular to a principal crystallographic axis in the plane of the surface is measured experimentally as a function of photon energy. The technique is surface-sensitive due to the fact that the ordinarily dominant bulk contributions are largely isotropic and cancel on subtraction, while the surface component is highly anisotropic.

Figure 18.20 presents a typical reflectance difference arrangement [54]. A schematic of a single-wavelength RD system adapted for MOCVD growth monitoring is shown in Fig. 18.21 [56]. A mathematical analysis of different RDS configurations has been presented [54]. Recently, a double-

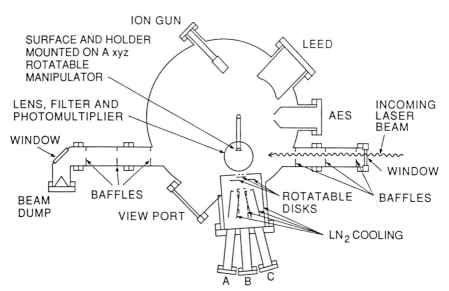

Fig. 18.18 Schematic of LIF attachment to MBE reactor for optical probing of epitaxial species. (From [48].)

modulation extension of RDS has been developed [55]. Figure 18.22 presents RD signals for GaInAs–InP heterojunctions grown at different temperatures [56]. An increase in temperature leads to a significant increase in interface quality as observed from the RDS signatures.

18.5.3   Silicide formation

Thin-film refractory metal silicides [65] are of considerable interest for use as polycide gates [66], interconnect material in VLSI technology and as the

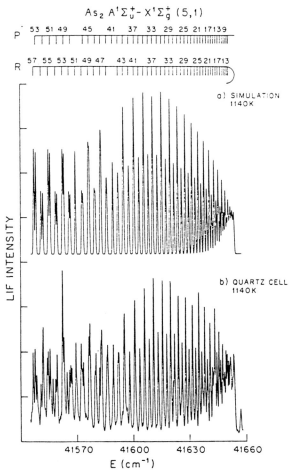

Fig. 18.19   LIF analysis: (a) 1140 K simulation of the (5, 1) band with rotational line assignments; (b) experimental spectrum of As₂ formed by thermal dissociation of As₄ in a quartz cell heated to 1140 K. [From [48].)

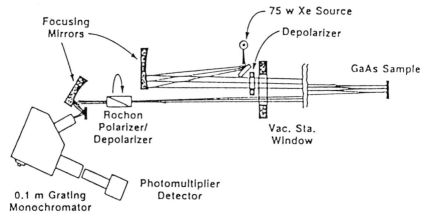

Fig. 18.20 Schematic of basic reflectance difference spectrometer. (From [54].)

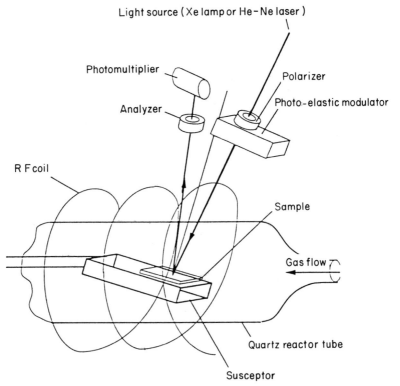

Fig. 18.21 Schematic of single-wavelength reflectance difference spectrometer adapted to a MOCVD reactor tube. (From [56].)

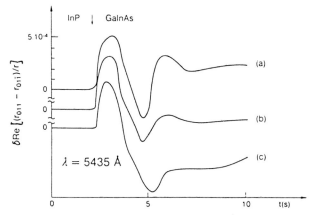

Fig. 18.22   Measured RDS spectra (wavelength 5435 Å) of GaInAs–InP lattice-matched heterostructures grown at different temperatures: (a) 500°C; (b) 520°C; (c) 540°C. (From [56].)

key component of novel heterostructure devices such as the metal-based transistor [67]. However, while the solid state reactions that occur upon deposition of a transition metal film on a silicon substrate and subsequent thermal treatment are well documented [68], the increasing use of very thin silicide films and novel thermal processing has posed questions concerning the microstructure of the silicide films that result. Future developments will increasingly depend on how well these materials can be characterized and controlled. This has led to the investigation of *in situ* photon probes for monitoring the microstructure and thickness of such films. Several probes have been evaluated for this application, including time-resolved optical reflectivity [69, 70], differential optical reflection spectroscopy [71] and photothermal reflectance (PR) measurements [72–74].

18.5.3.1   Differential optical reflectance spectroscopy

In reflection spectroscopy the technique is employed to avoid the uncertainty of absolute reflectance measurements. A typical instrumental configuration attached to a UHV chamber is shown in Fig. 18.23 [71]. In order to obtain the microstructure and thickness of the thin-film silicide, calculated reflectance spectra must be fitted to the measured spectra with the aid of a physical model. The development of such a model for ultrathin-film (<10 nm) titanium silicide ($TiSi_2$) has been discussed [71]. A comparison between measured and best-fit differential reflectance spectra for the ultrathin-film $TiSi_2$ samples is presented in Fig. 18.24. The results suggest that microstructural parameters can be evaluated *in situ* for silicide layers as thin as 1 nm.

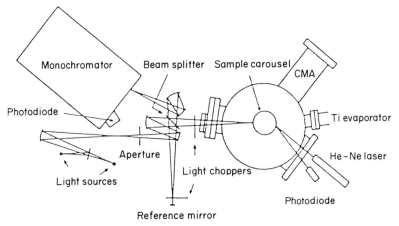

Fig. 18.23 Schematic of differential optical reflectance system attached to an UHV chamber. (From [71].)

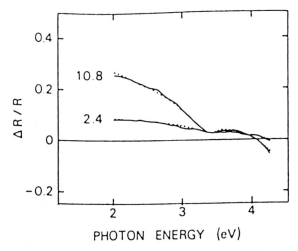

Fig. 18.24 Measured (———) and best-fit (·····) differential reflectance spectra for thin-film (10.8 and 2.4 nm) TiSi₂ on thermally cleaned Si(111). (From [71].)

18.5.3.2 Time-resolved optical reflectivity

This technique has been used in several thermal kinetics studies, and has recently been applied to the investigation of silicide formation kinetics in a rapid thermal anneal (RTA) system [70]. The technique is similar to differential reflectance and ellipsometry in that material parameters are extracted via the use of an optical response model for the layer structure

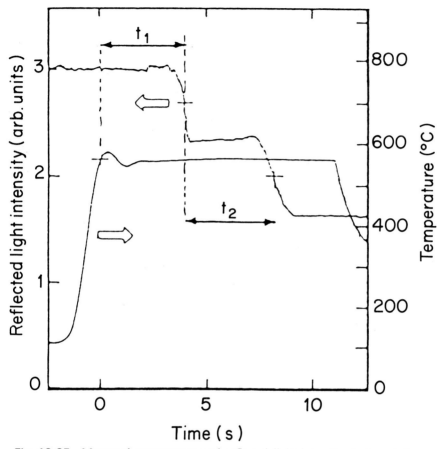

Fig. 18.25   Measured temperature and reflected light intensity during platinum silicide formation; t_1 and t_2 refer to Pt_2Si and $PtSi$ formation. (From [70].)

under study. Figure 18.25 presents temperature and reflected light intensity variations during the formation of platinum silicide, and demonstrates the sensitivity of the reflectance measurement to microstructural phase changes. Such a system could therefore be employed not only for *in situ* investigation of thin-film silicide kinetics but also for silicide endpoint detection.

18.5.3.3   Photothermal reflectance

Photothermal reflectance wave probing of both semiconductor and nonsemiconductor materials has been demonstrated. The technique is based upon monitoring local laser-induced variations in the thermo-elastic

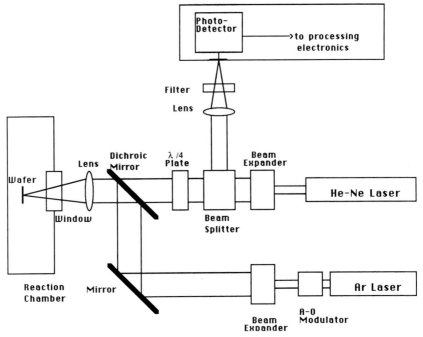

Fig. 18.26 Schematic of photothermal reflectance system.

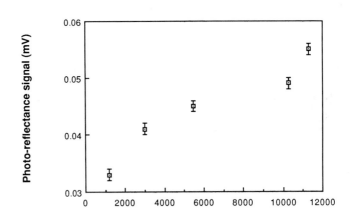

Tantalum Silicide Thickness (Å)

Fig. 18.27 Measured variation in PR amplitude as a function of tantalum silicide thickness. (From [76].)

and optical properties of the surface region of a material via a second probe laser [75]. An example of an *in situ* photothermal reflectance probe designed at the NMRC is presented in Fig. 18.26.

Recently, this technique has gained considerable attention for the characterization of thin-film silicide layers. Work by research groups at National Semiconductor Corporation, USA [72], Toshiba Corporation, Japan [73] and in Europe under the ESPRIT SEDESES project Siemens, Germany and NMRC, Ireland [76] has demonstrated the potential for in-process control of silicide parameters such as stoichiometry and film thickness. Figure 18.27 demonstrates the variation of photothermal reflectance amplitude as a function of tantalum silicide film thickness [76].

18.6 CONCLUSIONS

Stringent process tolerances on material parameters such as layer composition and geometrical dimensions are necessary for successful ULSI fabrication. To ensure that these and other process criteria (e.g. uniformity and reproducibility) are satisfied requires noncontact and nondestructive process control. As discussed in Section 18.1, photon-based probes are particularly suitable for this application.

Current work is focused on developing specific quantitative *in situ* photon characterization tool for specific semiconductor fabrication steps, as demonstrated in Section 18.5. However, this requires an analytical strategy that must address the choice of photon probe, optimized instrumental configuration and sophisticated material parameter extraction decision tree.

The optimal solution requires collaboration between semiconductor technologists and materials characterization scientists working in multi-disciplinary teams to respond to the above issues. This work is rapidly advancing, and it seems inevitable that within the next few years *in situ* process control will become an integral part of semiconductor manufacturing.

ACKNOWLEDGEMENTS

I am grateful to my postgraduate students, I. Little, S. Lynch, J. Beechinor, M. Murtagh and to NMRC staff members T. Flaherty, P. A. F. Herbert, A. Mathewson and M. O'Sullivan for many interesting discussions and their contribution to this chapter.

REFERENCES

[1] D. H. Phillips, in *ULSI Science and Technology* (ed. S. Broydo and G. M. Osburn), p. 752 (Proceedings Vol. 87-11, The Electrochemical Society, 1987).

[2] Commission of the European Communities, Brussels, 1989. Workprogramme for Microelectronics.

[3] W. Vandervorst and H. Bender, in *Analytical Techniques for Semiconductor Materials and Process Characterisation* (ed. B. Kolbesen, D. V. McCaughan and W. Vandervorst), p. 139 (Proceedings Vol. 90-11, The Electrochemical Society, 1990).

[4] W. Bergholz, V. Penka and G. Zoth, in *Analytical Techniques for Semiconductor Materials and Process Characterisation* (ed. B. Kolbesen, D. V. McCaughan and W. Vandervorst), p. 29 (Proceedings Vol. 90-11, The Electrochemical Society, 1990).

[5] D. Nudelman and S. S. Mitra (eds), *Optical Properties of Solids* (Plenum Press, New York, 1969).

[6] N. Bloembergen, *Nonlinear Optics* (Benjamin, Reading, Mass., 1977).

[7] M. Born and E. Wolf, *Principles of Optics* (Pergamon Press, Oxford, 1980).

[8] D. E. Aspnes, *Thin Solid Films* **89**, 262 (1982).

[9] A. J. Auberton-Herve, in *Silicon-on-Insulator Technology and Devices* (ed. D. N. Schmidt), p. 478 (Proceedings Vol. 90-6, The Electrochemical Society, 1990).

[10] M. T. Duffy, G. W. Cullen, A. Ipri, L. Jastrzebski and W. Sheed, in *Silicon-on-Insulator Technology and Devices* (ed. D. N. Schmidt), p. 298 (Proceedings Vol. 90-6, The Electrochemical Society, 1990).

[11] G. W. Cullen and M. T. Duffy, in *Silicon-on-Insulator Technology and Devices* (ed. D. N. Schmidt), p. 10 (Proceedings Vol. 90-6, The Electrochemical Society, 1990).

[12] R. M. Azzam and N. M. Bashara, *Ellipsometry and Polarized Light* (North-Holland, Amsterdam, 1976).

[13] S. Lynch, R. Greff, J. Margail and G. M. Creen, To be published.

[14] J. C. Maxwell Garnett, *Phil. Trans. R. Soc. Lond.* **A203**, 385 (1904).

[15] J. C. Maxwell Garnett, *Phil. Trans. R. Soc. Lond.* **A205**, 237 (1906).

[16] L. Lorentz, *Ann. Phys. Chem.* **11**, 70 (1880).

[17] H. A. Lorentz, *Theory of Electrons*, 2nd edn (Teubner, Leipzig, 1916).

[18] D. A. G. Bruggeman, *Ann. der Phys.* **24**, 636 (1935).

[19] A. A. Studna, D. E. Aspnes, L. T. Florez, B. J. Wilkens, J. P. Harbison and R. E. Ryan, *J. Vac. Sci. Technol.* **A7**, 3291 (1989).

[20] A. R. Reinberg, in *VLSI Electronics Microstructure Science* (ed. N. G. Einspruch) (Academic Press, New York, 1981).

[21] S. W. Pang, W. D. Goodhue, T. Lyszczarz, D. J. Ehrlich, R. B. Goodman and G. D. Johnson, *J. Vac. Sci. Technol.* **B6**, 1916 (1988).

[22] E. M. Clausen, H. G. Craighead, J. P. Harbison, A. Scherer, L. M. Schiavone, B. Van der Gaag and L. T. Florez, *J. Vac. Sci. Technol.* **B7**, 2011 (1989).

[23] R. R. Chang, R. Iyer and D. L. Lile, *J. Appl. Phys.* **61**, 1995 (1987).

[24] C. Harris, W. D. Sawyer, M. Konuma and J. Weber, in *Proceedings of European Materials Research Society Symposium on Science and Technology of Defects in Silicon* (North-Holland, Amsterdam, 1989).

[25] H. Weman, J. L. Lindstrom and G. S. Oerhlein, *Proceedings of European Materials Research Society Symposium on Science and Technology of Defects in Silicon* (North-Holland, Amsterdam, 1989).

[26] K. Akita, M. Taneya, Y. Sugimoto, H. Hidaka and M. Tajima, *J. Vac. Sci. Technol.* **A8**, 3274 (1990).

[27] A. Mitchell, R. A. Gottscho, S. J. Pearton and G. R. Scheller, *Appl. Phys. Lett.* **56**, 821 (1990).

[28] J. C. Tsang, G. S. Oehrlein, I. Haller and J. S. Custer, *Appl. Phys. Lett.* **46**, 589 (1985).

[29] J. Wagner and C. Hoffman, *Appl. Phys. Lett.* **50**, 682 (1987).

[30] D. G. Lishan, H. F. Wong, D. L. Green, E. L. Hu, J. L. Merz and D. Kirillov, *J. Vac. Sci. Technol.* **B7**, 556 (1989).

[31] J. L. Buckner, D. J. Vitkavage, E. A. Irene and T. M. Mayer, *J. Electrochem. Soc.* **133**, 1729 (1986).

[32] S. C. Vitkavage and E. A. Irene, *J. Appl. Phys.* **64**, 1983 (1988).

[33] U. Rossow, T. Fieseler, J. Geurts, D. R. T. Zahn, W. Richter, M. S. Puttock and K. P. Hilton, *J. Phys. Condens. Matter* **1**, SB231 (1989).

[34] M. Haverlag, G. M. Kroesen, C. de Zeeuw, Y. Creyghton, T. Bisschop and F. de Hoog, *J. Vac. Sci. Technol.* **B7**, 529 (1989).

[35] D. J. Thomas, P. Southworth, M. C. Flowers and R. Greef, *J. Vac. Sci. Technol.* **B7**, 1325 (1989).

[36] S. G. Hanson, G. Luckman, G. C. Nieman and S. D. Colson, *J. Vac. Sci. Technol.* **B8**, 128 (1990).

[37] G. M. Crean, I. Little and P. A. F. Herbert, *Appl. Phys. Lett.* (1990).

[38] P. A. F. Herbert, G. M. Crean, I. Little, W. M. Kelly, G. Hughes and M. Henry, *Microcircuit Engng* (1990).

[39] G. M. Crean, P. A. F. Herbert, I. Little, W. M. Kelly, J. Y. Marzin, A. Izrael and B. Jusserand, in *Proceedings of European Materials Research Society Symposium on Analytical Techniques for the Characterization of Compound Semiconductors, Strasbourg, 1990* (North-Holland, Amsterdam, 1991).

[40] W. E. Neal, *Surf. Technol.* **23**, 1 (1984).

[41] T. P. Chow, P. A. Mociel and G. M. Farelli, *J. Electrochem. Soc.* **134**, 1281 (1987).

[42] S. J. Sheard and M. G. Somekh, *Infrared Phys.* **28**, 287 (1988).

[43] I. P. Little, G. M. Crean and S. J. Sheard, *Mater. Sci. Engng* **B5**, 89 (1990).

[44] G. M. Crean, I. Little, C. Jeynes, M. Murtagh and R. P. Webb, in *Proceedings of 6th International Symposium on Silicon Materials Science and Technology* (ed. H. Hoff, K. Barraclough and J. Chikawa), p. 983 (Proceedings Vol. 9-7, The Electrochemical Society, 1990).

[45] E. H. C. Parker (ed.), *The Technology and Physics of Molecular Beam Epitaxy* (Plenum Press, New York, 1985).

[46] L. L. Chang and K. Ploog (eds), *Molecular Beam Epitaxy and Heterostructures* (Martinus Nijhoff, Dordrecht, 1987).

[47] M. Razeghi, *The MOCVD Challenge* (Adam Hilger, Bristol, 1989).

[48] R. V. Smilgys and S. R. Leone, *J. Vac. Sci. Technol.* **B8**, 416 (1990).

[49] A. J. Pidduck, D. J. Robbins, I. M. Young and G. Patel, *Thin Solid Films*, **183**, 255 (1989).

[50] A. J. Pidduck, D. J. Robbins, A. G. Cullis, D. B. Gasson and J. L. Glasper, *J. Electrochem. Soc.* **136**, 3083 (1989).

[51] D. E. Aspnes, *J. Vac. Sci. Technol.* **B3**, 1498 (1985).

[52] D. E. Aspnes, J. P. Harbison, A. A. Studna and L. T. Florez, *Phys. Rev. Lett.* **59**, 1687 (1987).

[53] J. P. Harbison, D. E. Aspnes, A. A. Studna and L. T. Florez, *J. Vac. Sci. Technol.* **B6**, 740 (1988).

[54] D. E. Aspnes, J. P. Harbison, A. A. Studna and L. T. Florez, *J. Vac. Sci. Technol.* **A6**, 1327 (1988).

[55] D. E. Aspnes, Y. C. Chang, A. A. Studna, L. T. Florez, H. H. Farrell and J. P. Harbison, *Phys. Rev. Lett.* **64**, 192 (1990).

[56] O. Acher, F. Omnes, M. Razeghi and M. Drevillon, in *Proceedings of European Acoustic, Thermal Wave and Optical Characterisation Materials Research Society Symposium of Materials* (eds G. M. Creen, M. Locatelli and J. McGilp), p. 277 (North-Holland, Amsterdam, 1990).

[57] H. W. K. Tom, T. F. Heinz and Y. R. Shen, *Phys. Rev. Lett.* **51**, 1983 (1983).

[58] J. A. van Vechten, in *Semiconductors Probed by Ultrafast Laser Spectroscopy*, Vol. II (ed. R. R. Alfano) (Academic Press, New York, 1984).

[59] T. F. Heinz, M. M. T. Loy and S. S. Iyer, *Mater. Res. Soc. Symp. Proc.* **75**, 697 (1987).

[60] G. L. Richmond, J. M. Robinson and V. L. Shannon, *Prog. Surf. Sci.* **28**, 1 (1988).

[61] R. W. Hollering, A. J. Hoeven and J. M. Lenssinck, *J. Vac. Sci. Technol.* **A8**, 3194 (1990).

[62] N. Bloembergen and P. S. Pershan, *Phys. Rev.* **128**, 606 (1962).

[63] P. G. Sionnest, W. Chen and Y. R. Shen, *Phys. Rev.* **B33**, 8254 (1986).

[64] T. F. Heinz, C. K. Chen, D. Richard and Y. R. Shen, *Phys. Rev. Lett.* **48**, 478 (1982).

[65] S. P. Murarka, in *Silicides For VLSI Applications* (Academic Press, New York, 1989).

[66] W. L. Wong, T. C. Holloway, R. F. Pinizzotto and A. F. Tasch, *IEEE Trans. Electron Devices* **29**, 597 (1982).

[67] E. Rosencher, S. Delage, Y. Campidelli and F. Arnaud d'Avitaya, *Electron. Lett.* **20**, 762 (1984).

[68] K. N. Tu and J. W. Mayer, in *Thin Films, Interdiffusion and Reactions* (eds J. M. Poate, K. N. Tu and J. M. Mayer) (Wiley-Interscience, New York, 1978).

[69] J. T. Pan and I. A. Blech, *Thin Solid Films* **113**, 129 (1984).

[70] J. M. Dilhac, N. Nolhier and C. Ganibal, *Proc. ESSDERC* **90**, 65 (1990).

[71] M. Tanaka, P. E. Schmid, A. Piaggi and F. Levy, *J. Vac. Sci. Technol.* **A7**, 3287 (1989).

[72] W. L. Smith, J. Opsal, A. Rosencwaig, J. B. Stimmell, J. C. Allison and A. S. Bhandia, *J. Vac. Sci. Technol.* **B2**, 710 (1984).

[73] N. Uchitomi, M. Nagaoka and N. Toyoda, *J. Appl. Phys.* **65**, 1743 (1989).

[74] P. Alpern and S. Wurm, *J. Appl. Phys.* **66**, 1676 (1989).

[75] A. Rosencwaig, in *Photoacoustic and Thermal Wave Phenomena in Semiconductors* (ed. A. Mandelis) (North-Holland, Amsterdam, 1987).

[76] T. Flaherty and G. M. Crean, *ESPRIT Project 5004 SEDESES: First Six Monthly Progress Report.*

Index